Path Integrals and Anomalies in Curved Space

Path integrals provide a powerful method for describing quantum phenomena, first introduced in physics by Dirac and Feynman. This book introduces the quantum mechanics of particles that move in curved space by employing the path integral method, and uses this formalism to compute anomalies in quantum field theories.

The authors start by deriving path integrals for particles moving in curved space (one-dimensional nonlinear sigma models), and their supersymmetric generalizations. Coherent states are used for fermionic particles. They then discuss the regularization and renormalization schemes essential to constructing and computing these path integrals.

In the second part of the book, the authors apply these methods to discuss and calculate anomalies in quantum field theories, with external gravitational and/or (non) abelian gauge fields. Anomalies constitute one of the most important aspects of quantum field theory; requiring that there are no anomalies is an enormous constraint in the search for physical theories of elementary particles, quantum gravity and string theories. In particular, the authors include explicit calculations of the gravitational anomalies, reviewing the seminal work of Alvarez-Gaumé and Witten in an original way, and their own work on trace anomalies.

FIORENZO BASTIANELLI is Professor of Theoretical Physics at the Department of Physics, University of Bologna. Following his Ph.D., from SUNY at Stony Brook in New York, he has worked at ITP, Stockholm University, and at the Niels Bohr Institute, University of Copenhagen. In 1995 he moved to Italy, initially at the University of Modena, and since 1999 at the University of Bologna. His research interests include quantum field theory and string theory.

PETER VAN NIEUWENHUIZEN is a Distinguished Professor at the C. N. Yang Institute of SUNY, Stony Brook. After his Ph.D. from Utrecht University he worked at CERN (Geneva), the Ecole Normale Superieure (Paris) and Brandeis University (Mass.) In 1975 he moved to Stony Brook, where he co-discovered supergravity in 1976, for which he received the Dirac award in 1993, and the Heinemann prize in 2005. His research interests are in supersymmetry, supergravity, quantum field theory, modern general relativity and string theory, about which he has published over 300 articles in scientific journals.

CAMBRIDGE MONOGRAPHS ON MATHEMATICAL PHYSICS

General editors: P. V. Landshoff, D. R. Nelson, S. Weinberg

S. J. Aarseth *Gravitational N-Body Simulations*
J. Ambjørn, B. Durhuus and T. Jonsson *Quantum Geometry: A Statistical Field Theory Approach*
A. M. Anile *Relativistic Fluids and Magneto-Fluids*
J. A. de Azcárrage and J. M. Izquierdo *Lie Groups, Lie Algebras, Cohomology and Some Applications in Physics*†
O. Babelon, D. Bernard and M. Talon *Introduction to Classical Integrable Systems*
F. Bastianelli and P. van Nieuwenhuizen *Path Integrals and Anomalies in Curved Space*
V. Belinkski and E. Verdaguer *Gravitational Solitons*
J. Bernstein *Kinetic Theory in the Expanding Universe*
G. F. Bertsch and R. A. Broglia *Oscillations in Finite Quantum Systems*
N. D. Birrell and P.C.W. Davies *Quantum Fields in Curved Space*†
M. Burgess *Classical Covariant Fields*
S. Carlip *Quantum Gravity in 2+1 Dimensions*
J. C. Collins *Renormalization*†
M. Creutz *Quarks, Gluons and Lattices*†
P. D. D'Eath *Supersymmetric Quantum Cosmology*
F. de Felice and C. J. S. Clarke *Relativity on Curved Manifolds*†
B. S. DeWitt *Supermanifolds*, 2nd edition†
P. G. O. Freund *Introduction to Supersymmetry*†
J. Fuchs *Affine Lie Algebras and Quantum Groups*†
J. Fuchs and C. Schweigert *Symmetries, Lie Algebras and Representations: A Graduate Course for Physicists*†
Y. Fujii and K. Maeda *The Scalar–Tensor Theory of Gravitation*
A. S. Galperin, E. A. Ivanov, V. I. Orievetsky and E. S. Sokatchev *Harmonic Superspace*
R. Gambini and J. Pullin *Loops, Knots, Gauge Theories and Quantum Gravity*†
M. Göckeler and T. Schücker *Differential Geometry, Gauge Theories and Gravity*†
C. Gómez, M. Ruiz Altaba and G. Sierra *Quantum Groups in Two-dimensional Physics*
M. B. Green, J. H. Schwarz and E. Witten *Superstring Theory, volume 1: Introduction*†
M. B. Green, J. H. Schwarz and E. Witten *Superstring Theory, volume 2: Loop Amplitudes, Anomalies and Phenomenology*†
V. N. Gribov *The Theory of Complex Angular Momenta*
S. W. Hawking and G. F. R. Ellis *The Large-Scale Structure of Space-Time*†
F. Iachello and A. Arima *The Interacting Boson Model*
F. Iachello and P. van Isacker *The Interacting Boson–Fermion Model*
C. Itzykson and J.-M. Drouffe *Statistical Field Theory, volume 1: From Brownian Motion to Renormalization and Lattice Gauge Theory*†
C. Itzykson and J.-M. Drouffe *Statistical Field Theory, volume 2: Strong Coupling, Monte Carlo Methods, Conformal Field Theory, and Random Systems*†
C. Johnson *D-Branes*
J. I. Kapusta *Finite-Temperature Field Theory*†
V. E. Korepin, A. G. Izergin and N. M. Boguliubov *The Quantum Inverse Scattering Method and Correlation Functions*†
M. Le Bellac *Thermal Field Theory*†
Y. Makeenko *Methods of Contemporary Gauge Theory*
N. Manton and P. Sutcliffe *Topological Solitons*
N. H. March *Liquid Metals: Concepts and Theory*
I. M. Montvay and G. Münster *Quantum Fields on a Lattice*†
L. O' Raifeartaigh *Group Structure of Gauge Theories*†
T. Ortín *Gravity and Strings*
A. Ozorio de Almeida *Hamiltonian Systems: Chaos and Quantization*†
R. Penrose and W. Rindler *Spinors and Space-Time, volume 1: Two-Spinor Calculus and Relativistic Fields*†
R. Penrose and W. Rindler *Spinors and Space-Time, volume 2: Spinor and Twistor Methods in Space-Time Geometry*†
S. Pokorski *Gauge Field Theories*, 2nd edition
J. Polchinski *String Theory, volume 1: An Introduction to the Bosonic String*†
J. Polchinski *String Theory, volume 2: Superstring Theory and Beyond*†
V. N. Popov *Functional Integrals and Collective Excitations*†
R. J. Rivers *Path Integral Methods in Quantum Field Theory*†
R. G. Roberts *The Structure of the Proton*†
C. Rovelli *Quantum Gravity*
W. C. Saslaw *Gravitational Physics of Stellar and Galactic Systems*†
H. Stephani, D. Kramer, M. A. H. MacCallum, C. Hoenselaers and E. Herlt *Exact Solutions of Einstein's Field Equations, 2nd edition*
J. M. Stewart *Advanced General Relativity*†
A. Vilenkin and E. P. S. Shellard *Cosmic Strings and Other Topological Defects*†
R. S. Ward and R. O. Wells Jr *Twistor Geometry and Field Theories*†
J. R. Wilson and G. J. Mathews *Relativistic Numerical Hydrodynamics*

†Issued as a paperback

Path Integrals and Anomalies in Curved Space

FIORENZO BASTIANELLI
Università degli Studi,
Bologna, Italy

PETER VAN NIEUWENHUIZEN
State University of New York at Stony Brook

CAMBRIDGE
UNIVERSITY PRESS

CAMBRIDGE UNIVERSITY PRESS
Cambridge, New York, Melbourne, Madrid, Cape Town, Singapore, São Paulo

Cambridge University Press
The Edinburgh Building, Cambridge CB2 2RU, UK

Published in the United States of America by Cambridge University Press, New York

www.cambridge.org
Information on this title: www.cambridge.org/9780521847612

First published 2006

Printed in the United Kingdom at the University Press, Cambridge

A catalog record for this publication is available from the British Library

ISBN-13 978-0-521-84761-2 hardback
ISBN-10 0-521-84761-3 hardback

To Bryce S. DeWitt (1923–2004)
who pioneered quantum mechanics in curved space

Contents

Preface

In 1983, L. Alvarez-Gaumé and E. Witten (AGW) wrote a fundamental article in which they calculated the one-loop gravitational anomalies (anomalies in the local Lorentz symmetry of $(4k + 2)$-dimensional Minkowskian quantum field theories coupled to external gravity) of complex chiral spin-$\frac{1}{2}$ and spin-$\frac{3}{2}$ fields and real self-dual antisymmetric tensor fields[1] [1]. They used two methods: a straightforward Feynman graph calculation in $4k + 2$ dimensions with Pauli–Villars regularization, and a quantum mechanical (QM) path integral method in which corresponding nonlinear sigma models appeared. The former has been discussed in detail in an earlier book [3]. The latter method is the subject of this book. AGW applied their formulas to $N = 2B$ supergravity in 10 dimensions, which contains precisely one field of each kind, and found that the sum of the gravitational anomalies cancels. Soon afterwards, M. B. Green and J. H. Schwarz [4] calculated the gravitational anomalies in one-loop string amplitudes, and concluded that these anomalies cancel in string theory, and therefore should also cancel in $N = 1$ supergravity in 10 dimensions with suitable gauge groups for the $N = 1$ matter couplings. Using the formulas of AGW, one can indeed show that the sum of anomalies in $N = 1$ supergravity coupled to super Yang–Mills theory with gauge group $SO(32)$ or $E_8 \times E_8$, though nonvanishing, is in the technical sense exact:

[1] Just as one can always shift the axial anomaly from the vector current to the axial current by adding a suitable counterterm to the action or by using a different regularization scheme, one can also shift the gravitational anomaly from the general coordinate symmetry to the local Lorentz symmetry [2]. Conventionally one chooses to preserve general coordinate invariance. AGW chose the symmetric vielbein gauge, so that the symmetry for which they computed the anomalies was a linear combination of a general coordinate transformation and a compensating local Lorentz transformation. However, they used a regulator that manifestly preserved general coordinate invariance, so that their calculation yielded the anomaly in the local Lorentz symmetry.

it can be removed by adding a local counterterm to the action. These two papers led to an explosion of interest in string theory.

We discussed these two papers in a series of internal seminars for advanced graduate students and faculty at Stony Brook (the "Friday seminars"). Whereas the basic philosophy and methods of the paper by AGW were clear, we stumbled on numerous technical problems and details. Some of these became clearer upon closer reading, some became more baffling. In a desire to clarify these issues we decided to embark on a research project: the AGW program for trace anomalies. Since gravitational and chiral anomalies only contribute at the one-worldline-loop level in the QM method, one need not be careful with definitions of the measure for the path integral, choice of regulators, regularization of divergent graphs, etc. This is explicitly discussed in [1]. However, we soon noticed that for the trace anomalies the opposite is true: if the field theory is defined in $n = 2k$ dimensions, one needs $(k+1)$-loop graphs on the worldline in the QM method. Consequently, every detail in the calculation matters. Our program of calculating trace anomalies turned into a program of studying path integrals for nonlinear sigma models in phase space and configuration space, a notoriously difficult and controversial subject. As already pointed out by AGW, the QM nonlinear sigma models needed for spacetime fermions (or self-dual antisymmetric tensor fields in spacetime) have $N = 1$ (or $N = 2$) worldline supersymmetry (susy), even though the original field theories were not spacetime supersymmetric. Thus, we also had to wrestle with the role of susy in the careful definitions and calculations of these QM path integrals.

Although it only gradually dawned upon us, we have come to recognize the problems with these susy and nonsusy QM path integrals as problems one should expect to encounter in any quantum field theory (QFT), the only difference being that these particular field theories have a one-dimensional (finite) spacetime, as a result of which infinities in the sum of Feynman graphs for a given process cancel. However, individual Feynman graphs are power-counting divergent (because these models contain double-derivative interactions just like quantum gravity). This cancellation of infinities in the sum of graphs is perhaps the psychological reason why there is no systematic discussion of regularization issues in the early literature on the subject (in the 1950s and 1960s). With the advent of the renormalization of gauge theories in the 1970s, issues of regularization of nonlinear sigma models were also studied. It was found that the regularization schemes used at that time (the time slicing method and the mode regularization method) broke general coordinate invariance at intermediate stages, but it was also noted that by adding noncovariant counterterms [5–9], the final physical results were still general coordinate invariant (we shall use the shorter term Einstein invariance for this symmetry in this book). The question thus arose as to how to determine those counterterms, and to understand the relation between the counterterms in one regularization scheme and those in other schemes. Once again, the answer to this question could be found in the general literature on QFT: the imposition of suitable renormalization conditions.

As we tackled more and more difficult problems (four-loop graphs for trace anomalies in six dimensions) it became clear to us that a scheme which needed only covariant counterterms would be very welcome. Dimensional regularization (DR) is such a scheme [10]. It had been used by Kleinert and Chervyakov [11] for the QM of a one-dimensional target space on an infinite worldline time interval (with a mass term added to regulate infrared divergences). For our purposes we have developed instead a version of dimensional regularization on a compact space; because the space is compact we do not need to add by hand a mass term to regulate the infrared divergences due to massless fields. The counterterms needed in such an approach are indeed covariant (both Einstein and locally Lorentz invariant).

The quantum mechanical path integral formalism can be used to compute anomalies in quantum field theories. This application forms the second part of this book. Chiral spin-$\frac{1}{2}$ and spin-$\frac{3}{2}$ fields and selfdual antisymmetric tensor (SAT) fields can produce anomalies in loop graphs with external gravitons and/or external gauge (Yang–Mills) fields. The treatment of the spin $\frac{3}{2}$ and SAT fields formed a major obstacle. For example, in the article by AGW the SAT fields are described by a bispinor $\psi_{\alpha\beta}$. However, the vector index of the spin-$\frac{3}{2}$ field and the β index of $\psi_{\alpha\beta}$ are treated differently from the spinor index of the spin-$\frac{1}{2}$ and spin-$\frac{3}{2}$ fields and the α index of $\psi_{\alpha\beta}$. In [1] one finds the following transformation rule for the spin-$\frac{3}{2}$ field (in their notation):

$$-\delta_\eta \psi_A = \eta^i D_i \psi_A + D_a \eta_b (T^{ab})_{AB} \psi_B \tag{1}$$

where $\eta^i(x)$ parametrizes an infinitesimal coordinate transformation $x^i \to x^i + \eta^i(x)$, and $A = 1, 2, \ldots, n$ is the flat vector index of the spin-$\frac{3}{2}$ (gravitino) field, while $(T^{ab})_{AB} = -i(\delta^a_A \delta^b_B - \delta^b_A \delta^a_B)$ are the matrix elements of the Euclidean Lorentz group $SO(n)$ in the vector representation. One would expect that this transformation rule is a linear combination of an Einstein transformation $\delta_E \psi_{A\alpha} = \eta^i \partial_i \psi_{A\alpha}$ (the vector index A of $\psi_{A\alpha}$ is flat and α is the spin index) and a local Lorentz rotation $\delta_{lL} \psi_{A\alpha} = \frac{1}{4} \eta^i \omega_{iBC} (\gamma^B \gamma^C)_\alpha{}^\beta \psi_{A\beta} + \eta^i \omega_{iA}{}^B \psi_{B\alpha}$. However, on top of this Lorentz rotation with parameter $\eta^i \omega_{iAB}$, one finds the second term in (1) which describes a local Lorentz rotation with parameter $(D_a \eta_b - D_b \eta_a)$ and this local Lorentz transformation only acts on the vector index of the gravitino. If one assumes (1), one finds a beautiful simple relation between the gravitational contribution to the axial (γ_5) anomaly in $4k+4$ dimensions and the gravitational (local Lorentz) anomaly in $4k+2$ dimensions. We shall derive (1) from first principles, and show that it is correct, but only if one uses a particular regulator $\mathcal{R}$.

The regulators for the spin-$\frac{1}{2}$ field λ, for the gravitino ψ_A, and for the bispinor $\psi_{\alpha\beta}$ are in all cases the square of the field operators for the nonchiral spinors $\tilde{\lambda}$, $\tilde{\psi}_A$ and $\tilde{\psi}_{\alpha\beta}$, where the "twiddled fields" $\tilde{\lambda}$, $\tilde{\psi}_A$ and $\tilde{\psi}_{\alpha\beta}$ are obtained from λ, ψ_A and $\psi_{\alpha\beta}$ by multiplication by $g^{1/4} = (\det e_\mu{}^m)^{1/2}$. These regulators are covariant regulators, not consistent regulators, and the anomalies we will obtain are covariant anomalies, not consistent anomalies [2]. However, when we come to the cancellation of anomalies, we shall use the descent equations to convert these covariant anomalies to consistent anomalies, and then construct counterterms whose variations cancel these consistent anomalies.

The twiddled fields were used by Fujikawa, who pioneered the path integral approach to anomalies [12]. An ordinary Einstein transformation of $\tilde{\lambda}$ is given by $\delta\tilde{\lambda} = \frac{1}{2}(\xi^\mu\partial_\mu + \partial_\mu\xi^\mu)\tilde{\lambda}$, where the second derivative ∂_μ can also act on $\tilde{\lambda}$, and if one evaluates the corresponding regulated anomaly $An_E = \mathrm{Tr}\frac{1}{2}(\xi^\mu\partial_\mu + \partial_\mu\xi^\mu)\mathrm{e}^{-\beta\mathcal{R}}$ by inserting a complete set of eigenfunctions $\tilde{\varphi}_k$ of $\mathcal{R}$ with non-negative eigenvalues λ_k, one finds

$$An_E = \lim_{\beta\to 0}\sum_k \int d^n x\, \tilde{\varphi}_k^*(x)\frac{1}{2}(\xi^\mu\partial_\mu + \partial_\mu\xi^\mu)\mathrm{e}^{-\beta\lambda_k}\tilde{\varphi}_k(x)\,. \tag{2}$$

Thus, the Einstein anomaly vanishes (partially integrate the second ∂_μ) as long as the regulator is self-adjoint with respect to the inner product $\langle\tilde{\lambda}_1|\tilde{\lambda}_2\rangle = \int dx\, \tilde{\lambda}_1^*(x)\tilde{\lambda}_2(x)$ (so that $\tilde{\varphi}_k$ form a complete set), and as long as both $\tilde{\varphi}_k(x)$ and $\tilde{\varphi}_k^*(x)$ belong to the same complete set of eigenstates, as in the case of plane waves e^{ikx}. One can always make a unitary transformation from $\tilde{\varphi}_k$ to the set e^{ikx}, and using these plane waves, the calculation of anomalies in the framework of quantum field theory is reduced to a set of n-dimensional Gaussian integrals over k. We shall use the regulator $\mathcal{R}$ discussed above, and twiddled fields, but then cast the calculation of anomalies in terms of quantum mechanics and path integrals. Calculating anomalies using quantum mechanics is much simpler than evaluating the Gaussian integrals of quantum field theory. Using path integrals simplifies the calculations even further.

When we first started studying the problems discussed in this book, we used the shortcuts and plausible arguments which are used by researchers and sometimes mentioned in the literature. However, the more we tried to clarify and complete these shortcuts and arguments, the more we were driven to basic questions and theoretical principles. We have been studying these issues now for over 15 years, and have accumulated a wealth of facts and insights. We decided to write a book in which all ideas and calculation were developed from scratch, with all intermediate steps worked out. The result looks detailed, and at places technical. We have made every effort to keep the text readable by providing verbal descriptions next to formulas, and providing introductory sections and historical reviews. In the end, however, we felt there is no substitute for a complete and fundamental treatment.

We end this preface by summarizing the content of this book. In the first part of this book we give a complete derivation of the path integrals for supersymmetric and nonsupersymmetric nonlinear sigma models describing bosonic and fermionic point particles (commuting coordinates $x^i(t)$ and anticommuting variables $\psi^a(t) = e_i^a(x(t))\psi^i(t)$) in a curved target space with metric $g_{ij}(x) = e_i^a(x)e_j^b(x)\delta_{ab}$. All of our calculations are performed in Euclidean target space. We consider a finite time interval because this is what is needed for the applications to anomalies. As these models contain double-derivative interactions, they are divergent according to power-counting, just as in quantum gravity, but ghost loops arising from the path integral measure cancel the divergences. Only the one- and two-loop graphs are power-counting divergent, hence in general the action may contain extra finite local one- and two-loop counterterms, the coefficients of which should be fixed. They are fixed by imposing suitable renormalization conditions. To regularize individual diagrams we use three different regularization schemes:

(i) time slicing (TS), known from the work of Dirac and Feynman;

(ii) mode regularization (MR), known from instanton and soliton physics;[2] and

(iii) dimensional regularization on a finite time interval (DR), discussed in this book.

The renormalization conditions relate a given quantum Hamiltonian $\hat{H}$ to a corresponding quantum action S, by which we mean the action that appears in the exponent of the path integral. The particular finite one- and two-loop counterterms in S thus obtained are different for each regularization scheme. In principle, any $\hat{H}$ with a definite ordering of the operators can be taken as the starting point, and gives a corresponding path integral (with different counterterms for different regularization schemes), but for our physical applications we shall consider quantum Hamiltonians that maintain reparametrization and local Lorentz invariance in target space (i.e. commute with the quantum generators of these

[2] Actually, the mode expansion had already been used by Feynman and Hibbs to compute the path integral for the harmonic oscillator.

symmetries. The chiral anomaly is then due to the chirality matrix in the Jacobians). Then there are no one-loop counterterms in the three schemes, but only two-loop counterterms. Having defined the regulated path integrals, the continuum limit can be taken and reveals the correct "Feynman rules" (the rules of how to evaluate the integrals over products of distributions and equal-time contractions) for each regularization scheme. All three regularization schemes give the same final answer for the transition amplitude, although the Feynman rules are different.

In the second part of this book we apply our methods to the evaluation of anomalies in n-dimensional relativistic quantum field theories with bosons and fermions in the loops (spin $0, \frac{1}{2}, 1, \frac{3}{2}$ and self-dual antisymmetric tensor fields) coupled to external gauge fields and/or gravity. We regulate the field-theoretical Jacobian for the symmetries whose anomalies we want to compute with a factor of $\exp(-\beta\mathcal{R})$, where $\mathcal{R}$ is the covariant regulator which follows from the corresponding quantum field theory, as discussed before, and β tends to zero only at the end of the calculation. Next, we introduce a quantum mechanical representation of the operators which enter in the field-theoretical calculation. The regulator $\mathcal{R}$ yields a corresponding quantum mechanical Hamiltonian $\hat{H}$. We rewrite the quantum mechanical operator expression for the anomalies as a path integral on the finite time interval $-\beta \leq t \leq 0$ for a linear or nonlinear sigma model with action S. For given spacetime dimension n, in the limit $\beta \to 0$ only graphs with a finite number of loops on the worldline contribute. In this way the calculation of the anomalies is transformed from a field-theoretical problem to a problem in quantum mechanics. We give details of the derivation of the chiral and gravitational anomalies as first given by Alvarez-Gaumé and Witten, and discuss our own work on trace anomalies. For the former one only needs to evaluate one-loop graphs on the worldline, but for the trace anomalies in two dimensions we need two-loop graphs, and for the trace anomalies in four dimensions we compute three-loop graphs. Here a technical but important problem was settled: using time-slicing or mode regularization, counterterms proportional to the product of two Christoffel symbols were found, but it is incorrect to invoke normal coordinates and to ignore these counterterms. Their expansion produces products of two Riemann curvatures which do contribute at 3 loops to trace anomalies. We obtain complete agreement with the results for these anomalies obtained from other methods. We conclude with a detailed analysis of the gravitational anomalies in 10-dimensional supergravities, both for classical and for exceptional gauge groups.

Twenty years have passed since AGW wrote their renowned article. We believe we have solved all major and minor problems we initially ran into.[3] The quantum mechanical approach to quantum field theory can be applied to more problems than only anomalies. If future work on such problems will profit from the detailed account given in this book, our scientific and geographical Odyssey has come to a good ending.

No book is everywhere totally clear and without misprints. We collect and correct them at the following webpage: http://insti.physics.sunysb.edu/itp/books/anomalies/ and we ask readers who have questions or spot misprints or errors to contact us by email at bastianelli@bo.infn.it and vannieu@insti.physics.sunysb.edu.

We would like to thank our respective universities of Bologna and Stony Brook, and the Istituto Nazionale di Fisica Nucleare (INFN) of Italy, the Teyler Foundation of the Netherlands, and the National Science Foundation (NSF) of the USA for financial support. Over the years part of our work was presented at lecture series, summer schools and conferences in Berlin, Brussels, Leiden, Leuven, Los Angeles, Marseille, Rio de Janeiro and Vienna, and we thank the organizers for these opportunities.

Bologna and Stony Brook, January 2005

[3]Except one problem: a rigorous derivation, based only on quantum mechanical path integrals, of the overall normalization of the gravitational anomaly of self-dual antisymmetric tensor fields, see Chapter 8. We fix this normalization by requiring agreement with bosonization formulas of two-dimensional quantum field theories.

Part I

Path integrals for quantum mechanics in curved space

1
Introduction to path integrals

Path integrals play an important role in modern quantum field theory. One usually first encounters them as useful formal devices to derive Feynman rules. For gauge theories they yield straightforwardly the Ward identities. Namely, if BRST symmetry (the "quantum gauge invariance" discovered by Becchi, Rouet, Stora and Tyutin [14]) holds at the quantum level, certain relations between Green functions can be derived from path integrals, but details of the path integral (for example, the precise form of the measure) are not needed for this purpose.[1] Once the BRST Ward identities for gauge theories have been derived, unitarity and renormalizability can be proven, and at this point one may forget about path integrals if one is only interested in perturbative aspects of quantum field theories. One can compute higher-loop Feynman graphs without ever using path integrals.

However, for nonperturbative aspects, path integrals are essential. The first place where one encounters path integrals in nonperturbative quantum field theory is in the study of instantons and solitons. Here advanced methods based on path integrals have been developed. For example, in the case of instantons the correct measure for integration over their collective coordinates (corresponding to the zero modes) is needed. In particular, for supersymmetric nonabelian gauge theories, there are only contributions from these zero modes, while the contributions from the nonzero modes cancel between bosons and fermions. Another area where the path integral

[1] To prove that the BRST symmetry is free from anomalies, one may either use regularization-free cohomological methods, or one may perform explicit loop graph calculations using a particular regularization scheme. When there are no anomalies, but the regularization scheme does not preserve the BRST symmetry, one can always add local counterterms to the action at each loop level to restore the BRST symmetry. In these manipulations the path integral measure is usually not taken into account.

measure is important is quantum gravity. In particular, in modern studies of quantum gravity based on string theory, the measure is crucial in obtaining the correct correlation functions.

One can compute path integrals at the nonperturbative level by going to Euclidean space, discretizing the path integrals on lattices and using powerful computers. In this book we use a continuum approach. We study a class of simple models which lead to path integrals in which no infinite renormalization is needed, but some individual diagrams are divergent and need be regulated, and subtle issues of regularization and measures can be studied explicitly. These models are the quantum mechanical (one-dimensional) nonlinear sigma models. The one- and two-loop diagrams in these models are power-counting divergent, but the infinities cancel in the sum of diagrams for a given process at a given loop level.

Quantum mechanical (QM) nonlinear sigma models can be described by path integrals and are toy models for realistic path integrals in four dimensions. They describe curved target spaces and contain double-derivative interactions (quantum gravity has also double-derivative interactions). The formalism for path integrals in curved space has been discussed in great generality in several books and reviews [15–26]. In the first half of this book we define the path integrals for these models and discuss various subtleties. However, quantum mechanical nonlinear sigma models can also be used to compute anomalies of realistic four- and higher-dimensional quantum field theories, and this application is thoroughly discussed in the second half of this book. Furthermore, quantum mechanical path integrals can be used to compute correlation functions and effective actions. For references in flat space see [27], and for some work in curved space see [28–30].

The study of path integrals in curved space was pioneered by DeWitt [15]. He first extended to curved space a result of Pauli [16] for the transition element for infinitesimal times which was the product of the exponent of the classical action evaluated for a classical trajectory, times the Van Vleck–Morette determinant [17]. He verified that this transition element satisfied a Schrödinger equation with Hamiltonian $\hat{H} + \frac{1}{12}\hbar^2 R$ ($-\frac{1}{12}\hbar^2 R$ in our conventions for R), where $\hat{H} = \frac{1}{2}\hat{g}^{-1/4}\hat{p}_i\hat{g}^{ij}\hat{g}^{1/2}\hat{p}_j\hat{g}^{-1/4}$. He also claimed that this transition element could be written as a path integral with a modified action, which was the sum of the classical action and a term $+\frac{\hbar^2}{12}R$. The latter term comes from the Van Vleck determinant.[2] His work has led to an enormous literature on this subject, with many authors proposing various ingenuous definitions or approximations of the

[2]There exists some confusion in the literature about the coefficient of R in the action in the path integral for the transition element related to the minimal hamiltonian operator $\hat{H}$ ("the counter term with R"). Initially DeWitt obtained $\frac{1}{6}$ [15]. However, recently in [26] he rectified this to $\frac{1}{8}$, a result with which we agree, at least if one uses the regularization schemes discussed in this book, see eqs. (2.81), (3.73), (4.28) and Appendix B. (Note: some of these schemes have additional noncovariant $\Gamma\Gamma$ terms.)

infinitesimal transition element, and various proposals for iterations which should produce the finite transition amplitude, see for example [31–34].

In Part I of this book we show how to define and compute the transition element for finite times using path integrals. This yields, in particular, the transition element for infinitesimal times in a series expansion. Path integrals are of course just one of many ways of computing the transition element, but for the calculation of anomalies the path integral method is far superior as we hope to demonstrate in this book.

1.1 The simplest case: a particle in flat space

Before considering path integrals in curved space, we first review the simple case of a nonrelativistic particle moving in an n-dimensional flat space and subject to a scalar potential $V(x)$. We are going to derive the path integral from the canonical (operatorial) formulation of quantum mechanics. We will also compute the transition amplitude in the free case (i.e. with vanishing potential), a useful result to compare with when we deal with the more complicated case of curved space.

Thus, let us consider a particle with coordinates x^i, conjugate momenta p_i and mass m. As the quantum Hamiltonian we take

$$H(\hat{x},\hat{p}) = \frac{1}{2m}\hat{p}_i\hat{p}^i + V(\hat{x}) \tag{1.1}$$

where, as usual, hats denote quantum mechanical operators. We are interested in deriving a path integral representation of the transition amplitude

$$T(z,y;\beta) \equiv \langle z|\mathrm{e}^{-\frac{\beta}{\hbar}\hat{H}}|y\rangle \tag{1.2}$$

for the particle to propagate from the point y^i to the point z^i in a Euclidean time β. We use a language appropriate to quantum mechanics ("transition amplitude", etc.) even though we consider a Euclidean approach. The usual quantum mechanics in Minkowskian time is obtained by the substitution $\beta \to it$, which corresponds to the so-called Wick rotation, an analytical continuation in the time coordinate that relates statistical mechanics to quantum mechanics, and vice versa.

We use eigenstates $|x\rangle$ and $|p\rangle$ of the position operator $\hat{x}^i$ and momentum operator $\hat{p}_i$, respectively,

$$\hat{x}^i|x\rangle = x^i|x\rangle, \quad \hat{p}_i|p\rangle = p_i|p\rangle, \tag{1.3}$$

together with the completeness relations

$$I = \int d^n x\, |x\rangle\langle x| = \int d^n p\, |p\rangle\langle p| \tag{1.4}$$

and the scalar products

$$\langle x_1|x_2\rangle = \delta^n(x_1 - x_2), \quad \langle p_1|p_2\rangle = \delta^n(p_1 - p_2), \quad \langle x|p\rangle = \frac{1}{(2\pi\hbar)^{n/2}}\mathrm{e}^{\frac{i}{\hbar}p_i x^i}. \tag{1.5}$$

It is easy to show that the transition amplitude should satisfy the Schrödinger equation (see (2.229) and (2.230))

$$-\hbar\frac{\partial}{\partial\beta}T(z,y;\beta) = H(z)T(z,y;\beta) \tag{1.6}$$

with the boundary condition

$$T(z,y;0) = \delta^n(z-y) \tag{1.7}$$

where the Hamiltonian in the coordinate representation is, of course, given by

$$H(z) = -\frac{\hbar^2}{2m}\frac{\partial}{\partial z^i}\frac{\partial}{\partial z_i} + V(z). \tag{1.8}$$

A similar equation holds at the point y^i.

The derivation of a path integral representation for the transition amplitude is rather standard. The transition amplitude can be split into N factors

$$\begin{aligned} T(z,y;\beta) &= \langle z|\left(\mathrm{e}^{-\frac{\beta}{\hbar N}\hat{H}}\right)^N|y\rangle = \langle z|\underbrace{\mathrm{e}^{-\frac{\epsilon}{\hbar}\hat{H}}\mathrm{e}^{-\frac{\epsilon}{\hbar}\hat{H}}\cdots\mathrm{e}^{-\frac{\epsilon}{\hbar}\hat{H}}}_{N\,\text{times}}|y\rangle \\ &= \int\left(\prod_{k=1}^{N-1}d^n x_k\right)\prod_{k=1}^{N}\langle x_k|\mathrm{e}^{-\frac{\epsilon}{\hbar}\hat{H}}|x_{k-1}\rangle \end{aligned} \tag{1.9}$$

where we have denoted $x_0^i = y^i$, $x_N^i = z^i$, $\epsilon = \beta/N$, and used $N-1$ times the completeness relations with position eigenstates. Then one can use N times the completeness relations with momentum eigenstates and obtain

$$T(z,y;\beta) = \int\left(\prod_{k=1}^{N-1}d^n x_k\right)\left(\prod_{k=1}^{N}d^n p_k\right)\prod_{k=1}^{N}\langle x_k|p_k\rangle\langle p_k|\mathrm{e}^{-\frac{\epsilon}{\hbar}\hat{H}}|x_{k-1}\rangle. \tag{1.10}$$

This is still an exact formula, but we are now going to evaluate it using approximations which are correct in the limit $N\to\infty$ ($\epsilon\to 0$). The key point for deriving the path integral is to evaluate the following matrix element

$$\begin{aligned} \langle p|\mathrm{e}^{-\frac{\epsilon}{\hbar}\hat{H}(\hat{x},\hat{p})}|x\rangle &= \langle p|\left[1-\frac{\epsilon}{\hbar}\hat{H}(\hat{x},\hat{p})+\cdots\right]|x\rangle \\ &= \langle p|x\rangle - \frac{\epsilon}{\hbar}\langle p|\hat{H}(\hat{x},\hat{p})|x\rangle + \cdots \\ &= \langle p|x\rangle\left[1-\frac{\epsilon}{\hbar}H(x,p)+\cdots\right] \\ &= \langle p|x\rangle\,\mathrm{e}^{-\frac{\epsilon}{\hbar}H(x,p)+\cdots}. \end{aligned} \tag{1.11}$$

The replacement $\langle p|\hat{H}(\hat{x},\hat{p})|x\rangle = \langle p|x\rangle\, H(x,p)$ follows from the simple structure of the Hamiltonian in (1.1), which allows to act with the position and momentum operators on the corresponding eigenstates, so that these operators are simply replaced by the corresponding eigenvalues. In this way the Hamiltonian operator $\hat{H}(\hat{x},\hat{p})$ is replaced by the Hamiltonian function $H(x,p) = p^2/2m + V(x)$. These approximations are justified in the limit $N \to \infty$ for many physically interesting potentials (i.e. the "dots" in (1.11) can be neglected in this limit), in which cases a rigorous mathematical proof is also available, and goes under the name of the "Trotter formula" [21]. Finally, using the expression for $\langle x|p\rangle$ given in (1.5), and recalling that $\langle p|x\rangle = \langle x|p\rangle^*$, one obtains

$$\langle x_k|p_k\rangle\langle p_k|\mathrm{e}^{-\frac{\epsilon}{\hbar}\hat{H}}|x_{k-1}\rangle = \frac{1}{(2\pi\hbar)^n}\mathrm{e}^{\frac{i}{\hbar}p_k\cdot(x_k-x_{k-1})-\frac{\epsilon}{\hbar}H(x_{k-1},p_k)} \tag{1.12}$$

which can now be inserted into (1.10). At this point the expression of the transition amplitude does not contain any more operators, and reads as

$$\begin{aligned} T(z,y;\beta) &= \lim_{N\to\infty}\int\left(\prod_{k=1}^{N-1} d^n x_k\right)\left(\prod_{k=1}^{N}\frac{d^n p_k}{(2\pi\hbar)^n}\right) \\ &\quad\times\exp\left\{-\frac{\epsilon}{\hbar}\sum_{k=1}^{N}\left[-ip_k\cdot\frac{(x_k-x_{k-1})}{\epsilon}+H(x_{k-1},p_k)\right]\right\} \\ &= \int Dx\,Dp\,\mathrm{e}^{-\frac{1}{\hbar}S[x,p]}. \end{aligned} \tag{1.13}$$

This is the path integral in phase space. We recognize in the exponent a discretization of the classical Euclidean phase space action

$$\begin{aligned} S[x,p] &= \int_0^\beta dt\,[-ip\cdot\dot{x}+H(x,p)] \\ &\to \epsilon\sum_{k=1}^{N}\left[-ip_k\cdot\frac{(x_k-x_{k-1})}{\epsilon}+H(x_{k-1},p_k)\right] \end{aligned} \tag{1.14}$$

where again $\beta = N\epsilon$. The last line in (1.13) is symbolic and indicates a formal sum over paths in phase space weighted by the exponential of minus their classical action.

The configuration space path integral is easily derived by integrating out the momenta in (1.13). Completing squares and using Gaussian integration one obtains

$$\begin{aligned} T(z,y;\beta) &= \lim_{N\to\infty}\int\left(\prod_{k=1}^{N-1} d^n x_k\right)\left(\frac{m}{2\pi\hbar\epsilon}\right)^{nN/2} \\ &\quad\times\exp\left\{-\frac{\epsilon}{\hbar}\sum_{k=1}^{N}\left[\frac{m}{2}\left(\frac{x_k-x_{k-1}}{\epsilon}\right)^2+V(x_{k-1})\right]\right\} \\ &= \int Dx\,\mathrm{e}^{-\frac{1}{\hbar}S[x]}. \end{aligned} \tag{1.15}$$

This is the path integral in configuration space. In the exponent one finds a discretization of the classical Euclidean configuration space action

$$S[x] = \int_0^\beta dt \left[\frac{m}{2}\dot{x}^2 + V(x)\right]$$
$$\to \epsilon \sum_{k=1}^{N} \left[\frac{m}{2}\Big(\frac{x_k - x_{k-1}}{\epsilon}\Big)^2 + V(x_{k-1})\right]. \tag{1.16}$$

Again the last line in (1.15) is symbolic, and indicates a sum over paths in configuration space.

For the case of a vanishing potential, the path integral can be evaluated exactly [45, 46, 21]. Performing successive Gaussian integrations one obtains

$$T(z, y; \beta) = \Big(\frac{m}{2\pi\hbar\beta}\Big)^{n/2} \mathrm{e}^{-m(z-y)^2/2\beta\hbar}. \tag{1.17}$$

This final result is very suggestive. Up to a prefactor, it consists of the exponential of the classical action evaluated on the classical trajectory. This is typical for the cases where the semiclassical approximation is exact. The prefactor can be considered as containing the "one-loop" corrections which make up the full result (thus "semiclassical" = "classical + one-loop").

The preceding approach is called time slicing, and will be applied to nonlinear sigma models (models in curved target space) in Chapter 2. In Chapters 3 and 4 we shall use two other equivalent methods of computing path integrals: mode regularization and dimensional regularization.

We shall actually use a somewhat different way to evaluate path integrals, by decomposing $x^i(t)$ as follows. We expand the continuous paths $x^i(t)$ into a fixed classical "background" part $x^i_{bg}(t)$ plus "quantum fluctuations" $q^i(t)$

$$x^i(t) = x^i_{bg}(t) + q^i(t). \tag{1.18}$$

Here $x^i_{bg}(t)$ is a fixed function: it solves the classical equations of motion and takes into account the boundary conditions ($x^i(0) = y^i$ and $x^i(\beta) = z^i$)

$$x^i_{bg}(t) = y^i + (z^i - y^i)\frac{t}{\beta}, \tag{1.19}$$

while the arbitrary fluctuations $q^i(t)$ vanish at the boundaries. One may interpret $x^i_{bg}(t)$ as the origin and $q^i(t)$ as the coordinates of the "space of paths".

Now one can compute the path integral (1.15) for a vanishing potential

$$T(z, y; \beta) = \int Dx\, \mathrm{e}^{-\frac{1}{\hbar}S[x]} = \int D(x_{bg} + q)\, \mathrm{e}^{-\frac{1}{\hbar}S[x_{bg}+q]}$$

$$= \int Dq\, \mathrm{e}^{-\frac{1}{\hbar}(S[x_{bg}]+S[q])} = \mathrm{e}^{-\frac{1}{\hbar}S[x_{bg}]} \int Dq\, \mathrm{e}^{-\frac{1}{\hbar}S[q]}$$
$$= A\mathrm{e}^{-\frac{1}{\hbar}S[x_{bg}]} = A\mathrm{e}^{-\frac{m(z-y)^2}{2\beta\hbar}} \tag{1.20}$$

where we have used the translational invariance of the path integral measure $Dx = D(x_{bg} + q) = Dq$ (at the discretized level this is evident from writing $d^n x_k = d^n(x_{k,bg} + q_k) = d^n q_k$) and the fact that in the action there is no term linear in q^i (the action is quadratic in q^i, but the term linear in q^i must also be linear in x_{bg}, but then this term must vanish by the equations of motion). Finally, the constant $A = \int Dq \exp(-\frac{1}{\hbar}S[q])$ is not determined by this method, but it can be fixed by requiring that (1.20) solves the Schrödinger equation (1.6) with the boundary condition in (1.7). The value $A = (m/2\pi\hbar\beta)^{n/2}$ is sometimes called the Feynman measure.

1.2 Quantum mechanical path integrals in curved space require regularization

The path integrals for the quantum mechanical systems we shall discuss have a Hamiltonian $\hat{H}(\hat{x}, \hat{p})$ which is more general than $\hat{T}(\hat{p}) + \hat{V}(\hat{x})$. We shall typically be considering models with a Euclidean Lagrangian of the form $L = \frac{1}{2}g_{ij}(x)\frac{dx^i}{dt}\frac{dx^j}{dt} + iA_i(x)\frac{dx^i}{dt} + V(x)$, where $i, j = 1, \ldots, n$. These systems are one-dimensional quantum field theories with double-derivative interactions, and hence they are not ultraviolet finite by power counting; rather, the one- and two-loop diagrams are divergent as we shall discuss in detail in the next section. The ultraviolet infinities cancel in the sum of diagrams, but one needs to regularize individual diagrams which are divergent. The results of individual diagrams are then regularization-scheme dependent, and also the results for the sum of diagrams are finite but scheme dependent. One must then add finite counterterms which are also scheme dependent, and which must be chosen such that certain physical requirements are satisfied (renormalization conditions). Of course, the final physical answers should be the same, no matter which scheme one uses. Since we shall be working with actions defined on a compact time-interval, there are no infrared divergences. We shall also discuss nonlinear sigma models with fermionic point particles $\psi^a(t)$ with again $a = 1, \ldots, n$. Also one- and two-loop diagrams containing fermions can be power-counting divergent. For applications to chiral and gravitational anomalies the most important cases are the rigidly supersymmetric models, in particular the quantum mechanical models with $N = 1$ and $N = 2$ supersymmetry, but nonsupersymmetric models with or without fermions will also be used as they are needed for applications to trace anomalies.

Quantum mechanical path integrals can be used to compute anomalies of n-dimensional quantum field theories. This was first shown by Alvarez-Gaumé and Witten (AGW) [1, 35, 36], who studied various chiral and gravitational anomalies (see also [37, 38]). Subsequently, Bastianelli and van Nieuwenhuizen [39, 40] extended their approach to trace anomalies. With the formalism developed below one can now, in principle, compute any anomaly, and not only chiral anomalies. In the work of Alvarez-Gaumé and Witten, the chiral anomalies themselves were written directly as a path integral in which the fermions have periodic boundary conditions. Similarly, the trace anomalies lead to path integrals with antiperiodic boundary conditions for the fermions. These are, however, only special cases, and in our approach any Jacobian will lead to a corresponding set of boundary conditions.

Because chiral anomalies have a topological character, one would expect details of the path integral to be unimportant and only one-loop graphs on the worldline to contribute. In fact, in the approach of AGW this is indeed the case.[3] On the other hand, for trace anomalies, which have no topological interpretation, the details of the path integral do matter and higher loops on the worldline contribute. In fact, it was precisely because three-loop calculations of the trace anomaly based on quantum mechanical path integrals initially did not agree with results known from other methods, that we started a detailed study of path integrals for nonlinear sigma models. These discrepancies have been resolved in the meantime, and the resulting formalism is presented in this book.

The reason that we do not encounter infinities in loop calculations for QM nonlinear sigma models is different from a corresponding statement for QM linear sigma models. For a linear sigma model with a kinetic term $\frac{1}{2}\dot{x}^i\dot{x}^i$ on an infinite t-interval, the propagator behaves as $1/k^2$ for large momenta, and vertices from $V(x)$ do not contain derivatives, hence loops $\int dk[\cdots]$ will always be finite. For nonlinear sigma models with $L = \frac{1}{2}g_{ij}(x)\dot{x}^i\dot{x}^j$, propagators still behave like k^{-2} but vertices now behave like k^2 (as in ordinary quantum gravity), hence single loops are linearly divergent by power counting and double loops are logarithmically

[3]Their approach uses a particular linear combination of general coordinate and local Lorentz transformations, and for this symmetry one only needs to evaluate single loops on the worldline. However if one directly computes the anomaly of the Lorentz operator $\gamma^{\mu\nu}\gamma_5$, using the same steps as in the case of the chiral operator γ_5 for gauge fields in flat space, one needs higher loops on the worldline. We discuss this at the end of Section 6.3.

divergent. It is clear by inspection of

$$\langle z|e^{-\frac{\beta}{\hbar}\hat{H}}|y\rangle = \int_{-\infty}^{\infty} \langle z|e^{-\frac{\beta}{\hbar}\hat{H}}|p\rangle\langle p|y\rangle\, d^n p \tag{1.21}$$

that no infinities should be present: the matrix element $\langle z|\exp(-\frac{\beta}{\hbar}\hat{H})|y\rangle$ is **finite** and **unambiguous**. Indeed, we could in principle insert a complete set of momentum eigenstates as indicated, and then expand the exponent and move all $\hat{p}$ operators to the right and all $\hat{x}$ operators to the left, taking commutators into account. The integral over $d^n p$ is Gaussian and converges. To any given order in β we would then find a finite and well-defined expression.[4] Hence, also **the path integrals should be finite**.

The mechanism by which loops based on path integrals are finite is different in phase space and configuration space path integrals. In the phase space path integrals the momenta are independent variables and the vertices contained in $H(x,p)$ are without derivatives. (The only derivatives are due to the term $p\dot{x}$, whereas the term $\frac{1}{2}p^2$ is free from derivatives.) The propagators and vertices are nonsingular functions (containing at most step functions but no delta functions) which are integrated over the finite domain $[-\beta, 0]$, hence no infinities arise. (We use the interval $[-\beta, 0]$ instead of $[0, \beta]$, but it is easy to change notation to go from one to the other.) In the configuration space path integrals, on the other hand, there are divergences in individual loops, as we mentioned. The reason for this is that although one still integrates over the finite domain $[-\beta, 0]$, single derivatives of the propagators are discontinuous and double derivatives are divergent (they contain delta functions).

However, since the results of configuration-space path integrals should be the same as those of phase-space path integrals, these infinities should not be there in the first place. The resolution of this paradox is that **configuration-space path integrals contain a new kind of ghost.** These ghosts are needed to exponentiate the factors $(\det g_{ij})^{1/2}$ which are produced when one integrates out the momenta. Historically, the cancellation of divergences at the one-loop level was first found by Lee and Yang [41], who studied nonlinear deformations of harmonic oscillators, and who wrote these determinants as new terms in the action of the form

$$\frac{1}{2}\sum_t \ln\det g_{ij}(x(t)) = \frac{1}{2}\delta(0)\int \mathrm{tr}\,\ln\, g_{ij}(x(t))\, dt. \tag{1.22}$$

To obtain the right-hand side one may multiply the left-hand side by $\Delta t/\Delta t$ and replace $1/\Delta t$ by $\delta(0)$ in the continuum limit. For higher loops,

[4] This program is executed in Section 2.5 to order β. For reasons explained there, we count the difference $(z - y)$ as being of order $\beta^{1/2}$.

it is inconvenient to work with $\delta(0)$; rather, we shall use the new ghosts in precisely the same manner as one uses the Faddeev–Popov ghosts in gauge theories: they occur in all possible manners in Feynman diagrams and have their own Feynman rules. These ghosts for quantum mechanical path integrals were first introduced by Bastianelli [39].

In configuration space, loops with ghost particles cancel divergences in corresponding loop graphs without ghost particles. Generically one has

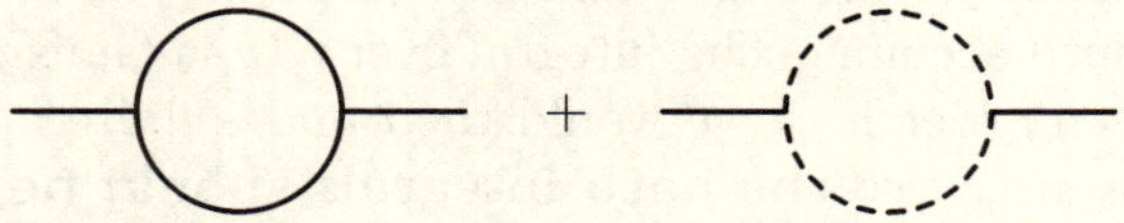

= finite.

However, the fact that the infinities cancel does not mean that the remaining finite parts are unambiguous. One must regularize the divergent graphs, and different regularization schemes can lead to different finite parts, as is well known from field theory. Since our actions are of the form $\int_{-\beta}^{0} L\,dt$, we are dealing with one-dimensional quantum field theories in a finite "spacetime". If one is not dealing with a circle, translational invariance is broken, and propagators depend on t and s, not only on $t-s$. In configuration space the propagators contain singularities. For example, the propagator for a free quantum particle $q(t)$ corresponding to $L=\frac{1}{2}\dot{q}^2$ with boundary conditions $q(-\beta)=q(0) = 0$ is proportional to $\Delta(\sigma,\tau)$, where $\sigma = s/\beta$ and $\tau = t/\beta$, with $-\beta \leq s,t \leq 0$ and $-1 \leq \sigma,\tau \leq 0$

$$\langle q(\sigma)q(\tau)\rangle \approx \Delta(\sigma,\tau) = \sigma(\tau+1)\theta(\sigma-\tau) + \tau(\sigma+1)\theta(\tau-\sigma). \quad (1.23)$$

It is easy to check that $\partial_\sigma^2\Delta(\sigma,\tau) = \delta(\sigma-\tau)$ and $\Delta(\sigma,\tau) = 0$ at $\sigma = -1, 0$ and $\tau = -1, 0$ (use $\partial_\sigma\Delta(\sigma,\tau) = \tau + \theta(\sigma-\tau)$).

It is clear that Wick contractions of $\dot{q}(\sigma)$ with $q(\tau)$ will contain a factor of $\theta(\sigma-\tau)$, and $\dot{q}(\sigma)$ with $\dot{q}(\tau)$ a factor $\delta(\sigma-\tau)$. Also the propagators for the ghosts contain factors of $\delta(\sigma-\tau)$. Thus one needs a consistent, unambiguous and workable regularization scheme for products of the distributions $\delta(\sigma-\tau)$ and $\theta(\sigma-\tau)$. In mathematics the products of distributions are ill-defined [42]. Thus, it comes as no surprise that in physics different regularization schemes give different answers for such integrals. For example, consider the following two familiar ways of evaluating the product of distributions: smoothing of distributions and using Fourier transforms. Suppose one is required to evaluate

$$I = \int_{-1}^{0}\int_{-1}^{0} \delta(\sigma-\tau)\theta(\sigma-\tau)\theta(\sigma-\tau)\,d\sigma\,d\tau. \quad (1.24)$$

Smoothing of a distribution can be achieved by approximating $\delta(\sigma-\tau)$ and $\theta(\sigma-\tau)$ by some smooth functions and requiring that at the regulated level one still has the relation $\delta(\sigma-\tau) = \frac{\partial}{\partial\sigma}\theta(\sigma-\tau)$. One then obtains

$I = \frac{1}{3}\int_{-1}^{0}\int_{-1}^{0}\frac{\partial}{\partial\sigma}[\theta(\sigma-\tau)]^3\,d\sigma\,d\tau = \frac{1}{3}$. On the other hand, if one were to interpret the delta function $\delta(\sigma-\tau)$ to mean that one should evaluate the function $\theta(\sigma-\tau)^2$ at $\sigma=\tau$ one obtains $\frac{1}{4}$. One could also decide to use the representations

$$\delta(\sigma-\tau) = \int_{-\infty}^{\infty}\frac{d\lambda}{2\pi}\,e^{i\lambda(\sigma-\tau)}$$

$$\theta(\sigma-\tau) = \int_{-\infty}^{\infty}\frac{d\lambda}{2\pi i}\frac{e^{i\lambda(\sigma-\tau)}}{\lambda-i\epsilon} \qquad \text{with } \epsilon>0. \tag{1.25}$$

Formally, $\partial_\sigma\theta(\sigma-\tau) = \delta(\sigma-\tau) - \epsilon\,\theta(\sigma-\tau)$, and upon taking the limit ϵ tending to zero one would again expect to obtain the value $\frac{1}{3}$ for I. However, if one first integrates over σ and τ, one finds

$$I = \left[\int_{-\infty}^{\infty}\frac{dy}{2\pi}\frac{(2-2\cos y)}{y^2}\right]\left(\int_{-\infty}^{\infty}\frac{d\lambda}{2\pi i}\frac{1}{\lambda-i\epsilon}\right)^2. \tag{1.26}$$

Depending on the prescription used to evaluate the last integral, one could obtain different results. Clearly, using different methods to evaluate I leads to different answers. Without further specifications, integrals such as I are indeed ambiguous and make no sense.

In the applications we are going to discuss, we sometimes choose a regularization scheme that reduces the path integral to a finite-dimensional integral. For example, for time slicing one chooses a finite set of intermediate points, and for mode regularization one begins with a finite number of modes for each one-dimensional field. Another scheme we use is dimensional regularization: here one regulates the various Feynman diagrams by moving away from $d=1$ dimensions, and performing partial integrations which make the integral manifestly finite at $d=1$. Afterwards one returns to $d=1$ and computes the values of these finite integrals. One omits boundary terms in the extra dimensions; this can be justified by noting that there are factors of $e^{i\mathbf{k}(\mathbf{t}-\mathbf{s})}$ in the propagators due to translation invariance in the extra D dimensions. They yield the Dirac delta functions $\delta^D(\mathbf{k}_1+\mathbf{k}_2+\cdots+\mathbf{k}_n)$ upon integration over the extra space coordinates. A derivative with respect to the extra space coordinate which yields, for example, a factor $\mathbf{k}_1$ can be replaced by $-\mathbf{k}_2-\mathbf{k}_3-\cdots-\mathbf{k}_n$ due to the presence of the delta function, and this replacement is equivalent to a partial integration without boundary terms. These are formal manipulations which should be viewed as specifying the regularization scheme.

In time slicing we find the value $I=\frac{1}{4}$ for (1.24): in fact, as we shall see, in this case the delta function is a Kronecker delta which gives the product of the θ functions at the point $\sigma=\tau$. In mode regularization, one finds $I=\frac{1}{3}$ because now $\delta(\sigma-\tau)$ is indeed $\partial_\sigma\theta(\sigma-\tau)$ at the regulated level. In dimensional regularization one must first decide which derivatives are contracted with which derivatives in $D+1$ dimensions (for example, $({}_\mu\Delta_\nu)({}_\mu\Delta)(\Delta_\nu)$). This follows from the form of the action in $D+1$ dimensions. Then one applies the usual manipulations of dimensional regularization in $D+1$ dimensions until one reaches a convergent integral which can directly be evaluated in one dimension.[5]

As we have seen, different regularization schemes lead to finite well-defined results for a given diagram which are in general different, but there are also ambiguities in the vertices: the finite one- and two-loop counterterms have not been fixed. The physical requirement that the theory be based on a given quantum Hamiltonian removes the ambiguities in the counterterms: for time slicing Weyl ordering of $\hat{H}$ directly produces the counterterms, while for the other schemes the requirement that the transition element satisfies the Schrödinger equation with a given Hamiltonian $\hat{H}$ fixes the counterterms. Thus in all of these schemes the renormalization condition is that the transition element should be derived from the same particular Hamiltonian $\hat{H}$.

The first scheme, time slicing (TS), has the advantage that one can deduce it directly from the operatorial formalism of quantum mechanics. This regularization can be considered to be equivalent to lattice regularization of standard quantum field theories. It is the approach followed by Dirac and Feynman. One must specify the Hamiltonian $\hat{H}$ with an a priori fixed operator ordering; this ordering corresponds to the renormalization conditions in this approach. All further steps are finite and unambiguous. This approach breaks general coordinate invariance in target space which is then recovered by a specific finite counterterm ΔV_{TS} in the action of the path integral. (To simplify the notation, we denote these counterterms in later sections by V_{TS} instead of ΔV_{TS}.) This counterterm also follows unambiguously from the initial Hamiltonian and is itself not coordinate invariant either. However, if the initial Hamiltonian is general coordinate invariant (as an operator, see Section 2.5) then the final result (the transition element) will also be general coordinate invariant.

The second scheme, mode regularization (MR), will be constructed directly without referring to the operatorial formalism. It can be thought

[5] For an example of an integral where dimensional regularization is applied, consider

$$
\begin{aligned}
J &= \int_{-1}^{0} d\sigma \int_{-1}^{0} d\tau\, ({}^{\bullet}\Delta^{\bullet})\,({}^{\bullet}\Delta)\,(\Delta^{\bullet}) \\
&= \int_{-1}^{0} d\sigma \int_{-1}^{0} d\tau\, [1-\delta(\sigma-\tau)]\,[\tau+\theta(\sigma-\tau)]\,[\sigma+\theta(\tau-\sigma)] \qquad (1.27)
\end{aligned}
$$

where dots on the left and right denote derivatives with respect to the first and second variable. One finds $J = -\frac{1}{6}$ for time slicing, see (2.270). Furthermore, $J = -\frac{1}{12}$ for mode regularization, see (3.82). In dimensional regularization one rewrites the integrand as $({}_\mu\Delta_\nu)({}_\mu\Delta)(\Delta_\nu)$ and one finds $J = -\frac{1}{24}$, see (4.24).

of as the equivalent of momentum cut-off in QFT.[6] It is close to the intuitive notion of path integrals, that are meant to give a global picture of the quantum phenomena by summing over entire paths (while one may view the time discretization method as being closer to the local picture of the differential Schrödinger equation, since one imagines the particle propagating by small time steps). Mode regularization gives, in principle, a nonperturbative definition of path integrals in that one does not have to expand the exponential of the interaction part of the action. However, this regularization also breaks general coordinate invariance, and one needs a different finite noncovariant counterterm ΔV_{MR} to recover it.

Finally, the third regularization scheme, dimensional regularization (DR), is based on the dimensional continuation of the ambiguous integrals appearing in the loop expansion. It is inherently a perturbative regularization, but it is the optimal one for perturbative computations in the following sense. It does not break general coordinate invariance at intermediate stages and the counterterm ΔV_{DR} is Einstein and local Lorentz invariant.

All of these different regularization schemes will be presented in separate chapters. Since our derivation of the path integrals contains several steps, each requiring a detailed discussion, we have decided to put all of these special discussions in separate sections after the main derivation. This has the advantage that one can read each section independently. The structure of our discussions is summarized by the flow chart in Fig. 1.1.

We shall first discuss time slicing, the lower part of the flow chart. This discussion is first given for bosonic systems with $x^i(t)$ and afterwards for systems with fermions. In the bosonic case, we first construct discretized phase-space path integrals, then discretized configuration-space path integrals, to be followed by the continuous configuration-space path integrals, and finally the continuous phase-space path integrals. We show that after Weyl ordering of the Hamiltonian operator $\hat{H}(\hat{x}, \hat{p})$ one obtains a path integral with a midpoint rule (Berezin's theorem). Then we repeat the analysis for fermions.

Next, we consider mode number regularization (the upper part of the flow chart). Here we define the path integrals *ab initio* in configuration space with the naive classical action and a counterterm ΔV_{MR} which is at first left unspecified. We then proceed to fix ΔV_{MR} by imposing the

[6] In more complicated cases, such as path integrals in spaces with a topological vacuum (for example, the kink background in Euclidean quantum mechanics), the mode regularization scheme and the momentum regularization scheme with a sharp cut-off are not equivalent (for example, they give different answers for the quantum mass of the kink). However, if one replaces the sharp energy cut-off by a smooth cut-off, those schemes become equivalent [43]. We do not consider such topologically nontrivial backgrounds.

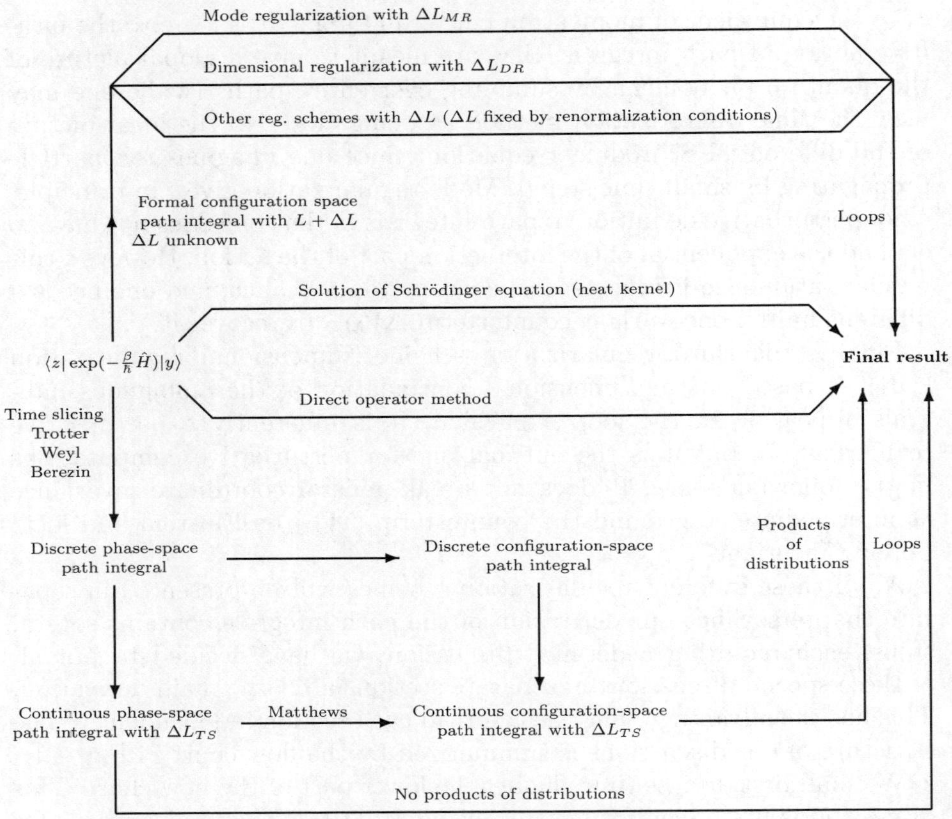

Fig. 1.1 Flow chart of Part I of the book.

requirement that the Schrödinger equation be satisfied with a specific Hamiltonian $\hat{H}$. Having fixed ΔV_{MR}, one can proceed to compute loops at any desired order.

Finally, we present dimensional regularization along similar lines. The counterterm 5 is now denoted by ΔV_{MR}. Each section can be read independently of the previous ones.

In all three cases we define the theory by the Hamiltonian $\hat{H}$ and then construct the path integrals and Feynman rules which correspond to $\hat{H}$. The choice of $\hat{H}$ defines the physical theory. One may be prejudiced about which $\hat{H}$ makes physical sense (for example, many physicists require that $\hat{H}$ preserves general coordinate invariance), but in our work one does not have to restrict oneself to these particular $\hat{H}$. Any $\hat{H}$, no matter how unphysical, leads to a corresponding path integral and corresponding Feynman rules. We repeat that the path integral and Feynman rules depend on the regularization scheme chosen, but the final result for the transition element and correlation functions are the same in each scheme.

In the time-slicing approach we shall solve some of the following basic problems: **given** a Hamiltonian operator $\hat{H}(\hat{x}, \hat{p})$ with arbitrary but a-priori fixed operator ordering, find a path integral expression for the matrix element[7] $\langle z|\exp(-\frac{\beta}{\hbar}\hat{H})|y\rangle$. (The bra $\langle z|$ and ket $|y\rangle$ are eigenstates of the position operator $\hat{x}^i$ with eigenvalues z^i and y^i, respectively. For fermions we shall use coherent states as bra and ket.) One way to obtain such a path integral representation is, as we have discussed, to insert complete sets of x- and p-eigenstates (namely N sets of p-eigenstates and $N-1$ sets of x-eigenstates), in the manner first studied by Dirac [44] and Feynman [45, 46], and leads to the following result:

$$\langle z|\mathrm{e}^{-\frac{\beta}{\hbar}\hat{H}}|y\rangle \approx \int Dx\, Dp\, \mathrm{e}^{-\frac{1}{\hbar}\int_{-\beta}^{0} L\, dt} \tag{1.28}$$

where $L = -ip_i(t)\frac{dx^i}{dt} + H(x, p)$ in our Euclidean phase-space approach. However, several questions arise if one studies (1.28).

(i) What is the precise relation between the operator $\hat{H}(\hat{x}, \hat{p})$ and the function $H(x, p)$? Different operator orderings of $\hat{H}$ are expected to lead to different functions $H(x, p)$. Are there special orderings of $\hat{H}$ for which $H(x, p)$ is particularly simple? And if so, are these special orderings consistent with general coordinate invariance?

(ii) What is the precise meaning of the measures $Dx\, Dp$ in phase space and Dx in configuration space in theories with external gravitational fields? Is there a normalization constant in front of the path integral? Does the measure depend on the metric? The measure $Dx\, Dp = \prod_{i=1}^{N-1} dx^i \prod_{i=1}^{N} dp_i$ is not a canonically invariant measure (not equal to the Liouville measure) because there is one more dp than dx. Does this have implications?

(iii) What are the boundary conditions one must impose on the paths over which one sums? One expects that all paths must satisfy the Dirichlet boundary conditions $x^i(-\beta) = y^i$ and $x^i(0) = z^i$, but are there also boundary conditions on $p_i(t)$? Is it possible to consider classical paths in phase space which satisfy boundary conditions both at $t = -\beta$ and at $t = 0$?

(iv) How does one compute such path integrals in practice? Performing the integrations over dx^i and dp_i for finite N and then taking

[7] The results in this book are for Euclidean path integrals with $L = -ip\dot{x} + H(x, p)$. However, they hold equally well in Minkowskian time, at least at the level of perturbation theory, with operators $\exp(-\frac{i}{\hbar}\hat{H}t)$ and path integrals with $\exp(\frac{i}{\hbar}\int L_M\, dt)$, where L_M is the Lagrangian in Minkowskian time, related to the positive-definite Euclidean Lagrangian L by an inverse a Wick rotation ($t \to +it$) and an extra overall minus sign.

the limit $N \to \infty$ is in practice hardly possible. Is there a simpler scheme by which one can compute the path integral loop-by-loop, and what are the precise Feynman rules for such an approach? Does the measure contribute to the Feynman rules?

(v) It is often advantageous to use a background formalism and to decompose bosonic fields $x(t)$ into background fields $x_{bg}(t)$ and quantum fluctuations $q(t)$. One can then require that $x_{bg}(t)$ satisfies the boundary conditions so that $q(t)$ vanishes at the endpoints. However, inspired by string theory, one can also compactify the interval $[-\beta, 0]$ to a circle, and then decompose $x(t)$ into a center of mass coordinate x_c and quantum fluctuations about it. What is the relation between both approaches?

(vi) When one is dealing with $N = 1$ supersymmetric systems, one has real (Majorana) particles $\psi^a(t)$. How does one define the Hilbert space in which $\hat{H}$ is supposed to act? Must one also impose an initial and a final condition on $\psi^a(t)$, even though the Dirac equation is only linear in (time) derivatives? We shall introduce operators $\hat{\psi}^a$ and $\hat{\psi}_a^\dagger$ and construct coherent states by contracting them with Grassmann variables $\bar{\eta}_a$ and η^a. If $\hat{\psi}_a^\dagger$ is the hermitian conjugate of $\hat{\psi}^a$, then is $\bar{\eta}_a$ the complex conjugate of η^a?

(vii) In certain applications, for example the calculation of trace anomalies, one must evaluate path integrals over fermions with antiperiodic boundary conditions. In the work of AGW the chiral anomalies were expressed in terms of integrals over the zero modes of the fermions. For antiperiodic boundary conditions there are no zero modes. How then should one compute trace anomalies from quantum mechanics?

These are some of the questions which come to mind if one contemplates (1.28) for some time. In the literature one can find discussions of some of these questions [47, 48], but we have made an effort to give a consistent discussion of all of them. Answers to these questions can be found in Chapter 8. New material in this book is an exact evaluation of all discretized expressions in the TS scheme as well as the derivation of the MR and DR schemes in curved space.

1.3 Power counting and divergences

Let us now give some examples of divergent graphs. The precise form of the vertices is given later, in (2.85), but for the discussion in this section we only need the qualitative features of the action. The propagators we are going to use later in this book are not of the simple form $1/k^2$ for a scalar, rather they have the form $\sum_{n=1}^{\infty}(2/\pi^2 n^2)\sin(\pi n\tau)\sin(\pi n\sigma)$ due to boundary conditions. (Even the propagator for time slicing can be cast into this form by Fourier transformation.) However, for ultraviolet divergences the sum of $1/n^2$ is equivalent to an integral over $1/k^2$, and in this section we analyze Feynman graphs with $1/k^2$ propagators. The physical justification is that ultraviolet divergences should not feel the boundaries.

Consider first the self-energy. At the one-loop level the self-energy without external derivatives receives contributions from the following two graphs

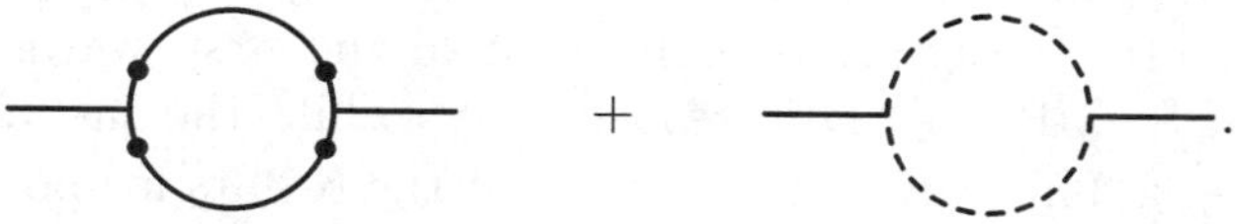

We used the vertices from $\frac{1}{2}[g_{ij}(x) - g_{ij}(z)](\dot{q}^i\dot{q}^j + a^i a^j + b^i c^j)$, where $x^i = x^i(\tau) = z^i + q^i(\tau)$ and $z^i = x^i(0)$. Dots indicate derivatives and dashed lines denote the ghost particles a^i, b^i, c^i. The two divergences are proportional to $\delta^2(\sigma - \tau)$ and cancel, but there are ambiguities in the finite part which must be fixed using suitable conditions. (In quantum field theories with divergences one calls these conditions "renormalization conditions".) In momentum space both graphs are linearly divergent, but the linear divergence $\int dk$ cancels in the sums of the graphs and the two remaining logarithmic divergences $\int dk\, k/k^2$ cancel by symmetric integration leaving in general a finite but ambiguous result.

Another example is the self-energy with one external derivative

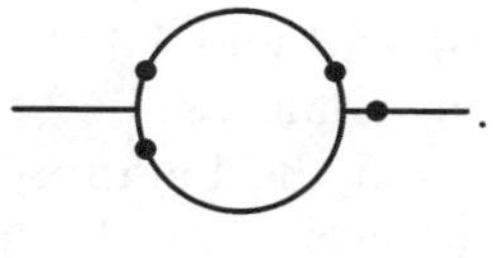

This graph is logarithmically divergent, $\int dk\, k^3/(k^2)^2$, but using symmetric integration it again leaves a finite but ambiguous part.

All three regularization schemes give the same answer for all one-loop graphs, so the one-loop counterterms are the same; in fact, there are no one-loop counterterms at all in any of the schemes if one starts with an Einstein-invariant Hamiltonian.[8]

At the two-loop level, there are similar cancellations and ambiguities. Consider the following vacuum graphs (vacuum graphs will play an

[8] If one were to use the Einstein-noninvariant Hamiltonian $g^{1/4-\alpha}\hat{p}_i\sqrt{g}g^{ij}\hat{p}_j g^{1/4+\alpha}$, one would obtain in the TS scheme a one-loop counterterm proportional to $\hbar p_i g^{ij}\partial_j \ln g$ in phase space or $\hbar\dot{x}^i\partial_i \ln g$ in configuration space (see Appendix B).

important role in the applications to anomalies)

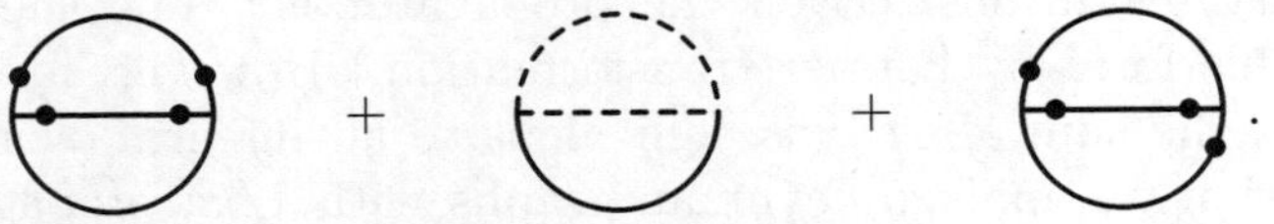

Again the infinities in the upper loop of the first two graphs cancel, but the finite part is ambiguous. The last graph is logarithmically divergent by power counting, and also the two subdivergences are logarithmically divergent by power counting, but actual calculation shows that it is finite but ambiguous (the leading singularities are of the form $\int \frac{dk\,k}{k^2}$ and cancel due to symmetric integration). The sum of the first two graphs yields $(\frac{1}{4}, \frac{1}{4}, \frac{1}{8})$ in TS, MR and DR, respectively, while the last graph yields $(-\frac{1}{6}, -\frac{1}{12}, -\frac{1}{24})$. This explicitly proves that the results for power counting logarithmically divergent graphs are ambiguous, even though the divergences cancel.

It is possible to use standard power-counting methods as used in ordinary quantum field theory to determine all possibly ultraviolet-divergent graphs. Let us interpret our quantum mechanical nonlinear sigma model as a particular QFT in one Euclidean time dimension. We consider a toy model of the type

$$S = \int dt \left[\frac{1}{2} g(\phi)\dot{\phi}\dot{\phi} + A(\phi)\dot{\phi} + V(\phi) \right] \tag{1.29}$$

where the functions $g(\phi), A(\phi)$ and $V(\phi)$ describe the various couplings. For simplicity we omit the indices i and j.

The choice $g(\phi) = 1$, $A(\phi) = 0$ and $V(\phi) = \frac{1}{2}m^2\phi^2$ reproduces a free massive theory, namely a harmonic oscillator of "mass" (frequency) m. The action is dimensionless and the Lagrangian then has the dimension of a mass. From this one deduces that the field ϕ has mass dimension $M^{-1/2}$. Next, let us consider general interactions and expand them in Taylor series

$$V(\phi) = \sum_{n=0}^{\infty} V_n \phi^n, \quad A(\phi) = \sum_{n=0}^{\infty} A_{n+1}\phi^n, \quad g(\phi) = \sum_{n=0}^{\infty} g_{n+2}\phi^n. \tag{1.30}$$

These expansions define the coupling constants V_n, A_n and g_n. We easily deduce the following mass dimensions for such couplings:

$$[V_n] = M^{n/2+1}; \quad [A_n] = M^{n/2}; \quad [g_n] = M^{n/2-1}. \tag{1.31}$$

The interactions correspond to the terms with $n \geq 3$ in (1.31), so all coupling constants have positive mass dimensions. This implies that the theory is super-renormalizable. Namely, from a certain loop level onwards, there are no more superficial divergences by power counting. We can work this out in more detail. Given a Feynman diagram, let us indicate by L the number of loops, I the number of internal lines, V_n, A_n and g_n the numbers of corresponding vertices present in the diagram. One can assign to the diagram a superficial degree of divergence D by

$$D = L - 2I + \sum_n (A_n + 2g_n) \tag{1.32}$$

reflecting the fact that each loop gives a momentum integration $\int dk$, the propagators give factors of k^{-2}, and the A_n and g_n vertices bring in at worst one and two momenta, respectively. Also, the number of loops is given by

$$L = I - \sum_n (V_n + A_n + g_n) + 1. \tag{1.33}$$

Combining these two equations we find that the degree of divergence D is given by

$$D = 2 - L - \sum_n (2V_n + A_n). \tag{1.34}$$

Let us analyze the consequences of this formula by considering first the case with nontrivial $V(\phi)$ couplings only (linear sigma models). Then (1.34) shows that no divergences can ever arise. Consequently, no ambiguities are expected in the path integral quantization of the model either. This is the class of models with $H = T(p) + V(x)$ which is extensively discussed in many textbooks, see for example [9, 21–25, 46, 48–51].

Next, let us consider a nontrivial $A(\phi)$. From (1.34) we see that there is now a possible logarithmic superficial divergence in the one-loop graphs with a single vertex A_n (n can be arbitrary since the extra fields that are not needed to construct the loop can be taken as being external)

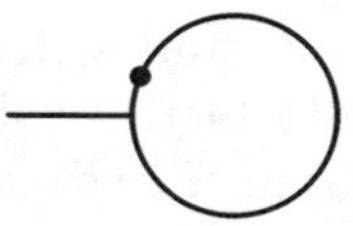
.

The logarithmic singularity actually cancels by symmetric integration, but the finite part which is left over must be fixed unambiguously by specifying a renormalization condition. If A corresponds to an electromagnetic field, gauge invariance can be used as a renormalization condition that fixes the ambiguity completely. In the continuum theory, the action $\int A_i \dot{x}^i \, dt$ is invariant under the gauge transformation $\delta A_i = \partial_i \lambda(x)$. Feynman [45] found that with TS one must evaluate A_i at the midpoints $\frac{1}{2}(x_{k+1} + x_k)$

in order to obtain the Schrödinger equation with a gauge-invariant Hamiltonian[9] $\hat{H} = \frac{1}{2m}(\hat{p} - \frac{e}{c}\hat{A})^2 + \hat{V}$. For further discussion, see for example, chapters 4 and 5 of [21]. If the regularization scheme chosen to define the above graphs does not respect gauge invariance, one must add local finite counterterms by hand to restore the gauge invariance.

Finally, consider the most general case with $g(\phi)$. There can be linear and logarithmic divergences in one-loop graphs as in

and logarithmic divergences at two-loops

Notice that (1.34) is independent of g_n. This implies that at the one- and two-loop level one can construct an infinity of divergent graphs from a given divergent graph by inserting g_n vertices. The following diagrams illustrate this fact

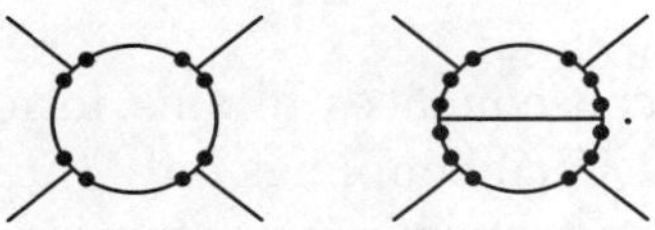

As we shall see, the ghost cancels the leading divergences, but we repeat that finite ambiguities remain which must be fixed by renormalization conditions. Of course, general coordinate invariance must also be imposed, but this symmetry requirement is not enough to fix all of the renormalization conditions. One can understand this from the following observation. In the canonical approach different orderings of the Hamiltonian $g_{ij}p^i p^j$ lead to ambiguities proportional to $(\partial_i g_{jk})^2$ and $\partial_i \partial_j g_{kl}$, and from them one can form the scalar curvature R. So one can always add to the Hamiltonian a term proportional to R and still maintain general coordinate invariance in the target space. In fact, we should distinguish between an explicit R term in the Hamiltonian $\hat{H}$ and an explicit R term in the action which appears in the path integrals. In all three schemes we shall discuss,

[9] To avoid confusion we repeat that in our treatment of path integrals there are no ambiguities. If one takes a Hamiltonian operator which is gauge invariant (commutes with the generator of gauge transformations at the operator level), then the corresponding path integral evaluated with time slicing uses the midpoint rule, but when using another Hamiltonian the midpoint rule does not hold.

one always produces a term $\frac{1}{8}R$ in the action as one proceeds from $\hat{H}$ to the path integral. So for a free scalar particle with $\hat{H}$ without an R term, the path integral contains a term $\frac{1}{8}R$ in the potential. However, in supersymmetric theories $\hat{H}$ is obtained by evaluating the susy anticommutator $\{\hat{Q}, \hat{Q}\}$, and one finds that this $\hat{H}$ contains a term $-\frac{1}{8}R$, but then in the corresponding path integral one does not obtain any R term.

1.4 Brief summary of the three regularization schemes

For experts who want a quick review of the main technical issues covered in the first part of this book, we give here a brief summary of the three regularization schemes we are going to use; namely, time slicing (TS), mode regularization (MR) and dimensional regularization (DR).

Time slicing

We begin with bosonic systems with arbitrary Hamiltonians $\hat{H}$, quadratic in momenta. Starting from the matrix element $\langle z|\exp(-\frac{\beta}{\hbar}\hat{H})|y\rangle$ (which we call the transition amplitude or transition element) with arbitrary but a-priori fixed operator ordering in $\hat{H}$, we insert complete sets of position and momentum eigenstates, and obtain the discretized propagators and vertices in *closed* form. These results tell us how to evaluate equal-time contractions in the corresponding continuum Euclidean path integrals, as well as products of distributions which are present in Feynman graphs, such as

$$I = \int_{-1}^{0}\int_{-1}^{0} \delta(\sigma-\tau)\theta(\sigma-\tau)\theta(\tau-\sigma)\, d\sigma\, d\tau. \tag{1.35}$$

It is found that $\delta(\sigma-\tau)$ should be viewed as a Kronecker delta function, even in the continuum limit, and the step functions as functions with $\theta(0) = \frac{1}{2}$ (yielding $I = \frac{1}{4}$). Here a Kronecker delta function means that $\int \delta(\sigma-\tau) f(\sigma)\, d\sigma = f(\tau)$, even when $f(\sigma)$ is a product of distributions.

We show that the kernel $\langle x_{k+1}|\exp(-\frac{\epsilon}{\hbar}\hat{H})|x_k\rangle$ with $\epsilon = \beta/N$ may be approximated by $\langle x_{k+1}|(1-\frac{\epsilon}{\hbar}\hat{H})|x_k\rangle$ for large N. For linear sigma models this result is well known and can be rigorously proven ("the Trotter formula"). For nonlinear sigma models, the Hamiltonian $\hat{H}$ is **rewritten** in Weyl ordered form (which leads to extra terms in the action for the path integrals of order $\hbar$ and $\hbar^2$), and the midpoint rule follows automatically (so not because we require gauge invariance). The continuum path integrals thus obtained are phase-space path integrals. By integrating out the momenta we obtain configuration-space path integrals. We discuss the relation between both of them (Matthews' theorem), both for our quantum mechanical nonlinear sigma models and for four-dimensional Yang–Mills theories.

The configuration-space path integrals contain new ghosts (anticommuting $b^i(\tau)$, $c^i(\tau)$ and commuting $a^i(\tau)$), obtained by exponentiating the factors $[\det g_{ij}(x(\tau))]^{1/2}$ which result when one integrates out the momenta. At the one-loop level these ghosts merely remove the overall $\delta(\sigma - \tau)$ singularity in the $\dot{x}\dot{x}$ propagator, but at higher loops they are as useful as in quantum chromodynamics (QCD) and electroweak gauge theories. In QCD one can choose a unitary gauge without ghosts, but loop calculations become horrendous. Similarly, one could start without ghosts and try to renormalize the theory in a consistent manner, but this is far more complicated than working with ghosts. Since the ghosts arise when we integrate out the momenta, it is natural to keep them. We stress that at any stage all expressions are finite and unambiguous once the operator $\hat{H}$ has been specified. As a result we do not have to fix normalization constants at the end by physical arguments, but the "measure" is unambiguously derived in explicit form. Several two- and three-loop examples are worked out, and confirm our path integral formalism in the sense that the results agree with a direct evaluation using operator methods for the canonical variables $\hat{p}$ and $\hat{x}$.

We then extend our results to fermionic systems. We define and use coherent states, define Weyl ordering and derive a fermionic midpoint rule, and also obtain the fermionic discretized propagators and vertices in closed form, with similar conclusions as for the continuum path integral for the bosonic case.

Particular attention is paid to the operatorial treatment of Majorana fermions. It is shown that "fermion-doubling" (by adding a full set of noninteracting Majorana fermions) and "fermion-halving" (by combining pairs of interacting Majorana fermions into Dirac fermions) yield different propagators and vertices but the same physical results such as anomalies.

Mode regularization

As quantum mechanics can be viewed as a one-dimensional quantum field theory (QFT), we can follow the same approach in quantum mechanics as that familiar from four-dimensional quantum field theories. One way to formulate quantum field theory is to expand fields into a complete set of functions and integrate in the path integral over the coefficients of these functions. One could try to derive this approach from first principles, starting, for example, from canonical methods for operators, but we shall follow a simpler approach for mode regularization. Namely we first write down formal rules for the path integral in mode regularization without derivation, and a posteriori fix all ambiguities and free coefficients using consistency conditions.

We start from the formal sum over paths weighted by the phase factor containing the classical action (which is like the Boltzmann factor of statistical mechanics in our Euclidean treatment), and next we suitably define the space of paths. We parametrize all paths as a background trajectory, which takes into account the boundary conditions and quantum fluctuations, which vanish at the time boundaries. Quantum fluctuations are expanded into a complete set of functions (the sines) and path integration is generated by integration over all Fourier coefficients appearing in the mode expansion of the quantum fields. General covariance demands a nontrivial measure $\mathcal{D}x = \prod_t \sqrt{\det g_{ij}(x(t))}\, d^n x(t)$. This measure is formally a scalar under general coordinate (Einstein) transformations, but it is not translationally invariant under $x^i(t) \to x^i(t) + \epsilon^i(t)$. To derive propagators it is more convenient to exponentiate the nontrivial part of the measure by using ghost fields $\prod_t \sqrt{\det g_{ij}(x(t))} \sim \int Da\, Db\, Dc \exp[-\int dt\, \frac{1}{2} g_{ij}(x)(a^i a^j + b^i c^j)]$. At this stage the construction is still formal, and one regulates it by integrating over only a finite number of modes, i.e. by cutting off the Fourier sums at a large mode number M. This makes all expressions well defined and finite. For example, in a perturbative expansion all Feynman diagrams are unambiguous and give finite results. This regularization is in spirit equivalent to a standard momentum cut-off in QFT. The continuum limit is achieved by sending M to infinity. Thanks to the presence of the ghost fields (i.e. of the nontrivial measure) there is no need to cancel infinities (i.e. to perform infinite renormalization). This procedure defines a consistent way of performing path integration, but it cannot determine the overall normalization of the path integral (in QFT it is generically infinite). More generally one would like to know how MR is related to the other regularization schemes. As is well known, in QFT different regularization schemes are related to each other by local counterterms. Defining the necessary renormalization conditions introduces a specific set of counterterms of order $\hbar$ and $\hbar^2$, and fixes all of these ambiguities. We do this last step by requiring that the transition amplitude computed in the MR scheme satisfies the Schrödinger equation with an a-priori fixed Hamiltonian $\hat{H}$ (the same one as used in the approach based on time slicing). The fact that one-dimensional nonlinear sigma models are super-renormalizable guarantees that no counterterms needed to match MR with other regularization schemes (and also needed to recover general coordinate invariance, which is broken by the TS and MR regularizations) are generated beyond two-loops.

Dimensional regularization

The dimensionally regulated path integral can be defined following steps similar to those used in the definition of the MR scheme, but the regularization of the ambiguous Feynman diagrams is achieved differently. One extends the one-dimensional compact time coordinate $-\beta \leq t \leq 0$ by adding D extra noncompact flat dimensions. The propagators on the worldline are now a combined sum–integral, where the integral is a momentum integral as usual in dimensional regularization. At this stage these momentum space integrals define expressions where the variable D can be analytically continued into the complex plane. We are not able to perform these momentum integrals explicitly, but we assume that for arbitrary D all expressions are regulated and define analytic functions, possibly with poles only at integer dimensions, as in usual dimensional regularization. Feynman diagrams are written in coordinate space (t-space), with propagators which contain momentum integrals. Time derivatives d/dt become derivatives $\partial/\partial t^\mu$, but how the indices μ get contracted follows directly from writing the action in $D+1$ dimensions. We perform operations which are valid in dimensional regularization (such as partial integration without boundary terms) to cast the integrals into alternative forms. Dropping the boundary terms in partial integration is always allowed in the extra D dimension, as in ordinary dimensional regularization, but it is only allowed in the original compact time dimension when the boundary term explicitly vanishes because of the boundary conditions. Using partial integrations one rewrites the integrands such that undifferentiated $(D+1)$-dimensional delta functions $\delta^{D+1}(t,s)$ appear, and these allow us to reduce the original integrals to simpler integrals which are finite and unambiguous, and can by computed even after removing the regulator, i.e. in the limit $D=0$. This procedure makes calculations quite easy, and at the same time frees one from the task of computing the analytical continuation of the momentum integrals at arbitrary D. In this way one can compute all Feynman diagrams. As in MR one determines all remaining finite ambiguities by imposing suitable renormalization conditions, namely requiring that the transition amplitude computed with dimensional regularization satisfies the Schrödinger equation with an a-priori given ordering for the Hamiltonian operator $\hat{H}$ (the same as used in mode regularization and time slicing). There are only covariant finite counterterms. Thus dimensional regularization also preserves general coordinate invariance at intermediate steps, and is the most convenient scheme for higher loop calculations. When extended to $N=1$ susy sigma models, dimensional regularization also preserves worldline supersymmetry, as we show explicitly: there are then no counterterms at all in the quantum action.

1.5 A brief history of path integrals

Path integrals yield a third approach to quantum physics, in addition to Heisenberg's operator approach and Schrödinger's wave function approach. They are due to Feynman [45], who in the 1940s developed an approach Dirac had briefly considered in 1932 [44]. In this section we discuss the motivations which led Dirac and Feynman to associate path integrals (with $i/\hbar$ times the action in the exponent) with quantum mechanics. In mathematics Wiener had already studied path integrals in the 1920s but these path integrals contained (-1) times the free action for a point particle in the exponent. Wiener's path integrals were Euclidean path integrals which are mathematically well defined but Feynman's path integrals do not have a similarly solid mathematical foundation. Nevertheless, path integrals have been successfully used in almost all branches of physics: particle physics, atomic and nuclear physics, optics and statistical mechanics [21].

In many applications one uses path integrals for perturbation theory, in particular for semiclassical approximations, and in these cases there are no serious mathematical problems. In other applications one uses Euclidean path integrals, and in these cases they coincide with Wiener's path integrals. However, for the nonperturbative evaluations of path integrals in Minkowski space a completely rigorous mathematical foundation is lacking. The problems increase in dimensions higher than four [52]. Feynman was well aware of these problems, but the physical ideas which stem from path integrals are so convincing that he (and other researchers) considered this not to be worrisome.

Our brief history begins with Dirac who in 1932 wrote an article in a USSR physics journal [44] in which he tried to find a description of quantum mechanics which was based on the Lagrangian instead of the Hamiltonian approach. Dirac was making a trip with Heisenberg around the world, and took the trans-Siberian railway to arrive in Moscow. In those days all work in quantum mechanics (including the work on quantum field theory) started with the Schrödinger equation or operator methods, and in both of these the Hamiltonian played a central role. For quantum mechanics this was fine, but for relativistic field theories an approach based on the Hamiltonian had the drawback that manifest Lorentz invariance was lost (although for QED it had been shown that physical results were nevertheless relativistically invariant). Dirac considered the transition element

$$\langle x_2, t_2 | x_1, t_1 \rangle = K(x_2, t_2 | x_1, t_1) = \langle x_2 | \mathrm{e}^{-\frac{i}{\hbar}\hat{H}(t_2 - t_1)} | x_1 \rangle \tag{1.36}$$

(for time-independent H), and asked whether one could find an expression for this matrix element in which the action was used instead of the Hamiltonian. (The notation $\langle x_2, t_2|x_1, t_1\rangle$ is due to Dirac who called this element a transformation function. Feynman introduced the notation $K(x_2, t_2|x_1, t_1)$ because he used it as the kernel in an integral equation which solved the Schrödinger equation.) Dirac knew that in classical mechanics the time evolution of a system could be written as a canonical transformation, with Hamilton's principal function $S(x_2, t_2|x_1, t_1)$ as the generating functional [53]. This function $S(x_2, t_2|x_1, t_1)$ is the classical action evaluated along the classical path that begins at the point x_1 at time t_1 and ends at the point x_2 at time t_2. In his 1932 article Dirac wrote that $\langle x_2, t_2|x_1, t_1\rangle$ **corresponds to** $\exp \frac{i}{\hbar} S(x_2, t_2|x_1, t_1)$. He used the words "corresponds to" to express that at the quantum level there were presumably corrections so that the exact result for $\langle x_2, t_2|x_1, t_1\rangle$ was different from $\exp \frac{i}{\hbar} S(x_2, t_2|x_1, t_1)$. Although Dirac wrote these ideas down in 1932, they were largely ignored until Feynman started his studies on the role of the action in quantum mechanics.

Towards the end of the 1930s Feynman started studying how to formulate an approach to quantum mechanics based on the action. (Here we follow the biography of Feynman by Mehra [54].) The reason he tackled this problem was that with Wheeler he had developed a theory of quantum electrodynamics from which the electromagnetic field had been eliminated. In this way they hoped to avoid the problems of the self-acceleration and infinite self-energy of an electron which are due to the interactions of an electron with the electromagnetic field and which Lienard, Wiechert, Abraham and Lorentz had tried in vain to solve. The resulting "Wheeler–Feynman theory" arrived at a description of the interactions between two electrons in which no reference was made to any field. It is a so-called action-at-a-distance theory. These theories were nonlocal in space and time. (In modern terminology one might say that the fields A_μ had been integrated out from the path integral by completing squares.) Fokker and Tetrode had found a classical action for such a system, given by [54]

$$\begin{aligned} S = & -\sum_i m_{(i)} \int \left[\frac{dx^\mu_{(i)}}{ds_{(i)}} \frac{dx^\nu_{(i)}}{ds_{(i)}} \eta_{\mu\nu} \right]^{1/2} ds_{(i)} \\ & - \frac{1}{2} \sum_{i \neq j} e_{(i)} e_{(j)} \int\!\!\int \delta \left[\left(x^\mu_{(i)} - x^\mu_{(j)} \right)^2 \right] \frac{dx^\rho_{(i)}}{ds_{(i)}} \frac{dx^\sigma_{(i)}}{ds_{(i)}} \eta_{\rho\sigma} \, ds_{(i)} \, ds_{(j)}. \end{aligned} \tag{1.37}$$

Here the sum over (i) denotes a sum over different electrons. So, two electrons only interact when the relativistic four-distance vanishes, and by taking $i \neq j$ in the second sum, the problem of infinite self-energy was eliminated. Wheeler and Feynman set out to quantize this system, but Feynman noticed that a Hamiltonian treatment was hopelessly complicated.[10] Thus Feynman was looking for an approach to quantum mechanics in which he could avoid the Hamiltonian. The natural object to use was the action.

At this moment in time, an interesting discussion helped him further. A physicist from Europe, Herbert Jehle, who was visiting Princeton, mentioned to Feynman (spring 1941) that Dirac had already (in 1932) studied the problem of how to use the action in quantum mechanics. Together they looked up Dirac's paper, and of course Feynman was puzzled by the ambiguous phrase "corresponds to" in it. He asked Jehle whether Dirac meant that they were equal or not. Jehle did not know, and Feynman decided to take a very simple example and to check. He considered the case where $t_2 - t_1 = \epsilon$ was very small, and wrote the time evolution of the Schrödinger wave function $\psi(x,t)$ as follows:

$$\psi(x, t+\epsilon) = \frac{1}{\mathcal{N}} \int \exp\left[\frac{i}{\hbar}\epsilon L(x, t+\epsilon; y, t)\right] \psi(y,t)\, dy. \tag{1.38}$$

With $L = \frac{1}{2}m\dot{x}^2 - V(x)$ one obtains, as we now know very well, the Schrödinger equation, provided the constant $\mathcal{N}$ is given by

$$\mathcal{N} = \left(\frac{2\pi i\hbar\epsilon}{m}\right)^{1/2} \tag{1.39}$$

(the combination $dy/\mathcal{N}$ is nowadays often called the Feynman measure). Thus, as Dirac correctly guessed, $\langle x_2, t_2 | x_1, t_1\rangle$ was analogous to $\exp(\frac{i}{\hbar}\epsilon L)$ for small $\epsilon = t_2 - t_1$; however, they were not equal but rather proportional.

There is an amusing continuation of this story [54]. In the fall of 1946 Dirac was giving a lecture at Princeton, and Feynman was asked to introduce Dirac and comment on his lecture afterwards. Feynman decided to simplify Dirac's rather technical lecture for the benefit of the audience, but senior physicists such as Bohr and Weisskopf did not much appreciate this watering down of the work of the great Dirac by the young and relatively unknown Feynman. Afterwards people were discussing Dirac's lecture and Feynman who (in his own words) felt a bit let down happened to look out of the window and saw Dirac lying on his back on a lawn and looking at the sky. So Feynman went outside and sitting down near Dirac asked him whether he could ask him a question concerning his 1932 paper. Dirac consented. Feynman said "Did you know that the

[10] By expanding expressions such as $1/(\partial_x^2 + \partial_t^2 - m^2)$ in a power series in ∂_t, and using Ostrogradsky's approach to a canonical formulation of systems with higher-order ∂_t derivatives, one can give a Hamiltonian treatment, but one must introduce infinitely many new fields $B, C, \ldots$ of the form $\partial_t A = B, \partial_t B = C, \ldots$. All of these new fields are, of course, equivalent to the oscillators of the original electromagnetic field.

two functions do not just 'correspond to' each other, but are actually proportional?" Dirac said "Oh, that's interesting". And that was the total reaction that Feynman got from Dirac.

Feynman then asked himself how to treat the case where $t_2 - t_1$ is not small. This Dirac had already discussed in his paper: by inserting a complete set of x-eigenstates one obtains

$$\langle x_{\mathrm{f}}, t_{\mathrm{f}} | x_{\mathrm{i}}, t_{\mathrm{i}} \rangle = \int \langle x_{\mathrm{f}}, t_{\mathrm{f}} | x_{N-1}, t_{N-1} \rangle \langle x_{N-1}, t_{N-1} | x_{N-2}, t_{N-2} \rangle \cdots \\ \cdots \langle x_1, t_1 | x_{\mathrm{i}}, t_{\mathrm{i}} \rangle \, dx_{N-1} \cdots dx_1. \tag{1.40}$$

Taking $t_j - t_{j-1}$ small and using the fact that for small $t_j - t_{j-1}$ one can use $\mathcal{N}^{-1} \exp \frac{i}{\hbar}(t_j - t_{j-1})L$ for the transformation function, Feynman arrived at

$$\langle x_{\mathrm{f}}, t_{\mathrm{f}} | x_{\mathrm{i}}, t_{\mathrm{i}} \rangle = \int \exp\left[\frac{i}{\hbar} \sum_{j=0}^{N-1} (t_{j+1} - t_j) L(x_{j+1}, t_{j+1}; x_j, t_j)\right] \frac{dx_{N-1} \cdots dx_1}{\mathcal{N}^N}. \tag{1.41}$$

At this point Feynman recognized that one obtains the action in the exponent and that by first summing over j and then integrating over x one is summing over paths. Hence $\langle x_{\mathrm{f}}, t_{\mathrm{f}} | x_{\mathrm{i}}, t_{\mathrm{i}} \rangle$ is equal to a sum over all paths of $\exp(\frac{i}{\hbar}S)$ with each path beginning at x_{i}, t_{i} and ending at x_{f}, t_{f}.

Of course, only one of these paths is the classical path, but by summing over all other paths (arbitrary paths not satisfying the classical equation of motion) quantum mechanical corrections are introduced. The tremendous result was that all quantum corrections were included if one summed the action over all paths. Dirac had entertained the possibility that in addition to summing over paths one would have to replace the action S by a generalization which contained terms with higher powers in $\hbar$.

Reviewing this development more than half a century later, when path integrals have largely superseded operators methods and the Schrödinger equation for relativistic field theories, one notices how close Dirac came to the solution of using the action in quantum mechanics, and how different Feynman's approach was to solving the problem. Dirac anticipated that the action had to play a role, and by inserting a complete set of states he did obtain (1.41). However, he did not pursue the observation that the sum of terms in (1.41) is the action because he anticipated for large $t_2 - t_1$ a more complicated expression. Feynman, on the other hand, started by working out a few simple examples, curious to see whether Dirac was correct that the complete result would need a more complicated expression than the action, and in this way found that the truth lies in between: Dirac's transformation functions (Feynman's transition kernel K) is equal to the exponent of the action up to a constant. This constant diverges as ϵ tends to zero, but for $N \to \infty$ the result for K (and other quantities) is finite.

Feynman initially believed that in his path integral approach to quantum mechanics ordering ambiguities of the p and x operators of the operator approach would be absent (as he wrote in his PhD thesis of May 1942). However, later in his fundamental 1948 paper in *Review of Modern Physics* [45], he realized that the same ambiguities would be present. For our work the existence of these ambiguities is very important and we shall discuss in great detail how to fix them. Schrödinger [55] had already noticed that ordering ambiguities occur if one tries to promote a classical function $F(x,p)$ to an operator $\hat{F}(\hat{x},\hat{p})$. Furthermore, one can in principle add further terms that are linear and of higher order in $\hbar$ to such operators $\hat{F}$. These are further ambiguities which have to be fixed before one can make definite predictions.

Feynman evaluated the kernels $K(x_{j+1},t_{j+1}|x_j,t_j)$ for small $t_{j+1}-t_j$ by inserting complete sets of **momentum eigenstates** $|p_j\rangle$ in addition to position eigenstates $|x_j\rangle$. In this way he constructed **phase-space path integrals**. We shall follow the same approach for the nonlinear sigma models we consider. It has been claimed in [21] that "...phase space path integrals have more troubles than merely missing details. On this basis they should have been left out [from the book]...". We have instead arrived at a different conclusion: they are well defined and can be used to **derive** the usual configuration-space path integrals from the operatorial approach by adding integrations over intermediate momenta. A continuous source of confusion is the notation $Dx(t)\,Dp(t)$ for these phase-space path integrals. Many authors, who attribute more meaning to the symbol than $dx_1\cdots dx_{N-1}\,dp_1\cdots dp_N$, assume that this measure is invariant under canonical transformations, and apply the powerful methods developed in classical mechanics for the Liouville measure. However, the measure $Dx(t)\,Dp(t)$ in path integrals is not invariant under canonical transformations of the x and the p because there is one more p integration then x integration in $\prod dx_j \prod dp_j$.

Another source of confusion for phase-space path integrals arises if one tries to interpret them as integrals over paths around classical solutions in phase space. Consider Feynman's expression

$$\begin{aligned} K(x_j,t_j|x_{j-1},t_{j-1}) &= \langle x_j|\mathrm{e}^{-\frac{i}{\hbar}\hat{H}(t_j-t_{j-1})}|x_{j-1}\rangle \\ &= \int \frac{dp_j}{2\pi}\langle x_j|\mathrm{e}^{-\frac{i}{\hbar}\hat{H}(t_j-t_{j-1})}|p_j\rangle\langle p_j|x_{j-1}\rangle. \qquad (1.42) \end{aligned}$$

For $\langle x_j|\mathrm{e}^{-\frac{i}{\hbar}\hat{H}(t_j-t_{j-1})}|x_{j-1}\rangle$ one can substitute $\exp[\frac{i}{\hbar}S(x_j,t_j|x_{j-1},t_{j-1})]$, where in S one uses the classical path from x_j, t_j to x_{j-1}, t_{j-1}. In a similar way some authors have tried to give meaning to $\langle x_j | \mathrm{e}^{-\frac{i}{\hbar}\hat{H}(t_j - t_{j-1})} | p_j \rangle$ by considering a classical path in phase space. However, several proposals have been shown to be inconsistent or impractical [21]. We shall not try to interpret the transition elements in phase space in terms of classical paths, but only do what we are supposed to do: integrate over p_j and x_j.

Yet another source of confusion has to do with path integrals over fermions for which one needs Grassmann numbers and Berezin integration [56]

$$\int d\theta = 0\ , \qquad \int d\theta\, \theta = 1. \tag{1.43}$$

Some authors claim that the notion of anticommuting classical fields makes no sense and that only quantized fermionic fields are consistent. However, the notion of Grassmann variables is completely consistent if one uses it only at the intermediate stages to construct, for example, fermionic coherent states: all one does is make use of mathematical identities. One begins with fermionic harmonic oscillator operators $\hat{\psi}$ and $\hat{\psi}^\dagger$ and constructs coherent bra and kets states $|\eta\rangle$ and $\langle\bar{\eta}|$ in Hilbert space. In applications traces are taken over these coherent states using Berezin rules for the integrations over η and $\bar{\eta}$. One ends up with physical results which are independent of the Grassmann variables, and since all intermediary steps are mathematical identities [19], defined by Berezin [56], at no point are there any conceptual problems in the treatment of path integrals for fermions.

2
Time slicing

In this chapter we discuss quantum mechanical path integrals defined by time slicing. Our starting point is an arbitrary but fixed Hamiltonian operator $\hat{H}$. We obtain the Feynman rules for nonlinear sigma models, first for bosonic point particles $x^i(t)$ with curved indices $i = 1, \ldots, n$ and then for fermionic point particles $\psi^a(t)$ with flat indices $a = 1, \ldots, n$. In the bosonic case we first discuss in detail configuration-space path integrals, and then return to the corresponding phase-space integrals. In the fermionic case we use coherent states to define bras and kets, and we discuss the proper treatment of Majorana fermions, both in the operatorial and in the path integral approach. Finally, we compute directly the transition element $\langle z|e^{-(\beta/\hbar)\hat{H}}|y\rangle$ to order β (two-loop order) using operator methods, and compare the answer with the results of a similar calculation based on the perturbative evaluation of the path integral with time slicing regularization. Complete agreement is found. These results were obtained in [57–59]. Additional discussions are found in [60–65].

The quantum action, i.e. the action to be used in the path integral, is obtained from the quantum Hamiltonian by mathematical identities, and the quantum Hamiltonian is fixed by the quantum field theory, the anomalies of which we study in Part II of the book. Hence, there is no ambiguity in the quantum action. It contains local finite counterterms of order $\hbar^2$. They were discussed in detail by Gervais and Jevicki [5]. Earlier Schwinger [6] and later Christ and Lee [8] found by the same method that four-dimensional Yang–Mills theory in the Coulomb gauge has such counterterms.

2.1 Configuration-space path integrals for bosons from time slicing

Consider a quantum Hamiltonian $\hat{H}(\hat{x},\hat{p})$ with a definite ordering of the operators $\hat{x}^i$ and $\hat{p}_i$. We will mostly focus on the operator

$$\hat{H}(\hat{x},\hat{p}) = \frac{1}{2} g^{-1/4} \hat{p}_i g^{ij} g^{1/2} \hat{p}_j g^{-1/4} \tag{2.1}$$

where $g = \det g_{ij}(x)$ and we omitted hats on $\hat{x}$ in the metric for notational simplicity. This Hamiltonian is Einstein invariant. A simple way to prove this is to use the x-representation for $\hat{p}_i$, namely $\hat{p}_i = \frac{\hbar}{i} g^{-1/4} \partial/\partial x^i g^{1/4}$ for the inner product in (2.4). One finds then the Laplacian in curved space, $g^{-1/2}\partial_i g^{1/2} g^{ij} \partial_j$. One can also give a more form proof by showing that $\hat{H}$ commutes with the generator of general coordinate transformations (see Section 2.5, in particular (2.198)). Our methods also apply to other Hamiltonians; for example, $\hat{H} = \frac{1}{2}\hat{p}_i g^{ij} \hat{p}_j$ or the nonhermitian operator $\frac{1}{2} g^{ij} \hat{p}_i \hat{p}_j$. The reason we focus on (2.1) is that this Hamiltonian is general coordinate invariant, and thus describes the motion of a particle in a curved space in an arbitrary coordinate system. Also, this same Hamiltonian leads to the regulators which we use in the second part of this book to compute anomalies by quantum mechanical methods.

The essential object from which all other quantities can be calculated, is the transition element (also called the transition amplitude)

$$T(z,y;\beta) = \langle z | \mathrm{e}^{-\frac{\beta}{\hbar}\hat{H}} | y \rangle \tag{2.2}$$

where $|y\rangle$ and $\langle z|$ are eigenstates of the position operator $\hat{x}^i$,

$$\hat{x}^i |y\rangle = y^i |y\rangle, \quad \langle z | \hat{x}^i = \langle z | z^i \tag{2.3}$$

with y^i and z^i being real numbers. We normalize the x and p eigenstates as follows:

$$\int |x\rangle \sqrt{g(x)}\, \langle x|\, d^n x = I \quad \rightarrow \quad \langle x | y \rangle = \frac{\delta^n(x-y)}{\sqrt{g(x)}} \tag{2.4}$$

where I is the identity operator and

$$\int |p\rangle \langle p|\, d^n p = I \quad \rightarrow \quad \langle p | p' \rangle = \delta^n(p-p'). \tag{2.5}$$

The delta function $\delta^n(x-y)$ is defined by $\int \delta^n(x-y) f(y)\, d^n y = f(x)$. Since $\hat{x}^i$ and $\hat{p}_i$ are diagonal and real on these complete sets of orthonormal states, they are both hermitian. The Hamiltonian in (2.1) is then also hermitian, but we could in principle also allow nonhermitian Hamiltonians. However, we stress that $\hat{x}^i$ and $\hat{p}_i$ are always hermitian.

We have chosen the normalization in (2.4) in order that $T(z,y;\beta)$ will be a bi-scalar (a scalar under general coordinate transformations of z and y separately). With this inner product one obtains $\langle\varphi|\psi\rangle = \int dx\, g^{1/2}(x)\, \varphi(x)\, \psi(x)$ where $\varphi(x)$ and $\psi(x)$ are scalars. There is no need to choose the normalization in (2.4) and one could also use $\int |x\rangle\langle x|\, d^n x = I$, for example. However, (2.4) leads to simpler formulas. For example, the inner product of two states $\langle\varphi|\psi\rangle$ takes the familiar form $\int \sqrt{g(x)}\varphi^*(x)\psi(x)\, d^n x$. Consequently, wave functions $\psi(x) = \langle x|\psi\rangle$ are scalar functions under a change of coordinates.

The inner product between x- and p-eigenstates yields plane waves with an extra factor of $g^{-1/4}$,

$$\langle x|p\rangle = \frac{e^{\frac{i}{\hbar}p_j x^j}}{(2\pi\hbar)^{n/2} g^{1/4}(x)}. \tag{2.6}$$

As a check, note that $\int \langle p|x\rangle \sqrt{g(x)}\, \langle x|p'\rangle\, d^n x = \delta^n(p - p')$, in agreement with the completeness relations.

We now insert N complete sets of momentum eigenstates and $N-1$ complete sets of position eigenstates into the transition element. Defining $\beta = N\epsilon$ we obtain

$$\begin{aligned} T(z, y; \beta) &= \langle z|\left(e^{-\frac{\epsilon}{\hbar}\hat{H}}\right)^N |y\rangle \\ &= \langle z|e^{-\frac{\epsilon}{\hbar}\hat{H}}|p_N\rangle \int d^n p_N\, \langle p_N|x_{N-1}\rangle \int \sqrt{g(x_{N-1})}\, d^n x_{N-1} \\ &\quad \times \langle x_{N-1}|e^{-\frac{\epsilon}{\hbar}\hat{H}}|p_{N-1}\rangle \int d^n p_{N-1}\, \langle p_{N-1}|x_{N-2}\rangle \\ &\quad \times \int \sqrt{g(x_{N-2})}\, d^n x_{N-2} \\ &\quad \vdots \\ &\quad \cdots\ \langle x_1|e^{-\frac{\epsilon}{\hbar}\hat{H}}|p_1\rangle \int d^n p_1\, \langle p_1|y\rangle. \end{aligned} \tag{2.7}$$

We have written the integration symbols between the bras and kets to which they belong in order to simplify the notation. It is natural to denote z by x_N and y by x_0. The order in which x_i and p_i appear can be indicated as follows:

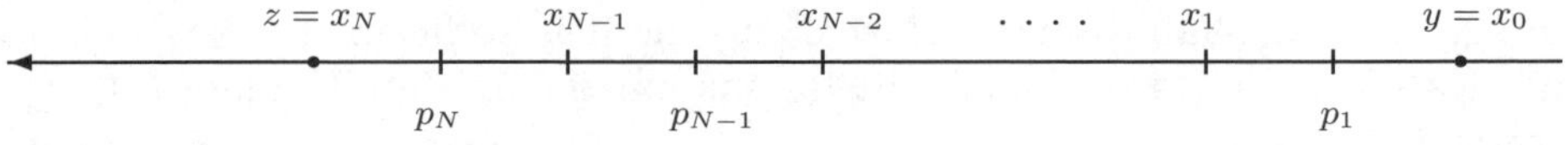

Although the p's occur between x's, we do not imply that a kind of mid-point rule holds. Only the ordering of the p's and x's matters.

We now **rewrite** the operators $\exp(-\frac{\epsilon}{\hbar}\hat{H}(\hat{x}, \hat{p}))$ in Weyl-ordered form [66]. This means that after the rewriting this operator is symmetric in all $\hat{x}$ and $\hat{p}$ it contains. Weyl ordering is discussed in Appendix B. As an example of such a rewriting consider the operator $\hat{x}\hat{p}$. We rewrite it as $\frac{1}{2}(\hat{x}\hat{p} + \hat{p}\hat{x}) + \frac{1}{2}(\hat{x}\hat{p} - \hat{p}\hat{x}) = \frac{1}{2}(\hat{x}\hat{p} + \hat{p}\hat{x}) + \frac{1}{2}i\hbar$. It is useful to denote the symmetrized expression $\frac{1}{2}(\hat{x}\hat{p}+\hat{p}\hat{x})$ by $(\hat{x}\hat{p})_S$. So in this example $\hat{x}\hat{p}=(\hat{x}\hat{p})_S+\frac{1}{2}i\hbar$. We shall refer to the form of the operator O in which $\hat{x}$ and $\hat{p}$ have been symmetrized as O_W. Hence, $\hat{x}\hat{p}=(\hat{x}\hat{p})_S+\frac{1}{2}i\hbar=(\hat{x}\hat{p})_W$. In more general cases, one has the formula $O(\hat{x},\hat{p})=O_S(\hat{x},\hat{p})+\textit{more}=O_W(\hat{x},\hat{p})$, where "*more*" may contain further operators that depend on $\hat{x}$ and $\hat{p}$, and which we again rewrite in a symmetrical way.

The reason we rewrite operators in Weyl-ordered form is that Weyl ordering leads to the midpoint rule in the following way [67]:

$$\begin{aligned}&\int \langle x_k|\, O_W(\hat{x},\hat{p})\, |p_k\rangle\langle p_k|x_{k-1}\rangle\, d^n p_k\\&\quad=\int \langle x_k|p_k\rangle\, O_W\left(\frac{1}{2}(x_k+x_{k-1}),\, p_k\right)\langle p_k|x_{k-1}\rangle\, d^n p_k. \qquad (2.8)\end{aligned}$$

This formula is also proven in Appendix B. For further discussions of this correspondence rule between operators and path integrals, see [68, 69]. The meaning of this equation is that one may extract the Weyl-ordered operator from the matrix element and replace it by a function by replacing each $\hat{p}_i$ by $p_{k,i}$ and each $\hat{x}^i$ by $\frac{1}{2}(x_k^i+x_{k-1}^i)$. A more precise notation for the function in the second line would have been

$$O_W\left(\hat{x}\to\frac{1}{2}(x_k+x_{k-1}),\, \hat{p}\to p_k\right). \qquad (2.9)$$

For simplicity we use the notation in (2.8).

Thus, in (2.7) we may use

$$\begin{aligned}&\int \langle x_k|\left(\mathrm{e}^{-\frac{\epsilon}{\hbar}\hat{H}}\right)_W |p_k\rangle\langle p_k|x_{k-1}\rangle\, d^n p_k\\&\quad=\int \langle x_k|p_k\rangle\left(\mathrm{e}^{-\frac{\epsilon}{\hbar}H(\frac{1}{2}(x_k+x_{k-1}),p_k)}\right)_W\langle p_k|x_{k-1}\rangle\, d^n p_k. \qquad (2.10)\end{aligned}$$

Note that we should first Weyl order the whole operator $\exp(-\frac{\epsilon}{\hbar}\hat{H})$, and not only $\hat{H}$, and then replace $\hat{x}^i$ by $\frac{1}{2}(x_k^i+x_{k-1}^i)$ and $\hat{p}_i$ by $p_{k,i}$.

In general, we cannot write down a closed expression for $(\exp(-\frac{\epsilon}{\hbar}\hat{H}))_W$. However, for path integrals one may replace the Weyl-ordered expression $(\exp(-\frac{\epsilon}{\hbar}\hat{H}))_W$ by $\exp(-\frac{\epsilon}{\hbar}\hat{H}_W)$, because the difference cancels in the path integral, as we shall discuss below (2.19). So it is sufficient to Weyl-order the Hamiltonian itself. The result for the particular Hamiltonian in (2.1) reads as follows:

$$\hat{H}_W(\hat{x},\hat{p})=\left(\frac{1}{2}g^{ij}\hat{p}_i\hat{p}_j\right)_S+\frac{\hbar^2}{8}(R+g^{ij}\Gamma^k_{il}\Gamma^l_{jk}). \qquad (2.11)$$

The definition of the scalar curvature R is given in Appendix A and the symmetrized operator $\left(\frac{1}{2}g^{ij}\hat{p}_i\hat{p}_j\right)_S$ is according to Appendix B given by

$$\left(\frac{1}{2}g^{ij}\hat{p}_i\hat{p}_j\right)_S=\frac{1}{8}g^{ij}\hat{p}_i\hat{p}_j+\frac{1}{4}\hat{p}_i g^{ij}\hat{p}_j+\frac{1}{8}\hat{p}_i\hat{p}_j g^{ij}. \qquad (2.12)$$

Note that $\hat{H}_W$ in (2.11) is still the same operator $\hat{H}$ of (2.1), but $\hat{H}_W$ is subsequently converted into a function H_W according to (2.9).

The extra terms of order $\hbar^2$ in (2.11) are crucial for our further discussions. One might at this point also ask whether in ordinary quantum field theory there are such further terms in the action of order $\hbar$ and higher, which more general studies have missed. There are such terms but only in unusual cases. For example, choosing the Coulomb gauge in Yang–Mills theory, elimination of the longitudinal part of the gauge field leads to a complicated nonlinear sigma model (of course a field-theoretical sigma model, not a quantum mechanical one), and the methods of this chapter yield straightforwardly extra $\hbar^2$ terms [6–8]. However, it has been argued that there are no extra order $\hbar^2$ terms in the usual higher-than-one-dimensional quantum field theories because, in that case, one employs a different regularization scheme than in quantum mechanics [70].

In the discretized transition element we then encounter the functions

$$\left(\mathrm{e}^{-\frac{\epsilon}{\hbar}H(\bar{x}_{k-1/2},p_k)}\right)_W = \exp\left[-\frac{\epsilon}{\hbar}\left\{\frac{1}{2}g^{ij}(\bar{x}_{k-1/2})p_{k,i}p_{k,j}\right.\right.$$
$$\left.\left.+\frac{\hbar^2}{8}\left[R(\bar{x}_{k-1/2})+g^{ij}(\bar{x}_{k-1/2})\Gamma^m_{il}(\bar{x}_{k-1/2})\Gamma^l_{jm}(\bar{x}_{k-1/2})\right]\right\}\right] \quad (2.13)$$

where $\bar{x}_{k-1/2} \equiv \frac{1}{2}(x_k+x_{k-1})$. Note that the term with the scalar curvature R and its coefficient $\frac{1}{8}$ as well as the $\Gamma\Gamma$ term are a mathematical consequence of rewriting the particular Hamiltonian in (2.1) in Weyl-ordered form. If we were to have started from another Hamiltonian (in particular, another operator ordering), we would have found different coefficients in (2.13). To summarize, we end up with the following path integral representation for the transition element:

$$T(z,y;\beta) = \lim_{N\to\infty}[g(z)g(y)]^{-1/4}\int\left(\prod_{k=1}^{N}\frac{d^n p_k}{(2\pi\hbar)^n}\prod_{l=1}^{N-1}d^n x_l\right)$$
$$\times\exp\left\{\sum_{k=1}^{N}\left[\frac{i}{\hbar}p_k\cdot(x_k-x_{k-1})-\frac{\epsilon}{\hbar}H_W(\bar{x}_{k-1/2},p_k)\right]\right\} \quad (2.14)$$

because the $g^{1/2}$ factors in (2.7) compensate exactly the $g^{-1/4}$ factors from the plane waves (2.6), except at the initial and final points. For the Hamiltonian (2.1) the function H_W is then

$$H_W(\bar{x}_{k-1/2},p_k) = \frac{1}{2}g^{ij}(\bar{x}_{k-1/2})p_{k,i}p_{k,j} + V_{TS}(\bar{x}_{k-1/2}) \quad (2.15)$$

where

$$V_{TS} = \frac{\hbar^2}{8}(R + g^{ij}\Gamma^m_{il}\Gamma^l_{jm}) \quad (2.16)$$

is the so-called "counterterm" of the time slicing regularization scheme.

Up to this point we have presented the standard approach to path integrals. In the rest of this chapter we evaluate the transition element without making any of the usual approximations. We must do this because for nonlinear sigma models we need propagators of the quantum fields $q(t)$ with double time derivatives, $\langle \dot{q}(t_1)\dot{q}(t_2)\rangle$. As a result, in the evaluation of the transition element in terms of Feynman graphs we shall encounter expressions which contain products of distributions, for example $\delta(t_1-t_2)\theta(t_1-t_2)\theta(t_2-t_1)$, and equal-time contractions. Such expressions are ambiguous, one can obtain different answers depending on how one regulates the distributions. One could at this point introduce further physical requirements that fix the ambiguities. For example, one could require that the Green functions satisfy certain symmetries. If symmetries are not enough, one could impose some further renormalization conditions; for example, that certain Green functions have certain values at certain momenta. This procedure has been used before for linear sigma models, namely if one tries to define the path integral for $L=\frac{1}{2}(\dot{x}^i)^2+\dot{x}^i A_i(x)+V(x)$. It makes a difference whether one discretizes to $(x^i_k-x^i_{k-1})A_i(x_k)$ or $(x^i_k-x^i_{k-1})A_i(x_{k-1})$ [46, 21]. One way to fix this ambiguity is to require that the transition element be gauge-invariant, and one discovers that this is achieved by using the midpoint rule $(x^i_k-x^i_{k-1})A_i(\frac{1}{2}(x_k+x_{k-1}))$. Conversely, one could have started with a Hamiltonian operator $\hat{H}$, for which the operator ordering is gauge-invariant, namely $\hat{H}$ commutes with the operator $\hat{G}$ of gauge transformations. Using this particular $\hat{H}$, the time slicing method – with complete p and x eigenstates – produces the midpoint rule automatically without any further terms. Similarly, by taking the particular $\hat{H}$ in (2.1) which is invariant under general coordinate transformations, all ambiguities are fixed from the start, and the time slicing method – with complete p and x eigenstates – leads to well-defined and unambiguous expressions, including definite rules for the products of distributions. Taking at the very end the limit $\epsilon \to 0$, $N \to \infty$ and $N\epsilon = \beta$, one **derives** the rules of how to evaluate integrals over products of distributions in the continuum theory.

Since the subsequent discussions in this chapter must be precise and therefore technical in order not to miss subtleties in the product of distributions, it may help the reader if we first give a short nontechnical summary of the results to be obtained. Such a summary follows in the next three paragraphs.

We begin by integrating over the N momenta $p_1, \ldots, p_N$. This leads to a product of N determinants $[\det g_{ij}(\bar{x}_{k-1/2})]^{1/2}$, where $\bar{x}_{k-1/2}=\frac{1}{2}(x_k+x_{k-1})$. We exponentiate these determinants using ghost fields $a_{k-1/2}, b_{k-1/2}$ and $c_{k-1/2}$. This yields discretized configuration path integrals for the transition element. We decompose x_k into background fields $x_{bg,k}$ and quantum fields q_k, and decompose the discretized action into a part $S^{(0)}$ quadratic in q_k and an interaction part $S^{(int)}$. We require that $x_{bg,k}$ satisfy the field equation of $S^{(0)}$ and the boundary conditions that x_{bg} be equal to z or y at the boundaries. Because $x_{bg,k}$ satisfy the field equation of $S^{(0)}$ and not of the full action S, there will be terms in $S^{(int)}$ linear in q. This poses no problems, because in a perturbative calculation to a given order in β only a finite number of tadpoles contribute. We introduce external sources F and G which couple to $\frac{1}{2}(q_k + q_{k-1})$ and $(q_k - q_{k-1})$, and extract $S^{(int)}$ from the path integral, as is usual in quantum field theory. Then we want to integrate over q but the action $S^{(0)}$ is not diagonal in q. In order to diagonalize $S^{(0)}$ we make an orthogonal change of integration variables with a unit Jacobian. The actual integration over q can then be performed in closed form, but it requires a few relatively unknown identities for products involving sines and cosines. The final result is given in (2.49)–(2.51). By differentiation with respect to the sources F and G and similar external sources A, B, C for the ghosts, we find the discretized propagators in closed form. The results are given in (2.54) for $\langle \dot{q}\dot{q} \rangle$, in (2.70) for $\langle q\dot{q} \rangle$, in (2.76) for $\langle qq \rangle$, and in (2.79) and (2.80) for the ghosts. These result can also be written in the continuum limit, see (2.83), but if one computes diagrams in the continuum limit, one should use the integration formulas for the products of distributions which we **derive** from the discretized approach.

For fermionic models with operators $\hat{\psi}^a$ and $\hat{\psi}_a^\dagger$ we begin by introducing bras $\langle \bar{\eta}|$ and kets $|\eta\rangle$ in terms of coherent states which depend on Grassmann variables $\bar{\eta}_a$ and η^a. We treat η and $\bar{\eta}$ as independent variables. This is the approach to be used for complex (Dirac) fermions, and hence for $N = 2$ models. We could equally well have defined $\bar{\eta}$ to be the complex conjugate of η because the only property we need is integration over the Grassmann variables, and this integration is the same whether $\bar{\eta}$ is an independent variable or the complex conjugate of η. The transition element we wish to compute is $\langle \bar{\eta}|e^{-\frac{\beta}{\hbar}\hat{H}}|\eta\rangle$ and again we insert complete sets $|\chi_k\rangle\langle\bar{\chi}_k|$ of coherent states to arrive at a discretized path integral. After defining Weyl ordering for anticommuting operators $\hat{\psi}^a$ and $\hat{\psi}_a^\dagger$, we again obtain a midpoint rule for the variables χ_k^a. We decompose the variables χ_k^a and $\bar{\chi}_{ka}$ again into background parts ξ_k^a and $\bar{\xi}_{ka}$, and quantum parts ψ_k^a and $\bar{\psi}_{ka}$. We decompose S into a part $S^{(0)}$ that is quadratic in ψ_k^a and $\bar{\psi}_{ka}$, and the rest $S^{(int)}$. We require that the background variables ξ and $\bar{\xi}$ satisfy the field equation of $S^{(0)}$, and the boundary conditions that $\xi = \eta$ at the right and $\bar{\xi} = \bar{\eta}$ at the left. Because ξ and $\bar{\xi}$ are constant (do not depend on k), the decomposition reduces to $\chi_{k-1/2}^a = \eta^a + \psi_{k-1/2}^a$ and $\chi_{ka} = \bar{\eta}_a + \bar{\psi}_{ka}$. We also treat ξ, $\bar{\xi}$ and ψ, $\bar{\psi}$, as independent Grassmann variables; again it makes no difference whether they are related by complex conjugation or not. We couple the quantum variables ψ_k^a and $\bar{\psi}_{ka}$ to external sources $\bar{K}_{ka}$ and K_k^a, complete squares and integrate over ψ_k^a and $\bar{\psi}_{ka}$. For the propagators of ψ and $\bar{\psi}$ we find the following exact discretized result:

$$\langle \psi_k^a \bar{\psi}_{lb} \rangle = \begin{cases} \delta_a^b & \text{if } k \geq l \\ 0 & \text{if } k < l. \end{cases} \tag{2.17}$$

(Time ordering is always understood, so the case $k < l$ refers to $-\langle \bar{\psi}_{lb} \psi_k^a \rangle$.) Owing to the midpoint rule, we rather need the propagator for $\psi_{k-1/2}^a \equiv \frac{1}{2}(\psi_k^a + \psi_{k-1}^a)$. It reads as

$$\langle \psi_{k-1/2}^a \bar{\psi}_{lb} \rangle = \begin{cases} \delta_a^b & \text{if } k > l \\ \frac{1}{2}\delta_a^b & \text{if } k = l \\ 0 & \text{if } k < l. \end{cases} \tag{2.18}$$

It becomes $\theta(t - t')$ in the continuum limit, but $\theta(0)$ is now equal to 1/2.

Finally, we consider Majorana fermions $\hat{\psi}_1^a$. We add another set of **free** Majorana fermions $\hat{\psi}_2^a$, and define the operators $\hat{\psi}^a = (\hat{\psi}_1^a + i\hat{\psi}_2^a)/\sqrt{2}$ and $\hat{\psi}_a^\dagger = (\hat{\psi}_1^a - i\hat{\psi}_2^a)/\sqrt{2}$. We can then apply the results for Dirac spinors. We again consider the transition element $\langle \bar{\eta} | \mathrm{e}^{-\frac{\beta}{\hbar}\hat{H}} | \eta \rangle$. The operator $\hat{\psi}_1^a$ in the Hamiltonian operator $\hat{H}$ can be written in terms of $\hat{\psi}$ and $\hat{\psi}^\dagger$, and after a Weyl ordering the midpoint rule yields the function $H(\bar{\chi}_{ka}, \chi_{k-1/2}^a)$, where $\chi_{k-1/2}^a \equiv (\chi_k^a + \chi_{k-1}^a)/2$. Because initially $\hat{H}$ depended only on $\hat{\psi}_1^a$, the action in the final path integral contains H_W, which depends on $(\chi_{k-1/2}^a + \bar{\chi}_k^a)/\sqrt{2}$. The Grassmann variables $\chi_{k-1/2}^a$, which are introduced by inserting unity, are decomposed into a background η^a and a quantum part $\psi_{k-1/2}^a$, according to $\chi_{k-1/2}^a = \eta^a + \psi_{k-1/2}^a$. Similarly, $\bar{\chi}_{ka} = \bar{\eta}_a + \bar{\psi}_{ka}$. (The background fields η^a and $\bar{\eta}_a$ are constant, hence they do not depend on k.) The propagator for $\psi_{1k}^a \equiv (\psi_{k-1/2}^a + \bar{\psi}_{ka})/\sqrt{2}$ follows from the propagator $\langle \psi_k^a \bar{\psi}_{lb} \rangle$ for the Dirac spinors and reads as

$$\langle \psi_{1k}^a \psi_{1l}^b \rangle = \frac{1}{2}\delta^{ab} \begin{cases} 1 & \text{if } k > l \\ 0 & \text{if } k = l \\ -1 & \text{if } k < l. \end{cases} \tag{2.19}$$

In the continuum limit it becomes $\langle \psi_1^a(t) \psi_1^b(t') \rangle = \frac{1}{2}\delta^{ab}[\theta(t-t') - \theta(t'-t)]$. With this propagator we can compute the transition element in a loop expansion, and the transition element will be used to compute anomalies. We now return to a detailed derivation of these results.

We first present the proof that one may replace $(\exp(-\frac{\epsilon}{\hbar}H))_W$ by $\exp(-\frac{\epsilon}{\hbar}H_W)$ in the path integral. In general, Weyl ordering and exponentiation do not commute, $(\exp(-\frac{\epsilon}{\hbar}H))_W \neq \exp(-\frac{\epsilon}{\hbar}H_W)$ and whereas H_W was easy to write down, a closed expression for $(\exp(-\frac{\epsilon}{\hbar}H))_W$ cannot be written down. One expects, however, that a suitable approximation of the kernels, containing only terms of order ϵ, suffices. It might seem that p is of order $\epsilon^{-1/2}$ due to the term $\exp(-\frac{1}{2}\epsilon p^2)$ in the action, see (2.13). Expansion of $\exp(-\frac{\epsilon}{\hbar}H_W)$ would contain terms of the form $\epsilon^s p^r f(x)$ for which $s \geq 2$ and such terms could still be of order ϵ if r is sufficiently large. We are now going to give an argument that p is actually of order unity, and therefore only the terms with one explicit ϵ need be retained. Hence, we may use as a kernel $\exp[-\frac{\epsilon}{\hbar}H_W(\frac{1}{2}(x_k + x_{k-1}), p_k)]$. In other words, the Trotter-like approximation

$$\begin{aligned}\langle x|\exp\Big(-\frac{\epsilon}{\hbar}\hat{H}\Big)|p\rangle &\simeq \langle x|\Big(1-\frac{\epsilon}{\hbar}\hat{H}\Big)|p\rangle = \Big(1-\frac{\epsilon}{\hbar}H_W\Big)\langle x|p\rangle \\ &\simeq \exp\Big(-\frac{\epsilon}{\hbar}H_W\Big)\langle x|p\rangle \end{aligned} \tag{2.20}$$

is still correct if used inside (2.10), but we repeat that H_W is not simply $\langle x|\hat{H}|p\rangle$ as in the usual models with $H = T(p) + V(x)$, but rather it is evaluated at the midpoints.

To prove this claim, we note that the kernels in (2.10) are proportional to

$$\int d^n p_k \, \mathrm{e}^{\frac{i}{\hbar}p_k\cdot\Delta x_{k-1/2}} \Big(\mathrm{e}^{-\frac{\epsilon}{\hbar}H(\bar{x}_{k-1/2},p_k)}\Big)_W \tag{2.21}$$

$$\Delta x_{k-1/2} \equiv x_k - x_{k-1}, \quad \bar{x}_{k-1/2} \equiv \frac{1}{2}(x_k + x_{k-1}).$$

The difference between $(\exp(-\frac{\epsilon}{\hbar}H))_W$ and $\exp(-\frac{\epsilon}{\hbar}H_W)$ consists of two kinds of terms.

(i) Terms without a p. These are certainly of higher order in ϵ and can be omitted.

(ii) Terms with at least one p.

In order to evaluate (2.21) one has to proceed as follows, as will be discussed in detail below. One extracts the interaction part of H from the path integral, while the terms quadratic in p and x yield the propagators. One then constructs Feynman graphs with $H^{(int)}(\bar{x}, p)$ as vertices and phase-space propagators for p and $\bar{x}$. The crucial observation is now that the phase-space propagators $\langle p_{k,i}p_{l,j}\rangle$ and $\langle p_{k,i}\bar{x}^j_{l-1/2}\rangle$ are both of order unity, and not of order ϵ^{-1} and $\epsilon^{-1/2}$, respectively.[1] An explicit proof is given later when we construct the discretized propagators for $\langle pp\rangle$ and $\langle px\rangle$, see (2.99). However, already at this point one might note that the pp propagator is not only determined by the term $-\frac{1}{2}\epsilon gpp$ contained in H but also by $ip\Delta x$. Completing squares, it is $p' = (p - i\Delta x/\epsilon)$, which is of

[1] We shall actually decompose $x(t) = x_{bg}(t) + q(t)$, and then we obtain propagators $\langle p_k\bar{q}_{l-1/2}\rangle$ instead of $\langle p_k\bar{x}_{l-1/2}\rangle$.

order $\epsilon^{-1/2}$. In the pp propagator the singularities of the $p'p'$ and $\Delta x \Delta x$ propagators cancel each other. (The origin of the more singular nature of the $\dot{x}\dot{x}$ propagator can be understood from canonical formalism: the xx propagator contains a time-ordering step function $\theta(t-t')$, and differentiation yields $\delta(t-t')$.) As a consequence, the pp and pq propagators are of order one, and this also proves the Trotter formula for nonlinear sigma models. (Already for linear sigma models with $H = T + V$ a completely rigorous proof of the Trotter formula uses Banach spaces [21], so for our nonlinear sigma models a completely rigorous proof is probably very complicated. However, we have identified the essential reason why p can be treated as being of order unity, and this is enough to justify (2.20).)

Using Weyl ordering and the midpoint rule to replace the operators $\exp(-\frac{\epsilon}{\hbar}\hat{H})$ by functions, we substitute the value of the various inner products. We arrive at the following discretized transition element T_N:

$$T_N(z, y; \beta) = \int \cdots \int \frac{e^{[\frac{i}{\hbar} p_N \cdot (z - x_{N-1}) - \frac{\epsilon}{\hbar} H_W(\frac{1}{2}(z + x_{N-1}), p_N)]}}{(2\pi\hbar)^n g^{1/4}(z) g^{1/4}(x_{N-1})} \sqrt{g(x_{N-1})}$$

$$\times \frac{e^{[\frac{i}{\hbar} p_{N-1} \cdot (x_{N-1} - x_{N-2}) - \frac{\epsilon}{\hbar} H_W(\frac{1}{2}(x_{N-1} + x_{N-2}), p_{N-1})]}}{(2\pi\hbar)^n g^{1/4}(x_{N-1}) g^{1/4}(x_{N-2})} \sqrt{g(x_{N-2})}$$

$$\vdots$$

$$\times \frac{e^{[\frac{i}{\hbar} p_1 \cdot (x_1 - y) - \frac{\epsilon}{\hbar} H_W(\frac{1}{2}(x_1 + y), p_1)]}}{(2\pi\hbar)^n g^{1/4}(x_1) g^{1/4}(y)} \left(\prod_{j=1}^{n} dp_1^j \cdots dp_N^j \, dx_1^j \cdots dx_{N-1}^j \right). \quad (2.22)$$

We note that all factors g cancel except an overall factor

$$[g(z)g(y)]^{-1/4}. \quad (2.23)$$

Furthermore we find in the exponents either coordinate differences $x_k - x_{k-1}$ or coordinate averages $\frac{1}{2}(x_k + x_{k-1})$, but the integration measure is $\prod_{j=1}^{n} dx_{N-1}^j \ldots dx_1^j$. In the continuum limit the exponent takes on the form $\exp(-\frac{1}{\hbar}S)$, where

$$S = \int dt \, [-ip\dot{x} + H_W(x, p)] \quad (2.24)$$

but we shall not yet take the continuum limit.

We shall now go from phase space to configuration space by integrating over the momenta. In the previous derivation we could have chosen to work with quite general Hamiltonians, but to be able to eliminate the momenta we must assume that the Hamiltonian contains at most two momenta, and consider the Hamiltonian function (not an operator)

$$H(\bar{x}, p) = \frac{1}{2} g^{ij}(\bar{x}) p_i p_j + V(\bar{x}). \tag{2.25}$$

We could also allow terms linear in p, but terms quartic or cubic in p we shall not consider. Of course the potential $V(\bar{x})$ contains the extra $\hbar^2$ terms displayed in (2.11) that arose from Weyl ordering. We find N Gaussian integrals (for $k = 1, \dots, N$)

$$\int \exp\left[-\frac{\epsilon}{2\hbar} g^{ij} \left(\frac{x_k + x_{k-1}}{2} \right) p_{k,i} p_{k,j} + \frac{i}{\hbar} p_{k,j} (x_k^j - x_{k-1}^j) \right] \prod_{j=1}^{n} dp_{k,j} \tag{2.26}$$

which yield N determinants upon completing squares

$$\left(\frac{2\pi\hbar}{\epsilon} \right)^{n/2} [\det\, g^{ij}(\bar{x}_{k-1/2})]^{-1/2} \tag{2.27}$$

where we recall the notation

$$\bar{x}^i_{k-1/2} \equiv \frac{1}{2}(x^i_k + x^i_{k-1}). \tag{2.28}$$

The discretized configuration space path integral then becomes

$$\begin{aligned} T_N(z, y; \beta) = &\left\{ [g(z) g(y)]^{-1/4} (2\pi\hbar)^{-nN} \right\} \\ &\times \int d^n x_{N-1} \cdots d^n x_1 \left\{ \left(\frac{2\pi\hbar}{\epsilon} \right)^{nN/2} \prod_{k=1}^{N} [\det g_{ij}(\bar{x}_{k-1/2})]^{1/2} \right\} \\ &\times \exp\left[-\sum_{k=1}^{N} \frac{1}{2\hbar\epsilon} g_{ij}(\bar{x}_{k-1/2}) \left(x^i_k - x^i_{k-1} \right) \left(x^j_k - x^j_{k-1} \right) - \frac{\epsilon}{\hbar} V(\bar{x}_{k-1/2}) \right]. \end{aligned} \tag{2.29}$$

We recall the definitions $x_N = z$ and $x_0 = y$.

Before we introduce external sources to compute discretized propagators, we remove the factors $g^{1/2}(\bar{x}) = [\det g_{ij}(\bar{x}_{k-1/2})]^{1/2}$ from the measure by exponentiating them with new ghost fields. If there had been factors of $g^{-1/2}(\bar{x})$ instead of $g^{1/2}(\bar{x})$ we could have replaced them by $\int da \exp a^i g_{ij} a^j$ with commuting ghosts. However, we have $g^{1/2}$ instead of $g^{-1/2}$. By writing this as $g^{1/2} = g^{-1/2} g$, we can still exponentiate if **we introduce two anticommuting real ghosts b^i and c^i and one commuting real ghost a^i.**[2] Then the Berezin integral over b^i and c^i yields

[2]One might think that one could simply use one real anticommuting ghost α^i to obtain a result like $\int \prod_{j=1}^{n} d\alpha^j e^{-\alpha^k g_{kl} \alpha^l} \sim (\det g_{kl})^{+1/2}$. However, since $\alpha^k g_{kl} \alpha^l$ vanishes for symmetric g_{kl} and anticommuting α^k, one must use the slightly more complicated approach with a, b and c ghosts.

$\det g_{ij}$ and the ordinary Gaussian integral over a^i yields $(\det g_{ij})^{-1/2}$. Altogether one finds the following result:

$$\Big[\det g_{ij}(\bar{x}_{k-1/2})\Big]^{1/2} = \alpha \int \left(\prod_{j=1}^{n} da^j_{k-1/2}\, db^j_{k-1/2}\, dc^j_{k-1/2}\right)$$
$$\times \exp\left[-\frac{\epsilon}{2\beta^2\hbar} g_{ij}(\bar{x}_{k-1/2})\left(b^i_{k-1/2}\, c^j_{k-1/2} + a^i_{k-1/2}\, a^j_{k-1/2}\right)\right]. \quad (2.30)$$

We define the normalization constant α such that the integral precisely yields $g^{1/2}(\bar{x}_{k-1/2})$; since later we shall perform the integral over a^i, b^i, c^i, see (2.46), the normalization constant α will cancel and for that reason we shall not bother to determine its value. The reason we have inserted the coefficients $-\epsilon(2\beta^2\hbar)^{-1}$ is that we obtain then in the continuum limit the same normalization for the ghost action as for the nonghost part, see (2.81). (The factor ϵ/β becomes $d\tau$ and the overall factor becomes $1/2\beta\hbar$.)

We have given the ghosts a^i, b^i and c^i which belong to $g(\bar{x}_{k-1/2})$ the subscripts $k-1/2$. This brings out clearly the fact that the ghosts are defined by integrating out the momenta which were located between the coordinates. Similarly, we could have written the momenta as $p^j_{k-1/2}$ instead of p^j_k to indicate that they occur between x^j_k and x^j_{k-1}.

To proceed, we decompose the action for the x_k and the ghosts into a free and an interacting part

$$S = S^{(0)} + S^{(int)}. \quad (2.31)$$

The results should not depend on how one makes this split, but for practical purposes we take $S^{(0)}$ as simple as possible. Since we obtain the answer for $T(z, y; \beta)$ defined in (2.2) as an expansion about z (we perform this calculation in Section 2.5), we find it convenient for the purposes of comparison to also take the metric in $S^{(0)}$ at the point z. We could also have taken the metric in $S^{(0)}$ at, for example, the midpoint $\frac{1}{2}(z+y)$ or perhaps at geodesic midpoints.

Next, we decompose x^i_k into a background part $x^i_{bg,k}$ and a quantum part q^i_k,

$$x^i_k = x^i_{bg,k} + q^i_k \quad (k = 0, \ldots, N). \quad (2.32)$$

For $k=0$ we define $x^i_0 = y^i$ and for $k=N$ one has $x^i_N = z^i$. We shall assume that the background $x^i_{bg,k}$ satisfies the boundary conditions $x^i_{bg,0} = y^i$ and $x^i_{bg,N} = z^i$, hence the quantum part q^i_k vanishes for $k=0$ and $k=N$. Furthermore, we assume that $x^i_{bg,k}$ is a solution of the $N-1$ equations of motion of $S^{(0)}$,

$$g_{ij}(z)\left(x^j_{k+1} - 2x^j_k + x^j_{k-1}\right) = 0 \quad (k = 1, \ldots, N-1). \quad (2.33)$$

In the continuum limit one obtains

$$x^i_{bg}(t) = z^i + \frac{t}{\beta}(z^i - y^i) = z^i + \tau(z^i - y^i) \tag{2.34}$$

where $\tau = t/\beta$ and τ runs from -1 to 0. At this point the time coordinate t appears. We take it to run from $-\beta$ to 0 in order that the point z corresponds to $t = 0$, but other definitions are of course also possible.

We thus define

$$S^{(0)} = \sum_{k=1}^{N} \left[\frac{1}{2\epsilon} g_{ij}(z)(q^i_k - q^i_{k-1})(q^j_k - q^j_{k-1}) \right.$$
$$\left. + \frac{\epsilon}{2\beta^2} g_{ij}(z)(b^i_{k-1/2}\, c^j_{k-1/2} + a^i_{k-1/2}\, a^j_{k-1/2}) \right] . \tag{2.35}$$

By definition then, $S^{(int)} = S - S^{(0)}$. Note that $S^{(int)}$ contains terms linear in q, see (2.85). As already mentioned, this is due to the fact that $x_{bg}(t)$ is a solution of the field equations of $S^{(0)}$, and not of the field equation of S. Also note that $S^{(0)}$ does not depend on V so that the propagators we obtain are model independent (V independent).

We should comment on why we require $x^i_{bg}(t)$ to satisfy the equations of motion of $S^{(0)}$. The reason is that the complete solution of the equations of motion of S cannot be given in closed form, so we settle for $S^{(0)}$. This has the drawback that terms linear in q will be produced if we expand $S^{(int)}$ in terms of q about x_{bg} and these give rise to tadpole diagrams. However, as we shall discuss, these tadpole diagrams are of order $\beta^{1/2}$, so to a given order in β, only a few tadpoles contribute.

The decomposition of S into $S^{(0)} + S^{(int)}$ is standard in perturbation theory, but in most cases one puts all terms proportional to q^2 into $S^{(0)}$, and not only the q^2 term with $g_{ij}(z)$. For example, in instanton physics $S^{(0)}$ contains all terms proportional to two quantum fields in the background of the full instanton, and not, for example, the instanton at a particular point. Because one is dealing then with a particular background (the instanton) instead of an arbitrary metric $g_{ij}(x)$, one can determine the propagator in that background, and with this choice of $S^{(0)}$ there are no tadpoles. In our case it is impossible to determine explicitly the exact propagator in an arbitrary background, hence we settle for $S^{(0)}$ in which we use $g_{ij}(x)$ at the point $x = z$.

To obtain discretized propagators, we could couple q^i_k to external sources. However, since the discretized action only depends on $q^i_{k-1/2} = \frac{1}{2}(q^i_k + q^i_{k-1})$ and $\frac{1}{\epsilon}(q^i_k - q^i_{k-1})$, we couple these combinations to independent real discretized external sources,

$$-\frac{1}{\hbar}S_{(sources,\ nonghost)} = \sum_{k=1}^{N}\left(F_{k-1/2,j}\frac{q_k^j - q_{k-1}^j}{\epsilon} + G_{k-1/2,j}q_{k-1/2}^j\right). \tag{2.36}$$

We should now complete squares in $S^{(0)} + S_{(sources,\ nonghost)}$ and then integrate over $dx_k^i = dq_k^i$. However, the action $S^{(0)}$ is not diagonal in q_k^i. Therefore we first make an orthogonal transformation which diagonalizes $S^{(0)}$.

We introduce modes for the quantum fluctuations by the orthogonal transformation

$$q_k^j = \sum_{m=1}^{N-1} r_m^j \sqrt{\frac{2}{N}}\sin\left(\frac{km\pi}{N}\right); \qquad k = 1,\dots,N-1. \tag{2.37}$$

The orthogonality of the real $(N-1)\times(N-1)$ matrix $O_k^m = \sqrt{(2/N)}\sin(km\pi/N)$ follows from the trigonometric formula $2\sin\alpha\sin\beta = \cos(\alpha-\beta) - \cos(\alpha+\beta)$. One finds

$$\sum_{m=1}^{N-1} O_j^m O_k^m = \frac{1}{N}\sum_{m=1}^{N-1}\left[\cos\frac{(j-k)m\pi}{N} - \cos\frac{(j+k)m\pi}{N}\right]. \tag{2.38}$$

The sum over the cosines is easy if one writes them as exponentials. For integer p one has

$$\begin{aligned}\sum_{m=1}^{N-1}\cos\frac{pm\pi}{N} &= \frac{1}{2}\sum_{m=1}^{N-1}\left(\mathrm{e}^{ipm\pi/N} + \mathrm{e}^{-ipm\pi/N}\right)\\ &= \frac{1}{2}\left[\sum_{m=-N+1}^{N}\mathrm{e}^{ipm\pi/N} - 1 - (-)^p\right]\\ &= \frac{1}{2}\sum_{m=1}^{2N}\mathrm{e}^{ipm\pi/N} - \frac{1}{2} - \frac{1}{2}(-)^p = N\delta_{p,0} - \frac{1}{2} - \frac{1}{2}(-)^p.\end{aligned} \tag{2.39}$$

Using this result in (2.38) we find

$$\sum_{m=1}^{N-1} O_j^m O_k^m = \delta_{j,k} \tag{2.40}$$

since $(-)^{j-k}$ equals $(-)^{j+k}$. Because O_j^m is orthogonal, we can replace $\prod_{k=1}^{N-1} dq_k^j$ by $\prod_{m=1}^{N-1} dr_m^j$.

The orthogonality of O_j^m implies that $S^{(0)}$ is diagonal in r_m^j. To demonstrate this we just evaluate $S^{(0)}$

$$S^{(0)} = \frac{1}{2\epsilon}\sum_{k=1}^{N} g_{ij}(z)(q_k^i - q_{k-1}^i)(q_k^j - q_{k-1}^j)$$

$$= \frac{1}{2\epsilon} g_{ij}(z) \sum_{k=1}^{N}\sum_{m,n=1}^{N-1} (O_k^m - O_{k-1}^m) r_m^i (O_k^n - O_{k-1}^n) r_n^j$$

$$= \frac{1}{2\epsilon} g_{ij}(z) \left[\sum_{m=1}^{N-1} 2 r_m^i r_m^j - \sum_{k=1}^{N-1}\sum_{m,n=1}^{N-1} O_k^m (O_{k-1}^n + O_{k+1}^n) r_m^i r_n^j \right]. \tag{2.41}$$

(We shifted $k \to k+1$ in the last term, which is allowed since O_k^m vanishes for $k = 0$ and $k = N$.) Using $\sin\alpha + \sin\beta = 2\sin\frac{1}{2}(\alpha+\beta)\cos\frac{1}{2}(\alpha-\beta)$ we obtain $O_{k-1}^n + O_{k+1}^n = 2O_k^n \cos(n\pi/N)$. Using again the orthogonality of O_j^m we find

$$S^{(0)} = \frac{1}{\epsilon}\sum_{m=1}^{N-1} g_{ij}(z) r_m^i r_m^j \left(1 - \cos\frac{m\pi}{N}\right). \tag{2.42}$$

We denote the path integral with $S^{(0)}$ and the external sources by $Z_N^{(0)}$. In the nonghost sector it becomes

$$Z_N^{(0)}(F,G) = (\text{factor in (2.29)}) \int \left(\prod_{j=1}^{n}\prod_{m=1}^{N-1} dr_m^j \right) \exp E$$

$$E = -\frac{1}{\epsilon\hbar}\sum_{m=1}^{N-1} g_{ij}(z) r_m^i r_m^j \left(1 - \cos\frac{m\pi}{N}\right)$$

$$+ \sum_{k=1}^{N-1} \left[\frac{1}{\epsilon}(F_{k-1/2,j} - F_{k+1/2,j}) + \frac{1}{2}(G_{k-1/2,j} + G_{k+1/2,j}) \right]$$

$$\times \left(\sum_{m=1}^{N-1} \sqrt{\frac{2}{N}} r_m^j \sin\frac{km\pi}{N} \right). \tag{2.43}$$

By "factor in (2.29)" we mean both factors in curly brackets in front of the exponential. Summations over $i, j = 1, \ldots, n$ are always understood. For vanishing F and G one recovers the discretized transition element $T_N^{(0)}(z, y; \beta)$.

Completing squares is now straightforward and performing the integration over dr^i_m yields

$$Z_N^{(0)}(F,G) = (\text{factor in (2.29)}) \left(\prod_{m=1}^{N-1} \frac{(\pi\epsilon\hbar)^{n/2}}{\sqrt{g(z)}(1-\cos\frac{m\pi}{N})^{n/2}} \right)$$
$$\times \exp\left\{ \sum_{m=1}^{N-1} \frac{\epsilon\hbar}{4\,(1-\cos\frac{m\pi}{N})} \right.$$
$$\times \left[\frac{2}{\epsilon}\sqrt{\frac{2}{N}}\sin\frac{m\pi}{2N}\sum_{k=0}^{N-1}\cos\left\{\left(k+\frac{1}{2}\right)\frac{m\pi}{N}\right\} F_{k+1/2,j} \right.$$
$$\left.\left. + \sqrt{\frac{2}{N}}\cos\frac{m\pi}{2N}\sum_{k=0}^{N-1}\sin\left\{\left(k+\frac{1}{2}\right)\frac{m\pi}{N}\right\} G_{k+1/2,j} \right]^2 \right\}. \tag{2.44}$$

The square denoted by $[\ldots]^2$ is taken with $g^{ij}(z)$, so written out in full it reads as $g^{ij}(z)[\ldots]_i[\ldots]_j$.

We similarly couple the ghosts $a^i_{k+1/2}$, $b^i_{k+1/2}$ and $c^i_{k+1/2}$ to external (commuting or anticommuting) sources as follows:

$$-\frac{1}{\hbar}S_{(sources,\, ghosts)} = \sum_{k=0}^{N-1}(A_{k+1/2,i}a^i_{k+1/2} + b^i_{k+1/2}B_{k+1/2,i}$$
$$+ C_{k+1/2,i}c^i_{k+1/2}). \tag{2.45}$$

Completing squares and integrating over a, b, c we find, using (2.30),

$$Z_N^{(0)}(A,B,C) = g^{N/2}(z)\exp\sum_{k=0}^{N-1}\frac{\beta^2\hbar}{\epsilon}g^{ij}(z)$$
$$\times\left(2C_{k+1/2,i}B_{k+1/2,i} + \frac{1}{2}A_{k+1/2,i}A_{k+1/2,j}\right). \tag{2.46}$$

The factor of $g^{N/2}(z)$ is due to integration over a, b, c and corresponds to the N factors $g^{1/2}(x_{k-1/2})$ which we exponentiated in (2.30). The integration over a, b, c then cancels the normalization constant α in (2.30), which we never computed for this reason.

The complete discretized transition element with external sources now becomes

$$Z_N(z,y;\beta) = [g(z)g(y)]^{-1/4}(2\pi\hbar)^{-nN}\left[\left(\frac{2\pi\hbar}{\epsilon}\right)^{nN/2} g^{N/2}(z)\right]$$
$$\times\left[(\pi\epsilon\hbar)^{n(N-1)/2}g^{-N/2+1/2}(z)\prod_{m=1}^{N-1}\left(1-\cos\frac{m\pi}{N}\right)^{-n/2}\right]$$
$$\times\left[\mathrm{e}^{-(1/\hbar)S^{int}}\mathrm{e}^{-(1/\hbar)S(F,G,A,B,C)}\right]. \tag{2.47}$$

The first line is due to the factor in (2.29), the various inner products and the integration over momenta and a, b and c, the second line is due to the second factor in (2.44) and accounts for the integration over r^j_m, while $-\frac{1}{\hbar}S[F,G,A,B,C]$ denotes the terms bilinear in external sources. In the interaction term S^{int} the quantum fields $q^j_k - q^j_{k-1}$, $q^j_{k-1/2}$, $a^j_{k-1/2}$, $b^j_{k-1/2}$ and $c^j_{k-1/2}$, should be replaced by derivatives with respect to the corresponding sources $F^j_{k-1/2}$, $G^j_{k-1/2}$, $A^j_{k-1/2}$, $B^j_{k-1/2}$ and $C^j_{k-1/2}$ as is usual in quantum field theory.

Using the identity[3]

$$\prod_{m=1}^{N-1} 2\left(1 - \cos\frac{m\pi}{N}\right) = N \tag{2.48}$$

the factors $1/\varepsilon$ are converted into factors $1/\beta$. The final result for (2.2) with Hamiltonian (2.1) is very simple

$$Z_N(z,y;\beta) = \left[\frac{g(z)}{g(y)}\right]^{1/4} \frac{1}{(2\pi\hbar\beta)^{n/2}} \left[\mathrm{e}^{-(1/\hbar)S^{int}} \mathrm{e}^{-(1/\hbar)S[F,G,A,B,C]}\right] \tag{2.49}$$

[3] To prove this identity, consider the function

$$f(x) = \prod_{k=0}^{2N-1}\left(x - \cos\frac{k\pi}{N}\right) = (x^2-1)\left[\prod_{k=1}^{N-1}\left(x - \cos\frac{k\pi}{N}\right)\right]^2 .$$

The function $p(x) = -1 + (x + i\sqrt{1-x^2})^{2N}$ has zeros at the roots of unity, hence at $x = \cos(k\pi/N)$ for $k = 0, 1, \ldots, 2N-1$. In particular, its real part vanishes there. Since *Re* $p(x)$ is a polynomial in x of degree $2N$, we see that $f(x)$ and Re $p(x)$ are proportional. Since Re $p(x) = a_{2N}x^{2N} + \cdots$ with

$$a_{2N} = \sum_{k\ \text{even}} \binom{2N}{k} = \lim_{x\to 1}\left[\frac{1}{2}(1+x)^{2N} + \frac{1}{2}(1-x)^{2N}\right] = 2^{2N-1},$$

we find $f(x) = 2^{1-2N}\mathrm{Re}\ p(x)$. Furthermore, near $x = 1$ we have

$$\begin{aligned}\mathrm{Re}\ p(x) &= x^{2N} - \binom{2N}{2} x^{2N-2}(1-x^2) + \mathcal{O}(1-x^2)^2 - 1 \\ &= (x^2-1)\left[(x^{2N}-1)/(x^2-1) + \binom{2N}{2}x^{2N-2}\right] + \mathcal{O}(1-x^2)^2.\end{aligned}$$

Hence

$$2^{2N-1}\prod_{k=1}^{N-1}\left(x - \cos\frac{k\pi}{N}\right)^2 = \frac{\mathrm{Re}\ p(x)}{x^2-1} \to \frac{x^{2N}-1}{x^2-1} + \binom{2N}{2} = N + \binom{2N}{2} = 2N^2$$

as $x \to 1$. This proves the identity.

where we recall for completeness

$$
\begin{aligned}
&-\frac{1}{\hbar}S[F,G,A,B,C] \\
&= \sum_{m=1}^{N-1} \frac{\epsilon\hbar}{4\left(1-\cos\frac{m\pi}{N}\right)} \left\{ \frac{2}{\epsilon}\sqrt{\frac{2}{N}} \sin\frac{m\pi}{2N} \right. \\
&\times \sum_{k=0}^{N-1} \cos\left[\left(k+\frac{1}{2}\right)\frac{m\pi}{N}\right] F_{k+1/2,j} + \sqrt{\frac{2}{N}} \cos\frac{m\pi}{2N} \\
&\times \left. \sum_{k=0}^{N-1} \sin\left[\left(k+\frac{1}{2}\right)\frac{m\pi}{N}\right] G_{k+1/2,j} \right\}^2 + \sum_{k=0}^{N-1} \frac{\beta^2\hbar}{\epsilon} g^{ij}(z) \\
&\times \left(2C_{k+1/2,i}B_{k+1/2,j} + \frac{1}{2}A_{k+1/2,i}A_{k+1/2,j} \right). \qquad (2.50)
\end{aligned}
$$

The interactions are given by

$$
\begin{aligned}
-\frac{1}{\hbar}S^{int} = \sum_{k=1}^{N} \Big\{ &-\frac{1}{2\epsilon\hbar}[g_{ij}(\bar{x}_{k-1/2}) - g_{ij}(z)](x_k^i - x_{k-1}^i)(x_k^j - x_{k-1}^j) \\
&- \frac{1}{2\epsilon\hbar} g_{ij}(z)[(x_k^i - x_{k-1}^i)(x_k^j - x_{k-1}^j) - (q_k^i - q_{k-1}^i)(q_k^j - q_{k-1}^j)] \\
&- \frac{\epsilon}{2\beta^2\hbar}[g_{ij}(\bar{x}_{k-1/2}) - g_{ij}(z)](a_{k-1/2}^i a_{k-1/2}^j + b_{k-1/2}^i c_{k-1/2}^j) \\
&- \frac{\epsilon}{\hbar} V(\bar{x}_{k-1/2}) \Big\} \qquad (2.51)
\end{aligned}
$$

where $V(\bar{x}_{k-1/2})$ is the order $\hbar^2$ counterterm given in (2.13). As already discussed, the quantum fields q_k^i, $a_{k-1/2}^i$, $b_{k-1/2}^i$ and $c_{k-1/2}^i$ in S^{int} should be replaced by the corresponding differential operators with respect to the external sources; this is, of course, standard practice in quantum field theory.

The expression for $Z_N(z,y;\beta)$ in (2.49) is a correct expression to order $\epsilon = \beta/N$. It was derived with much labor but the final result is very simple. It contains the Feynman measure $(2\pi\beta\hbar)^{-n/2}$, and the factor $[g(z)/g(y)]^{1/4}$ is due to expanding the metric in S^{int} about z. If we had expanded about the midpoint this factor would even have been absent. Using this exact expression we can now find unambiguous Feynman rules at the discretized level.

We obtain the discretized propagators by twice differentiating the expression $\exp(-\frac{1}{\hbar}S[F,G,A,B,C])$ with respect to F, G, A, B, C and then setting external sources to zero. An easy case is the $\dot{q}\dot{q}$ propagator, by which we mean

$$\left\langle\left(\frac{q^i_{k+1}-q^i_k}{\epsilon}\right)\left(\frac{q^j_{k'+1}-q^j_{k'}}{\epsilon}\right)\right\rangle; \quad 0\leq k,k'\leq N. \tag{2.52}$$

According to (2.36) it is given by

$$\begin{aligned}\langle \dot{q}^i_{k+1/2}\dot{q}^j_{k'+1/2}\rangle &= \frac{\partial}{\partial F_{k+1/2,i}}\frac{\partial}{\partial F_{k'+1/2,j}}\exp\left(-\frac{1}{\hbar}S[F,G,A,B,C]\right)\Big|_0 \\ &= 2\sum_{m=1}^{N-1}\frac{\epsilon\hbar}{4(1-\cos\frac{m\pi}{N})}g^{ij}(z)\left(\frac{2}{\epsilon}\sqrt{\frac{2}{N}}\sin\frac{m\pi}{2N}\right)^2 \\ &\quad\times\cos\left(k+\frac{1}{2}\right)\frac{m\pi}{N}\cos\left(k'+\frac{1}{2}\right)\frac{m\pi}{N}.\end{aligned} \tag{2.53}$$

Twice the square of the sine cancels the factor $1-\cos m\pi/n$ in the denominator, and using $2\cos\alpha\cos\beta=\cos(\alpha+\beta)+\cos(\alpha-\beta)$ one finds from (2.39)

$$\begin{aligned}\langle \dot{q}^i_{k+1/2}\dot{q}^j_{k'+1/2}\rangle &= \frac{\hbar}{N\epsilon}g^{ij}(z)\sum_{m=1}^{N-1}\left[\cos(k+k'+1)\frac{m\pi}{N}+\cos(k-k')\frac{m\pi}{N}\right] \\ &= \frac{\hbar}{N\epsilon}g^{ij}(z)(-1+N\delta_{k,k'}).\end{aligned} \tag{2.54}$$

Since in the continuum limit $\frac{1}{\epsilon}\delta_{k,k'}$ becomes $\delta(t-t')$, we find in the continuum limit with $t=k\epsilon-\beta$

$$\langle \dot{q}^i(t)\dot{q}^j(t')\rangle = \hbar g^{ij}(z)\left[-\frac{1}{\beta}+\delta(t-t')\right]. \tag{2.55}$$

Let us compare this with the result we would have obtained naively (i.e. disregarding all subtleties involving discretizations). The naive continuum propagator is obtained from

$$\exp\left(-\frac{1}{2\hbar}\int_{-\beta}^{0} g_{ij}\dot{q}^i\dot{q}^j\,dt+\int_{-\beta}^{0} q^iJ_i\,dt\right) \tag{2.56}$$

by the usual steps

$$\begin{aligned}\langle q^i(t)q^j(t')\rangle &= \frac{\delta}{\delta J_i(t)}\frac{\delta}{\delta J_j(t')}\exp\left[\left(-\frac{\hbar}{2}\right)g^{ij}(z)\int_{-\beta}^{0} J_i(t'')\frac{1}{\partial^2/\partial t^2}J_j(t'')\,dt''\right] \\ &= -\hbar g^{ij}(z)\frac{1}{\partial^2/\partial t^2}\delta(t-t').\end{aligned} \tag{2.57}$$

Differentiating with respect to t and t' yields only the delta function; one misses the term with $-1/\beta$ in (2.55). However, this is due to not having taken into account the boundary conditions. Imposing the boundary condition $q(0)=q(-\beta)=0$, one must add suitable terms linear in t and t' to

(2.57) so that the propagator vanishes at $t = 0, -\beta$ and $t' = 0, -\beta$ while still maintaining $\partial^2/\partial t^2 \langle q^i(t) q^j(t')\rangle = -\hbar g^{ij}(z)\delta(t-t')$. The naive result, as one may check, is

$$\langle q^i(t) q^j(t')\rangle = -\hbar g^{ij}(z)\frac{1}{\beta}[t(t'+\beta)\theta(t-t') + t'(t+\beta)\theta(t'-t)]. \quad (2.58)$$

This is the expression for $\frac{1}{\partial^2/\partial t^2}\delta(t-t')$ in the space of functions which vanish at the boundaries. It follows that naively

$$\langle \dot{q}^i(t)\dot{q}^j(t')\rangle = -\hbar g^{ij}(z)\left[\frac{1}{\beta} - \delta(t-t')\right]. \quad (2.59)$$

This agrees with our discretized expression in (2.54), but note that **the symbol $\boldsymbol{\delta(t-t')}$ is proportional to a Kronecker delta function in the discretized case**. Thus when one evaluates Feynman graphs, this Kronecker $\delta(t-t')$ instructs one to set everywhere in the integrand $t = t'$, and not to replace $\delta(t-t')$ by some smooth function. This will be crucial when we evaluate Feynman graphs with equal-time contractions.

Next, we evaluate the $q\dot{q}$ propagator. By this we mean $\langle \frac{1}{2}(q_{k+1} + q_k)\frac{1}{\epsilon}(q_{k'+1} - q_{k'})\rangle$, of course. It is given by

$$\begin{aligned}
\langle q^i_{k+1/2}\dot{q}^j_{k'+1/2}\rangle &= \frac{\partial}{\partial G_{k+1/2,i}}\frac{\partial}{\partial F_{k'+1/2,j}}\exp\left(-\frac{1}{\hbar}S[F,G,A,B,C]\right)\Big|_0 \\
&= 2\sum_{m=1}^{N-1}\frac{\epsilon\hbar g^{ij}(z)}{4\left(1-\cos\frac{m\pi}{N}\right)}\frac{4}{N\epsilon}\sin\frac{m\pi}{2N}\cos\frac{m\pi}{2N} \\
&\quad\times\sin\left[\left(k+\frac{1}{2}\right)\frac{m\pi}{N}\right]\cos\left[\left(k'+\frac{1}{2}\right)\frac{m\pi}{N}\right] \\
&= \hbar g^{ij}(z)\frac{1}{N}\sum_{m=1}^{N-1}\cos\frac{m\pi}{2N}\left[\frac{\sin\left(k+\frac{1}{2}\right)\frac{m\pi}{N}}{\sin\frac{m\pi}{2N}}\right] \\
&\quad\times\cos(k'+1/2)\frac{m\pi}{N}.
\end{aligned} \quad (2.60)$$

The symbol $|_0$ indicates that one should set the external sources to zero after the differentiations. To evaluate this series, we introduce the notation

$$\zeta = \exp\frac{i\pi}{2N} \quad (2.61)$$

and then find the following result for the product of the four trigonometric functions

$$\frac{1}{4N}\sum_{m=1}^{N-1}[\zeta^m + \zeta^{-m}][\zeta^{2km} + \zeta^{(2k-2)m} + \cdots + \zeta^{-2km}][\zeta^{(2k'+1)m} + \zeta^{-(2k'+1)m}]. \quad (2.62)$$

(In the ratio of the two sines $(\zeta^{(2k+1)m} - \zeta^{-(2k+1)m})/(\zeta^m - \zeta^{-m})$ only powers of ζ^2 remain.) There are then four series to sum, which we write in the following four lines:

$$\frac{1}{4N}\sum_{m=1}^{N-1}\left[\begin{array}{l}\varsigma^{(2k+2k'+2)m}+\varsigma^{(2k+2k')m}+\cdots+\varsigma^{(-2k+2k'+2)m}\\ +\varsigma^{(2k+2k')m}+\varsigma^{(2k+2k'-2)m}+\cdots+\varsigma^{(-2k+2k')m}\\ +\varsigma^{-(2k+2k')m}+\varsigma^{-(2k-2+2k')m}+\cdots+\varsigma^{-(-2k+2k')m}\\ +\varsigma^{-(2k+2k'+2)m}+\varsigma^{-(2k+2k')m}+\cdots+\varsigma^{-(-2k+2k'+2)m}\end{array}\right]. \tag{2.63}$$

We have written the terms in the last two lines with increasing exponents, since this allows us to combine in the same column the terms in the first and fourth row, or the second and third row. Using (2.39)

$$\sum_{m=1}^{N-1}(\varsigma^{2pm}+\varsigma^{-2pm})=-1-(-)^p+2N\delta_{p,0} \tag{2.64}$$

we find for (2.63)

$$\frac{1}{4N}\sum_{p=-k+k'+1}^{k+k'+1}[-1-(-)^p+2N\delta_{p,0}]+\frac{1}{4N}\sum_{p=-k+k'}^{k+k'}[-1-(-)^p+2N\delta_{p,0}]. \tag{2.65}$$

The terms with $(-)^p$ cancel. In the remainder we distinguish the cases $k>k'$, $k<k'$ and $k=k'$. We then obtain

$$\begin{aligned}&\frac{1}{4N}\Big[-(2k+1)+2N\delta_{k>k'}-(2k+1)+2N\delta_{k\geq k'}\Big]\\ &=-\frac{\left(k+\frac{1}{2}\right)}{N}+\frac{1}{2}\delta_{k>k'}+\frac{1}{2}\delta_{k\geq k'}\end{aligned} \tag{2.66}$$

where $\delta_{k>k'}$ equals unity if $k>k'$ and zero otherwise. Therefore

$$\langle q^i_{k+1/2}\dot{q}^j_{k'+1/2}\rangle=\hbar g^{ij}(z)\left[-\frac{\left(k+\frac{1}{2}\right)}{N}+\left\{\begin{array}{ll}0 & \text{if } k<k'\\ \frac{1}{2} & \text{if } k=k'\\ 1 & \text{if } k>k'\end{array}\right\}\right]. \tag{2.67}$$

In the continuum limit this becomes

$$\langle q^i(t)\dot{q}^j(t')\rangle=\hbar g^{ij}(z)\left[-\frac{t+\beta}{\beta}+\theta(t-t')\right] \tag{2.68}$$

which agrees with the naive continuum result obtained by differentiating (2.58),

$$\langle q^i(t)\dot{q}^j(t')\rangle=[-\hbar g^{ij}(z)]\frac{1}{\beta}\left[t\theta(t-t')+(t+\beta)\theta(t'-t)\right]. \tag{2.69}$$

The discretized approach tells us that $\theta(t-t')=\frac{1}{2}$ at $t=t'$. However, at the point $t=t'=0$, the naive continuum propagator in (2.69) does not vanish and thus violates the boundary conditions. In the discretized approach one should write q_k as $q_{k-1/2} + \frac{1}{2}(q_k - q_{k-1})$. By combining (2.67) and (2.54) one finds then no problems at the boundary

$$\langle q^i_{k+1}\dot{q}^j_{k'+1/2}\rangle = \hbar g^{ij}(z)\left[-\frac{(k+1)}{N} + \frac{1}{2}\delta_{k,k'} + \theta_{k,k'}\right] \tag{2.70}$$

which vanishes for $k+1 = N$. The extra term $\frac{1}{2}\delta_{k,k'}$ saves the day. The reason that the continuum approach fails to give zero while the discretized approach yields the correct result is that $q^i(t)$ should be defined at midpoints, and thus never really reaches the endpoints.

Finally, we consider the qq propagators. This is the most complicated propagator. It is given by

$$\begin{aligned}\langle q^i_{k+1/2}q^j_{k'+1/2}\rangle &= \frac{\partial}{\partial G_{k+1/2,i}}\frac{\partial}{\partial G_{k'+1/2,j}}\exp\left(-\frac{1}{\hbar}S[G,F,A,B,C,0]\right)\Big|_0 \\ &= 2\sum_{m=1}^{N-1}\frac{\epsilon\hbar g^{ij}(z)}{4\,(1-\cos\frac{m\pi}{N})}\left(\sqrt{\frac{2}{N}}\cos\frac{m\pi}{2N}\right)^2 \\ &\quad\times\sin\left(k+\frac{1}{2}\right)\frac{m\pi}{N}\sin\left(k'+\frac{1}{2}\right)\frac{m\pi}{N} \\ &= \frac{\epsilon\hbar}{2N}g^{ij}(z)\sum_{m=1}^{N-1}\left(\cos\frac{m\pi}{2N}\right)^2 \\ &\quad\times\left[\frac{\sin\left(k+\frac{1}{2}\right)\frac{m\pi}{N}}{\sin\frac{m\pi}{2N}}\right]\left[\frac{\sin\left(k'+\frac{1}{2}\right)\frac{m\pi}{N}}{\sin\frac{m\pi}{2N}}\right].\end{aligned} \tag{2.71}$$

Again we write the ratios of sines as polynomials in ζ^2, with $2k+1$ and $2k'+1$ terms, respectively. This leads to the series

$$\begin{aligned}&\frac{1}{4}\sum_{m=1}^{N-1}(\zeta^m+\zeta^{-m})^2(\zeta^{2km}+\zeta^{(2k-2)m}+\dots+\zeta^{-2km}) \\ &\quad\times(\zeta^{2k'm}+\zeta^{(2k'-2)m}+\dots+\zeta^{-2km'}) \\ &=\frac{1}{4}\sum_{m=1}^{N-1}\sum_{\alpha=-k}^{k}\sum_{\beta=-k'}^{k'}[\zeta^{(2\alpha+2\beta+2)m}+2\zeta^{(2\alpha+2\beta)m}+\zeta^{(2\alpha+2\beta-2)m}].\end{aligned} \tag{2.72}$$

Replacing $\alpha \to -\alpha$ and $\beta \to -\beta$ in half of the terms (which yields the same result), we obtain cosines

$$\frac{1}{2}\sum_{m=1}^{N-1}\sum_{\alpha=-k}^{k}\sum_{\beta=-k'}^{k'}\left[\cos(\alpha+\beta+1)\frac{m\pi}{N}+\cos(\alpha+\beta)\frac{m\pi}{N}\right] \tag{2.73}$$

and using the formula in (2.39) for summing cosines we obtain

$$\frac{1}{2}\sum_{\alpha=-k}^{k}\sum_{\beta=-k'}^{k'}\left[-\frac{1}{2}-\frac{1}{2}(-)^{\alpha+\beta+1}+N\delta_{\alpha+\beta+1,0}-\frac{1}{2}-\frac{1}{2}(-)^{\alpha+\beta}+N\delta_{\alpha+\beta,0}\right]$$
$$=\frac{1}{2}\sum_{\alpha=-k}^{k}\sum_{\beta=-k'}^{k'}[-1+N(\delta_{\alpha+\beta+1,0}+\delta_{\alpha+\beta,0})]. \tag{2.74}$$

It is again easiest to consider the cases $k > k'$, $k < k'$ and $k = k'$ separately. We find

$$\langle q^i_{k+1/2} q^j_{k'+1/2}\rangle = \frac{\epsilon\hbar}{4N} g^{ij}(z) \times\left[-(2k+1)(2k'+1)+\begin{Bmatrix} 2N(2k'+1) & \text{for} & k>k' \\ N(4k+1) & \text{for} & k=k' \\ 2N(2k+1) & \text{for} & k<k' \end{Bmatrix}\right]. \tag{2.75}$$

The last term contains a discretized theta function

$$\langle q^i_{k+1/2} q^j_{k'+1/2}\rangle = \epsilon\hbar g^{ij}(z)\left[-\frac{\left(k+\frac{1}{2}\right)\left(k'+\frac{1}{2}\right)}{N} +\left(k'+\frac{1}{2}\right)\theta(k,k')+\left(k+\frac{1}{2}\right)\theta(k',k)-\frac{1}{4}\delta_{k,k'}\right]. \tag{2.76}$$

In the continuum limit this becomes

$$\langle q^i(t)q^j(t')\rangle = -\beta\hbar g^{ij}(z)\left[\frac{(\beta+t)}{\beta}\frac{(\beta+t')}{\beta} -\frac{(\beta+t')}{\beta}\theta(t-t')-\frac{(\beta+t)}{\beta}\theta(t'-t)\right] \tag{2.77}$$

which agrees with the naive continuum propagator $-\hbar g^{ij}(z)\frac{1}{\beta}[t(t'+\beta)\theta(t-t')+t'(t+\beta)\theta(t'-t)]$ except that the value $\theta(0)=\frac{1}{2}$ is now justified.

As a check we may combine the propagators for $q\dot{q}$ and qq, and check that $\langle q^i_{k+\frac{1}{2}}\, q^j_{k'}\rangle$ indeed vanishes for $k'=0$ or $k'=N$. Using $q_{k'+1}=q_{k'+1/2}+\frac{\epsilon}{2}\dot{q}_{k'+1/2}$ and combining (2.76) and (2.67) one finds the following result:

$$\langle q^i_{k+1/2}\, q^j_{k'+1}\rangle = \epsilon\hbar g^{ij}(z) \times\left[-\frac{(k+\frac{1}{2})(k'+1)}{N}+(k'+1)\theta(k,k')+\left(k+\frac{1}{2}\right)\theta(k',k)-\frac{1}{4}\delta_{k,k'}\right]. \tag{2.78}$$

This expression indeed vanishes at $k'+1=N$, while the naive continuum limit does not vanish at $t'=0$. Similar results hold for the other endpoint. One could try to improve the naive continuum results by extending the integration region beyond $[-\beta,0]$, and require that also for $t>0$ and $t<-\beta$ the propagator satisfies $\partial_t^2\langle q(t)q(t')\rangle \sim \delta(t-t')$ while still $q(0)=q(-\beta)=0$. This is possible, but the resulting function is quite complicated. Rather, we shall derive rules for products of continuum distributions which follow directly from the corresponding discretized expressions.

Finally, we determine the propagators of the ghosts. We find from (2.50) and (2.45),

$$\begin{aligned}\langle b^i_{k+1/2}c^j_{k'+1/2}\rangle &= -\frac{\partial}{\partial B_{k+1/2,i}}\frac{\partial}{\partial C_{k'+1/2,j}}\mathrm{e}^{-\frac{1}{\hbar}S[F,G,A,B,C]}\Big|_0 \\ &= -\frac{2\beta^2\hbar}{\epsilon}g^{ij}(z)\delta_{k,k'} \quad\rightarrow\quad -2\beta^2\hbar g^{ij}(z)\delta(t-t') \qquad (2.79)\\ \langle a^i_{k+1/2}a^j_{k'+1/2}\rangle &= \frac{\partial}{\partial A_{k+1/2,i}}\frac{\partial}{\partial A_{k'+1/2,j}}\mathrm{e}^{-\frac{1}{\hbar}S[F,G,A,B,C]}\Big|_0 \\ &= \frac{\beta^2\hbar}{\epsilon}g^{ij}(z)\delta_{k,k'} \quad\rightarrow\quad \beta^2\hbar g^{ij}(z)\delta(t-t'). \qquad (2.80)\end{aligned}$$

Again we note that $\delta(t-t')$ in the continuum limit should be interpreted as a Kronecker delta function; moreover, in the discretized approach the ghosts are only defined on midpoints.

Summary

We now summarize our results for the path integral representation of $\langle z|\exp(-\frac{\beta}{\hbar}\hat{H})|y\rangle$ for bosonic systems in configuration space with the Hamiltonian $\hat{H}$ given in (2.1). It can be written in terms of propagators and vertices from $S^{(int)}$ as follows. The vertices are given by

$$\langle z|\exp\Big(-\frac{\beta}{\hbar}\hat{H}\Big)|y\rangle = \left[\frac{g(z)}{g(y)}\right]^{1/4}\frac{1}{(2\pi\beta\hbar)^{n/2}}\langle \mathrm{e}^{-\frac{1}{\hbar}S^{(int)}}\rangle \qquad (2.81)$$

$$\begin{aligned}S^{(int)} &= S - S^{(0)}\\ -\frac{1}{\hbar}S &= -\frac{1}{\beta\hbar}\int_{-1}^{0}\frac{1}{2}g_{ij}(x)\left[\frac{dx^i}{d\tau}\frac{dx^j}{d\tau}+b^i(\tau)c^j(\tau)+a^i(\tau)a^j(\tau)\right]d\tau\\ &\quad -\frac{\beta\hbar}{8}\int_{-1}^{0}\left[R(x)+g^{ij}(x)\Gamma^l_{ik}(x)\Gamma^k_{jl}(x)\right]d\tau\\ -\frac{1}{\hbar}S^{(0)} &= -\frac{1}{\beta\hbar}\int_{-1}^{0}\frac{1}{2}g_{ij}(z)\left[\frac{dq^i}{d\tau}\frac{dq^j}{d\tau}+b^i(\tau)c^j(\tau)+a^i(\tau)a^j(\tau)\right]d\tau\end{aligned}$$

where

$$\begin{aligned}x^i(\tau) &= x^i_{bg}(\tau)+q^i(\tau)\\ x^i_{bg}(\tau) &= z^i+\tau(z^i-y^i) \qquad (2.82)\end{aligned}$$

The propagations are given by

$$\begin{aligned}
\langle q^i(\sigma)q^j(\tau)\rangle &= -\beta\hbar g^{ij}(z)\Delta(\sigma,\tau)\\
\langle q^i(\sigma)\dot{q}^j(\tau)\rangle &= -\beta\hbar g^{ij}(z)[\sigma+\theta(\tau-\sigma)]\\
\langle \dot{q}^i(\sigma)\dot{q}^j(\tau)\rangle &= -\beta\hbar g^{ij}(z)[1-\delta(\tau-\sigma)]\\
\langle a^i(\sigma)a^j(\tau)\rangle &= \beta\hbar g^{ij}(z)\partial_\sigma^2\Delta(\sigma,\tau)\\
\langle b^i(\sigma)c^j(\tau)\rangle &= -2\beta\hbar g^{ij}(z)\partial_\sigma^2\Delta(\sigma,\tau)
\end{aligned} \tag{2.83}$$

where

$$\begin{aligned}
\Delta(\sigma,\tau) &= \sigma(\tau+1)\theta(\sigma-\tau)+\tau(\sigma+1)\theta(\tau-\sigma)\\
\partial_\sigma^2\Delta(\sigma,\tau) &= \delta(\sigma-\tau).
\end{aligned} \tag{2.84}$$

The reader may add the contribution from a general $V(x)$, but we shall not need this. Since only the combination $\beta\hbar$ occurs, β counts the number of loops. To obtain a uniform overall factor of $(\beta\hbar)^{-1}$ in the action we normalized the ghost actions as in (2.30). By expanding $\exp[-\frac{1}{\hbar}S^{(int)}]$ and using these propagators, we can evaluate $\langle z|\exp(-\frac{\beta}{\hbar}\hat{H})|y\rangle$ to any order in loops. The transition element $T(z,y;\beta)$ corresponds to the vacuum expectation value of $\exp[-\frac{1}{\hbar}S^{(int)}]$: loops with internal quantum fields but no external quantum fields.

The interactions are more explicitly given by

$$\begin{aligned}
-\frac{1}{\hbar}S^{(int)} &= -\frac{1}{\beta\hbar}\int_{-1}^{0}\left\{\frac{1}{2}g_{ij}(x)[(z^i-y^i)(z^j-y^j)+2(z^i-y^i)\dot{q}^j]\right.\\
&\quad\left.+\frac{1}{2}[g_{ij}(x)-g_{ij}(z)](\dot{q}^i\dot{q}^j+b^ic^j+a^ia^j)\right\}d\tau \qquad (2.85)\\
&= -\frac{1}{\beta\hbar}\int_{-1}^{0}\left\{\frac{1}{2}g_{ij}(z)(z^i-y^i)(z^j-y^j)+g_{ij}(z)(z^i-y^i)\dot{q}^j\right.\\
&\quad+\frac{1}{2}\partial_k g_{ij}(z)(z^i-y^i)(z^j-y^j)[(z-y)^k\tau+q^k]\\
&\quad+\partial_k g_{ij}(z)(z^i-y^i)\dot{q}^j[(z^k-y^k)\tau+q^k]\\
&\quad+\frac{1}{4}\partial_k\partial_l g_{ij}(z)[(z^k-y^k)\tau+q^k][(z^l-y^l)\tau+q^l]\\
&\quad\times[(z^i-y^i)(z^j-y^j)+2(z^i-y^i)\dot{q}^j]\\
&\quad\left.+\cdots+\frac{1}{2}[g_{ij}(x)-g_{ij}(z)](\dot{q}^i\dot{q}^j+b^ic^j+a^ia^j)\right\}d\tau\\
&\quad-\frac{\beta\hbar}{8}\int_{-1}^{0}[R(x_{bg}+q)+g^{ij}(x_{bg}+q)\Gamma^l_{ik}(x_{bg}+q)\Gamma^k_{jl}(x_{bg}+q)]\,d\tau.
\end{aligned}$$

Checks

We now briefly discuss and check some of the terms in the action. We do this because all of our later calculations will be based on this action, so we should be absolutely sure it is correct. The classical terms (the terms without q or ghosts) yield in the path integral a factor

$$= \exp\left\{-\frac{1}{\beta\hbar}\left[\frac{1}{2}g_{ij}(z)(z^i-y^i)(z^j-y^j)\right.\right.$$
$$-\frac{1}{4}(z^k-y^k)\partial_k g_{ij}(z)(z^i-y^i)(z^j-y^j)$$
$$\left.\left.+\frac{1}{12}\partial_k\partial_l g_{ij}(z)(z^i-y^i)(z^j-y^j)(z^k-y^k)(z^l-y^l)+\cdots\right]\right\}. \quad (2.86)$$

These terms are not equal to an expansion of the classical action about z because $x_{bg}(\tau)$ is only a solution of $S^{(0)}$. As a consequence, tree graphs with vertices which are linear in q from $S^{(int)}$ also contribute to the order $1/\beta\hbar$ terms. Let us study this further. The term with $\dot{q}^j$ in the first line of $-\frac{1}{\hbar}S^{(int)}$ vanishes due to the boundary conditions, but the vertices

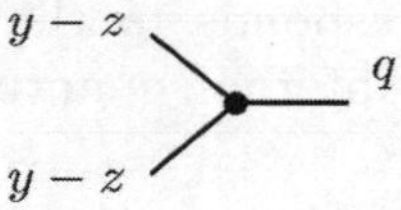

$$= -\frac{1}{\beta\hbar}\int_{-1}^{0}\frac{1}{2}\partial_k g_{ij}\Big[(z^i-y^i)(z^j-y^j)q^k$$
$$-2(z^i-y^i)(z^k-y^k)q^j)\Big]d\tau$$
$$=\frac{1}{\beta\hbar}\Gamma_{ij;k}(z-y)^i(z-y)^j\int_{-1}^{0}q^k\,d\tau \quad (2.87)$$

do contribute. (We partially integrated to obtain the second term.) Two of these vertices produce a tree graph which contributes a term

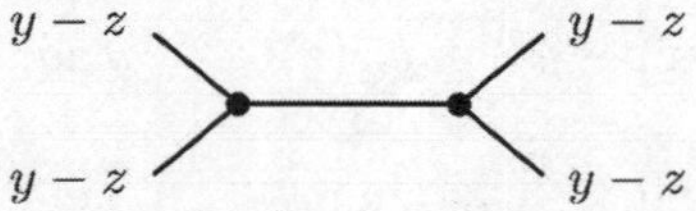

$$= -\frac{1}{\beta\hbar}\frac{1}{24}\Gamma_{ij;k}\Gamma_{i'j';k'}(z-y)^i(z-y)^{i'}(z-y)^j(z-y)^{j'}g^{kk'} \quad (2.88)$$

where we have used the fact that $\int_{-1}^{0}\int_{-1}^{0}d\sigma\,d\tau\,\Delta(\sigma-\tau)=-\frac{1}{12}$. Recalling that $y-z$ is of order $\beta^{1/2}$, this indeed completes the classical action to this order in β (see (2.220)). Tree graphs with two q-propagators contribute at the $\beta^{3/2}$ level, and so on. Hence, the tree graph part is in good shape (see Fig. 2.1)

Next, consider the one-loop part (the part independent of $\hbar$). Expanding the measure $[g(z)/g(y)]^{1/4}$ in (2.81) one finds a factor $1+\frac{1}{4}g^{ij}\times(z-y)^k\partial_k g_{ij}(z)$ multiplying $\exp(-\frac{1}{\hbar}S_{cl}[z,y;\beta])$. This factor is canceled

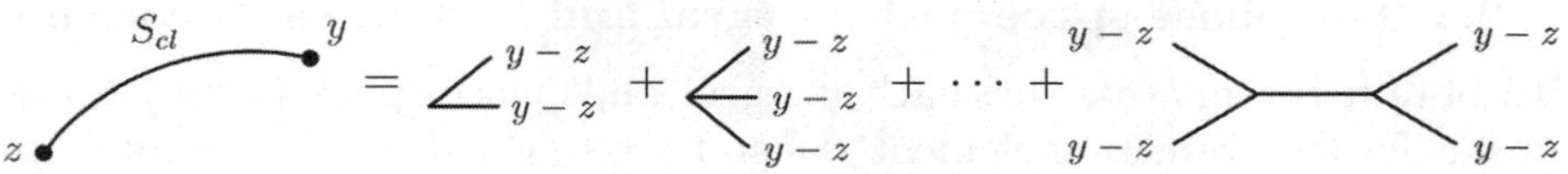

Fig. 2.1 The expansion of the classical action evaluated for a geodesic from y to z, expanded in terms of $y-z$. The internal propagators come from the quantum fields q, and external lines denote factors of $y-z$.

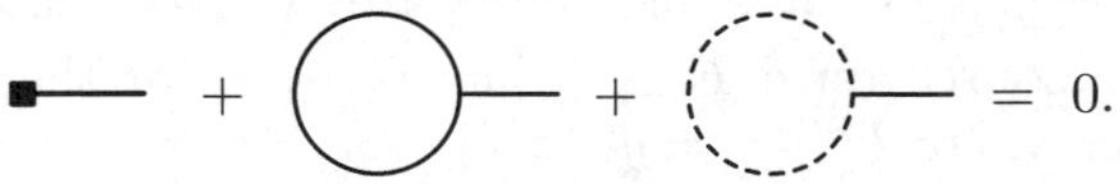

Fig. 2.2 At the one-loop level the contributions of the measure, denoted by a black box, cancel loops with q and ghost loops. External lines again denote factors of $y-z$.

by the one-loop equal-time contractions from the vertices in the last-but-one line of $-\frac{1}{\hbar}S^{int}$,

$$-\frac{1}{\beta\hbar}(z^k-y^k)\frac{1}{2}\partial_k g_{ij}(z)\int_{-1}^{0}\tau\,\langle\dot{q}^i\dot{q}^j+b^ic^j+a^ia^j\rangle\,d\tau \tag{2.89}$$

and from the vertex in the third line

$$-\frac{1}{\beta\hbar}(\partial_k g_{ij})(z^i-y^i)\int_{-1}^{0}\langle\dot{q}^jq^k\rangle\,d\tau. \tag{2.90}$$

The latter does not contribute since

$$\langle\dot{q}^jq^k\rangle\sim(\tau+1)\theta(\sigma-\tau)+\tau\theta(\tau-\sigma)=\tau+\frac{1}{2}\qquad\text{at }\ \sigma=\tau \tag{2.91}$$

which integrates to zero. In the former all $\delta(\sigma-\tau)$ cancel, as the discretized approach rigorously shows, and with

$$\langle\dot{q}^i\dot{q}^j+b^ic^j+a^ia^j\rangle_{\sigma=\tau}=-\hbar\beta g^{ij}(z) \tag{2.92}$$

it yields $-\frac{1}{4}(z^k-y^k)\partial_k g_{ij}(z)g^{ij}(z)$, canceling the factor from the measure (see Fig. 2.2).

There are many other one- and two-loop graphs, and the contribution from each corresponds to a particular term in the expansion of $\langle z|\exp(-\frac{\beta}{\hbar}\hat{H})|y\rangle$ about z. In particular, the two-loop graph with one $\dot{q}\dot{q}$, one $\dot{q}q$ and one $q\dot{q}$ propagator only agrees with the transition element if

$$\int_{-1}^{0}\int_{-1}^{0}\delta(\sigma-\tau)\theta(\sigma-\tau)\theta(\tau-\sigma)\,d\sigma\,d\tau=\frac{1}{4}. \tag{2.93}$$

This result immediately follows from the discretized approach, where $\delta(\sigma-\tau)$ is a Kronecker delta and $\theta(\sigma-\tau)=\frac{1}{2}$ at $\sigma=\tau$. We shall give a complete analysis of all two-loop graphs in Section 2.6.

2.2 The phase-space path integral and Matthews' theorem

To obtain the phase-space path integral and phase-space Feynman diagrams for the transition element, we go back to the discretized expression for $T_N(z,y;\beta)$ in (2.22) with the momenta not yet integrated out, and add external sources for $q_{k-1/2} \equiv \frac{1}{2}(q_k + q_{k-1})$ and p_k, because these are the variables on which H depends. Since the free equation of motion for p_k reads $g^{ij}(z)p_{k,j} = i(q_k^i - q_{k-1}^i)/\epsilon$ (see below), we denote the sources for $p_{k,j}$ by $-iF^j_{k-1/2}$. If one then integrates out the momenta p_k one again finds the source term $F_{k-1/2,j}(q_k^j - q_{k-1}^j)/\epsilon$ of the configuration-space approach, where $F^j_{k-1/2} = g^{ij}(z)F_{k-1/2,i}$. Hence

$$Z_N(F,G,z,y;\beta) = [g(z)g(y)]^{-1/4}(2\pi\hbar)^{-nN} \int \prod_{j=1}^{n} \left(\prod_{k=1}^{N} dp_{k,j} \prod_{l=1}^{N-1} dx_l^j \right)$$
$$\times \exp \left\{ \sum_{k=1}^{N} \left[\frac{i}{\hbar} p_k \cdot (x_k - x_{k-1}) - \frac{\epsilon}{\hbar} H(x_{k-1/2}, p_k) \right.\right.$$
$$\left.\left. - iF^j_{k-1/2} p_{k,j} + G_{k-1/2,j} q^j_{k-1/2} \right] \right\}. \tag{2.94}$$

Next, we decompose H into $H^{(0)} + H^{(int)}$, where

$$H^{(0)} = \sum_{k=1}^{N} \frac{1}{2} g^{ij}(z) p_{k,i} p_{k,j}, \tag{2.95}$$

and we again decompose $x = x_{bg} + q$, but we add the term $\frac{i}{\hbar} p_{k,j}(x^j_{bg,k} - x^j_{bg,k-1})$ to $-\frac{\epsilon}{\hbar} H^{(int)}$. The term $\frac{i}{\hbar} p_{k,j}(q^j_k - q^j_{k-1})/\epsilon$ is part of $S^{(0)}$. Note that we do not decompose p into a background part and a quantum part. We discuss boundary conditions in phase-space further below (2.101).

We then complete squares in the terms depending on p in the sum of $ip_k(q_k - q_{k-1}) - \epsilon H^{(0)}$ and the source terms, and perform the p-integrals. This yields

$$Z(F,G,z,y;\beta) = [g(z)g(y)]^{-1/4}(2\pi\hbar)^{-nN}$$
$$\times \left[\left(\frac{2\pi\hbar}{\epsilon} \right)^{\frac{1}{2}nN} g(z)^{\frac{1}{2}N} \right] \int \prod_{j=1}^{n} dx^j_{N-1} \cdots dx^j_1$$
$$\times \exp \left[-\frac{\epsilon}{\hbar} H^{(int)} \left(x^j_{k-1/2} \to x^j_{bg,k-1/2} + \frac{\partial}{\partial G_{k-1/2,j}},\ p_{k,j} \to i \frac{\partial}{\partial F^j_{k-1/2}} \right) \right]$$
$$\times \exp \left\{ \sum_{k=1}^{N} \frac{\epsilon}{2\hbar} \left[-\frac{i\hbar}{\epsilon} F^j_{k-1/2} + \frac{i}{\epsilon}(q^j_k - q^j_{k-1}) \right]^2 + G_{k-1/2,j} q^j_{k-1/2} \right\}. \tag{2.96}$$

The square in the last line of (2.96) is again taken with $g_{ij}(z)$; expanding this square we again find the terms of the configuration-space path integral, multiplied by the factor

$$\exp\left\{-\frac{\hbar}{2\epsilon}\left(\sum_{k=1}^{N} g_{ij}(z)F^i_{k-1/2}F^j_{k-1/2}\right)\right\}. \tag{2.97}$$

The propagators are again obtained by differentiation with respect to the sources in (2.94). It follows that the discrete qq propagators are the same, while the pq propagator in the phase-space approach is equal to i times the $\dot{q}q$ propagator in the configuration-space approach, in agreement with the linearized field equations $g^{ij}p_{k,j} = i(q^i_k - q^i_{k-1})/\epsilon$. However, the pp propagator is not equal to minus the $\dot{q}\dot{q}$ propagator; rather, there is an extra term proportional to $\delta_{k,k'}$ which comes from the term with F^2,

$$\langle p_{k,i}p_{l,j}\rangle = -g_{ii'}(z)g_{jj'}(z)\langle \dot{q}^{i'}_{k-1/2}\dot{q}^{j'}_{l-1/2}\rangle + \frac{\hbar}{\epsilon}g_{ij}(z)\delta_{k,l}. \tag{2.98}$$

The last term cancels the singularity $\frac{\hbar}{\epsilon}g^{ij}(z)\delta_{k,l}$ which appears in the term with $\langle \dot{q}^i_{k-1/2}\dot{q}^j_{l-1/2}\rangle$, see (2.54). Hence, as is well known, the phase-space propagator is nonsingular for short distances. The continuum limit reads

$$\begin{aligned}
\langle p_i(\sigma)p_j(\tau)\rangle &= \frac{1}{\beta}\hbar g_{ij}(z) \\
\langle q^i(\sigma)p_j(\tau)\rangle &= -i\hbar\delta^i_j(\sigma + \theta(\tau-\sigma)) = \langle p_j(\tau)q^i(\sigma)\rangle \\
\langle q^i(\sigma)q^j(\tau)\rangle &= -\beta\hbar g^{ij}(z)\Delta(\sigma,\tau).
\end{aligned} \tag{2.99}$$

Using

$$\begin{aligned}
-\frac{1}{\hbar}H^{(int)} = \sum_{k=1}^{N}\Big\{&-\frac{\epsilon}{2\hbar}[g^{ij}(x_{k-1/2}) - g^{ij}(z)]p_{k,i}p_{k,j} \\
&+ \frac{i}{\hbar}p_{k,j}(x^j_{bg,k} - x^j_{bg,k-1}) - \epsilon\frac{\hbar}{8}(R + g^{ij}\Gamma^l_{ik}\Gamma^k_{jl})\Big\}
\end{aligned} \tag{2.100}$$

we can again compute the transition element loop-by-loop. In the continuum limit we find

$$\begin{aligned}
-\frac{1}{\hbar}\int_{-\beta}^{0} H^{(int)}dt = &-\frac{\beta}{\hbar}\int_{-1}^{0}\frac{1}{2}[g^{ij}(x) - g^{ij}(z)]p_ip_j\,d\tau \\
&+ \frac{i}{\hbar}\int_{-1}^{0}p_j(z^j - y^j)\,d\tau - \frac{\beta\hbar}{8}\int_{-1}^{0}(R + g^{ij}\Gamma^k_{il}\Gamma^l_{jk})\,d\tau.
\end{aligned} \tag{2.101}$$

The tree graph with two $p(z-y)$ vertices now yields the leading term $-\frac{1}{2\beta\hbar}g_{ij}(z)(z^i - y^i)(z^j - y^j)$ in the classical action (see Section 2.5, in particular (2.218)), and the reader can check a few other graphs. Of course, the propagators as well as the vertices differ in the phase-space approach (the latter contain $g^{ij}(x)$ instead of $g_{ij}(x)$ and there are no ghosts), but the result for $\langle z|\exp(-\frac{\epsilon}{\hbar}H)|y\rangle$ should be the same. This equality is known as the Matthews' theorem.[4] We shall later check this in a few examples.

In the phase-space approach, we imposed only boundary conditions on $x^j(t)$, namely $x^j(0) = z^j$ and $x^j(-\beta) = y^j$. This is the correct number of boundary conditions, both in configuration and phase-space, and we need only boundary conditions on x because we consider $T(z, y; \beta)$. We decomposed $x^j(t)$ into a background solution of the field equation $\ddot{x} = 0$, which took care of the boundary conditions, and a quantum part. Note that the complete field equations for a free particle with $S^{(0)} \sim \int(ip\dot{x} - \frac{1}{2}p^2)$ are $p = i\dot{x}$ and $\dot{p} = 0$. For a harmonic oscillator with $H = \frac{1}{2}(p^2 + x^2)$ one obtains $p = i\dot{x}$ and $x = -i\dot{p}$, leading to $\ddot{x} = x$ which is solved in terms of sinh and cosh. So in these cases one can find a background trajectory x_{cl}, but the background solution for p would have to be imaginary ($p = i\dot{q}$, we are in Euclidean space). It is easier to decompose only x but not p; the background trajectory with $p_{cl} = 0$ is then no longer a solution of $S^{(0)}$, but one can still calculate with it (as we have done).

One could also consider transition elements with a p-eigenstate at $t = 0$ and an x-eigenstate at $t = -\beta$. Then one could introduce background trajectories x_{cl} and p_{cl} satisfying the field equations $p = i\dot{x}$ and $\dot{p} = 0$ and the boundary conditions $x_{cl}(t = -\beta) = y$ and $p_{cl}(t = 0) = p$. A simpler method is to start from $\langle z|\mathrm{e}^{-\frac{\beta}{\hbar}\hat{H}}|y\rangle$ and Fourier transform to obtain

$$\langle z|\mathrm{e}^{-\frac{\beta}{\hbar}\hat{H}}|p\rangle = \int dy\, \langle z|\mathrm{e}^{-\frac{\beta}{\hbar}\hat{H}}|y\rangle \sqrt{g(y)}\langle y|p\rangle. \qquad (2.102)$$

For $T(z, y; \beta)$ no boundary conditions on $p_j(t)$ are needed to make the p integrals convergent because $p_j(t)$ has no zero modes. (A zero mode is a mode which drops out of the action.) There are no zero modes for $p_j(t)$ because it appears without derivatives in the action, with a leading term p^2. For $q^j(t)$, only differences $q^j_k - q^j_{k-1}$ appear in the discretized expressions of $T(z, y; \beta)$, hence we must fix the zero mode of $q^j(t)$ by suitable boundary conditions. It follows that one cannot impose boundary conditions on p alone both at $t = -\beta$ and $t = 0$. The technical reason is that the field equation $\dot{p} = 0$ does not allow the boundary conditions $p(t = -\beta) = p_1$ and $p(t = 0) = p_2$. The deeper reason is that in our model $q(t)$ has a zero mode which must be fixed by a boundary condition

[4]The original Matthews' theorem only applied to meson field theories with at most one time derivative in the interaction [71]. It was extended to quantum mechanical models with $\dot{q}\dot{q}$ interactions and higher time derivatives by Nambu [72]. Provided one adds the new ghosts as we have done, the equivalence between the Lagrangian and Hamiltonian approach also holds for nonlinear sigma models [73]. For a general proof of Matthews' theorem based on path integrals see [74].

on $q(t)$. Only by adding a mass term for $q(t)$ would this zero mode become a nonzero model and in that case one could impose boundary conditions on $p(t)$ at both ends. The complete sets of p_k states were inserted between x-eigenstates, so one can view them as being defined at midpoints, not at the endpoints, and this explains that for the computation of $T(z, y; \beta)$ one need not impose boundary conditions on the momenta as well. The most compelling reason is that there are no classical trajectories in phase-space which connect two arbitrary points in phase-space. (In the so-called holomorphic approach [48], one introduces variables $z \sim x + ip$ and $\bar{z} \sim x - ip$, and fixing boundary conditions for z at $t = 0$ and $\bar{z}$ at $t = -\beta$ would seem to imply that one does impose separate boundary conditions at both endpoints for x and p, but this is not true as the classical z_{cl} and $\bar{z}_{cl}$ are to be considered as independent, and not as complex conjugates of each other.)

It is instructive to see how the p-propagators are obtained in the continuum approach. The kinetic terms in the phase-space approach are given by

$$-\frac{1}{\hbar}S^{(0)} = \frac{i}{\hbar}\int_{-\beta}^{0} p_j \dot{q}^j \, dt - \frac{1}{\hbar}\int_{-\beta}^{0} \frac{1}{2} g^{ij}(z) p_i p_j \, dt. \tag{2.103}$$

The kinetic matrix for (p_i, q^j) is thus (replacing $g^{ij}(z)$ and $\hbar$ by unity for notational simplicity)

$$K = \begin{pmatrix} 1 & -i\partial_t \\ i\partial_t & 0 \end{pmatrix}, \tag{2.104}$$

and the Feynman (translationally invariant) propagator is its inverse

$$G = \begin{pmatrix} 0 & -\frac{1}{2}i\,\epsilon(t-t') \\ \frac{1}{2}i\,\epsilon(t-t') & -\frac{1}{2}(t-t')\theta(t-t') + (t \leftrightarrow t') \end{pmatrix}. \tag{2.105}$$

It satisfies $KG = \delta(t-t')$. To satisfy the boundary condition $q(0) = q(-\beta) = 0$, while still satisfying $KG = \delta(t-t')$, we add to G a polynomial in t, t' which is annihilated by K

$$P(t,t') = \begin{pmatrix} p_1(t') & p_2(t') \\ -itp_1(t') - iq_1(t') & -itp_2(t') - iq_2(t') \end{pmatrix}. \tag{2.106}$$

We then require that $(G+P)_{12}$ vanishes at $t' = 0, -\beta$, $(G+P)_{21}$ at $t = 0, -\beta$, and $(G+P)_{22}$ at all $t = 0, -\beta$ and $t' = 0, -\beta$. (These entries correspond to $\langle pq \rangle$, $\langle qp \rangle$ and $\langle qq \rangle$, respectively.) The solution is

$$P(t,t') = \begin{pmatrix} \frac{1}{\beta} & -i\left(\frac{t'}{\beta} + \frac{1}{2}\right) \\ -i\left(\frac{t}{\beta} + \frac{1}{2}\right) & -\frac{tt'}{\beta} - \frac{1}{2}(t+t') \end{pmatrix}. \tag{2.107}$$

Adding P to G, it is clear that the naive continuum results in (2.99) agree with the discretized propagators, again except when both t and t'

lie on the boundaries. For these values one must again use the discretized propagators.

The Feynman propagator in (2.105) is given in position space, but for an analysis of the divergences in loops it is more convenient to give it in momentum space. The Fourier transform of (2.105) may not seem obvious, but it helps to first add a small mass term $\frac{1}{2}m^2q^2$ to $S^{(0)}$. The kinetic operator now becomes

$$K = \begin{pmatrix} 1 & -i\partial_t \\ i\partial_t & m^2 \end{pmatrix}, \tag{2.108}$$

and the Feynman propagator becomes

$$G(t-t';m^2) = \begin{pmatrix} \frac{m}{2}\mathrm{e}^{-m|t-t'|} & -\frac{i}{2}\epsilon(t-t')\mathrm{e}^{-m|t-t'|} \\ \frac{i}{2}\epsilon(t-t')\mathrm{e}^{-m|t-t'|} & \frac{1}{2m}\mathrm{e}^{-m|t-t'|} \end{pmatrix}. \tag{2.109}$$

For small m one recovers (2.105), except that one finds in the qq propagator the constant $1/2m$. It cancels in KG, but if one were to evaluate Feynman graphs for a massive theory (with, for example, $V = \frac{1}{2}m^2q^2 + \lambda q^4$) in the infinite t-interval one would need to include the contributions from this $1/2m$ term. The limit $m \to 0$ would then lead to infrared divergences which makes the theory ill-defined. On a finite t-interval one must specify boundary conditions, similarly to what we did in the massless theory with the matrix P in (2.107), and the mass singularity is removed by the boundary conditions. The Fourier transform of $G(t-t';m^2)$ is easily found

$$G(t-t';m^2) = \int_{-\infty}^{\infty} \frac{dk_0}{2\pi} \mathrm{e}^{ik_0(t-t')} \begin{pmatrix} \frac{m^2}{k_0^2+m^2} & \frac{-k_0}{k_0^2+m^2} \\ \frac{k_0}{k_0^2+m^2} & \frac{1}{k_0^2+m^2} \end{pmatrix}. \tag{2.110}$$

By decomposing m^2 into $(k_0^2 + m^2) - k_0^2$ we see once again how the delta function singularity in the $\dot{q}\dot{q}$ propagator is canceled in the pp propagator: the propagator in (2.110) is clearly nonsingular as $t' \to t$, but the $\dot{q}\dot{q}$ propagator corresponds to the numerator k_0^2 and is singular as $t' \to t$. We can now study divergences in phase-space.

All loops computed with phase-space Feynman diagrams are finite because in t-space all propagators are bounded and all integration regions are finite. One can also explain by power counting methods (which are formulated in momentum space) why phase-space path integrals are convergent, while configuration-space path integrals are only convergent after taking the ghosts into account. The kinetic matrix has entries of unity (from the p^2 term) and k (from the $p\dot{q}$ term). Disregarding $P(t,t')$, on an infinite interval the pp propagators vanish, while the qq propagators behave like $1/k^2$ for large k but the pq propagators go only like $\int dk\, k/k^2$ and would seem to lead to an ultraviolet divergence in a tadpole graph. However, the integral over k/k^2 vanishes since it is odd in k. These results do not change if there is a mass term of the form $\frac{1}{2}m^2q^2$ present.

To illustrate the calculations in phase-space by another example, consider the following Hamiltonian:

$$\hat{H} = \frac{1}{2} g^{\alpha} p_i g^{1/2} g^{ij} p_j g^{-1/2-\alpha}. \tag{2.111}$$

For $\alpha = -\frac{1}{4}$, this is just (2.1), but for $\alpha \neq -\frac{1}{4}$ there are extra terms proportional to $\alpha + \frac{1}{4}$,

$$\begin{aligned} \hat{H} &= \hat{H}\left(\alpha = -\frac{1}{4}\right) + \Delta\hat{H}\left(\alpha + \frac{1}{4}\right), \\ \Delta\hat{H}\left(\alpha + \frac{1}{4}\right) &= \frac{1}{2}\left(\alpha + \frac{1}{4}\right) i\hbar \left\{p_i, g^{ij}\partial_j \ln g\right\} \\ &\quad - \frac{1}{2}\left(\alpha + \frac{1}{4}\right)^2 \hbar^2 g^{ij} (\partial_i \ln g)(\partial_j \ln g). \end{aligned} \tag{2.112}$$

Since the extra terms are Weyl-ordered, we can at once go to the phase-space path integral. Suppose we were to compute

$$\operatorname{Tr} \sigma(x) \mathrm{e}^{-\frac{\beta}{\hbar} H} = \int dx_0 \sqrt{g(x_0)} \sigma(x_0) \langle x_0 | \mathrm{e}^{-\frac{\beta}{\hbar} H} | x_0 \rangle. \tag{2.113}$$

(These kinds of expressions are found when one evaluates trace anomalies, see Chapter 7, but these interpretations do not concern us at this point.) The path integral leads to

$$\left\langle \mathrm{e}^{-\frac{\beta}{\hbar}\int_{-1}^{0} H^{(int)}\, d\tau}\, \mathrm{e}^{-\frac{\beta}{\hbar}\int_{0}^{1} \Delta H^{(int)}(\alpha+\frac{1}{4}) d\tau} \right\rangle \tag{2.114}$$

with $H^{(int)}$ given by (2.101) and

$$\begin{aligned} -\frac{\beta}{\hbar} \Delta H^{(int)}\left(\alpha + \frac{1}{4}\right) &= -i\left(\alpha + \frac{1}{4}\right)(\beta p_i) g^{ij} \partial_j \ln g \\ &\quad + \frac{1}{2}(\beta\hbar)\left(\alpha + \frac{1}{4}\right)^2 g^{ij} (\partial_i \ln g)(\partial_j \ln g). \end{aligned} \tag{2.115}$$

Since the trace is cyclic, the result for the path integral should be α-independent. To order $(\alpha + \frac{1}{4})$ there are no contributions since p_i is a quantum field for which the vacuum expectation value vanishes.[5] However, to order $(\alpha + \frac{1}{4})^2$ there are two contributions: a tree graph with two $(\alpha + \frac{1}{4})$ vertices and a *pp* propagator, and, further more, the vertex is proportional to $(\alpha + \frac{1}{4})^2$. Using the *pp* propagators from (2.99), the sum of both contributions clearly cancels, as it should.

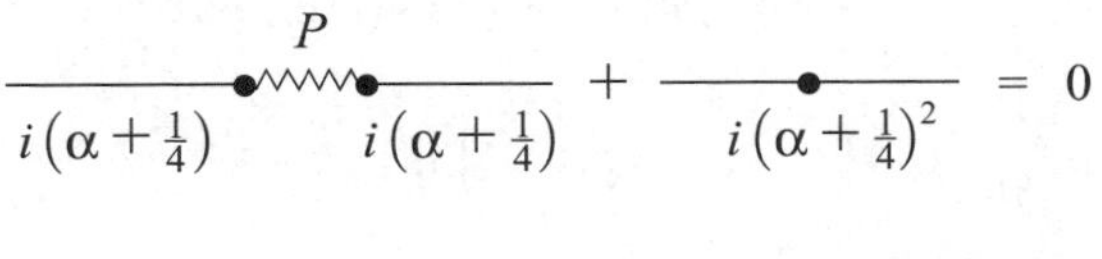

[5]There is also a term linear in p in $H(\alpha = -\frac{1}{4})$, namely the term $-i\int_{-1}^{0} p_j(z^j - y^j)\, d\tau$ in (2.100), but it does not contribute to the trace in (2.113) because $z = y = x_0$ in the trace.

Upon eliminating the momenta p_j, the phase-space path integral becomes a configuration-space path integral, and infinities are introduced which are canceled by new ghosts, as we have discussed. The phase-space approach should yield the same finite answers as the configuration-space approach. That this indeed happens is called Matthews' theorem. We illustrate it with a few examples, although a formal path integral proof can also be given, see [74].

The interaction part of the action for the phase-space path integral differs from that for the configuration-space path integral by the following terms

$$
\begin{aligned}
-\frac{1}{\hbar}\int_{-\beta}^{0}\Big[H^{(int)}_{phase}(p,q)-H^{(int)}_{conf}(q)\Big]\,dt &= -\frac{\beta}{\hbar}\int_{-1}^{0}\frac{1}{2}\Big\{g^{ij}(x)-g^{ij}(z)\Big\}p_ip_j\,d\tau \\
&+\frac{1}{\beta\hbar}\int_{-1}^{0}\frac{1}{2}\Big\{g_{ij}(x)-g_{ij}(z)\Big\}\Big(\dot q^i\dot q^j+b^ic^j+a^ia^j\Big)\,d\tau \\
&+\frac{i}{\hbar}\int_{-1}^{0}p_j(z^j-y^j)\,d\tau+\frac{1}{\beta\hbar}\int_{-1}^{0}\frac{1}{2}g_{ij}(x)(z^i-y^i)(z^j-y^j)\,d\tau \\
&+\frac{1}{\beta\hbar}\int_{-1}^{0}g_{ij}(x)(z^i-y^i)\dot q^j\,d\tau \qquad (2.116)
\end{aligned}
$$

where $x=z+\tau(z-y)+q$.

Inserting the complete field equation for p, namely $p_i=\frac{i}{\beta}g_{ij}(x)\dot x^i$, into the complete action for p, which reads $-\frac{1}{2}g^{ij}(x)p_ip_j+ip\dot x$, one obtains, of course, the complete action in configuration space, namely $-\frac{1}{2}g_{ij}(x)\dot x^i\dot x^j$ plus ghosts. However, if one calculates in perturbation theory, one decomposes p and q into a free part (in-and-out fields) and the rest. These free parts satisfy free field equations which differ, of course, from the full field equations and for p they read $p_i^{(0)}=\frac{i}{\beta}g_{ij}(z)\dot q^j$. Substituting these free field equations into the kinetic part of the phase-space action (of the form $ip\dot q-\frac{1}{2}p^2$) one finds the free part of the configuration-space action ($-\frac{1}{2}\dot q^2$), but **the interaction parts of the phase-space and configuration-space actions differ after substituting the free p field equations.** The difference is easily calculated in our case

$$
\begin{aligned}
-\frac{1}{\hbar}\Delta S &= \frac{1}{\beta\hbar}\int_{-1}^{0}\frac{1}{2}\Big\{g_{ij}(z)g^{jk}(x)g_{kl}(z)+g_{il}(x)-2g_{il}(z)\Big\}\,\dot q^i\dot q^l\,d\tau \\
&+\frac{1}{\beta\hbar}\int_{-1}^{0}\frac{1}{2}\left\{g_{ij}(x)-g_{ij}(z)\right\}(b^ic^j+a^ia^j)\,d\tau \\
&+\frac{1}{\beta\hbar}\int_{-1}^{0}\left\{g_{ij}(x)-g_{ij}(z)\right\}\dot q^i(z^j-y^j)\,d\tau \\
&+\frac{1}{\beta\hbar}\int_{-1}^{0}\frac{1}{2}g_{ij}(x)(z^i-y^i)(z^j-y^j)\,d\tau. \qquad (2.117)
\end{aligned}
$$

Clearly, for linear sigma models with $g_{ij} = \delta_{ij}$ and no background fields ($z=y$), the actions are the same, $\Delta S=0$. The claim of Matthews' theorem is now that the effects of the extra term $\hbar g^{ij}(z)\delta(\sigma-\tau)$ in the $\dot{q}^i\dot{q}^j$ propagator cancel the effects due to ΔS. In other words, the phase-space approach and the configuration-space approach should give the same result.

To avoid confusion, we spell out the procedure in detail. In the phase-space approach one has an interaction $L^{(int)}_{phase}(p,q)$. The interactions in the configuration-space approach are given by $L^{(int)}_{conf}(\dot{q},q)$. The statement that these interactions are different means that $L^{(int)}_{phase}(p,q) \neq L^{(int)}_{conf}(\dot{q},q)$, where p should be eliminated using the field equations of motion of $S^{(0)}$. This field equation reads as $p_i = g_{ij}(z)\dot{q}^j$. If one changes the notation in $L^{(int)}_{phase}$ and writes $L^{(int)}_{phase}(i\dot{q},q)$, then one may use the same propagators for $\langle qq\rangle$ and $\langle q\dot{q}\rangle$ as in the configuration-space approach, but for $\langle\dot{q}\dot{q}\rangle$ the propagators in the phase-space approach and the configuration-space approach are different. The claim of Matthews' theorem is then that one may either work with $L^{(int)}_{phase}$ and the nonsingular propagator for $\langle\dot{q}\dot{q}\rangle$, or with $L^{(int)}_{conf}$ but with singular propagators for $\langle\dot{q}\dot{q}\rangle$. Green functions (with p identified with $i\dot{q}$ on external lines) should be the same.

To bring out the essentials, we consider a simplified model [39], in which $z=y$ and $g_{ij}(z) = \delta_{ij}$ and $g_{ij}(x) - \delta_{ij} \equiv A_{ij}(q)$ and we choose an ordering of $\hat{H}$ such that there are no extra terms of order $\hbar^2$. Furthermore, the model is one-dimensional, so $i,j=1$. This leads to $L = \frac{1}{2\beta}\dot{q}(1+A)\dot{q}$ with $A = A(q)$. We define H in the Euclidean case by $i\dot{q}p - H = -L$. This yields $H = \frac{1}{2}\beta p \frac{1}{1+A} p$. Then the interaction Hamiltonian and Lagrangian read as, respectively,

$$-\frac{1}{\hbar}H^{(int)} = \frac{\beta}{2\hbar}p\frac{A}{1+A}p \quad \Leftrightarrow \quad -\frac{1}{\hbar}L^{(int)} = -\frac{1}{2\beta\hbar}(\dot{q}A\dot{q} + bAc + aAa). \tag{2.118}$$

The free field equation for p reads $\beta p = i\dot{q}$, and substitution into H^{int} produces a result which differs from $-L^{int}$ because instead of A one finds $A/(1+A)$ at vertices (and because there are no ghosts in H^{int}). However, the $\langle\dot{q}\dot{q}\rangle$ and $\langle pp\rangle$ propagators are also different. In the phase-space approach the qq one-loop self-energy receives contributions from a p-loop and a seagull graph with a pp loop:

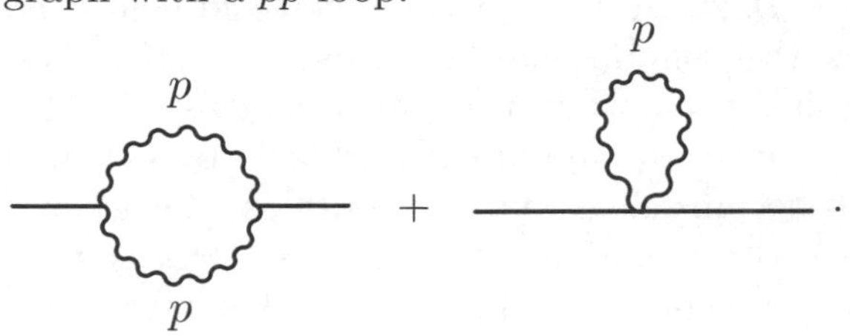

The external lines denote $A(q)$. In the configuration-space approach there are $\dot{q}\dot{q}$ loops and ghost loops but no seagull graph:

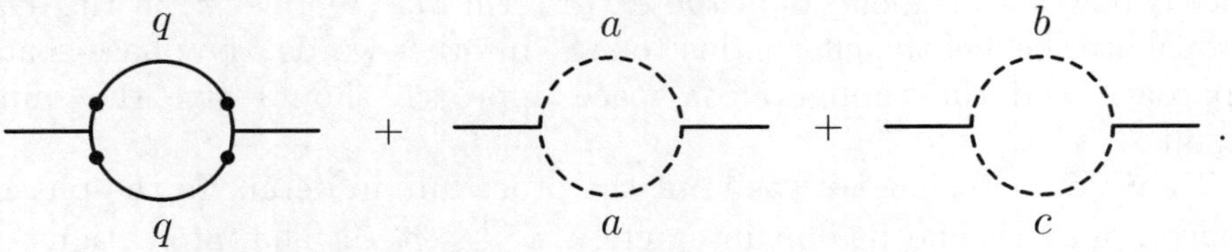

Matthews' theorem claims that both results are equal. Comparing both results, we see that in the configuration-space approach one is left with the integrand

$$\frac{1}{4}\left\{({}^{\bullet}\!\Delta^{\bullet})^2 - ({}^{\bullet\bullet}\!\Delta)^2\right\} = \frac{1}{4}\left\{1 - 2\delta(\sigma - \tau)\right\}. \tag{2.119}$$

The factor of 1 agrees with the result one obtains in the Hamiltonian approach from the p-loop, whereas the factor of $\frac{1}{4}(-2)\delta(\sigma - \tau)$ agrees with result one obtains in the Hamiltonian approach from the seagull graph. (To write the integration $\int d\tau$ of the seagull graph as a double integral $\int d\tau \int d\sigma$ we added the factor $\delta(\sigma - \tau)$. This $\delta(\sigma - \tau)$ contracts the self-energy graph to a seagull graph.) Hence, the extra term in the $\dot{q}\dot{q}$ propagator (the $\delta(\sigma - \tau)$ term) gives the same contribution as the extra vertex ($ppAA$).

The reader may verify that other Green functions also give the same results. For example, the $p - p$ self-energy gives the same result as minus the $\dot{q}\dot{q}$ self-energy (the minus sign comes from the factor of i in $p = i\dot{q}$) because the mixed loops (with Ap and pA propagators or with $A\dot{q}$ and $\dot{q}A$ propagators) agree, whereas in the phase-space case the loop with an AA and a pp propagator plus the seagull graph with an AA propagator gives the same result as in the configuration case of a loop with an AA and a $\dot{q}\dot{q}$ propagator. There are no ghost contributions to the one-loop pp self-energy.

Historically the difference between the Hamiltonian and Lagrangian approach to quantum field theories with derivative interactions first became a source of confusion in the 1940s when "mesotron theories" (theories with scalar fields) were studied with gradient couplings. (QED was in this respect simpler because it had no derivatives interactions, but scalar QED with $L = -|\partial_\mu \varphi - ieA_\mu \varphi|^2$ was studied and it has the same difficulties.) Matthews' theorem [71] clarified the situation for these theories, and a general analysis for quantum mechanical models with double-derivative interactions was given by Nambu [72] and later by Lee and Yang [41]. By the 1970s it had become clear that one could use the action itself to obtain the interaction vertices, and the propagators were "covariant", by which it is meant that propagators of derivatives fields were equal to derivatives of propagators of the fields

$$\langle \partial_\mu A_\nu(x) \partial_\rho A_\sigma(y) \rangle = \frac{\partial}{\partial x^\mu} \frac{\partial}{\partial y^\rho} \langle A_\nu(x) A_\sigma(y) \rangle. \tag{2.120}$$

To retain part of the canonical methods so that one can work out the radiative corrections to current algebra, one introduces the notion of a T^* product, so one can write $\langle B(x)C(y)\rangle = \langle \Omega | T^* B(x) C(y) | \Omega \rangle$, where $|\Omega\rangle$ denotes the vacuum. This T^* operator commutes with derivatives, and is thus different from the usual time-ordering symbol T which involves theta functions $\theta(x^0 - y^0)$ and $\theta(y^0 - x^0)$ and thus does not commute with $\partial/\partial x^0$ and $\partial/\partial y^0$. In fact, for one time derivative the results of using T^* or T are still the same because, for example in QED,

$$\begin{aligned} &\frac{\partial}{\partial x^0} \langle \Omega | T A_\mu(x) A_\nu(y) | \Omega \rangle \\ &\quad = \langle \Omega | T \partial_0 A_\mu(x) A_\nu(y) | \Omega \rangle + \delta(x^0 - y^0)[A_\mu(x), A_\nu(y)] \\ &\quad = \langle \Omega | T \partial_0 A_\mu(x) A_\nu(y) | \Omega \rangle \\ &\quad = \langle \Omega | T^* \partial_0 A_\mu(x) A_\nu(y) | \Omega \rangle \end{aligned} \tag{2.121}$$

since $[A_\mu(x), A_\nu(y)] = 0$. But for two time derivatives the Hamiltonian (canonical) propagator with T and the Lagrangian (covariant) propagator with T^* differ. For example, for QED

$$\begin{aligned} &\frac{\partial}{\partial x^0} \frac{\partial}{\partial y^0} \langle \Omega | T A_\mu(x) A_\nu(y) | \Omega \rangle \\ &\quad = \frac{\partial}{\partial x^0} \langle \Omega | T A_\mu(x) \partial_0 A_\nu(y) | \Omega \rangle \\ &\quad = \langle \Omega | T \partial_0 A_\mu(x) \partial_0 A_\nu(y) | \Omega \rangle + \delta(x^0 - y^0)[A_\mu(x), \partial_0 A_\nu(y)] \\ &\quad = \langle \Omega | T^* \partial_0 A_\mu(x) \partial_0 A_\nu(y) | \Omega \rangle + i\hbar \delta^4(x - y) \eta_{\mu\nu} \end{aligned} \tag{2.122}$$

where we have used the fact that $\partial_0 A_\nu = P(A_\nu) + \cdots$.

As we have already seen in the case of nonlinear sigma models, in the canonical approach there are extra vertices and extra terms in the propagators, but all of these extra effects cancel if one computes Green functions. In the 1960s "current algebra" was developed as a tool to deal with the strong interactions in a nonperturbative way. This was an operator formalism, which therefore used T products, but complicated noncovariant extra terms ("Schwinger terms") were found to be present in the commutation relations of (in particular, the space components of) currents. To simplify the current algebras, the T^* product was introduced, and relations between the current algebra with T ordering and with T^* ordering were developed. Here, of course, the theorems by Matthews and

Nambu were of some use. We shall not enter into a discussion of current algebras, but instead we now study the same problems in nonabelian gauge theory. We do this in Minkowski space to facilitate comparison with the literature.

Consider the nonghost sector. After adding the gauge-fixing term $\mathcal{L}_{(fix)} = -\frac{1}{2}(\partial_\mu A^\mu)^2$ the Lagrange density reads as

$$\begin{aligned}\mathcal{L}(q) &= -\frac{1}{2}(\partial_\mu A^\mu)^2 - \frac{1}{4}F_{\mu\nu}^2 \\ &= \frac{1}{2}(\partial_0 A_j)^2 - \frac{1}{2}(\partial_0 A_0)^2 + \frac{1}{2}(\partial_j A_0)^2 - \frac{1}{2}(\partial_i A_j)^2 \\ &\quad + \partial_0 A_j A_0 \wedge A_j - \partial_j A_0 A_0 \wedge A_j - \frac{1}{4}(F_{ij}^2)^{int} + \frac{1}{2}(A_0 \wedge A_j)^2 \quad (2.123)\end{aligned}$$

where $(F_{ij}^2)^{int} = F_{ij}^2 - (\partial_i A_j - \partial_j A_i)^2$ and $A \wedge B$ denotes $f^{abc} A^b B^c$. Group contractions with the Killing metric δ^{ab} are not indicated explicitly. The conjugate momenta are

$$\begin{aligned}p^j &\equiv p(A_j) = \partial_0 A_j + A_0 \wedge A_j = D_0 A_j = D_0 A^j \\ p^0 &\equiv p(A_0) = -\partial_0 A_0 = \partial_0 A^0. \end{aligned} \quad (2.124)$$

The Hamiltonian density $\mathcal{H} = \dot{A}_\mu p^\mu - \mathcal{L}$ becomes

$$\begin{aligned}\mathcal{H}(p,q) &= \frac{1}{2}(p^j)^2 + p^j A_j \wedge A_0 - \frac{1}{2}(p^0)^2 - \frac{1}{2}(\partial_j A_0)^2 \\ &\quad + \partial_j A_0 A_0 \wedge A_j + \frac{1}{2}(\partial_i A_j)^2 + \frac{1}{4}(F_{ij}^2)^{int}. \end{aligned} \quad (2.125)$$

The Lagrangian density in the phase-space approach is then

$$\mathcal{L}(p,q) = p^j \dot{A}_j + p^0 \dot{A}_0 - \mathcal{H}(p,q). \quad (2.126)$$

We define $\mathcal{L}^{(0)}$ to be the terms quadratic in fields. The interactions in phase-space are given by

$$\begin{aligned}\mathcal{L}^{(int)}_{phase}(p,q) &= -\mathcal{H}^{(int)}(p,q) \\ &= -p^j A_j \wedge A_0 - \partial_j A_0 A_0 \wedge A_j - \frac{1}{4}(F_{ij}^2)^{int}. \end{aligned} \quad (2.127)$$

On the other hand, in configuration space the interactions follow from (2.123),

$$\begin{aligned}\mathcal{L}^{(int)}_{conf}(q) &= -\frac{1}{4}(F_{\mu\nu}^2)^{int} = \frac{1}{2}(F_{0j}^2)^{int} - \frac{1}{4}(F_{ij}^2)^{int} \\ &= \left[(\partial_j A_0 - \partial_0 A_j) A_j \wedge A_0 - \frac{1}{4}(F_{ij}^2)^{int}\right] + \frac{1}{2}(A_0 \wedge A_j)^2. \quad (2.128)\end{aligned}$$

The reason for grouping these terms in this way will become clear.

If one first eliminates $p^\mu = \partial S/\partial \dot{A}_\mu$ from the action $\int \mathcal{L}(p,q)$ by using the full nonlinear field equations $p^j = D_0 A^j$ and $p^0 = -\partial_0 A_0$, one recovers, of course, $\mathcal{L}(q)$. One may check this by replacing p^j in $\mathcal{L}^{(int)}_{phase}(p,q)$ by $D_0 A^j$, and further by substituting $p^j = A_0 \wedge A_j$ into the terms of $\mathcal{L}(p,q)$ which are quadratic in p and q. One then finds

$$\begin{aligned}\mathcal{L}^{(int)}_{phase} - \mathcal{L}^{(int)}_{conf} &= -A_0 \wedge A_j A_j \wedge A_0 + A_0 \wedge A_j \dot{A}_j \\ &- \left[\partial_j A_0 A_0 \wedge A_j + \frac{1}{2}(A_0 \wedge A_j)^2\right] - \frac{1}{2}(A_0 \wedge A_j)^2 = 0. \quad (2.129)\end{aligned}$$

Suppose one performs perturbation theory, using the interaction picture, both with the Hamiltonian theory in phase-space and with the Lagrangian theory in configuration space. One then has the vertices from $\mathcal{L}^{(int)}_{phase}$ and $\mathcal{L}^{(int)}_{conf}$, and the propagators as they are found from the terms bilinear in p and q, and linear in q, respectively

The propagators $\langle p^j A_\mu\rangle$ and $\langle p^0 A_\mu\rangle$ in phase-space are the same as the propagators $\langle \partial_0 A^j A_\mu\rangle$ and $\langle \partial_0 A^0 A_\mu\rangle$ in configuration space. Thus one may substitute, as far as the propagators are concerned, the linear field equation of p^μ into the propagators. However, as we have already discussed, the $\langle p^\mu p^\nu\rangle$ propagators are not equal to the $\langle \partial_0 A^\mu \partial_0 A^\nu\rangle$ propagators (they differ by contact terms with $\delta^4(x-y)$). To facilitate comparison, we may therefore replace everywhere (both in the phase-space action and in the phase-space propagators) p^μ by $\partial_0 A^\mu$, but then there are two sources of extra terms in the phase-space approach:

(i) the vertices $\mathcal{L}^{(int)}_{phase}(p^\mu \to \partial_0 A^\mu)$ differ from those in $\mathcal{L}^{(int)}_{conf}$ by the extra term $\mathcal{L}^{(int)\ extra}_{phase}(p,q) = -\frac{1}{2}(A_j \wedge A_0)^2$;

(ii) the propagators $\langle \partial_0 A^\mu \partial_0 A^\nu\rangle_{phase}$ in the phase-space theory differ from the propagators $\langle \partial_0 A^\mu \partial_0 A^\nu\rangle_{conf}$ of the covariant Lagrangian by the extra contact term

$$\langle \partial_0 A^\mu \partial_0 A^\nu\rangle_{phase} - \langle \partial_0 A^\mu \partial_0 A^\nu\rangle_{conf} = -i\hbar\delta^4(x-y)\eta^{\mu\nu}. \quad (2.130)$$

The content of Matthews' theorem is that all extra contributions cancel. We now check this with a few instructive examples.

Consider first the $A_0 A_0$ self-energy in the phase-space approach at the one-loop level. There is one extra diagram

$$\sim \frac{i}{\hbar}\int\left(-\frac{1}{2}\right)A_0\langle A_i A_i\rangle A_0 \quad (2.131)$$

where the cross denotes that this is an extra vertex. To find the extra contributions from the propagators one first determines all interactions with a time derivative. These are given by $\mathcal{L}^{(int)} = -\partial_0 A_j A_j \wedge A_0$. These are ordinary vertices (vertices in the covariant theory), so there are no cross contributions where both extra vertices and extra propagators contribute. Two such vertices yield the following extra contribution:

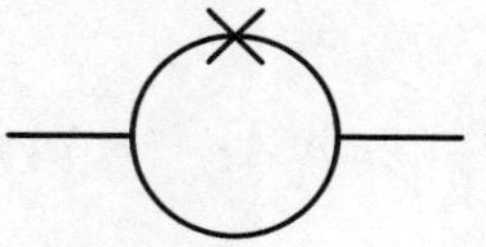

$$= \left(\frac{i}{\hbar}\right)^2 \frac{1}{2!} \iint A_0 \langle A_i A_i \rangle [-i\hbar \delta^4(x-y)] A_0$$

$$= \frac{i}{\hbar} \int \frac{1}{2} A_0 \langle A_i A_i \rangle A_0. \tag{2.132}$$

The cross denotes the extra term in the propagator. The extra contributions indeed exactly cancel

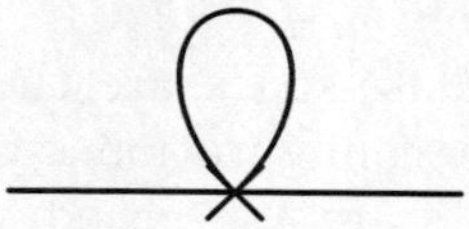

$$+$$

$$= 0.$$

The same cancellation follows for the $A_j A_j$ one-loop self-energy.

Let us now see if the ghost sector gives extra contributions. The ghost action is

$$\mathcal{L} = -\partial^\mu b(\partial_\mu c + A_\mu \wedge c) = \dot{b}(\dot{c} + A_0 \wedge c) - \partial_j b D_j c. \tag{2.133}$$

The conjugate momenta (we always use left-differentiation, so $p(b) = \frac{\partial}{\partial b} S$) are

$$p(b) = (\dot{c} + A_0 \wedge c) = D_0 c, \qquad p(c) = -\dot{b}. \tag{2.134}$$

The Lagrangian in phase-space (for left-differentiation) is

$$\mathcal{L}_{phase} = \dot{c} p(c) + \dot{b} p(b) - [p(b) - A_0 \wedge c] p(c) - \partial_j b D^j c. \tag{2.135}$$

The interactions in phase-space are

$$\mathcal{L}^{(int)}_{phase} = -p(c) A_0 \wedge c - \partial_j b(A_j \wedge c). \tag{2.136}$$

The interactions in configuration space are

$$\mathcal{L}^{(int)}_{conf} = \partial_0 b A_0 \wedge c - \partial_j b(A_j \wedge c). \tag{2.137}$$

Substituting the linearized field equation for $p(c)$ we find that there are no extra vertices

$$\mathcal{L}^{(int)}_{phase}(-p(c) = \dot{b}) - \mathcal{L}^{(int)}_{conf} = 0. \tag{2.138}$$

There are also no contributions with an extra propagator term since these would have to come from $\langle \partial_0 b \partial_0 c \rangle$, but there are no interactions with $\partial_0 c$. Thus, in the ghost sector there are no subtleties.

2.3 Path integrals for Dirac fermions

We shall now discuss the extension of the time slicing approach to fermions. We distinguish between complex (Dirac) fermions and real (Majorana) fermions. The latter have some special problems, so defined by we begin with Dirac fermions ψ. The action reads $\int i\psi^\dagger \dot{\psi}\, dt$, so the conjugate momentum $(\partial/\partial\dot{\psi})S = -i\psi^\dagger$ when we remove $\dot{\psi}$ from the left) is not proportional to ψ. In order to integrate in the path integral over fermionic fields ψ, one must introduce Grassmann variables. It is sometimes said that path integrals with Grassmann variables are mathematically not well founded. We have two answers to such criticisms.

(i) In our approach we begin with operators (such as $\hat{\psi}$ and $\hat{\psi}^\dagger$) without any Grassmann variables. Then when we convert expressions such as $\mathrm{Tr}\, \hat{J} \exp(-\beta \hat{H})$ into **discretized** path integrals, we introduce Grassmann variables by means of mathematical **identities**. Finally, one integrates over these Grassmann variables, and no Grassmann variables are left in the end. So the discretized path integrals are mathematically well defined.

(ii) The question of whether the continuum limit of the path integrals exists is easier to prove for fermions than bosons because there are no convergence problems with Berezinian integration: $\int d\theta\, \theta = 1$. In particular, in our applications to anomalies we need only graphs with a given number of loops. We are then working at the level of perturbation theory, and at this level path integrals with fermions are manifestly finite and well defined.

For fermions the problem is to evaluate expressions like $\mathrm{Tr}\, \hat{J} \exp(-\beta \hat{H})$, where $\hat{H}$ contains now also fermions. This Hamiltonian is constructed from the Minkowskian action as $H = \dot{q}p - L$, but once it is constructed, it is a well-defined operator which acts in a well-defined Hilbert space. Hermiticity and general coordinate invariance of $\hat{H}$ define $\hat{H}$ uniquely up to corrections proportional to $\hbar^2 R$. Supersymmetry ($\{\hat{Q}, \hat{Q}\} \sim \hat{H}$) even determines the coefficient of R because $\hat{Q}$ is unambiguous. When we insert complete sets of states in $\exp(-\frac{\beta}{\hbar}\hat{H})$ we will be led to a Euclidean path integral, but we arrive at this path integral by a series of identities, and not by a Wick rotation of the fermionic fields ψ and $\psi^\dagger$.

The Minkowskian action for a free complex (Dirac) fermion in n dimensions is

$$S = \int \left[-(\psi^\dagger i \gamma^0) \left(\gamma^0 \frac{\partial}{\partial t} + \gamma^k \frac{\partial}{\partial x^k} \right) \psi \right] d^n x\,, \quad (\gamma^0)^2 = -1. \quad (2.139)$$

For one-component spinors ψ in quantum mechanics this reduces to

$$S = \int i\psi^\dagger \dot{\psi}\, dt. \tag{2.140}$$

The conjugate momentum of ψ is defined by left-differentiation of the action with respect to $\dot{\psi}$ and is given by $-i\psi^\dagger$, and the equal-time anti-commutation relations yield $\{\hat{\psi}, \hat{\psi}\} = \{\hat{\psi}^\dagger, \hat{\psi}^\dagger\} = 0$ and, furthermore,

$$\{-i\hat{\psi}^\dagger, \hat{\psi}\} = \frac{\hbar}{i} \quad \rightarrow \quad \{\hat{\psi}, \hat{\psi}^\dagger\} = \hbar. \tag{2.141}$$

If there are more than one pair of $\hat{\psi}$ and $\hat{\psi}^\dagger$ we denote them by an index $a = 1, \ldots, n$, as in $\hat{\psi}^a$ and $\hat{\psi}_b^\dagger$. Then

$$\{\hat{\psi}^a, \hat{\psi}_b^\dagger\} = \hbar\, \delta_b^a. \tag{2.142}$$

In curved space this index a is a flat index, related to a curved index i by the vielbein fields $e^a{}_i(x)$ as usual in general relativity

$$\psi^a(t) = e^a{}_i(x(t))\, \psi^i(t). \tag{2.143}$$

It is convenient to work with flat indices for fermions and curved indices for bosons. Thus, we shall be using $x^i(t)$ and $\psi^a(t)$ where in both cases i and a run form 1 to n.

Having defined the basic operators $\hat{\psi}^a$ and $\hat{\psi}_b^\dagger$ we shall be using, we rewrite the trace as a path integral by inserting complete sets of states constructed from $\hat{\psi}^a$ and $\hat{\psi}_b^\dagger$. In this way we shall arrive at a path integral representation of $\mathrm{Tr}\, \hat{J} \exp(-\beta \hat{H})$. This path integral has the appearance of a Euclidean path integral (no $i/\hbar$ in front of the action). One might then, out of curiosity, wonder whether this Euclidean path integral could also have been obtained from a Minkowskian path integral by making a Wick rotation on the spinor fields ψ^a and $\psi_b^\dagger$ (and, of course, rotating the Minkowskian time t_M to $-it_E$, where t_E is the Euclidean time). This is a tricky question (especially in higher dimensions) which we do not need to answer because we are using well-defined operators with given (anti)commutation relations, which were derived from the Minkowski theory but which in our applications lead to Euclidean path integrals. We repeat that we do not need to make any Wick rotation on the fermionic fields. The Wick rotation on fermionic fields has been discussed in [75].

One could try to parallel the bosonic treatment, and introduce eigenstates of $\hat{\psi}$, namely bras and kets, as well as also introduce a complete set of "momentum" eigenstates, i.e. eigenstates of $\hat{\psi}^\dagger$ which are again bras and kets. These eigenstates are coherent states as we shall see, so this approach would lead to four kinds of coherent states, with several inner products to be specified. This is one of the approaches in the literature [76], but we shall follow a simpler approach [48, 51]. Namely, we shall only need one kind of coherent bras and one kind of coherent kets, and thus need only one inner product.

We begin with Dirac fermions $\hat{\psi}^a$ and $\hat{\psi}_b^\dagger$ with the equal-time canonical anticommutation relations derived previously,

$$\{\hat{\psi}^a, \hat{\psi}_b^\dagger\} = \hbar\delta_b^a; \qquad \{\hat{\psi}^a, \hat{\psi}^b\} = \{\hat{\psi}_a^\dagger, \hat{\psi}_b^\dagger\} = 0. \tag{2.144}$$

For an $N = 1$ supersymmetric system with Majorana fermions the conjugate momentum is proportional to ψ^a itself, and for certain purposes one can use Dirac brackets without having to distinguish between annihilation and creation operators. For our purposes, however, we need to be able to distinguish between $\hat{\psi}$ and $\hat{\psi}^\dagger$, and we shall later show how to do this for $N = 1$ models. (We shall either add another set of free Majorana fermions and then construct a larger Hilbert space, or combine pairs of Majorana spinors into $\hat{\psi}$ and $\hat{\psi}^\dagger$, and construct a smaller Hilbert space. Both approaches yield the same final results for physical quantities as we shall see.) We repeat that in this section we restrict our attention to Dirac fermions. This is enough for $N = 2$ models ($N = 2$ susy on the worldline).

To define fermionic coherent states without having to write factors of $\hbar^{\pm 1/2}$ all the time, it is useful to introduce rescaled variables $\hat{\psi}^a \to \hbar^{1/2}\hat{\psi}^a$ and $\hat{\psi}_a^\dagger \to \hbar^{1/2}\hat{\psi}_a^\dagger$, satisfying

$$\{\hat{\psi}^a, \hat{\psi}_b^\dagger\} = \delta_b^a. \tag{2.145}$$

We shall only use the rescaled variables with (2.145), and never scale back to (2.144). The coherent states we need are then defined by (dropping hats from now on)

$$|\eta\rangle = e^{\psi_a^\dagger \eta^a}|0\rangle; \qquad \psi^a|0\rangle = 0 \tag{2.146}$$

$$\langle\bar{\eta}| = \langle 0|e^{\bar{\eta}_a \psi^a}; \qquad \langle 0|\psi_a^\dagger = 0. \tag{2.147}$$

We choose η^a and $\bar{\eta}_a$ as **independent** complex (i.e. without reality conditions) Grassmann variables even though $\psi_b^\dagger = (\psi^b)^\dagger$. Therefore, we write $\bar{\eta}_a$ instead of $\eta_a^\dagger$. (We could equally well have chosen $\bar{\eta}_a$ to be given by $(\eta^a)^\dagger$ because this does not change the result of the Grassmann integration.) The state $|0\rangle$ is the Fock vacuum for the ψ, and by definition it commutes with the Grassmann numbers: $|0\rangle\eta^a = \eta^a|0\rangle$ and $|0\rangle\bar{\eta}_a = \bar{\eta}_a|0\rangle$. The same property holds by definition for $\langle 0|$.

It is clear that these coherent states satisfy the following relations

$$\psi^a|\eta\rangle = \eta^a|\eta\rangle; \qquad \langle\bar{\eta}|\psi_a^\dagger = \langle\bar{\eta}|\bar{\eta}_a. \tag{2.148}$$

To prove this one may expand the exponent. There are then only a finite number of terms because $\bar{\eta}_{a_1}\bar{\eta}_{a_2} = 0$ when $a_1 = a_2$.

The inner product is given by

$$\langle \bar{\eta}|\eta\rangle = \mathrm{e}^{\bar{\eta}_a \eta^a}. \tag{2.149}$$

This relation follows from $\mathrm{e}^A \mathrm{e}^B = \mathrm{e}^B \mathrm{e}^A \mathrm{e}^{[A,B]}$ with $A = \bar{\eta}_a \psi^a$ and $B = \psi_a^\dagger \eta^a$. Since $[A, B]$ commutes with A and B there are no further terms in the Baker–Campbell–Hausdorff formula.

Grassmann integration is defined by

$$\int d\eta^a\, \eta^b = \delta^{ab}, \quad \int d\bar{\eta}_a\, \bar{\eta}_b = \delta_{ab}, \quad \int d\eta^a = 0, \quad \int d\bar{\eta}_a = 0. \tag{2.150}$$

So, for example,

$$\int \prod_{a=1}^{n} d\bar{\eta}_a\, d\eta^a \left(1 + \sum_{b=1}^{n} \eta^b \bar{\eta}_b\right) = \delta_{n,1}. \tag{2.151}$$

Identities for one fermion are easily extended to the case of several fermions by observing that one can factorize into spaces with different a, for example,

$$\exp\left(\sum_{a=1}^{n} \psi_a^\dagger \eta^a\right) = \prod_{a=1}^{n} \mathrm{e}^{\psi_a^\dagger \eta^a} = \mathrm{e}^{\psi_1^\dagger \eta^1} \mathrm{e}^{\psi_2^\dagger \eta^2} \cdots \mathrm{e}^{\psi_n^\dagger \eta^n}. \tag{2.152}$$

The completeness relation reads

$$I = \int \left(\prod_{a=1}^{n} d\bar{\eta}_a\, d\eta^a\right) |\eta\rangle \mathrm{e}^{-\bar{\eta}_a \eta^a} \langle \bar{\eta}|. \tag{2.153}$$

For one pair of $\bar{\eta}$ and η the completeness relation is easily checked by expanding the exponent

$$\int d\bar{\eta}\, d\eta\, (1 + \psi^\dagger \eta)|0\rangle(1 - \bar{\eta}\eta)\langle 0|(1 + \bar{\eta}\psi) = |0\rangle\langle 0| + \psi^\dagger |0\rangle\langle 0|\psi. \tag{2.154}$$

The right-hand side is the identity operator in Fock space. Note the opposite sign in the exponent of the inner product and decomposition of unity. From now on by $d\bar{\xi}\, d\xi$ we shall mean the product

$$d\bar{\xi}\, d\xi \equiv \prod_{a=1}^{n} (d\bar{\xi}_a\, d\xi^a) = d\bar{\xi}_1\, d\xi^1 \cdots d\bar{\xi}_n\, d\xi^n = d\bar{\xi}_n \cdots d\bar{\xi}_1\, d\xi^1 \cdots d\xi^n. \tag{2.155}$$

We now consider the transition element between two coherent states

$$\langle \bar{\eta}|\mathrm{e}^{-\frac{\beta}{\hbar}\hat{H}}|\eta\rangle. \tag{2.156}$$

We assume that $\hat{H}$ depends on $\hat{\psi}^a$, $\hat{\psi}_a^\dagger$ (and on $\hat{x}^i, \hat{p}_i$ which we suppress writing) with again an arbitrary but definite a priori operator ordering. In order to compute traces like $\mathrm{Tr}\,\hat{J}\exp(-\frac{\beta}{\hbar}\hat{H})$, we shall use the completeness relation for coherent states to define the trace, and then all Grassmann variables are integrated over and disappear.

We begin by inserting $N-1$ complete sets of coherent states and obtain then (for notational clarity writing the integral signs between the coherent states to which they belong)

$$\begin{aligned}
&\langle\bar{\eta}|\mathrm{e}^{-\frac{\beta}{\hbar}\hat{H}}|\eta\rangle \\
&\quad = \langle\bar{\eta}|\mathrm{e}^{-\frac{\epsilon}{\hbar}\hat{H}}|\eta_{N-1}\rangle \int d\bar{\eta}_{N-1}\, d\eta_{N-1}\mathrm{e}^{-\bar{\eta}_{N-1}\eta_{N-1}}\langle\bar{\eta}_{N-1}|\mathrm{e}^{-\frac{\epsilon}{\hbar}\hat{H}}|\eta_{N-2}\rangle \\
&\qquad \cdots \langle\bar{\eta}_1|\mathrm{e}^{-\frac{\epsilon}{\hbar}\hat{H}}|\eta\rangle; \qquad \bar{\eta}\equiv\bar{\eta}_N, \quad \eta\equiv\eta_0, \quad \epsilon=\beta/N.
\end{aligned} \tag{2.157}$$

This is analogous to the insertion of $N-1$ sets of x eigenstates by Dirac. Next, we introduce N other complete sets of coherent states which are analogous to the N complete sets of p eigenstates of Feynman. We do this to obtain a fermionic midpoint rule. We thus consider

$$\langle\bar{\eta}_{k+1}|\mathrm{e}^{-\frac{\epsilon}{\hbar}\hat{H}}|\eta_k\rangle = \int d\bar{\chi}_k\, d\chi_k\, \langle\bar{\eta}_{k+1}|\mathrm{e}^{-\frac{\epsilon}{\hbar}\hat{H}}|\chi_k\rangle \mathrm{e}^{-\bar{\chi}_k\chi_k}\langle\bar{\chi}_k|\eta_k\rangle. \tag{2.158}$$

We now repeat our prescription of rewriting an operator with fermionic creation and annihilation operators into Weyl-ordered form. We rewrite the operator into a symmetrized part (or, rather, an antisymmetrized part) plus extra terms. A basis for the symmetrized operators is obtained by expanding $(\bar{\eta}_a\hat{\psi}^a+\eta^a\hat{\psi}_a^\dagger)^N$ and retaining all terms with a given number of operators in the order they come. For one fermion one has

$$(m+n)!(\hat{\psi}^m\hat{\psi}^{\dagger n})_S = \left(\frac{\partial}{\partial\bar{\eta}}\right)^m\left(\frac{\partial}{\partial\eta}\right)^n(\bar{\eta}\hat{\psi}+\eta\hat{\psi}^\dagger)^N; \;\; N=m+n \tag{2.159}$$

where m and n can only take the values 0 and 1. For several fermions one has, for example,

$$\begin{aligned}
(\hat{\psi}^a\hat{\psi}_b^\dagger)_S &= \frac{1}{2}(\hat{\psi}^a\hat{\psi}_b^\dagger - \hat{\psi}_b^\dagger\hat{\psi}^a) \\
(\hat{\psi}^a\hat{\psi}^b\hat{\psi}_c^\dagger)_S &= \frac{1}{6}(2\hat{\psi}^a\hat{\psi}^b\hat{\psi}_c^\dagger - \hat{\psi}^a\hat{\psi}_c^\dagger\hat{\psi}^b + \hat{\psi}^b\hat{\psi}_c^\dagger\hat{\psi}^a + 2\hat{\psi}_c^\dagger\hat{\psi}^a\hat{\psi}^b) \\
&= \frac{1}{2}(\hat{\psi}^a\hat{\psi}^b\hat{\psi}_c^\dagger + \hat{\psi}_c^\dagger\hat{\psi}^a\hat{\psi}^b)
\end{aligned} \tag{2.160}$$

where in the last line we have kept $\psi^a\psi^b$ together. Hence $\hat{\psi}^a\hat{\psi}_b^\dagger = (\hat{\psi}^a\hat{\psi}_b^\dagger)_S + \frac{1}{2}i\hbar\delta_b^a \equiv (\hat{\psi}^a\hat{\psi}_b^\dagger)_W$. Weyl ordering of fermions is further explained and worked out in Appendix C.

Given a Weyl-ordered operator $\hat{B}(\hat{\psi}^\dagger, \hat{\psi})$, the following midpoint rule holds

$$\begin{aligned}\langle \bar{\eta}|\hat{B}|\eta\rangle &= \int d\bar{\chi}\, d\chi\, \mathrm{e}^{-\bar{\chi}\chi} \langle \bar{\eta}|\chi\rangle B\Big(\bar{\chi}, \frac{1}{2}(\chi+\eta)\Big)\langle\bar{\chi}|\eta\rangle \\ &= \int d\bar{\chi}\, d\chi\, \mathrm{e}^{-\bar{\chi}\chi} \langle \bar{\eta}|\chi\rangle B\Big(\frac{1}{2}(\bar{\eta}+\bar{\chi}), \chi\Big)\langle\bar{\chi}|\eta\rangle. \qquad (2.161)\end{aligned}$$

Both formulas are true, but we shall only use the first one. The proof of this fermionic midpoint rule can either be given by following the same steps as in the bosonic case or by starting with an operator $(\hat{\psi}^\dagger)^k$ (or $(\hat{\psi})^k$) for which (2.161) is clear,[6] and then using the property that if $\hat{A}$ is Weyl-ordered then $\frac{1}{2}(\hat{\psi}\hat{A} \pm \hat{A}\hat{\psi})$ is also Weyl-ordered, where the + sign is needed if A is of commuting nature and the − sign if A is of anticommuting nature. Repeated application of this property to $\hat{A} = (\hat{\psi}^\dagger)^k$ then proves the fermionic midpoint rule for any operator $\hat{B}$ which is a polynomial in $\hat{\psi}$ and $\hat{\psi}^\dagger$.

Next, we again use the linear approximation. That is, we replace the Weyl-ordered operators $(\exp(-\frac{\epsilon}{\hbar}\hat{H}))_W$ by $\exp(-\frac{\epsilon}{\hbar}\hat{H}_W)$. In matrix elements (in particular, the kernels of the path integral) these two expressions differ by terms which are of order ϵ^2 and higher, and in the path integral these extra terms do not contribute. For example, if $\hat{H} = \hat{\psi}_a^\dagger\hat{\psi}^a$ then $(\hat{H})_W = \frac{1}{2}(\hat{\psi}_a^\dagger\hat{\psi}^a - \hat{\psi}^a\hat{\psi}_a^\dagger) + \frac{1}{2}n$ and the ϵ^2 terms in $(\exp(-\frac{\epsilon}{\hbar}\hat{H}))_W$ are given by $\frac{\epsilon^2}{2\hbar^2}(\hat{\psi}_a^\dagger\hat{\psi}^a\hat{\psi}_b^\dagger\hat{\psi}^b)_W$ while $\exp(-\frac{\epsilon}{\hbar}\hat{H}_W)$ yields $\frac{\epsilon^2}{2\hbar^2}(\hat{\psi}_a^\dagger\hat{\psi}^a)_W(\hat{\psi}_b^\dagger\hat{\psi}^b)_W$. The difference is $\epsilon^2 n/8\hbar^2$, see Appendix C, equation (C.14). These terms of order ϵ^2 do not contribute for $\epsilon \to 0$. In the bosonic case one had two terms $p\Delta q$ and ϵp^2 in the exponent, and whereas $p' = p - \frac{i}{\epsilon}\Delta q$ is of order $\epsilon^{-1/2}$, the phase-space variable p itself is of order ϵ^0. For fermions there are not two terms in the exponent, but only one term $\bar{\chi}\Delta\chi$. Thus, for fermions there are no corresponding subtleties.

After rewriting the Hamiltonian in Weyl-ordered form and applying the linear approximation we arrive at

$$\begin{aligned}\langle\bar{\eta}|\exp\Big(-\frac{\beta}{\hbar}\hat{H}\Big)|\eta\rangle &= \int \Big(\prod_{k=1}^{N-1} d\bar{\eta}_k\, d\eta_k\, \mathrm{e}^{-\bar{\eta}_k\eta_k}\Big)\Big(\prod_{k=0}^{N-1} d\bar{\chi}_k\, d\chi_k\, \mathrm{e}^{-\bar{\chi}_k\chi_k}\Big) \\ &\times \prod_{k=0}^{N-1} \langle\bar{\eta}_{k+1}|\chi_k\rangle \exp\Big[-\frac{\epsilon}{\hbar}H_W\Big(\bar{\chi}_k, \frac{1}{2}(\eta_k+\chi_k)\Big)\Big]\langle\bar{\chi}_k|\eta_k\rangle \qquad (2.162)\end{aligned}$$

where $\langle\bar{\eta}_N| = \langle\bar{\eta}|$ and $|\eta_0\rangle = |\eta\rangle$. Substituting the inner products for coherent states and using the lemma

$$\int d\bar{\eta}_k\, d\eta_k\, \mathrm{e}^{-\bar{\eta}_k(\eta_k-\chi_{k-1})} f(\eta_k) = f(\chi_{k-1}) \qquad (2.163)$$

[6]For example, $(\hat{\psi})^k$ becomes $(\chi)^k$, and the $\bar{\chi}$ integral yields the fermionic delta function $\Pi(\eta-\chi)$ which converts $(\chi)^k$ into $(\frac{\chi+\eta}{2})^k$.

we arrive at the following suggestive result:

$$\langle\bar{\eta}|\exp\Big(-\frac{\beta}{\hbar}\hat{H}\Big)|\eta\rangle = \int\left(\prod_{k=0}^{N-1} d\bar{\chi}_k\, d\chi_k\right)$$
$$\times\exp\left\{\bar{\eta}\chi_{N-1} - \epsilon\sum_{k=0}^{N-1}\left[\bar{\chi}_k\left(\frac{\chi_k-\chi_{k-1}}{\epsilon}\right)+\frac{1}{\hbar}H_W\left(\bar{\chi}_k,\frac{\chi_k+\chi_{k-1}}{2}\right)\right]\right\} \tag{2.164}$$

where $\chi_{-1}\equiv\eta$.

In the continuum limit one obtains the action $S=\int\bar{\chi}\dot{\chi}\,dt_E$ for χ in the exponent of $e^{-\frac{1}{\hbar}S}$. This action could have been obtained by starting from the Minkowski action $S=i\int\bar{\chi}\dot{\chi}\,dt_M$ and then continuing $t_M\to -it_E$ to Euclidean time t_E. However, we started from the Minkowski action, which we only used to derive the anticommutation relation of ψ^a and $\psi_a^\dagger$, then we took the Hamiltonian (the usual Hamiltonian of Minkowski space) and started computing $\langle\bar{\eta}|e^{-\frac{\beta}{\hbar}\hat{H}}|\eta\rangle$ using well-defined rules. The outcome is (2.164), with the action $S=\int\bar{\chi}\dot{\chi}\,dt_E$ for χ. We shall refer to this as the Euclidean action for χ.

The term $\bar{\eta}\chi_{N-1}$ is the extra term which one already encounters in path integrals with bosonic coherent states [48, 51]. It arises from the inner product $\langle\bar{\eta}|\chi_{N-1}\rangle$. In the continuum theory it becomes $\bar{\eta}\chi(0)$, and it is needed in the action $S=\bar{\eta}\chi(0)-\int_{-\beta}^0\bar{\chi}(t)\dot{\chi}(t)\,dt$ in order for the field equation for $\chi(t)$ to be given by $\frac{d}{dt}\bar{\chi}(t)=0$ without extra boundary terms. (Note that $\chi(t=-\beta)=\eta$ and $\bar{\chi}(t=0)=\bar{\eta}$, hence $\delta\chi(t)$ vanishes at $t=-\beta$ but not at $t=0$.) However, we shall go on with the discretized approach and not yet make this (or any other further) approximations.

Since $\bar{\eta}\chi_{N-1}$ is at most linear in quantum deviations, we get rid of this term by again introducing the background formalism. We decompose χ_k and $\bar{\chi}_k$ into a background part ξ_k and $\bar{\xi}_k$, and a quantum part ψ_k and $\bar{\psi}_k$,

$$\begin{aligned}\chi_k^a &= \xi_k^a+\psi_k^a && \text{with } k=-1,\dots,N-1\\ \bar{\chi}_{ka} &= \bar{\xi}_{ka}+\bar{\psi}_{ka} && \text{with } k=0,\dots,N.\end{aligned} \tag{2.165}$$

Again ξ_k^a and $\bar{\xi}_{ka}$ are independent complex Grassmann variables, and similarly for ψ_k^a and $\bar{\psi}_{ka}$. For $k=-1$ we have already defined $\chi_{-1}=\eta$. The background fermions are assumed to satisfy the boundary conditions, hence $\chi_{-1}=\xi_{-1}=\eta$ and $\psi_{-1}=0$. Similarly, $\bar{\chi}_N=\bar{\xi}_N=\bar{\eta}$ and $\bar{\psi}_N=0$. Of course, $d\bar{\chi}_{ka}\,d\chi_k^a=d\bar{\psi}_{ka}\,d\psi_k^a$, since Berezin integration is translationally invariant.

Next, we split off a free part $H^{(0)}$ from H_W. In our applications, we shall always choose $H^{(0)}=0$ for the fermions, so we concentrate on this case. However, nonvanishing $H^{(0)}$ can also be handled by our methods. (All of our applications are to massless fermions; had there been a mass term present, we would have put it into $H^{(0)}$.)

To be able to extract the interaction part of the action from the path integral, we introduce external sources $\bar{K}_{ka}$ and K_k^a which couple to ψ_k^a and $\bar{\psi}_{ka}$, and study the quadratic part of the path integral first,

$$Z_N^{(0)}(K,\bar{K}) = \int \left(\prod_{k=0}^{N-1} d\bar{\psi}_{ka}\, d\psi_k^a \right) \exp\left(-\frac{1}{\hbar} S^{(0)} \right) \tag{2.166}$$

$$-\frac{1}{\hbar} S^{(0)} = -\sum_{k=0}^{N-1} \bar{\psi}_{ka}(\psi_k^a - \psi_{k-1}^a) + \sum_{k=0}^{N-1} (\bar{K}_{ka}\psi_k^a + \bar{\psi}_{ka}K_k^a)$$

where we recall that $\psi_{-1}^a = 0$. All of the remaining parts of $\bar{\chi}\dot{\chi}$ and $\bar{\eta}\chi_{N-1}$ in (2.164) combine with H_W into what we shall call $H^{(int)}$. Just as in the bosonic case, the kinetic terms are not diagonal. In the bosonic case we therefore first made an orthogonal transformation on the q_k^j which diagonalized the kinetic terms, but the fermionic kinetic terms are sufficiently simple that we need not first diagonalize them. By completing squares one finds

$$\begin{aligned} -\bar{\psi}_{ka}A_{kl}\psi_l^a + \bar{K}_{ka}\psi_k^a + \bar{\psi}_{ka}K_k^a = &-(\bar{\psi}_{ka} - \bar{K}_{k'a}A_{k'k}^{-1})A_{kl}(\psi_l^a - A_{ll'}^{-1}K_{l'}^a) \\ &+ \bar{K}_{ka}(A^{-1})_{kl}K_l^a \end{aligned} \tag{2.167}$$

where A is the lower triangular matrix

$$A_{kl} = \delta_{kl} - \delta_{k,l+1}; \quad k,l = 0, N-1; \qquad A = \begin{pmatrix} 1 & 0 & 0 & . & 0 \\ -1 & 1 & 0 & . & 0 \\ 0 & -1 & 1 & . & . \\ . & . & . & . & 0 \\ 0 & . & 0 & -1 & 1 \end{pmatrix}. \tag{2.168}$$

The inverse of the matrix A is given by

$$\begin{aligned} A_{kl}^{-1} &= 0 \quad \text{if } k < l \\ A_{kl}^{-1} &= 1 \quad \text{if } k \geq l \end{aligned} \qquad A^{-1} = \begin{pmatrix} 1 & 0 & 0 & . & 0 \\ 1 & 1 & 0 & . & 0 \\ 1 & 1 & 1 & . & . \\ . & . & . & . & 0 \\ 1 & 1 & . & 1 & 1 \end{pmatrix} \tag{2.169}$$

(hence A^{-1} is also lower triangular). The integration over $\bar{\psi}_{ka}$ and ψ_k^a in

$Z^{(0)}(K, \bar{K})$ yields unity since $\det A_{kl} = 1$. Hence

$$Z_N^{(0)}(K, \bar{K}) = \exp \sum_{k,l=0}^{N-1} \bar{K}_{ka} A_{kl}^{-1} K_l^a. \tag{2.170}$$

The propagators follow by twice differentiating $Z^{(0)}$. We find

$$\langle \psi_k^a \bar{\psi}_{lb} \rangle = \frac{\partial}{\partial K_l^b} \frac{\partial}{\partial \bar{K}_{ka}} Z^{(0)} \Big|_{K=\bar{K}=0} = A_{kl}^{-1} \delta_b^a. \tag{2.171}$$

Since H_W depends on $\bar{\chi}_k$ and $\frac{1}{2}(\chi_k + \chi_{k-1})$, we rather need the propagators for the $\psi_{k-1/2}$ where

$$\psi_{k-1/2} \equiv \frac{1}{2}(\psi_k + \psi_{k-1}). \tag{2.172}$$

One clearly has

$$\langle \psi_{k-1/2}^a \bar{\psi}_{lb} \rangle = \frac{1}{2}(A_{k,l}^{-1} + A_{k-1,l}^{-1}) \delta_a^b = \left\{ \begin{array}{ll} 1 & \text{if } k > l \\ \frac{1}{2} & \text{if } k = l \\ 0 & \text{if } k < l \end{array} \right\} \delta_a^b. \tag{2.173}$$

The right-hand side contains the same discretized theta function as encountered in the bosonic case. The $\psi\psi$ and $\bar{\psi}\bar{\psi}$ propagators clearly vanish. In the continuum limit

$$\langle \psi^a(t) \bar{\psi}_b(t') \rangle = \theta(t - t') \delta_b^a \tag{2.174}$$

but when in doubt we shall go back to the discretized propagators with the discretized theta function.

The correlation functions obtained from the path integral formalism are always time-ordered, an automatic consequence of the time slicing in which the factors $\exp(-\epsilon \hat{H}/\hbar)$ move from one time to the next. Thus, the fermion propagator in (2.174) should be interpreted as

$$\begin{aligned} \langle \psi^a(t) \bar{\psi}_b(t') \rangle &= \delta_b^a && \text{if } t > t' \\ \langle \psi^a(t) \bar{\psi}_b(t') \rangle &= \frac{1}{2} \delta_b^a && \text{if } t = t' \\ \langle \bar{\psi}_b(t') \psi^a(t) \rangle &= 0 && \text{if } t' > t. \end{aligned} \tag{2.175}$$

The equal-time propagator is now well defined.

The fermionic path integral for the transition element in the discretized formulation reads as

$$T_N(\bar{\eta}, \eta; \beta) = \left[\exp\left(-\frac{1}{\hbar} H^{(int)} \right) \exp(\bar{K}_{ka} A_{kl}^{-1} K_l^a) \right]_{K=\bar{K}=0} \tag{2.176}$$

where $H^{(int)}$ follows from (2.164) and (2.165)

$$-\frac{1}{\hbar}H^{int} =$$

$$-\frac{\epsilon}{\hbar}\sum_{k=0}^{N-1} H_W\left(\bar{\psi}_{ka} \to -\frac{\partial}{\partial K_k^a}, \frac{1}{2}(\psi_k^a + \psi_{k-1}^a) \to \frac{1}{2}\left(\frac{\partial}{\partial \bar{K}_{ka}} + \frac{\partial}{\partial \bar{K}_{k-1,a}}\right)\right)$$

$$+\bar{\eta}_a \xi_{N-1}^a - \sum_{k=0}^{N-1} \bar{\xi}_{ka}(\xi_k^a - \xi_{k-1}^a). \tag{2.177}$$

We have suppressed the dependence of H_W on $\bar{\xi}_{ka}$ and $\frac{1}{2}(\xi_k^a + \xi_{k-1}^a)$ for notational simplicity. All terms linear in quantum fields except those in H_W cancel if we require that the background fermions ξ and $\bar{\xi}$ satisfy the (discretized) equations of motion of $S^{(0)}$. In particular, the term $\bar{\eta}\psi(0)$ cancels with the term $\int \bar{\chi}\dot{\psi}$ coming from the last but one term in (2.164) The background fermion fields are then all constant because $H^{(0)} = 0$. Hence $\xi_k^a = \eta^a$, for all $k = -1, 0, 1, \ldots, N-1$ and $\bar{\xi}_{ka} = \bar{\eta}_a$ for all $k = 0, 1, \ldots, N$. It is instructive to check that all terms linear in the quantum variables ψ_k^a and $\bar{\psi}_{k,a}$ cancel.

The path integral can now formally be written in the continuum limit as

$$\langle \bar{\eta}|\mathrm{e}^{-\frac{\beta}{\hbar}\hat{H}}|\eta\rangle = \int D\bar{\psi}\, D\psi\, \exp\left(-\int_{-\beta}^{0} \bar{\psi}\dot{\psi}\, dt - \frac{1}{\hbar}\int_{-\beta}^{0} H_W\, dt + \bar{\eta}_a\eta^a\right)$$
$$= \mathrm{e}^{\bar{\eta}_a\eta^a}\left\langle \exp\left(-\frac{1}{\hbar}\int_{-\beta}^{0} S^{(int)}\, dt\right)\right\rangle \tag{2.178}$$

with propagators $\langle \psi^a(t)\bar{\psi}_b(t')\rangle = \theta(t-t')\delta_b^a$, and $S^{(int)} = H_W$ because we took $H^{(0)} = 0$. The Hamiltonian H_W depends on $\bar{\eta}_a + \bar{\psi}_a(t)$ and $\eta^a + \psi^a(t)$, and $\psi(t)$ vanishes at $t = -\beta$ while $\bar{\psi}(t)$ vanishes at $t = 0$. The extra term $\mathrm{e}^{\bar{\eta}_a\eta^a}$ will play an important role in the computation of anomalies. It came from the extra term $\bar{\eta}\chi_{N-1}$ in (2.164), but in the continuum limit in (2.178) it simply corresponds to the inner product $\langle\bar{\eta}|\eta\rangle = \mathrm{e}^{\bar{\eta}_a\eta^a}$. When we consider both $x(t)$ and $\psi(t)$, we obtain the Feynman measure $(2\pi\beta\hbar)^{-n/2}$ from the bosomic sector in front of the path integral.

2.4 Path integrals for Majorana fermions

In the previous section we developed a path integral formalism for Dirac spinors $\hat{\psi}^a$ and $\hat{\psi}_b^\dagger$ satisfying $\{\hat{\psi}^a, \hat{\psi}_b^\dagger\} = \hbar\delta_a^b$. However, for many applications one needs Majorana spinors. The Dirac bracket for Majorana spinors ψ^a yields $\{\hat{\psi}^a, \hat{\psi}_b\} = \hbar\delta_a^b$, but one cannot directly construct a path integral formalism for Majorana spinors because we need separate operators $\hat{\psi}^a$ and $\hat{\psi}_b^\dagger$ in order to construct coherent states. There are two ways to achieve this objective: either by adding an extra set of free Majorana fermions, or by combining pairs of Majorana spinors into complex spinors $\hat{\psi}$ and $\hat{\psi}^\dagger$. We discuss these constructions separately. When one is dealing with one Majorana spinor (or an odd number of Majorana spinors), only the former procedure can be used.

Doubling of Majorana spinors

One way to construct separate operators $\hat{\psi}$ and $\hat{\psi}^\dagger$ is to extend the set of interacting Majorana spinors $\hat{\psi}_1^a$ with $a = 1, \ldots, n$ by adding another set of free Majorana spinors $\hat{\psi}_2^a$ with again $a = 1, \ldots, n$. The Hamiltonian depends only on $\hat{\psi}_1^a$ but not on $\hat{\psi}_2^a$. We then combine $\hat{\psi}_1^a$ and $\hat{\psi}_2^a$ into creation and annihilation operators as follows:

$$\hat{\psi}^a = \frac{1}{\sqrt{2}}(\hat{\psi}_1^a + i\hat{\psi}_2^a); \quad \hat{\psi}_a^\dagger = \frac{1}{\sqrt{2}}(\hat{\psi}_1^a - i\hat{\psi}_2^a). \tag{2.179}$$

For convenience we rescale the fermions such that there are no $\hbar$ terms in the brackets

$$\{\hat{\psi}_i^a, \hat{\psi}_j^b\} = \delta^{ab}\delta_{ij}; \quad \{\hat{\psi}^a, \hat{\psi}_b^\dagger\} = \delta_a^b. \tag{2.180}$$

Furthermore, of course, $\{\hat{\psi}^a, \hat{\psi}^b\} = \{\hat{\psi}_a^\dagger, \hat{\psi}_b^\dagger\} = 0$.

Given these operators $\hat{\psi}$ and $\hat{\psi}^\dagger$ we can now construct the path integral. The transition element is, according to (2.164),

$$\begin{aligned}\langle\bar{\eta}|\mathrm{e}^{-\frac{\beta}{\hbar}\hat{H}(\hat{\psi}_1)}|\eta\rangle &= \int \prod_{k=0}^{N-1} d\bar{\chi}_k\, d\chi_k\, \mathrm{e}^{\bar{\eta}\chi_{N-1}} \\ &\times \exp\left[-\epsilon\sum_{k=0}^{N-1} \bar{\chi}_k(\chi_k - \chi_{k-1})/\epsilon - \frac{\epsilon}{\hbar}H_W(\bar{\chi}_k, (\chi_k + \chi_{k-1})/2)\right]\end{aligned} \tag{2.181}$$

where the operators $\hat{\psi}_1^a$ in $\hat{H}$ are first written as $(\hat{\psi}^a + \hat{\psi}_a^\dagger)/\sqrt{2}$ and again $\chi_{-1} = \eta$. After Weyl reordering with respect to $\hat{\psi}$ and $\hat{\psi}^\dagger$ one then finds H_W as a function of $\bar{\chi}_k$ and $(\chi_k + \chi_{k-1})/2$ according to Berezin's theorem. In fact, because $\hat{H}$ originally only depended on $\hat{\psi}_1^a$ but not on $\hat{\psi}_2^a$, H_W depends only on the sum of $\bar{\chi}_k$ and $\frac{1}{2}(\chi_k + \chi_{k-1})$. One may then introduce a background/quantum split of χ and $\bar{\chi}$, and H_W then only depends on $\psi_{1,bg} + \psi_{1,qu}$. One constructs propagators for $\psi_{1,qu}$, and the path integral representation of the trace includes an integration over the boundary values η and $\bar{\eta}$ (i.e. over $\psi_{1,bg}$).

In applications one begins by considering a trace $\mathrm{Tr}\,\hat{J}\exp(-\frac{\beta}{\hbar}\hat{H})$ in a quantum field theory, and then converts this trace to a problem in quantum mechanics by representing the Dirac matrices γ^a by Majorana spinors $\sqrt{2}\,\hat{\psi}_1^a$ with the same anticommutation relations. One then adds free fermions $\hat{\psi}_2^a$ as explained above. Since $\hat{J}$ and $\hat{H}$ depend on $\hat{\psi}_1$, one combines $\hat{\psi}_1$ and $\hat{\psi}_2$ into $\hat{\psi} = (\hat{\psi}_1 + i\hat{\psi}_2)/\sqrt{2}$ and $\hat{\psi}^\dagger = (\hat{\psi}_1 - i\hat{\psi}_2)/\sqrt{2}$. The matrix elements of $\hat{\psi}_1$ are the same as the matrix element of γ^a. Thus the trace is still well-defined: the Hilbert space on which $\hat{\psi}_1$ acts is obtained by acting with $\hat{\psi}^\dagger$ on the vacuum annihilated by $\hat{\psi}$. Adding a free set of fermions implies that one is considering a larger Hilbert space. One must then afterwards divide by the dimension of the subspace which is due to $\hat{\psi}_2$. For an application where this construction is worked out in great detail, see Section 6.1.

Halving of Majorana spinors

The other way of constructing path integrals for Majorana spinors ψ^a is to combine pairs of Majorana spinors into complex Dirac spinors. This evidently is only possible if one has an even number of Majorana spinors. One defines then

$$\hat{\chi}^A = \frac{1}{\sqrt{2}}(\hat{\psi}^{2A-1} + i\hat{\psi}^{2A}); \ \ \hat{\chi}_A^\dagger = \frac{1}{\sqrt{2}}(\hat{\psi}^{2A-1} - i\hat{\psi}^{2A}) \tag{2.182}$$

where $A = 1, \ldots, n/2$ and $\{\hat{\chi}^A, \hat{\chi}_B^\dagger\} = \delta_B^A$. The inverse relations are given by

$$\begin{aligned} \hat{\psi}^a &= \frac{1}{\sqrt{2}}[\hat{\chi}^{(a+1)/2} + \hat{\chi}^\dagger_{(a+1)/2}] \quad \text{if } a \text{ is odd} \\ \hat{\psi}^a &= \frac{1}{\sqrt{2}}(-i\hat{\chi}^{a/2} + i\hat{\chi}^\dagger_{a/2}) \qquad \text{if } a \text{ is even.} \end{aligned} \tag{2.183}$$

We then again define bras $|\eta\rangle$ and kets $\langle\bar{\eta}|$ using the operators $\hat{\chi}^\dagger$ and $\hat{\chi}$, respectively, to construct coherent states. The Hamiltonian depends on $\hat{\psi}^a$, so we should first express $\hat{\psi}^a$ in terms of $\hat{\chi}^A$ and $\hat{\chi}_A^\dagger$, then Weyl-order this expression, and then go over to the path integral. Once again, we may then introduce a background/quantum split for the fermions and end up with a path integral for the transition element $\langle\bar{\eta}|\exp(-\frac{\beta}{\hbar}\hat{H})|\eta\rangle$ where one integrates over $\bar{\eta}$ and η, see (6.86). Then one changes integration variables from the $n/2 + n/2$ variables $\bar{\eta}$ and η to new variables ψ_1^a where $a = 1, \ldots, n$, and one ends up with a path integral over these ψ_1^a. For an application where all details are discussed see Section 6.2 below (6.44).

Boundary conditions for fermions

We conclude this section with a discussion of how antiperiodic boundary conditions (APB) and periodic boundary conditions (PBC) arise in the continuum path integrals. We will derive these results straightforwardly from our discretized path integrals. When we compute anomalies in the second part of this book we shall explicitly perform the integrals over Grassmann variables at the discretized level, and then we shall not need to know whether in the continuum limit fermionic fields are periodic or antiperiodic. However, in the approach of Alvarez-Gaumé and Witten the continuum limit is first taken, and then one must evaluate one-loop determinants with certain boundary conditions for the quantum fields.

Let us first go back to Dirac fermions and the transition element given in (2.156), namely $\langle\bar{\eta}|\exp(-\frac{\beta}{\hbar}\hat{H})|\eta\rangle$. Recall the expression in (2.181). We want to take the trace of this transition element, so first we discuss how to take the trace of an operator.

The trace of an operator $\hat{A}$ is given by

$$\operatorname{Tr}\hat{A} = \int \sqrt{g(x_0)}\prod_{i=1}^{n} dx_0^i \prod_{a=1}^{n}(d\chi^a\, d\bar{\chi}_a)\, \mathrm{e}^{\bar{\chi}\chi}\, \langle\bar{\chi}, x_0|\hat{A}|\chi, x_0\rangle \quad (2.184)$$

where $\prod_{a=1}^{n}(d\chi^a\, d\bar{\chi}_a)$ can be written as $d\chi^1\cdots d\chi^n\, d\bar{\chi}_n\cdots d\bar{\chi}_1$. The states $|\chi, x_0\rangle$ and $\langle\bar{\chi}, x_0|$ contain the fermionic coherent states. The factor of $\sqrt{g(x_0)}$ comes from the completeness relation $\int |x\rangle\sqrt{g(x)}\langle x|\, dx = I$ in (2.4). Here we take the Grassmann variables χ and $\bar{\chi}$ as independent and not related by complex conjugation. The only property they satisfy is Berezin integration $\int d\bar{\chi}_a\, \bar{\chi}_a = 1$ and $\int d\chi^a\, \chi^a = 1$ for fixed a. Note that the order of $d\chi\, d\bar{\chi}$ and the sign in the exponent $\exp(\bar{\chi}\chi)$ are different in the trace formula from the completeness relation in (2.153). To check this trace formula, consider one pair $\chi, \bar{\chi}$. Then $\int d\chi\, d\bar{\chi}\,(1+\bar{\chi}\chi)\,\langle 0|(1+\bar{\chi}\hat{\psi})\hat{A}(1+\hat{\psi}^\dagger\chi)|0\rangle$ should be equal to $\langle 0|\hat{A}|0\rangle + \langle 0|\hat{\psi}\hat{A}\hat{\psi}^\dagger|0\rangle = \langle 0|\hat{A}|0\rangle + \langle 1|\hat{A}|1\rangle$, where $|1\rangle = \psi^\dagger|0\rangle$. We assume here that χ commutes with $\hat{A}$, so $\hat{A}$ must have even statistics (one sometimes uses the not quite correct terminology that $\hat{A}$ is commuting). This is indeed equal to $\operatorname{Tr}\hat{A}$.[7]

[7] At the risk of confusing the nonexpert reader, let us mention that one can actually distinguish between a trace and a supertrace, where by supertrace we mean the usual trace but with a minus sign for the fermionic states. The Jacobian in quantum field theory leads to a superdeterminant and a supertrace; this is not a choice but can be proven [77]. In the quantum mechanical model one can also distinguish between an ordinary trace and a supertrace. Both traces are mathematically consistent operations, and different physical applications may require different choices. In the quantum mechanical case one needs the ordinary trace since one is taking the trace in spinor space. The $2^{n/2}$ states in spinor space split into two sets, one set with an even numbers of $\hat{\psi}^\dagger$ operators and the other set with an odd number. For two states $|0\rangle$ and $|1\rangle \equiv \hat{\psi}^\dagger|0\rangle$, the trace of an operator $\hat{A}$ is $\langle 0|\hat{A}|0\rangle + \langle 1|\hat{A}|1\rangle$ while the supertrace would be $\langle 0|\hat{A}|0\rangle - \langle 1|\hat{A}|1\rangle$. We need a trace in the QM case because in the original formulation in terms of quantum field theory we needed of course an ordinary trace over the spinor indices of the Dirac matrices. The issue of

We now return to the trace of the transition element

$$\mathrm{Tr}\, \mathrm{e}^{-\frac{\beta}{\hbar}\hat{H}} = \int d\eta^a\, d\bar{\eta}_a\, \mathrm{e}^{\bar{\eta}\eta}\, \langle \bar{\eta}|\mathrm{e}^{-\frac{\beta}{\hbar}\hat{H}}|\eta\rangle. \tag{2.185}$$

The transition element contains the expected path integral over $d\chi\, d\bar{\chi}$ with action $-\int_{-\beta}^{0} \bar{\chi}\dot{\chi}\, dt - \frac{1}{\hbar}\int_{-\beta}^{0} H_W\, dt$, plus the extra term $\bar{\eta}\chi(0)$. Performing the integration over $d\bar{\eta}$ of

$$\mathrm{e}^{\bar{\eta}\eta}\, \mathrm{e}^{\bar{\eta}\chi(0)} \tag{2.186}$$

leads to a factor $(\eta+\chi(0))$ which is a fermionic delta function $\delta(\eta+\chi(0))$ because $\int d\eta\, (\eta+\chi(0)) f(\eta) = f(-\chi(0))$. Subsequent integration over η leads then to

$$\chi(0) = -\eta \qquad \text{(ABC)}. \tag{2.187}$$

Recalling that $\chi(-\beta) = \eta$, this means that the path integral is over paths $\chi(t)$ with ABC.

On the other hand, consider the trace with a matrix γ^5, $\mathrm{Tr}\, \gamma^5 \mathrm{e}^{-\beta H}$. As we show in Section 6.1, the QM operator corresponding to γ^5 is $(-)^F$ where F is the fermion number operator. The path integral is as before, except that $\gamma^5|\eta\rangle = |-\eta\rangle$.[8] The extra terms are now

$$\mathrm{e}^{-\bar{\eta}\eta}\, \mathrm{e}^{\bar{\eta}\chi(0)}. \tag{2.188}$$

Integration over $d\bar{\eta}$ now yields a factor $(-\eta+\chi(0))$. Thus one obtains the same path integral as before, but now with periodic boundary conditions

$$\chi(0) = \eta \qquad \text{(PBC)}. \tag{2.189}$$

2.5 Direct evaluation of the transition element to order β

In this section we shall follow [57] and determine the transition element $\langle z|\exp(-\frac{\beta}{\hbar}\hat{H})|y\rangle$ to order β for the Hamiltonian

$$\hat{H} = \frac{1}{2} g^{-1/4} \hat{p}_i g^{1/2} g^{ij} \hat{p}_j g^{-1/4}. \tag{2.190}$$

We shall use operatorial methods rather than path integrals. We shall insert a complete set of p eigenstates, expand the exponent, move $\hat{p}$ operators to the p eigenstate and $\hat{x}$ to the x eigenstate, and in this

---

whether one should use a trace or a supertrace in finite-temperature physics arose in the 1980s. In [78] a trace had been used, but in [79] it was argued that one needs a supertrace. It was finally settled that one needs a trace [80].

[8] The proof is as follows. Consider for simplicity two dimensions. Then $\gamma^5 = -i\gamma^1\gamma^2 = \psi\psi^\dagger - \psi^\dagger\psi$ with $\psi = \frac{1}{2}(\gamma^1 + i\gamma^2)$ and $\psi^\dagger = \frac{1}{2}(\gamma^1 - i\gamma^2)$. Acting with this γ^5 on $|\eta\rangle = \mathrm{e}^{\psi^\dagger\eta}|0\rangle = (1+\psi^\dagger\eta)|0\rangle$ yields $|-\eta\rangle$.

way one obtains an answer that is completely unambiguous. We shall determine the order β corrections to the flat space transition element. Because the leading term in the transition element is proportional to $\exp[-(z-y)^2/2\beta\hbar]$ we take $z-y$ to be of order $\sqrt{\beta}$. At no stage in the calculation is there any ambiguity; we move operators $\hat{p}$ next to eigenstates $|p\rangle$ or $\langle p|$ where they become c-numbers p, and the operator $\hat{x}$ next to eigenstates $|x\rangle$ or $\langle x|$ where they become c-numbers x, taking commutators carefully into account. The path integral should exactly reproduce these results, and this is verified through two-loop orders in the next section.

We expand the exponent and take all terms in the expansion into account which contain none, one or two commutators. The final result will factorize into a classical part, a one-loop part which is given by the Van Vleck–Morette determinant and a two-loop part proportional to the scalar curvature. As expected the final result preserves Einstein invariance: it is a biscalar as we shall explain.

We first prove that the operator $\hat{H}$ in (2.190) is Einstein invariant. Here we follow DeWitt [19]. To demonstrate this, we note that infinitesimal general coordinate transformations $x^i \to x'^i \equiv x^i + \xi^i(x)$ are generated by the antihermitian operator

$$\hat{G}_E = \frac{1}{2i\hbar}[\hat{p}_k \xi^k(\hat{x}) + \xi^k(\hat{x})\hat{p}_k]. \tag{2.191}$$

Coordinates then transform as

$$\delta \hat{x}^j = [\hat{x}^j, \hat{G}_E] = \xi^j(\hat{x}). \tag{2.192}$$

The momenta transform as follows:

$$\delta \hat{p}_j = [\hat{p}_j, \hat{G}_E] = -\frac{1}{2}\left[\hat{p}_k \frac{\partial \xi^k(\hat{x})}{\partial x^j} + \frac{\partial \xi^k(\hat{x})}{\partial x^j}\hat{p}_k\right] \tag{2.193}$$

which agrees with the symmetrized tensor law $p'_j = \frac{1}{2}\left\{\frac{\partial x^k}{\partial x'^j}, p_k\right\}$ for $x'^j = x^j + \xi^j$. (There is, of course, no transport term $-\xi^k \partial_k p_j$ because p_j does not depend on x^k.) To simplify the notation we shall from now on omit the hats on operators. For what follows, it is useful to rewrite this result in factorized form as $p'_j = \frac{\partial x^i}{\partial x'^j}(p_i + \cdots)$. To this end we write

$$p'_j = \frac{\partial x^i}{\partial x'^j} p_i - \frac{1}{2}\left[\frac{\partial x^i}{\partial x'^j}, p_i\right] \tag{2.194}$$

and rewrite the last term as follows [15, 19]

$$\begin{aligned}\left[\frac{\partial x^i}{\partial x'^j}, p_i\right] &= i\hbar \frac{\partial x'^k}{\partial x^i}\frac{\partial^2 x^i}{\partial x'^k \partial x'^j} = i\hbar \frac{\partial}{\partial x'^j}\ln\det\frac{\partial x^i}{\partial x'^k} \\ &= \frac{\partial x^i}{\partial x'^j}\left(i\hbar\frac{\partial}{\partial x^i}\ln\left|\frac{\partial x}{\partial x'}\right|\right).\end{aligned}$$

Hence under a finite general coordinate transformation the momenta transform as follows

$$p'_j = \frac{\partial x^i}{\partial x'^j}\left(p_i - \frac{1}{2}i\hbar\frac{\partial}{\partial x^i}\ln\left|\frac{\partial x}{\partial x'}\right|\right). \tag{2.195}$$

Consider now the operator $g^{1/4}p_i g^{-1/4}$, where $g = \det g_{ij}$. Using $g'(x') = |\partial x/\partial x'|^2\, g(x)$, it is seen to transform as follows:[9]

$$\begin{aligned}
&(g')^{1/4}p'_i(g')^{-1/4} \\
&= g^{1/4}\left|\frac{\partial x}{\partial x'}\right|^{1/2}\frac{\partial x^j}{\partial x'^i}\left(p_j - \frac{1}{2}i\hbar\frac{\partial}{\partial x_j}\ln\left|\frac{\partial x}{\partial x'}\right|\right)\left|\frac{\partial x}{\partial x'}\right|^{-1/2} g^{-1/4} \\
&= g^{1/4}\frac{\partial x^j}{\partial x'^i}p_j g^{-1/4} = \frac{\partial x^j}{\partial x'^i}(g^{1/4}p_j g^{-1/4}).
\end{aligned} \tag{2.196}$$

Similarly,

$$(g')^{-1/4}p'_i(g')^{1/4} = \left(g^{-1/4}p_j g^{1/4}\right)\frac{\partial x^j}{\partial x'^i}. \tag{2.197}$$

Returning to the Hamiltonian, in (2.190), we obtain

$$\hat{H}' = \frac{1}{2}\left(g^{-1/4}p_k g^{1/4}\frac{\partial x^k}{\partial x'^i}\right)\left(\frac{\partial x'^i}{\partial x^m}\frac{\partial x'^j}{\partial x^n}g^{mn}\right)\left(\frac{\partial x^l}{\partial x'^j}g^{1/4}p_l g^{-1/4}\right) = \hat{H}. \tag{2.198}$$

Hence, we have demonstrated that $\hat{H}$ is Einstein invariant, $[\hat{H}, \hat{G}_E] = 0$. In a similar manner one can demonstrate that $\hat{H}$ is Lorentz invariant when fermions are present, provided one replaces p_i by a Lorentz covariant derivative π_i. (For $N=2$ models the Lorentz generator is given by $\hat{J} = \frac{1}{2}\lambda_{ab}(x)\psi^a_\alpha\psi^b_\alpha = \lambda_{ab}(x)\psi^{\dagger a}\psi^b$ with $\alpha = 1, 2$ and $\delta\psi^c = [\psi^c, \hat{J}] = \lambda^c{}_b\psi^b$. It leaves the coordinates x^i inert but transforms the momenta p_j.) It is

[9] The transformation rule $g'(x') = |\partial x/\partial x'|^2\, g(x)$ is of course well known from the tensor calculus of classical general relativity, but we should really derive this result by evaluating the commutator of $\hat{G}_E$ with $g(\hat{x})$. One can achieve this by writing $\hat{G}_E$ as the sum of the orbital part given in (2.191) and the following spin part: $\hat{G}^{spin}(\xi) = \int d^n x\, [\xi^\lambda\partial_\lambda g_{\mu\nu} + (\partial_\mu\xi^\lambda)g_{\lambda\nu} + (\partial_\nu\xi^\lambda)g_{\mu\lambda}]\partial/\partial g_{\mu\nu}$. The structure of a generator as a sum of an orbital part and a spin part is well known from the case of rigid Lorentz symmetry. The orbital and spin generators always commute with each other (they act in different spaces). The spin generators must therefore satisfy the same algebra as the orbital generators, $[\hat{G}_E(\xi_1), \hat{G}_E(\xi_2)] = \hat{G}_E(\xi_2^\nu\partial_\nu\xi_1 - \xi_1^\nu\partial_\nu\xi_2)$, and this fixes the spin generators [73]. In the commutator of $\hat{G}_E$ the transport term from the orbital part cancels the transport term of the spin part. This is easy to check for a scalar field, which satisfies $\phi'(x') = \phi(x)$ and $\phi'(x') = \phi(x) + [\phi(x), \hat{G}_E + \hat{G}^{spin}]$. It is then straightforward to check that $\hat{H}$ commutes with $\hat{G}_E$.

defined by $\pi_i = p_i - \frac{1}{2} i\hbar\omega_{iab}\psi^a_\alpha\psi^b_\alpha$ and it is Lorentz invariant if one also adds a spin term with $\partial/\partial\omega_{iab}$ to $\hat{J}$ which transforms ω_{iab}, similar to the spin term in $\hat{G}_E$ (see footnote 8). For $N=1$ models $\alpha=1$, and using Dirac brackets the same results are obtained.

We turn now to the task of evaluating

$$\langle x| \exp\left(-\frac{\beta}{\hbar}\hat{H}\right)|p\rangle \tag{2.199}$$

with

$$\hat{H} = \frac{1}{2} g^{-1/4}\hat{p}_i g^{1/2} g^{ij}\hat{p}_j g^{-1/4}. \tag{2.200}$$

Expanding the exponent in (2.199), we define

$$\langle x|\hat{H}^k|p\rangle \equiv \sum_{l=0}^{2k} A_l^k(x) p^l \langle x|p\rangle \tag{2.201}$$

where $A_l^k(x)$ is a c-number function and p^l denotes a homogeneous polynomial of order l in the momenta.

In order to compute the transition amplitude $\langle x|\exp\left(-\frac{\beta}{\hbar}\hat{H}\right)|y\rangle$ to order β compared to the leading terms, it will turn out that we only need the terms on the right-hand side of (2.201) with $l = 2k,\ 2k-1$ and $2k-2$. The proof will be given in (2.205). We find, defining $p^2 \equiv g^{ij}(x)p_ip_j$,

$$A_{2k}^k(x)p^{2k} = \left(\frac{1}{2}p^2\right)^k. \tag{2.202}$$

Since this is the term containing the maximal number of p, it can be easily computed because all $\hat{p}$ operators are just replaced by the corresponding c-numbers when acting on $|p\rangle$.

The next term is

$$\begin{aligned} A_{2k-1}^k(x)p^{2k-1} &= -i\hbar k\left(\frac{1}{2}p^2\right)^{k-1}\frac{1}{2}(\partial_i g^{ij})p_j \\ &\quad - i\hbar\binom{k}{2}\left(\frac{1}{2}p^2\right)^{k-2}\frac{1}{2}g^{ij}(\partial_i g^{kl})p_jp_kp_l. \end{aligned} \tag{2.203}$$

In this expression one of the $\hat{p}$ acts as a derivative, whereas the other $2k-1$ are replaced by the corresponding c-numbers. The first term in (2.203) comes about when the derivative acts within the same factor $\hat{H}$ in which it appears, and is multiplied by k since there are k factors of $\hat{H}$. The second term arises if this derivative acts on a different factor of $\hat{H}$. For this to occur there are $\binom{k}{2}$ possible combinations, and taking into account that there are two $\hat{p}$ in $\hat{H}$ we obtain an extra factor 2. Notice that in both cases the terms involving a derivative acting on the determinant g cancel.

The last term we have to calculate is obtained when two of the $\hat{p}$ act as derivatives,

$$
\begin{aligned}
&A^k_{2k-2}(x)p^{2k-2} \\
&= \hbar^2 k \left(\frac{1}{2}p^2\right)^{k-1} \left[\frac{1}{32} g^{ij}(\partial_i \log g)(\partial_j \log g) + \frac{1}{8} g^{ij}(\partial_i\partial_j \log g) \right.\\
&\qquad \left. + \frac{1}{8}(\partial_i g^{ij})(\partial_j \log g)\right] \\
&\quad - \hbar^2 \binom{k}{2} \left(\frac{1}{2}p^2\right)^{k-2} \left[\frac{1}{2} g^{ij}(\partial_i\partial_k g^{kl}) + \frac{1}{4}(\partial_i g^{ij})(\partial_k g^{kl}) \right.\\
&\qquad \left. + \frac{1}{4}(\partial_i g^{ik})(\partial_k g^{jl}) + \frac{1}{4} g^{ik}(\partial_i\partial_k g^{jl})\right] p_j p_l \\
&\quad - \hbar^2 \binom{k}{3} \left(\frac{1}{2}p^2\right)^{k-3} \left[\frac{1}{2} g^{ik} g^{jl}(\partial_i\partial_j g^{mn}) + \frac{3}{4} g^{im}(\partial_i g^{kl})(\partial_j g^{jn}) \right.\\
&\qquad \left. + \frac{1}{2} g^{jl}(\partial_j g^{ik})(\partial_i g^{mn}) + \frac{1}{4} g^{ij}(\partial_i g^{kl})(\partial_j g^{mn})\right] p_k p_l p_m p_n \\
&\quad - \hbar^2 \binom{k}{4} \left(\frac{1}{2}p^2\right)^{k-4} \left[\frac{3}{4} g^{ij} g^{mn}(\partial_i g^{kl})(\partial_n g^{pq})\right] p_j p_k p_l p_m p_p p_q. \qquad (2.204)
\end{aligned}
$$

The first set of terms appears when both derivatives act within the same factor $\hat{H}$; again there are k terms of this kind.

The next set of terms arises when only two of the factors $\hat{H}$ play a role. There are four possibilities: (i) one $\hat{p}$ from the left factor acts on the right factor, while another $\hat{p}$ from the right factor acts within the right factor; (ii) the first $\hat{p}$ acts within the first $\hat{H}$, while the second $\hat{p}$ acts within the second $\hat{H}$; (iii) both $\hat{p}$ come from the left $\hat{H}$, but one of them acts inside the left $\hat{H}$ while the other acts on the right $\hat{H}$; and (iv) both $\hat{p}$ from the left $\hat{H}$ act on the right $\hat{H}$. In all cases it is easy to see that again the derivatives on g cancel.

The following set of terms in (2.204) comes from combinations using three factors $\hat{H}$, hence its overall factor is $\binom{k}{3}$. There are again four cases: (i) a $\hat{p}$ from the first $\hat{H}$ and a $\hat{p}$ from the second $\hat{H}$ acts on the third $\hat{H}$; (ii) one $\hat{p}$ acts inside the factor $\hat{H}$ in which it appears, whereas a $\hat{p}$ from another $\hat{H}$ acts on the remaining $\hat{H}$ (there are three terms of this kind); (iii) a $\hat{p}$ from the first $\hat{H}$ hits the second $\hat{H}$ and a $\hat{p}$ from the second $\hat{H}$ acts on the third $\hat{H}$; and (iv) of the two $\hat{p}$ from the first $\hat{H}$ one acts on the second and one on the third $\hat{H}$.

Finally, the term with $\binom{k}{4}$ in (2.204) involves four factors $\hat{H}$, such that one $\hat{p}$ from one $\hat{H}$ acts on another $\hat{H}$ and the other $\hat{p}$ from one of the remaining factors $\hat{H}$ acts on the last $\hat{H}$.

The reason further terms do not contribute can be most easily seen if we rescale $q = \sqrt{\frac{\beta}{\hbar}}p$. Then the transition amplitude becomes

$$
\begin{aligned}
\langle z|\exp\left(-\frac{\beta}{\hbar}\hat{H}\right)|y\rangle &= \int d^n p \langle z|\exp\left(-\frac{\beta}{\hbar}\hat{H}\right)|p\rangle\langle p|y\rangle \\
&= g^{-1/4}(z)g^{-1/4}(y)(2\pi\hbar)^{-n} \\
&\quad \times \left(\frac{\hbar}{\beta}\right)^{n/2}\int d^n q \exp\left[i\frac{q_i(z-y)^i}{\sqrt{\beta\hbar}}\right] \\
&\quad \times \sum_{k=0}^{\infty}\frac{(-1)^k}{k!}\left(\frac{\beta}{\hbar}\right)^k \sum_{l=0}^{2k} A_l^k(z)q^l\left(\frac{\beta}{\hbar}\right)^{-l/2}
\end{aligned} \tag{2.205}
$$

where the first factor is due to the values of $\langle z|p\rangle$ and $\langle p|y\rangle$. If we consider $z-y$ of order $\sqrt{\beta}$, the q are of order β^0. Then only the A^k_{2k-1} and A^k_{2k-2} terms contribute through order β compared to the leading term A^k_{2k}.

The sum over k in (2.205) can be performed for fixed l. All terms in (2.202)–(2.204) have a prefactor $(p^2)^{k-s}/(k-s)!$ with $s = 0, 1, 2, 3, 4$ which leads to a factor of $\exp(-\frac{1}{2}\beta p^2\hbar)$ after summing over k. One is then left with factors $(-)^s\frac{1}{s!}(\beta\hbar)^{s-l/2}$ from the last line in (2.205). Integration over p is then straightforward and one obtains

$$
\begin{aligned}
&\langle z|\exp\left(-\frac{\beta}{\hbar}\hat{H}\right)|y\rangle = g^{-1/4}(z)g^{-1/4}(y)(4\pi^2\hbar\beta)^{-n/2}\int d^n q \\
&\times \exp\left[-\frac{1}{2}g^{ij}(z)q_iq_j + i\frac{q_i(z-y)^i}{\sqrt{\beta\hbar}}\right] \\
&\times \left[1 + i\sqrt{\beta\hbar}\left[\frac{1}{2}(\partial_i g^{ij})q_j - \frac{1}{4}g^{ij}(\partial_i g^{kl})q_jq_kq_l\right]\right. \\
&+\beta\hbar\left\{\left[-\frac{1}{32}g^{ij}(\partial_i\log g)(\partial_j\log g) - \frac{1}{8}g^{ij}(\partial_i\partial_j\log g)\right.\right. \\
&\left.-\frac{1}{8}(\partial_i g^{ij})(\partial_j\log g)\right] \\
&-\left[\frac{1}{4}g^{ij}(\partial_i\partial_k g^{kl}) + \frac{1}{8}(\partial_i g^{ij})(\partial_k g^{kl}) + \frac{1}{8}(\partial_i g^{ik})(\partial_k g^{jl})\right. \\
&\left.+\frac{1}{8}g^{ik}(\partial_i\partial_k g^{jl})\right]q_jq_l \\
&+\left[\frac{1}{12}g^{ik}g^{jl}(\partial_i\partial_j g^{mn}) + \frac{1}{8}g^{im}(\partial_i g^{kl})(\partial_j g^{jn})\right. \\
&\left.+\frac{1}{12}g^{jl}(\partial_j g^{ik})(\partial_i g^{mn}) + \frac{1}{24}g^{ij}(\partial_i g^{kl})(\partial_j g^{mn})\right]q_kq_lq_mq_n \\
&\left.\left.-\left[\frac{1}{32}g^{ij}g^{mn}(\partial_i g^{kl})(\partial_n g^{pq})\right]q_jq_kq_lq_mq_pq_q\right\} + \mathcal{O}(\beta^{3/2})\right].
\end{aligned} \tag{2.206}
$$

We combined $(2\pi\hbar)^{-n}$ and $(\hbar/\beta)^{n/2}$ into $(4\pi^2\hbar\beta)^{-n/2}$ The terms from (2.202) only give the leading exponential, but the terms from (2.203) give the two terms with $\sqrt{\beta\hbar}$, while the four sets of terms in (2.204) give the terms proportional to $\beta\hbar$. We have boldfaced the terms which are present if one does not take any commutators between different factors $\hat{H}$ into account. The last boldfaced term is clearly due to expanding the exponent of the first boldfaced term. The remaining terms, also of order β, are crucial to obtaining the correct result for $\langle z|\exp(-\frac{\beta}{\hbar}H)|y\rangle$ to order β. To avoid confusion: if one uses Weyl ordering to evaluate the path integral rather than directly evaluating the transition element, one need not take these commutators into account, but only those which follow from Weyl ordering $\hat{H}$ itself.

We can now complete the square in the exponent and integrate out the momenta q_i, since the integral becomes just a sum of Gaussian integrals which can easily be evaluated. The problem is then to factorize the result such that it is manifestly a scalar both in z and y (a 'bi-scalar') under general coordinate transformations. We expect, of course, to find at least the classical action integrated along a geodesic. In the expansion of this functional around $x(0) = z$, one finds many of the terms in (2.206). However, there are terms left over. They combine into R or R_{ij}, while expansion of $g(y)$ yields terms with $\partial \log g$ or derivatives thereof. With this in mind, we write the result in a factorized form, where in one factor we put all terms which possibly can come from expanding some power of $g(y)$, while into another factor we put the expanded action and curvature terms. It is quite nontrivial, and an excellent check on the results obtained so far, that this factorization is at all possible. The resulting expression is

$$\begin{aligned}
\langle z|\exp\left(-\frac{\beta}{\hbar}\hat{H}\right)|y\rangle &= g^{-1/4}(z)g^{-1/4}(y)(2\pi\hbar\beta)^{-n/2} \\
&\times\Big\{g^{1/2}(z) + g^{1/4}(z)(y-z)^i[\partial_i g^{1/4}(z)] \\
&+\frac{1}{2}g^{1/4}(z)(y-z)^i(y-z)^j[\partial_i\partial_j g^{1/4}(z)]\Big\} \\
&\times\exp\left[-\frac{1}{2\beta\hbar}g_{ij}(z)(y-z)^i(y-z)^j\right] \\
&\times\Big\{1 - \frac{1}{4}\frac{1}{\beta\hbar}[\partial_k g_{ij}(z)](y-z)^i(y-z)^j(y-z)^k \\
&+\frac{1}{2}\left[\frac{1}{4}\frac{1}{\beta\hbar}(\partial_k g_{ij}(z)(y-z)^i(y-z)^j(y-z)^k\right]^2 \\
&-\frac{1}{12}\frac{1}{\beta\hbar}\left[\partial_k\partial_l g_{ij}(z) - \frac{1}{2}g_{mn}(z)\Gamma^m_{ij}(z)\Gamma^n_{kl}(z)\right] \\
&\times(y-z)^i(y-z)^j(y-z)^k(y-z)^l \\
&-\frac{1}{12}\beta\hbar R(z) - \frac{1}{12}R_{ij}(z)(y-z)^i(y-z)^j + \mathcal{O}(\beta^{3/2})\Big\}
\end{aligned} \quad (2.207)$$

where the Ricci tensor is defined in Appendix A. For example, the term proportional to $(y-z)^i \partial_i g^{1/4}$ in the first pair of curly braces comes from the terms with $(\partial_j g^{ij}) q_j$ and $g^{ij}(\partial_i g^{kl}) q_j q_k q_k$ in (2.206) after integration over q.

Note that since the difference $(y-z)$ is of order $\sqrt{\beta}$, all terms are of order β or less. The terms within the first pair of curly braces are, through order β, equal to $g^{1/4}(z) g^{1/4}(y)$ and cancel the factors of $g^{-1/4}(z) g^{-1/4}(y)$ in front of the whole expression. The terms with $\partial_k g_{ij}$ and its square are clearly the first two terms in an expansion of an exponent. This suggests exponentiating all terms, yielding

$$
\begin{aligned}
\langle z | \exp\left(-\frac{\beta}{\hbar}\hat{H}\right) | y \rangle = (2\pi\hbar\beta)^{-n/2} \exp\Bigg[-\frac{\beta}{\hbar} \bigg\{ \frac{1}{2} g_{ij}(z) + \frac{1}{4} \partial_k g_{ij}(z) (y-z)^k \\
+ \frac{1}{12} \left[\partial_k \partial_l g_{ij}(z) - \frac{1}{2} g_{mn}(z) \Gamma_{ij}^m(z) \Gamma_{kl}^n(z) \right] (y-z)^k (y-z)^l \bigg\} \\
\times \frac{(y-z)^i}{\beta} \frac{(y-z)^j}{\beta} \\
- \frac{1}{12} \beta\hbar R(z) - \frac{1}{12} R_{ij}(z) (y-z)^i (y-z)^j + \mathcal{O}(\beta^{3/2}) \Bigg] . \qquad (2.208)
\end{aligned}
$$

We shall now show that all terms in the exponent except the last two just correspond to an expansion around z of the classical action, which is equal to the integral along the geodesic joining z and y of the invariant line element.

The classical action $S_{cl}[z, y; \beta]$ is given by

$$
S_{cl}[z, y; \beta] = \int_{-\beta}^{0} \frac{1}{2} g_{ij}(x_{cl}(t)) \frac{dx_{cl}^i(t)}{dt} \frac{dx_{cl}^j(t)}{dt} \, dt \qquad (2.209)
$$

where x_{cl}^i satisfies the equation of motion obtained from the Euler–Lagrange variational principle

$$
\frac{D}{dt} \frac{d}{dt} x^i \equiv \ddot{x}^i + \Gamma_{jk}^i \dot{x}^j \dot{x}^k = 0 \qquad (2.210)
$$

together with the boundary conditions

$$
x^i(-\beta) = y^i, \quad x^i(0) = z^i. \qquad (2.211)
$$

(To avoid confusion concerning the notation we consider functions $x^i(t)$ and endpoints y^i and z^i.) Expanding $x_{cl}^i(t)$ into a Taylor series

$$
x_{cl}^i(t) = \sum_{n=0}^{\infty} \frac{t^n}{n!} \frac{d^n}{dt^n} x_{cl}^i(0) \qquad (2.212)
$$

$$
x_{cl}^i(0) = z^i, \quad \ddot{x}_{cl}^i(0) = -\Gamma_{jk}^i(z) \dot{x}_{cl}^j(0) \dot{x}_{cl}^k(0), \quad \text{etc.} \qquad (2.213)
$$

we see that we can express $x^i_{cl}(t)$ into $x^i(0)$ and $\dot{x}^i(0)$. The value of $\dot{x}^i(0)$ follows from the boundary condition at $t = -\beta$. Namely, equation (2.212) at $t = -\beta$ yields

$$\begin{aligned} y^i &= z^i - \beta \dot{x}^i_{cl}(0) + \frac{1}{2}\beta^2 \ddot{x}^i_{cl}(0) - \frac{1}{6}\beta^3 \dddot{x}^i_{cl}(0) + \cdots \\ \dot{x}^i_{cl}(0) &= \frac{1}{\beta}(z^i - y^i) + \frac{1}{2}\beta \ddot{x}^i_{cl}(0) - \frac{1}{6}\beta^2 \frac{d}{dt}[\ddot{x}^i_{cl}(0)] + \cdots \\ &= \frac{1}{\beta}(z-y)^i - \frac{1}{2}\beta \Gamma^i_{jk}(z)\dot{x}^j_{cl}(0)\dot{x}^k_{cl}(0) + \frac{1}{6}\beta^2 \frac{d}{dt}(\Gamma^i_{kl}\dot{x}^k_{cl}\dot{x}^l_{cl}) + \cdots . \end{aligned} \tag{2.214}$$

Solving iteratively for $\dot{x}^i_{cl}(0)$ to order $(z-y)$ yields

$$\begin{aligned} \dot{x}^i_{cl}(0) &= \frac{1}{\beta}(z-y)^i - \frac{1}{2\beta}\Gamma^i_{jk}(z-y)^j(z-y)^k \\ &\quad + \frac{1}{6\beta}(\partial_l \Gamma^i_{jk} + \Gamma^i_{sj}\Gamma^s_{kl})(z-y)^j(z-y)^k(z-y)^l + \cdots . \end{aligned} \tag{2.215}$$

From these results we can obtain an expansion of the classical action in terms of z^i and $(z-y)^i$ by Taylor expanding the Lagrangian $L(t)$,

$$S_{cl}[z, y; \beta] = \int_{-\beta}^{0} \left[L(0) + t\frac{d}{dt}L(0) + \cdots \right] dt. \tag{2.216}$$

However, $L(t)$ is conserved because it coincides with the energy in the absence of a potential

$$\begin{aligned} \frac{d}{dt}\left(\frac{1}{2}g_{ij}\dot{x}^i\dot{x}^j\right) &= g_{ij}\ddot{x}^i\dot{x}^j + \frac{1}{2}\partial_k g_{ij}\dot{x}^k\dot{x}^i\dot{x}^j \\ &= g_{ij}(\ddot{x}^i + \Gamma^i_{kl}\dot{x}^k\dot{x}^l)\dot{x}^j = 0 \quad \text{for } x = x_{cl} \end{aligned} \tag{2.217}$$

hence only $L(0)$ contributes in (2.216). We then find

$$\begin{aligned} S_{cl}[z, y; \beta] = \beta L(0) &= \frac{\beta}{2} g_{ij}(z)\dot{x}^i_{cl}(0)\dot{x}^j_{cl}(0) \\ &= \frac{1}{2\beta}g_{ij}(z)\Big[(z-y)^i - \frac{1}{2}\Gamma^i_{kl}(z-y)^k(z-y)^l \\ &\quad + \frac{1}{6}(\partial_k\Gamma^i_{lm} + \Gamma^i_{sk}\Gamma^s_{lm})(z-y)^k(z-y)^l(z-y)^m\Big] \\ &\quad \times \Big[(z-y)^j - \frac{1}{2}\Gamma^j_{pq}(z-y)^p(z-y)^q \\ &\quad + \frac{1}{6}(\partial_p\Gamma^j_{qr} + \Gamma^j_{tp}\Gamma^t_{qr})(z-y)^p(z-y)^q(z-y)^r\Big] + \cdots . \end{aligned} \tag{2.218}$$

The terms quartic in $(z-y)$ have as coefficient

$$\left[\frac{1}{8}g_{ij}\Gamma^i_{kl}\Gamma^j_{mn} + \frac{1}{6}g_{sn}(\partial_k\Gamma^s_{lm} + \Gamma^s_{tk}\Gamma^t_{lm})\right]. \tag{2.219}$$

The last two terms in (2.219) yield actually $-\frac{1}{6}$ times the first $g\Gamma\Gamma$ plus a $\partial\partial g$ term. Hence, one finally arrives at

$$\begin{aligned} S_{cl}[z,y;\beta] = \frac{1}{\beta}\bigg\{ & \frac{1}{2}g_{ij}(z)(z-y)^i(z-y)^j \\ & -\frac{1}{4}\partial_k g_{ij}(z)(z-y)^i(z-y)^j(z-y)^k \\ & +\left[\frac{1}{12}\partial_k\partial_l g_{mn}(z) - \frac{1}{24}g_{ij}(z)\Gamma^i_{kl}\Gamma^j_{mn}(z)\right] \\ & \times (z-y)^k(z-y)^l(z-y)^m(z-y)^n\bigg\} + \mathcal{O}(z-y)^5. \end{aligned} \tag{2.220}$$

These terms agree perfectly with the first four terms in (2.208).

The transition amplitude can then be written as

$$\begin{aligned} \langle z|\mathrm{e}^{-\frac{\beta}{\hbar}\hat{H}}|y\rangle = & \frac{1}{(2\pi\hbar\beta)^{n/2}}\mathrm{e}^{-\frac{1}{\hbar}S_{cl}[z,y;\beta]}\bigg\{1 - \frac{1}{24}\beta\hbar(R(z)+R(y)) \\ & -\frac{1}{24}[R_{ij}(z)+R_{ij}(y)](z-y)^i(z-y)^j + \mathcal{O}(\beta^{3/2})\bigg\}. \end{aligned} \tag{2.221}$$

We have replaced $R(z)$ by $\frac{1}{2}[R(z)+R(y)]$ which is allowed to order β, to show that the result is symmetric under exchange of z and y. We have followed [57], but for an alternative derivation, see [14]. The transition amplitude is now known up to order $\beta^{5/2}$ (three-loops) and can be found in [81] where it is computed by path integral methods.

Of course, the composition rule should hold

$$\int \langle z|\mathrm{e}^{-\frac{\beta}{\hbar}\hat{H}}|x\rangle\sqrt{g(x)}\langle x|\mathrm{e}^{-\frac{\beta}{\hbar}\hat{H}}|y\rangle\, d^n x = \langle z|\mathrm{e}^{-\frac{2\beta}{\hbar}\hat{H}}|y\rangle. \tag{2.222}$$

A quick way to check this is to use normal coordinates around x since then only the leading term in the classical action survives (for normal coordinates $\partial_i g_{jk}(x) = \partial_{(i}\partial_j g_{kl)}(x) = 0$ while $g_{ij}(z) = g_{ij}(x) - \frac{1}{3}R_{iklj}(\Gamma)(x)(z-x)^k(z-x)^l$ through order β). Taking the opposite point of view, we can impose the composition rule and find then that this fixes the coefficient of the Ricci tensor, but not that of the scalar curvature. The latter terms yield the trace anomaly in $d=2$ dimensions, and its coefficient should not be fixed by requiring the composition rule to hold because we can view $\hbar^2 R$ as a potential term in the action, and the composition rule should hold for any potential.

The terms with R_{ij} can be expressed in terms of the classical action. One should expect this: they are one-loop terms and hence should be proportional to the determinant of the double derivative of the classical action ([17], see also the textbook by Schulman [21]). One may check that (2.220) yields

$$D_{ij} \equiv -\frac{\partial}{\partial z^i}\frac{\partial}{\partial y^j}S_{cl}[z,y;\beta] = \frac{1}{\beta}\Big\{g_{ij}(z) - \Gamma_{jk;i}(z-y)^k - \frac{3}{4}\partial_i\partial_{(k}g_{lj)}(z-y)^k(z-y)^l + \left[\partial_{(i}\partial_j g_{mn)} - \frac{1}{2}g_{st}\Gamma^s_{(ij}\Gamma^t_{mn)}\right](z-y)^m(z-y)^n\Big\}. \quad (2.223)$$

Hence, using the notation $D_{ij} = \frac{1}{\beta}(g_{ij} - \Delta g_{ij})$, we have

$$\det D_{ij} = \beta^{-n}g(z)\left[1 - g^{ij}\Delta g_{ij} - \frac{1}{2}g^{ij}\Delta g_{jk}g^{kl}\Delta g_{li} + \frac{1}{2}(g^{ij}\Delta g_{jk})^2 + \cdots\right]. \quad (2.224)$$

Since the first term in $-g^{ij}\Delta g_{ij}$ is equal to $-g^{ij}\Gamma_{jk;i}(z-y)^k$ which is equal to $-\frac{1}{2}g^{ij}\partial_k g_{ij}(z-y)^k \sim g^{-1/2}(z)g^{1/2}(y) - 1$, we can remove the term proportional to $(z-y)\Gamma$ from $\det D_{ij}$ by replacing $g(z)$ by $g^{1/2}(z)g^{1/2}(y)$

$$\det D_{ij} = \beta^{-n}g(z)^{1/2}g^{1/2}(y)(1 + \cdots). \quad (2.225)$$

This parametrization makes sense because $g(z)^{-1/2}\left(\det \frac{\partial}{\partial z^i}\frac{\partial}{\partial y^j}S\right) \times g^{-1/2}(y)$ is a biscalar. If we now work out the remaining terms in (2.225) denoted by $\cdots$, one finds a nice surprise,

$$g^{ij}\left[-\frac{3}{4}\partial_i\partial_{(k}g_{lj)} + \partial_{(i}\partial_j g_{kl)} - \frac{1}{2}g_{st}\Gamma^s_{(ij}\Gamma^t_{kl)} + \frac{1}{2}\Gamma_{jk,n}g^{nm}\Gamma_{ml,i}\right] \times (z-y)^k(z-y)^l = -\frac{1}{6}R_{kl}(z-y)^k(z-y)^l. \quad (2.226)$$

Hence the R_{ij} terms in (2.221) can be written as

$$\left(1 - \frac{1}{12}R_{ij}(z-y)^i(z-y)^j\right) = \beta^{n/2}g^{-1/4}(z)\left(\det -\frac{\partial}{\partial z^i}\frac{\partial}{\partial y^j}S_{cl}\right)^{1/2} g^{-1/4}(y) \equiv \tilde{D}^{1/2}. \quad (2.227)$$

The final result for the transition element becomes

$$\langle z|e^{-\frac{\beta}{\hbar}\hat{H}}|y\rangle = \frac{1}{(2\pi\hbar\beta)^{n/2}}e^{-\frac{1}{\hbar}S_{cl}[z,y;\beta]}\tilde{D}^{1/2} \times\left\{1 - \frac{1}{24}\beta\hbar\left[R(z) + R(y)\right] + \mathcal{O}(\beta^{3/2})\right\}. \quad (2.228)$$

The Einstein invariance is manifest: the transition element is a biscalar (it does not depend on the coordinates one chooses around z and y, nor on the choice of coordinates anywhere else).

The Van Vleck determinant $\tilde{D}$ is $\hbar$-independent and thus it yields the one-loop corrections. (Since $\frac{\partial}{\partial y^j} S_{cl}[z, y; \beta] = -p_j$, where p_j is the momentum conjugate to y^j, one could interpret $\det D_{ij}$ as the Jacobian for the change of variable $p(y) \to z$.) In the next section we shall directly calculate the one- and two-loop Feynman diagrams, and indeed obtain the Van Vleck determinant as part of the one-loop corrections. In flat space, it reduces to unity in our normalization. The factor $(2\pi\hbar\beta)^{-n/2}$ is the Feynman measure.

We have thus obtained the order β corrections to the transition element $\langle z|\exp(-\frac{\beta}{\hbar}\hat{H})|y\rangle$ by direct evaluation. Another way to obtain these corrections is to use the Schrödinger equation,

$$-\hbar\frac{\partial}{\partial\beta}\langle z|\exp\left(-\frac{\beta}{\hbar}\hat{H}\right)|y\rangle = \int \langle z|\hat{H}|x\rangle\sqrt{g(x)}\langle x|\exp\left(-\frac{\beta}{\hbar}\hat{H}\right)|y\rangle\, d^n x$$
$$= H(z)\langle z|\exp\left(-\frac{\beta}{\hbar}\hat{H}\right)|y\rangle = H(y)\langle z|\exp\left(-\frac{\beta}{\hbar}\hat{H}\right)|y\rangle. \tag{2.229}$$

In the last step we used the fact that the left-hand side of this equation is symmetric in z, y. This follows either from general arguments, or by looking at the explicit expression we obtained for $\langle z|\exp(-\frac{\beta}{\hbar}\hat{H})|y\rangle$. Since $\langle z|\exp(-\frac{\beta}{\hbar}\hat{H})|y\rangle$ is given by an expansion about z, it is evidently much easier to evaluate the action of $H(y)$ than that of $H(z)$.

The operator $H(z)$ is given by

$$H(z) = -\frac{\hbar^2}{2}g^{-1/2}\partial_i g^{1/2}g^{ij}\partial_j. \tag{2.230}$$

Let us show in some detail how this asymmetric-looking expression arises. Define $\hat{x}^i(t) = \exp(\frac{i}{\hbar}\hat{H}t)\hat{x}^i\exp(-\frac{i}{\hbar}\hat{H}t)$ and $|x, t\rangle = \exp(\frac{i}{\hbar}\hat{H}t)|x\rangle$ ("moving frames") as eigenstates of $\hat{x}^i(t)$. Similarly we introduce $|p, t\rangle = \exp(\frac{i}{\hbar}\hat{H}t)|p\rangle$. As before, $\langle x, t|x', t\rangle = g^{-1/2}(x)\delta^{(n)}(x - x')$ and $\langle p, t|p', t\rangle = \delta^{(n)}(p - p')$. Given a state $|\psi\rangle$ in the Hilbert space, $\psi(x, t) = \langle x, t|\psi\rangle$ is the Schrödinger wave function. In this x-representation, $\hat{p}_j(t)$ is represented when acting on $\psi(x, t)$ by

$$(p_x)_j = g^{-1/4}(x)\frac{\hbar}{i}\frac{\partial}{\partial x^j}g^{1/4}(x) \tag{2.231}$$

as follows from

$$\langle x, t|\hat{p}_j(t)|x', t\rangle = \int \langle x, t|p, t\rangle\langle p, t|\hat{p}_j(t)|x', t\rangle\, d^n p$$
$$= \int p_j\langle x, t|p, t\rangle\langle p, t|x', t\rangle\, d^n p$$

$$
\begin{aligned}
&= \int p_j \frac{\exp \frac{i}{\hbar}(x-x')\cdot p}{(2\pi\hbar)^n g^{1/4}(x) g^{1/4}(x')}\, d^n p \\
&= g^{-1/4}(x) g^{-1/4}(x') \frac{\hbar}{i}\frac{\partial}{\partial x^j}\delta^{(n)}(x-x') \\
&= g^{-1/4}(x) g^{-1/4}(x') \frac{\hbar}{i}\frac{\partial}{\partial x^j} g^{1/2}(x)\langle x,t|x',t\rangle \\
&= g^{-1/4}(x) \frac{\hbar}{i}\frac{\partial}{\partial x^j} g^{1/4}(x)\langle x,t|x',t\rangle. \qquad (2.232)
\end{aligned}
$$

The Dirac delta function is defined by $\int \delta^{(n)}(x-x')f(x')\, d^n x' = f(x)$. In the last step we moved $g^{-1/4}(x')$ past $\frac{\partial}{\partial x^j}$ and then converted the factor of $g^{-1/4}(x')\delta^{(n)}(x-x')$ to $g^{-1/4}(x)\delta^{(n)}(x-x')$. A quick argument to justify (2.231) is to note that with the $\sqrt{g}$ in the inner product in x-space the operator $(p_x)_j$ is hermitian. Similarly, one may derive

$$
\langle x,t|\hat{p}_j|x',t\rangle = -g^{-1/4}(x')\frac{\hbar}{i}\frac{\partial}{\partial x'^j} g^{1/4}(x')\langle x,t|x',t\rangle. \qquad (2.233)
$$

It then follows that

$$
\begin{aligned}
\langle x,t|\hat{H}(t)|x',t\rangle &= \langle x|\frac{1}{2} g^{-1/4}(\hat{x})\hat{p}_i g^{1/2}(\hat{x}) g^{ij}(\hat{x})\hat{p}_j g^{-1/4}(\hat{x})|x'\rangle \\
&= \frac{1}{2} g^{-1/4}(x)\left[g^{-1/4}(x)\frac{\hbar}{i}\frac{\partial}{\partial x^i} g^{1/4}(x)\right] g^{1/2}(x) g^{ij}(x)\langle x|\hat{p}_j g^{-1/4}(x)|x'\rangle \\
&= H(x)\langle x|x'\rangle. \qquad (2.234)
\end{aligned}
$$

From (2.233) one finds that this expression is also equal to $H(x')\langle x|x'\rangle$, and this proves the last step of (2.229). One may now check that the transition element given in (2.221) satisfies

$$
\left[H(y) + \hbar\frac{\partial}{\partial\beta}\right]\langle z|\mathrm{e}^{-\frac{\beta}{\hbar}\hat{H}}|y\rangle = 0. \qquad (2.235)
$$

Already at the level of the terms of the form $(z-y)\partial g$ this is quite a good check.

2.6 Two-loop path integral evaluation of the transition element to order β

In this section we shall explicitly verify through two-loops that the path integral corresponding to the Hamiltonian $\hat{H} = \frac{1}{2} g^{-1/4} p_i g^{1/2} g^{ij} p_j g_i^{-1/4}$ for path $x^i(t)$ satisfying the boundary conditions $x^i(-\beta) = y^i$, $x^i(0) = z^i$, reproduces to order β the results of the previous section for the matrix elements $\langle z|\exp\left(-\frac{\beta}{\hbar}H\right)|y\rangle$.

We recall from (2.81)

$$
\langle z|\exp\left(-\frac{\beta}{\hbar}H\right)|y\rangle = \left[\frac{g(z)}{g(y)}\right]^{1/4}\frac{1}{(2\pi\beta\hbar)^{n/2}}\langle \mathrm{e}^{-\frac{1}{\hbar}S^{int}}\rangle \qquad (2.236)
$$

where the brackets $\langle \cdots \rangle$ indicate that all quantum fields $q^i(\tau), b^i(\tau), c^i(\tau)$ and $a^i(\tau)$ are to be contracted using the propagators

$$\begin{aligned}
\langle q^i(\sigma)q^j(\tau)\rangle &= -\beta\hbar g^{ij}(z)\Delta(\sigma-\tau)\\
\langle q^i(\sigma)\dot{q}^j(\tau)\rangle &= -\beta\hbar g^{ij}(z)[\sigma+\theta(\tau-\sigma)]\\
\langle \dot{q}^i(\sigma)\dot{q}^j(\tau)\rangle &= -\beta\hbar g^{ij}(z)[1-\delta(\sigma-\tau)]\\
\langle b^i(\sigma)c^j(\tau)\rangle &= -2\beta\hbar g^{ij}(z)\delta(\sigma-\tau)\\
\langle a^i(\sigma)a^j(\tau)\rangle &= \beta\hbar g^{ij}(z)\delta(\sigma-\tau)\\
\Delta(\sigma,\tau) &= \sigma(\tau+1)\theta(\sigma-\tau)+\tau(\sigma+1)\theta(\tau-\sigma).
\end{aligned} \tag{2.237}$$

We also recall the definition of the interactions

$$\begin{aligned}
S^{int} &= \frac{1}{\beta}\int_{-1}^{0}\frac{1}{2}g_{ij}(x)(\dot{x}^i\dot{x}^j+b^ic^j+a^ia^j)\,d\tau\\
&\quad -\frac{1}{\beta}\int_{-1}^{0}\frac{1}{2}g_{ij}(z)(\dot{q}^i\dot{q}^j+b^ic^j+a^ia^j)\,d\tau\\
&\quad +\frac{\beta\hbar^2}{8}\int_{-1}^{0}(R+g^{ij}\Gamma^k_{il}\Gamma^l_{jk})\,d\tau,\\
x^i(t) &= z^i+(z-y)^i\tau+q^i(\tau),\qquad \tau=t/\beta.
\end{aligned} \tag{2.238}$$

We shall encounter at various points ill-defined expressions to which we shall give meaning by going back to the discretized approach. Since each Feynman graph corresponds in a one-to-one fashion to terms in the answer for $\langle z|\exp(-\frac{\beta}{\hbar}\hat{H})|y\rangle$, derived in the previous section without any ambiguities, this procedure will, in a very direct way, produce a list of continuum integrals for products of distributions. At the end of this section we shall check that our discretized Feynman rules produce these continuum integrals.

We shall organize the calculation as follows: first we compute all tree graphs (first from one vertex S^{int}, then from two vertices S^{int}), then all one-loop graphs (first from one S^{int}, then from two S^{int}) and finally all two-loop graphs (first those from one S^{int}, then the one-particle reducible ones from two S^{int} and finally the one-particle irreducible ones from two S^{int}). There will be no contributions from three S^{int} vertices to order β.

Tree graphs. They consist of the classical vertices themselves and tree graphs with q-propagators. The former are given by

$$\bullet = -\frac{1}{\hbar}S^{int} = -\frac{1}{2\beta\hbar}\begin{pmatrix} g_{ij}(z)-\frac{1}{2}(z-y)^k\partial_k g_{ij}(z)\\ +\frac{1}{6}(z-y)^k(z-y)^l\partial_k\partial_l g_{ij}(z)\end{pmatrix}(z-y)^i(z-y)^j. \tag{2.239}$$

These terms are part of the classical action $S_{cl}[z,y;\beta]$ in (2.220); the terms with two and three factors $(z-y)$ are already correct, while those with four $(z-y)$ and $\partial_k\partial_l g_{ij}$ are also correct, but the $\partial g \partial g (z-y)^4$ terms are lacking. They come from the tree graph with one propagator and two S^{int}. It yields (for terms with one Δ)

$$
\begin{aligned}
\bullet\!\!-\!\!\!-\!\!\bullet &= \left(\frac{-1}{\beta\hbar}\right)^2 \frac{1}{2!}\int_{-1}^{0}\int_{-1}^{0}\left\langle \frac{1}{2}g_{ij}(x)\dot{x}^i\dot{x}^j\frac{1}{2}g_{mn}(x)\dot{x}^m\dot{x}^n\right\rangle d\sigma\, d\tau \\
&= \frac{1}{8\beta^2\hbar^2}\int_{-1}^{0}\int_{-1}^{0} d\sigma\, d\tau \Big\{ (\partial_k g_{ij})(\partial_l g_{mn})(-\hbar\beta g^{kl})\Delta(\sigma,\tau) \\
&\times (z-y)^i(z-y)^j(z-y)^m(z-y)^n \\
&+ 4(\partial_k g_{ij}(z+(z-y)\sigma))(-\hbar\beta g^{km})\Delta^{\bullet}(\sigma,\tau) g_{nm}(z+(z-y)\tau) \\
&\times (z-y)^i(z-y)^j(z-y)^n \\
&+ 4 g_{ij}(z+(z-y)\sigma) g_{mn}(z+(z-y)\tau) \\
&\times [-\hbar\beta g^{im}(z)]\,{}^{\bullet}\Delta^{\bullet}(\sigma,\tau)(z-y)^j(z-y)^n \Big\}.
\end{aligned} \tag{2.240}
$$

We used the notation $\Delta^{\bullet}(\sigma,\tau) = \frac{\partial}{\partial\tau}\Delta(\sigma,\tau)$ and ${}^{\bullet}\Delta(\sigma,\tau) = \frac{\partial}{\partial\sigma}\Delta(\sigma,\tau)$ and ${}^{\bullet}\Delta^{\bullet}(\sigma,\tau) = \frac{\partial}{\partial\sigma}\frac{\partial}{\partial\tau}\Delta(\sigma,\tau)$. There are terms with three and four factors of $(z-y)$ in (2.240). Since the former were already accounted for, the $(z-y)^3$ terms above should vanish. This leads to a first condition on continuum integrals

$$
\int_{-1}^{0}\int_{-1}^{0}\Delta^{\bullet}(\sigma,\tau)\, d\sigma\, d\tau + 2\int_{-1}^{0}\int_{-1}^{0}\sigma\,{}^{\bullet}\Delta^{\bullet}(\sigma,\tau)\, d\sigma\, d\tau = 0. \tag{2.241}
$$

Since the terms of the form $\partial_i\partial_j g_{kl}(z-y)^4$ were already recovered in (2.239), the terms from the last line in (2.240), with g_{ij} or g_{mn} expanded to second order, or the terms from the second line expanded to first order, should also vanish. This leads to another condition

$$
\int_{-1}^{0}\int_{-1}^{0}\sigma^2\,{}^{\bullet}\Delta^{\bullet}(\sigma,\tau)\, d\sigma\, d\tau + \int_{-1}^{0}\int_{-1}^{0}\sigma\,\Delta^{\bullet}(\sigma,\tau)\, d\sigma\, d\tau = 0. \tag{2.242}
$$

The $\partial g \partial g (z-y)^4$ terms from (2.240) are given by

$$
\begin{aligned}
&\left(-\frac{1}{8\beta\hbar}\partial_k g_{ij}\partial_l g_{mn}\right)(z-y)^j(z-y)^n \\
&\quad\times\Bigg[g^{kl}(z-y)^i(z-y)^m\int_{-1}^{0}\int_{-1}^{0}\Delta\, d\sigma\, d\tau \\
&\quad+ (z-y)^i(z-y)^l g^{km} 4\int_{-1}^{0}\int_{-1}^{0}\tau\,\Delta^{\bullet}\, d\sigma\, d\tau \\
&\quad+ (z-y)^k(z-y)^l g^{im} 4\int_{-1}^{0}\int_{-1}^{0}\sigma\tau\,{}^{\bullet}\Delta^{\bullet}\, d\sigma\, d\tau\Bigg].
\end{aligned} \tag{2.243}
$$

We obtain the correct terms of the form $\partial g \partial g (z-y)^4$ which occur in the classical action in (2.220) provided the following integrals are correct:

$$\int_{-1}^{0}\int_{-1}^{0} \Delta(\sigma,\tau)\, d\sigma\, d\tau = -\frac{1}{12}$$
$$\int_{-1}^{0}\int_{-1}^{0} \tau\, \Delta^{\bullet}(\sigma,\tau)\, d\sigma\, d\tau = +\frac{1}{12}$$
$$\int_{-1}^{0}\int_{-1}^{0} \sigma\tau\, {}^{\bullet}\!\Delta^{\bullet}(\sigma,\tau)\, d\sigma\, d\tau = -\frac{1}{12}. \qquad (2.244)$$

If the integrals in (2.242)–(2.244) have the values indicated, the tree graph contributions correctly reproduce the classical action to order β. We first consider all other graphs and integrals, and then we shall discuss these integrals.

We considered only the connected tree graphs because we compared the result with S_{cl} instead of $\exp(-S_{cl}/\hbar)$. The reader may have wondered why we did not consider connected tree graphs with three or more vertices. The reason is that these graphs do not contribute at order β. The vertices at the end of such a tree graph contain one quantum field and at least two factors of $z-y$ (use the fact that $\int_{-1}^{0} \dot{q}\, d\sigma = 0$ since $q(\sigma)$ vanishes at the boundaries). Other vertices contain two quantum fields and at least one factor of $z-y$. Thus the total number of factors of $z-y$ is five or more, which leads to contributions of order $\beta^{3/2}$ or higher.

One-loop graphs. From the vertex S^{int} one finds, by expanding g_{ij} once, the following equal-time contractions:

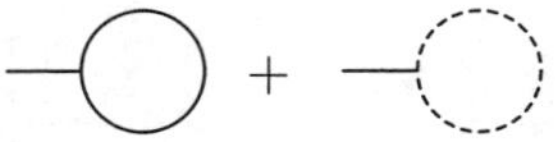

$$= -\frac{1}{2\beta\hbar}(z-y)^k \partial_k g_{ij}(z) \int_{-1}^{0} \tau\, ({}^{\bullet}\!\Delta^{\bullet} + \Delta^{\bullet\bullet})_{\sigma=\tau}\, d\tau\, [-\beta\hbar g^{ij}(z)]$$
$$-\frac{1}{\beta\hbar}\partial_k g_{ij}(z) \int_{-1}^{0} (\Delta^{\bullet})_{\sigma=\tau}\, d\tau\, (z-y)^j [-\beta\hbar g^{ik}(z)]. \qquad (2.245)$$

Since we already know from the previous section that the sum of all one-loop graphs is given by $-\frac{1}{12}R_{ij}(z)(z-y)^i(z-y)^j$, these equal-time contractions with ∂g should cancel the contribution from the measure $[g(z)/g(y)]^{1/4}$. This yields the conditions

$$\int_{-1}^{0} \tau\, ({}^{\bullet}\!\Delta^{\bullet} + \Delta^{\bullet\bullet})_{\sigma=\tau}\, d\tau = -\frac{1}{2}$$
$$\int_{-1}^{0} (\Delta^{\bullet})_{\sigma=\tau}\, d\tau = 0. \qquad (2.246)$$

These equal-time contractions are a priori ill-defined in field theory, but we deduce their value unambiguously as the limit from the discretized expressions.

By expanding g_{ij} to second order in one vertex S^{int}, one finds further equal-time one-loop contractions with $\partial\partial g$

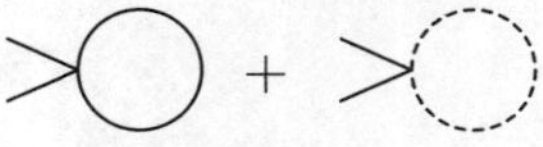

$$= \left(\frac{-1}{\beta\hbar}\right)\frac{1}{4}\partial_k\partial_l g_{ij}(z)\int_{-1}^{0} d\tau \Big[(z-y)^k\tau(z-y)^l\tau\,({}^{\bullet}\!\Delta^{\bullet}+\Delta^{\bullet\bullet})_{\sigma=\tau}\, g^{ij}(z)$$
$$+4(z-y)^k\tau(z-y)^i(\Delta^{\bullet})_{\sigma=\tau}g^{lj}(z)$$
$$+(z-y)^i(z-y)^j g^{kl}(z)\,(\Delta)_{\sigma=\tau}\Big]\, d\tau(-\beta\hbar). \tag{2.247}$$

There is also a term of the form $(\partial_k\partial_l g_{ij})(z-y)^k(z-y)^l g^{ij}$ coming from the measure factor $[g(z)/g(y)]^{1/4}$. Its coefficient is $-\frac{1}{8}$.

To obtain the linearized contribution

$$-\frac{1}{12}R_{iklj}(z-y)^k(z-y)^l g^{ij}$$
$$\sim \frac{1}{24}(-\partial_i\partial_j g_{kl}-\partial_k\partial_l g_{ij}+2\partial_i\partial_l g_{kj})g^{ij}(z-y)^k(z-y)^l \tag{2.248}$$

we need the following equal-time contractions:

$$\int_{-1}^{0}\tau^2\,({}^{\bullet}\!\Delta^{\bullet}+\Delta^{\bullet\bullet})\,d\tau = \frac{1}{3}$$
$$\int_{-1}^{0}\tau\,(\Delta^{\bullet})_{\sigma=\tau}\,d\tau = \frac{1}{12}$$
$$\int_{-1}^{0}(\Delta)_{\sigma=\tau}\,d\tau = -\frac{1}{6}. \tag{2.249}$$

Next, we consider the contributions with $\partial g\partial g$. They come from one-loop graphs with two vertices H^{int}. They can either consist of two unequal-time propagators, or one tree-graph propagator times an equal-time loop. We first consider the one-loop graphs with two unequal-time propagators. The contractions of two factors $(\dot{q}\dot{q}+bc+aa)$ yield (dots denote derivatives)

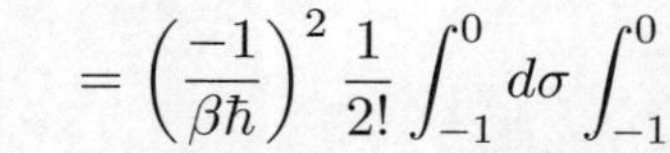

$$= \left(\frac{-1}{\beta\hbar}\right)^2\frac{1}{2!}\int_{-1}^{0}d\sigma\int_{-1}^{0}d\tau\,\frac{1}{2}\partial_k g_{ij}\frac{1}{2}\partial_l g_{mn}(z-y)^k\sigma(z-y)^l\tau$$
$$\times\{2g^{im}g^{jn}[{}^{\bullet}\!\Delta^{\bullet}(\sigma,\tau)\,{}^{\bullet}\!\Delta^{\bullet}(\sigma,\tau)-{}^{\bullet\bullet}\!\Delta(\sigma,\tau)\Delta^{\bullet\bullet}(\sigma,\tau)](-\beta\hbar)^2\}$$
$$= \frac{1}{4}\partial_k g_{ij}\partial_l g_{mn}g^{im}g^{jn}(z-y)^k(z-y)^l I \tag{2.250}$$

where I is fixed by requiring that these terms complete the Ricci tensor

$$I = \int ({}^{\bullet}\Delta^{\bullet}\,{}^{\bullet}\Delta^{\bullet} - {}^{\bullet\bullet}\Delta\,\Delta^{\bullet\bullet})\,\sigma\tau\, d\sigma\, d\tau = -\frac{5}{12}. \tag{2.251}$$

In addition there are contributions with $q\dot{q}$ propagators (indicated by putting a dot above them) and qq propagators (without a dot). Namely,

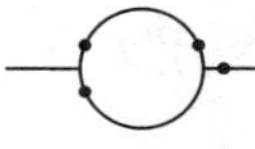

$$= \left(\frac{-1}{\beta\hbar}\right)^2 \frac{1}{2!}\left(\frac{1}{2}\partial_k g_{ij}\right)\left(\frac{1}{2}\partial_l g_{mn}\right) g^{im} g^{jl}\, 8$$

$$\int_{-1}^{0}\int_{-1}^{0} \sigma\, {}^{\bullet}\Delta^{\bullet}(\sigma,\tau)\, {}^{\bullet}\Delta(\sigma,\tau)(z-y)^k (z-y)^n (-\beta\hbar)^2\, d\sigma\, d\tau. \tag{2.252}$$

This should vanish, hence

$$\int_{-1}^{0}\int_{-1}^{0} \sigma\, {}^{\bullet}\Delta^{\bullet}(\sigma,\tau)\, {}^{\bullet}\Delta(\sigma,\tau)\, d\sigma\, d\tau = 0. \tag{2.253}$$

Further,

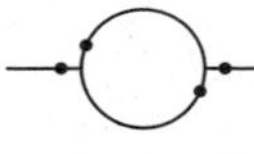

$$= \left(\frac{-1}{\beta\hbar}\right)^2 \frac{1}{2!}\frac{1}{2}\partial_k g_{ij}\frac{1}{2}\partial_l g_{mn} g^{km} g^{il}\, 4 (z-y)^j (z-y)^n$$

$$\int_{-1}^{0}\int_{-1}^{0} {}^{\bullet}\Delta(\sigma,\tau)\Delta^{\bullet}(\sigma,\tau)(-\beta\hbar)^2\, d\sigma\, d\tau. \tag{2.254}$$

We need

$$\int_{-1}^{0}\int_{-1}^{0} {}^{\bullet}\Delta(\sigma,\tau)\Delta^{\bullet}(\sigma,\tau)\, d\sigma\, d\tau = -\frac{1}{12}. \tag{2.255}$$

Finally,

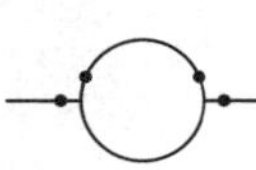

$$= \left(\frac{-1}{\beta\hbar}\right)^2 \frac{1}{2!}\frac{1}{2}\partial_k g_{ij}\frac{1}{2}\partial_l g_{mn} g^{kl} g^{im}$$

$$4\int_{-1}^{0}\int_{-1}^{0} \Delta(\sigma,\tau)\, {}^{\bullet}\Delta^{\bullet}(\sigma,\tau)\, d\sigma\, d\tau (z-y)^j (z-y)^n. \tag{2.256}$$

We need

$$\int_{-1}^{0}\int_{-1}^{0} \Delta(\sigma,\tau)\, {}^{\bullet}\Delta^{\bullet}(\sigma,\tau)\, d\sigma\, d\tau = \frac{1}{12}. \tag{2.257}$$

We now record the one-particle reducible one-loop graphs with one equal-time propagator and one tree propagator. We need one vertex with one q, and the other vertex with three q, or one q and two ghosts. In all cases we consider terms proportional to $\partial_k g_{ij}\partial_l g_{mn}$. We then find the following results provided the integrals have the values indicated (the symbol α denotes $\partial_k g_{ij} \partial_l g_{mn}$):

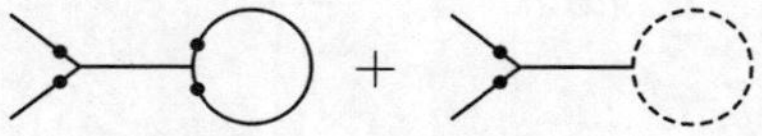

$$= -\frac{1}{48} \alpha g^{ij} g^{kl} (z-y)^m (z-y)^n I$$

$$I = \int_{-1}^{0} \int_{-1}^{1} \left[{}^{\bullet}\Delta^{\bullet}(\sigma,\sigma) + {}^{\bullet\bullet}\Delta(\sigma,\sigma) \right] \Delta(\sigma,\tau)\, d\sigma\, d\tau = -\frac{1}{12}, \tag{2.258}$$

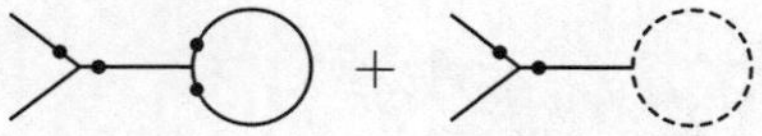

$$= \frac{1}{24} \alpha g^{ij} g^{km} (z-y)^l (z-y)^n I$$

$$I = \int_{-1}^{0} \int_{-1}^{0} \left[{}^{\bullet}\Delta^{\bullet}(\sigma,\sigma) + {}^{\bullet\bullet}\Delta(\sigma,\sigma) \right] \Delta^{\bullet}(\sigma,\tau)\, \tau\, d\sigma\, d\tau = \frac{1}{12}, \tag{2.259}$$

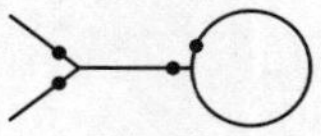

$$= \frac{1}{24} \alpha g^{ik} g^{jl} (z-y)^m (z-y)^n I$$

$$I = \int_{-1}^{0} \int_{-1}^{0} \Delta^{\bullet}(\sigma,\sigma)\, {}^{\bullet}\Delta(\sigma,\tau)\, d\sigma\, d\tau = \frac{1}{12}, \tag{2.260}$$

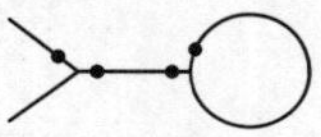

$$= -\frac{1}{12} \alpha g^{ik} g^{jm} (z-y)^l (z-y)^n I$$

$$I = \int_{-1}^{0} \int_{-1}^{0} \Delta^{\bullet}(\sigma,\sigma)\, {}^{\bullet}\Delta^{\bullet}(\sigma,\tau)\, \tau\, d\sigma\, d\tau = -\frac{1}{12}. \tag{2.261}$$

If all the integrals in (2.245)–(2.261) have the values indicated, the one-loop contributions correctly reproduce the Van Vleck–Morette determinant in (2.225).

Two-loop contributions. The two-loop graphs should reproduce the terms of order $\beta\hbar$ in the transition element. These were found to be given by

$$-\frac{\beta\hbar}{12} R(z). \tag{2.262}$$

We again quote the various graphs and below them the values which the corresponding integrals should have.

First there is the figure-of-eight graph due to one vertex

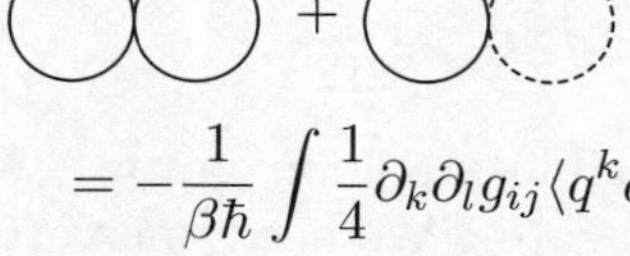

$$= -\frac{1}{\beta\hbar} \int \frac{1}{4} \partial_k \partial_l g_{ij} \langle q^k q^l (\dot{q}^i \dot{q}^j + b^i c^j + a^i a^j) \rangle \, d\tau \, (-\beta\hbar)^2$$

$$= \frac{1}{24} \beta\hbar \partial_k \partial_l g_{ij}(z) (g^{kl} g^{ij} I_1 - g^{ik} g^{jl} I_2)$$

$$I_1 = \int_{-1}^{0} \Delta(\tau,\tau)\,[{}^{\bullet}\Delta^{\bullet}(\tau,\tau) + {}^{\bullet\bullet}\Delta(\tau,\tau)]\,d\tau = -\frac{1}{6}$$

$$I_2 = \int_{-1}^{0} \Delta^{\bullet}(\tau,\tau)\,\Delta^{\bullet}(\tau,\tau)\,d\tau = \frac{1}{12}. \tag{2.263}$$

Next, there are the products of two equal-time loops connected by an unequal-time propagator

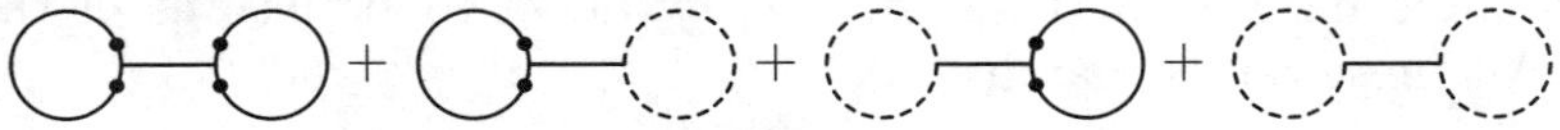

$$= \beta\hbar\frac{1}{96}\alpha g^{kl} g^{ij} g^{mn}\,(\partial_k g_{im})\,(\partial_l g_{jn})\,I$$

$$I = \int_{-1}^{0}\int_{-1}^{0} [{}^{\bullet}\Delta^{\bullet}(\sigma,\sigma) + \Delta^{\bullet\bullet}(\sigma,\sigma)]\,\Delta(\sigma,\tau)\,[{}^{\bullet\bullet}\Delta(\tau,\tau) + \Delta^{\bullet\bullet}(\tau,\tau)]\,d\sigma\,d\tau$$

$$= -\frac{1}{12}, \tag{2.264}$$

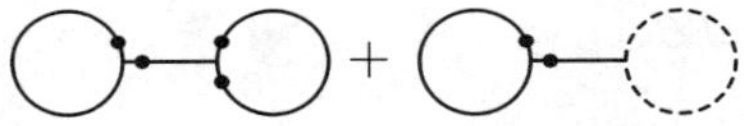

$$= -\beta\hbar\frac{1}{24}\alpha g^{ij} g^{lm} g^{kn}\,(\partial_m g_{ij})\,(\partial_n g_{kl})\,I$$

$$I = \int_{-1}^{0}\int_{-1}^{0} [{}^{\bullet}\Delta^{\bullet}(\sigma,\sigma) + {}^{\bullet\bullet}\Delta(\sigma,\sigma)]\Delta^{\bullet}(\sigma,\tau)\Delta^{\bullet}(\tau,\tau)\,d\sigma\,d\tau$$

$$= \frac{1}{12}, \tag{2.265}$$

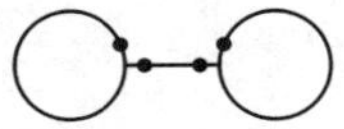

$$= \beta\hbar\frac{1}{24}\alpha g^{ik} g^{jm} g^{ln}\,(\partial_m g_{ij})\,(\partial_n g_{kl})\,I$$

$$I = \int_{-1}^{0}\int_{-1}^{0} {}^{\bullet}\Delta(\sigma,\sigma)\,{}^{\bullet}\Delta^{\bullet}(\sigma,\tau)\Delta^{\bullet}(\tau,\tau)\,d\sigma\,d\tau = -\frac{1}{12}. \tag{2.266}$$

Finally there are the two-loop graphs with the form of a setting sun. They come from all possible contractions of $q(\dot q\dot q + bc + aa)$ times $q(\dot q\dot q + bc + aa)$,

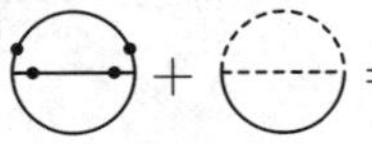

$$= -\beta\hbar\frac{1}{16}\alpha g^{kl} g^{im} g^{jn}\,(\partial_k g_{ij})\,(\partial_l g_{mn})\,I$$

$$I = \int_{-1}^{0}\int_{-1}^{0} \Delta(\sigma,\tau)\,[{}^{\bullet}\Delta^{\bullet}(\sigma,\tau)\,{}^{\bullet}\Delta^{\bullet}(\sigma,\tau) - {}^{\bullet\bullet}\Delta(\sigma,\tau)\Delta^{\bullet\bullet}(\sigma,\tau)]\,d\sigma\,d\tau$$

$$= \frac{1}{4}, \tag{2.267}$$

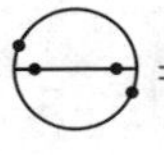

$$= \beta\hbar\frac{1}{12}\alpha g^{km} g^{in} g^{jl}\,(\partial_k g_{ij})\,(\partial_l g_{mn})\,I$$

$$I = \int_{-1}^{0}\int_{-1}^{0} {}^{\bullet}\Delta^{\bullet}(\sigma,\tau)\,\Delta^{\bullet}(\sigma,\tau)\,{}^{\bullet}\Delta(\sigma,\tau)\,d\sigma\,d\tau = -\frac{1}{6}. \tag{2.268}$$

The sum of all of these two-loop contributions plus the contribution from the counterterm should be equal to $-\beta\hbar\frac{1}{12}R$. Different regularization schemes lead to different counterterms, and thereby the one- and two-loop graphs give different results if one uses different regularization methods. In time slicing the counterterm is $-\frac{\beta}{\hbar}\frac{\hbar^2}{8}(R + \Gamma^k_{il}\Gamma^l_{jk}g^{ij})$. Hence, the sum of all two-loop graphs evaluated above should be equal to $\frac{1}{24}\beta\hbar R + \frac{1}{8}\beta\hbar(\Gamma\Gamma)$. Expanding the $\Gamma\Gamma$ term, one finds two structures of the form $\partial g \partial g$, while R contains three more structures with $\partial g \partial g$, and two structures of the form $\partial\partial g$. All these terms match:

$$
\begin{aligned}
\text{two-loop} &= \frac{1}{24}(\Box g - \partial^i g_i) + \frac{1}{96}\Big(\partial_i g \partial^i g - 4\partial_i g g^i + 4 g_i g^i \\
&\quad - 6(\partial_i g_{jk})^2 + 8\partial_i g_{jk}\partial_j g_{ik}\Big) \\
\left(\frac{1}{8} - \frac{1}{12}\right) R &= \frac{1}{24}\Big[\Box g - \partial^i g_i + \frac{1}{2}\partial_i g_{jk}\partial_j g_{ik} - \frac{3}{4}(\partial_i g_{jk})^2 \\
&\quad + \frac{1}{4}\partial_i g \partial^i g - \partial_i g g^i + g_i g^i\Big] \\
\frac{1}{8}\Gamma\Gamma &= \frac{1}{8}\left[\frac{1}{2}\partial_i g_{jk}\partial_j g_{ik} - \frac{1}{4}(\partial_i g_{jk})^2\right]. \qquad (2.269)
\end{aligned}
$$

Contractions are performed with the metric g^{ij}, for example $(\partial_i g_{jk})^2 = g^{ii'}g^{jj'}g^{kk'}(\partial_i g_{jk})(\partial_{i'} g_{j'k'})$.

We now discuss the integrals we have encountered. We first make a list. All integrals run over $-1 \le \sigma \le 0$ and $-1 \le \tau \le 0$.

Trees

$$
\begin{aligned}
&\iint \Delta^{\bullet}(\sigma,\tau) + 2\iint \sigma\, {}^{\bullet}\!\Delta^{\bullet}(\sigma,\tau) = 0 \\
&\iint \sigma^2\, {}^{\bullet}\!\Delta^{\bullet}(\sigma,\tau) + \iint \sigma\Delta^{\bullet}(\sigma,\tau) = 0 \\
&\iint \Delta(\sigma,\tau) = -\frac{1}{12}, \quad \iint \tau\Delta^{\bullet}(\sigma,\tau) = \frac{1}{12}, \quad \iint \sigma\, {}^{\bullet}\!\Delta^{\bullet}(\sigma,\tau)\tau = -\frac{1}{12}.
\end{aligned}
$$

One-loop

$$
\begin{aligned}
&\iint \left({}^{\bullet}\!\Delta^{\bullet}\, {}^{\bullet}\!\Delta^{\bullet} - {}^{\bullet\bullet}\!\Delta\Delta^{\bullet\bullet}\right)\sigma\tau = -\frac{5}{12}, \quad \iint \sigma\, {}^{\bullet}\!\Delta^{\bullet}\, {}^{\bullet}\!\Delta = 0 \\
&\iint {}^{\bullet}\!\Delta\Delta^{\bullet} = -\frac{1}{12}, \quad \iint \Delta\, {}^{\bullet}\!\Delta^{\bullet} = \frac{1}{12} \\
&\int \tau[{}^{\bullet}\!\Delta^{\bullet}(\tau,\tau) + \Delta^{\bullet\bullet}(\tau,\tau)] = -\frac{1}{2}, \quad \int \tau^2[{}^{\bullet}\!\Delta^{\bullet}(\tau,\tau) + \Delta^{\bullet\bullet}(\tau,\tau)] = \frac{1}{3} \\
&\int \Delta(\tau,\tau) = -\frac{1}{6}, \quad \int \Delta^{\bullet}(\tau,\tau) = 0, \quad \int \Delta^{\bullet}(\tau,\tau)\tau = \frac{1}{12}.
\end{aligned}
$$

Two-loop

$$
\begin{aligned}
&\int\!\!\int [{}^{\bullet}\Delta^{\bullet}(\sigma,\sigma) + {}^{\bullet\bullet}\Delta(\sigma,\sigma)]\Delta(\sigma,\tau) = -\frac{1}{12}\\
&\int\!\!\int [{}^{\bullet}\Delta^{\bullet}(\sigma,\sigma) + {}^{\bullet\bullet}\Delta(\sigma,\sigma)]\Delta^{\bullet}(\sigma,\tau)\tau = \frac{1}{12}\\
&\int\!\!\int [\Delta^{\bullet}(\sigma,\sigma)\,{}^{\bullet}\Delta(\sigma,\tau)] = \frac{1}{12}\\
&\int\!\!\int \Delta^{\bullet}(\sigma,\sigma)\,{}^{\bullet}\Delta^{\bullet}(\sigma,\tau)\tau = -\frac{1}{12}\\
&\int \Delta(\tau,\tau)[{}^{\bullet}\Delta^{\bullet}(\tau,\tau) + {}^{\bullet\bullet}\Delta(\tau,\tau)] = -\frac{1}{6}, \quad \int \Delta^{\bullet}(\tau,\tau)\Delta^{\bullet}(\tau,\tau) = \frac{1}{12}\\
&\int\!\!\int [{}^{\bullet}\Delta^{\bullet}(\sigma,\sigma) + {}^{\bullet\bullet}\Delta(\sigma,\sigma)]\Delta(\sigma,\tau)[{}^{\bullet}\Delta^{\bullet}(\tau,\tau) + {}^{\bullet\bullet}\Delta(\tau,\tau)] = -\frac{1}{12}\\
&\int\!\!\int [{}^{\bullet}\Delta^{\bullet}(\sigma,\sigma) + {}^{\bullet\bullet}\Delta(\sigma,\sigma)]\Delta^{\bullet}(\sigma,\tau)\Delta^{\bullet}(\tau,\tau) = \frac{1}{12}\\
&\int\!\!\int {}^{\bullet}\Delta(\sigma,\sigma)\,{}^{\bullet}\Delta^{\bullet}(\sigma,\tau)\Delta^{\bullet}(\tau,\tau) = -\frac{1}{12}\\
&\int\!\!\int \Delta(\sigma,\tau)[{}^{\bullet}\Delta^{\bullet}(\sigma,\tau)\,{}^{\bullet}\Delta^{\bullet}(\sigma,\tau) - {}^{\bullet\bullet}\Delta(\sigma,\tau)\Delta^{\bullet\bullet}(\sigma,\tau)] = \frac{1}{4} \qquad \text{(with TS)}\\
&\int\!\!\int {}^{\bullet}\Delta^{\bullet}(\sigma,\tau)\Delta^{\bullet}(\sigma,\tau)\,{}^{\bullet}\Delta(\sigma,\tau) = -\frac{1}{6} \qquad \text{(with TS)}. \qquad (2.270)
\end{aligned}
$$

Only if all of these integrals have the values indicated is there complete agreement between the Feynman diagram result and the operator approach result.

Using the naive continuum limits

$$
\begin{aligned}
&\Delta(\sigma,\tau) = \sigma(\tau+1)\theta(\sigma-\tau) + \tau(\sigma+1)\theta(\tau-\sigma)\\
&\Delta^{\bullet}(\sigma,\tau) = \sigma + \theta(\tau-\sigma), \quad \Delta^{\bullet}(\tau,\tau) = \tau + \frac{1}{2}\\
&{}^{\bullet}\Delta^{\bullet}(\sigma,\tau) = 1 - \delta(\sigma-\tau), \quad {}^{\bullet\bullet}\Delta(\sigma,\tau) = \delta(\sigma-\tau)\\
&{}^{\bullet}\Delta^{\bullet}(\sigma,\sigma) = 1 - \delta(\sigma-\sigma), \quad {}^{\bullet\bullet}\Delta(\sigma,\sigma) = \delta(\sigma-\sigma) \qquad (2.271)
\end{aligned}
$$

we find complete agreement for the transition element provided we interpret $\delta(\sigma-\tau)$ as a Kronecker delta and $\theta(0)=\frac{1}{2}$. The expressions $\delta(\sigma-\sigma)$ always cancel since they only appear in the combination ${}^{\bullet}\Delta^{\bullet}(\sigma,\sigma) + {}^{\bullet\bullet}\Delta(\sigma,\sigma)$. In the next chapters we discuss two other regularization schemes; these also lead to complete agreement.

3
Mode regularization

In this chapter we discuss path integrals defined by mode regularization (MR). Ideally, one would like to derive mode regularization from first principles, namely starting from the transition amplitude defined as the matrix element of the evolution operator $\langle x_{\rm f}^k|\exp(-\beta\hat{H})|x_{\rm i}^k\rangle$, as done in the time slicing regularization of the previous section (we set $\hbar = 1$ in this chapter). However, a derivation along those lines seems quite laborious and will not be attempted here. We find it easier to take a more pragmatic approach, and present a different construction of the mode-regulated path integral. This can be done by recalling general properties of quantum field theories (QFTs) in d dimensions as a guideline, and specializing those properties to the simpler context of one dimension.

General theorems for quantum field theories with local Lagrangians guarantee the possibility of constructing a consistent perturbative expansion by renormalizing the infinities away. Renormalization is usually achieved by adding local counterterms with infinite coefficients to the original Lagrangian. At the same time *finite local counterterms* relate different regularization schemes to each other. More precisely, the finite counterterms, left undetermined after the removal of divergences, are fixed by imposing a sufficient number of renormalization conditions. All regularization schemes should then produce the same physical results. The renormalization program through counterterms is performed iteratively, loop by loop.

Let us review the simple case of a scalar QFT, which is enough for our purposes. One can classify the interactions as nonrenormalizable, renormalizable and super-renormalizable according to the mass dimension of the corresponding coupling constant being negative, zero and positive, respectively. Coupling constants with negative mass dimensions render the theory nonrenormalizable, since one is forced to introduce an infinite number of counterterms (of a structure not contained in the original

Lagrangian) to cancel the divergences of the Feynman graphs. These theories are generically considered to be effective field theories, like the Fermi theory of the weak interactions, gravity and supergravity. Renormalizable interactions allow instead for infinities to be removed with the use of a finite number of counterterms, though at each loop the coefficients of the counterterms receive additional infinite contributions, as in QED and in the Standard Model. Finally, super-renormalizable interactions generate a perturbative expansion which can be made finite at any loop by counterterms that can appear only up to a finite loop order, like perturbative $\lambda\phi^3$ theory in four dimensions.

Our one-dimensional nonlinear sigma model is super-renormalizable (recall the explicit power-counting exercise presented in Section 1.2). Thus the QFT theorems guarantee that one needs to consider counterterms only up to a finite loop order. In addition, we will see that there is no need to cancel infinities thanks to the inclusion of extra vertices coming from the measure. Therefore only *finite* counterterms can appear. We will show that they appear only up to two-loops. As described above they are needed to satisfy the renormalization conditions. The precise renormalization conditions that we impose are contained in the following requirement: the transition amplitude computed with the regulated path integral must satisfy the Schrödinger equation with a given Hamiltonian operator. Without loss of generality, we choose this operator to be the one containing the covariant Laplacian without any additional coupling to the scalar curvature (if desired, extra couplings can always be introduced by including them into the potentials V and A_i). This renormalization condition completely fixes the counterterm V_{MR} as well as the overall normalization of the path integral. In this way, the MR scheme is also fully specified and can be used in applications.

Mode regularization for one-dimensional nonlinear sigma models was introduced in [39, 40, 73]. The latter reference contains the correct counterterm V_{MR}. This regularization was used in [81] to compute the transition amplitude at three-loops. Related references are [82, 83]. An early use of mode regularization for quantizing nonlinear sigma models was attempted in [84].

3.1 Mode regularization in configuration space

We start from a general classical action in Euclidean time for the fields x^i with $i = 1, \ldots, n$

$$S = \int_{t_\mathrm{i}}^{t_\mathrm{f}} dt \left[\frac{1}{2} g_{ij}(x) \frac{dx^i}{dt} \frac{dx^j}{dt} + i A_i(x) \frac{dx^i}{dt} + V(x) \right] \tag{3.1}$$

and try to define the transition amplitude directly as a path integral

$$\langle x_{\rm f}^k, t_{\rm f} | x_{\rm i}^k, t_{\rm i} \rangle = \int_{\rm BC} \mathcal{D}x \; {\rm e}^{-S} \tag{3.2}$$

$$\mathcal{D}x = \prod_{t_{\rm i}<t<t_{\rm f}} \sqrt{\det g_{ij}(x(t))}\; d^n x(t) \tag{3.3}$$

where BC indicates the Dirichlet boundary conditions at initial and final time $x^k(t_{\rm i}) = x_{\rm i}^k$ and $x^k(t_{\rm f}) = x_{\rm f}^k$. The measure $\mathcal{D}x$ is formally a scalar since it is the product of scalar measures. The action is also a scalar and the transition element should therefore be a scalar when properly defined.

Usually in QFT one considers the path integral representation for the transition amplitude from the in-vacuum to the out-vacuum, which corresponds to an infinite propagation time. In quantum mechanics one can be more general, and ask for the transition amplitude between an arbitrary initial state $|\Psi_{\rm i}\rangle$ at time $t_{\rm i}$ and an arbitrary final state $|\Psi_{\rm f}\rangle$ at time $t_{\rm f}$. For simplicity we consider initial and final states as eigenstates of the position operator, since a general transition amplitude is then given by

$$\langle \Psi_{\rm f}, t_{\rm f} | \Psi_{\rm i}, t_{\rm i} \rangle = \int d^n x_{\rm f} \sqrt{g(x_{\rm f})} \int d^n x_{\rm i} \sqrt{g(x_{\rm i})}\; \Psi_{\rm f}^*(x_{\rm f})\; \langle x_{\rm f}, t_{\rm f} | x_{\rm i}, t_{\rm i} \rangle \; \Psi_{\rm i}(x_{\rm i}). \tag{3.4}$$

Note again that the transition amplitude on the right-hand side, given in (3.2), is formally a scalar since the measure factors for integrating over the initial and final points are not included into (3.3). Therefore they appear in (3.4).

The nontrivial measure in (3.3) is not translationally invariant under $x^i(t) \to x^i(t) + \epsilon^i(t)$. This makes it difficult to generate the perturbative expansion: one cannot complete squares and shift integration variables to derive the propagators as usual. A standard trick to obtain a translationally invariant measure is to introduce ghost fields and exponentiate the nontrivial factor appearing in (3.3),

$$\prod_{t_{\rm i}<t<t_{\rm f}} \sqrt{\det g_{ij}(x(t))} = \int Da\, Db\, Dc\; {\rm e}^{-S_{gh}} \tag{3.5}$$

$$S_{gh} = \int_{t_{\rm i}}^{t_{\rm f}} dt\; \frac{1}{2} g_{ij}(x)(a^i a^j + b^i c^j) \tag{3.6}$$

where the translationally invariant measures for the ghosts are given by

$$Da = \prod_{t_{\rm i}<t<t_{\rm f}} d^n a(t), \quad Db = \prod_{t_{\rm i}<t<t_{\rm f}} d^n b(t), \quad Dc = \prod_{t_{\rm i}<t<t_{\rm f}} d^n c(t). \tag{3.7}$$

The ghosts a^i are commuting while the ghosts b^i and c^i are anticommuting, so they reproduce the same measure factor that is also obtained by integrating out the momenta in phase space.

Up to this point the whole construction is completely formal and we should try to give it a concrete meaning. Thus, we must introduce a regularization scheme to define the path integral and evaluate it unambiguously. The regularization will bring along a corresponding counterterm ΔV which will be used to satisfy the renormalization conditions mentioned previously. In particular, the counterterm will restore the symmetries which may be accidentally broken by the regularization (one may recall that no anomalies are expected in quantum mechanics). The regularization that we choose to present in this chapter is equivalent to a cut-off in the loop momenta. Since the momenta on a compact space are discrete this scheme is called mode regularization.

To get started it is convenient to shift and rescale the time parameter in order to extract the total propagation time β out of the action $S = \frac{1}{\beta}S'$. We do this by defining $t = t_{\rm f} + \beta\tau$ with $\beta = t_{\rm f} - t_{\rm i}$, so that $-1 \le \tau \le 0$. The full rescaled action reads as

$$S' = \int_{-1}^{0} d\tau \left\{ \frac{1}{2} g_{ij}(x)\dot{x}^i\dot{x}^j + i\beta A_i(x)\dot{x}^i + \beta^2[V(x) + V_{MR}(x)] \right\} \quad (3.8)$$

where $\dot{x}^i = dx^i/d\tau$. We have denoted by V_{MR} the counterterm required by mode regularization. Note that $\exp(-\frac{1}{\beta}S')$ is the weight factor for the sum over paths, so that here the total propagation time β plays a role similar to the Planck constant $\hbar$ (which we have set to one) and can be used to count the number of loops. In the loop expansion generated by β the potentials V and V_{MR} start contributing only at two-loops, while A_i starts at one-loop.[1] From now on we drop the prime on S.

For an arbitrary metric $g_{ij}(x)$ one is only able to calculate the path integral in a perturbative expansion in β and in the coordinate displacements $\xi^i \equiv x^i_{\rm i} - x^i_{\rm f}$. Thus we start by parametrizing

$$x^i(\tau) = x^i_{bg}(\tau) + q^i(\tau) \quad (3.9)$$

where $x^i_{bg}(\tau)$ is a background trajectory and $q^i(\tau)$ represents the quantum fluctuations. After choosing a coordinate system in which one carries out the computations, the background trajectory is most conveniently taken to satisfy the free equations of motion in the reference frame chosen. It is a function linear in τ connecting the initial point $x^i_{\rm i}$ to the final point $x^i_{\rm f}$, enforcing the correct boundary conditions

$$x^i_{bg}(\tau) = x^i_{\rm f} - \xi^i\tau, \quad \text{with} \quad \xi^i \equiv x^i_{\rm i} - x^i_{\rm f} \quad (3.10)$$

[1] Reintroducing $\hbar$ one can see that the classical potentials V and A_i are of order $\hbar^0$, while the counterterm V_{MR} will turn out to contribute only at the two-loop level (order $\hbar^2$). Thus, if one uses $\hbar$ to count loops, V appears at the tree level, but if one uses β then V starts contributing at two-loops. In Feynman graphs one might represent V by a cross to indicate that it is a term of order β^2.

where $x_{\rm f} = z$ and $x_{\rm i} = y$ was the notation used in the previous chapter. Note that by free equations of motion we mean those arising from (3.8) by neglecting the potentials $V + V_{MR}$ (which are explicitly of order β^2) and A_i (which is explicitly of order β), and by keeping the constant leading term in the expansion of the metric $g_{ij}(x)$ around the final point $x^i_{\rm f}$ (thus making the space effectively flat). Of course one could have taken any other point to expand about. Also, one could use the exact solution of the classical equations of motion as the background trajectory, but this cannot change the result of the computation. It would just correspond to a different parametrization of the space of paths.

The quantum fields $q^i(\tau)$ in (3.9) should vanish at the time boundaries since the boundary conditions are already included in $x^i_{bg}(\tau)$. Therefore they can be expanded in a sine series. For the ghosts we use the same Fourier expansion. This cannot be justified with the same rigor as for the fields x, but we can give the following arguments. First of all, in (3.3) there are no factors of $(\det g_{ij})^{1/2}$ at the end points; they do not appear because we want to introduce them explicitly later in (3.4) in order for the transition amplitude to be a biscalar. Since the factors $(\det g_{ij})^{1/2}$ were exponentiated with ghosts, we do not need ghosts at the end points, and the way to achieve this is to impose as boundary conditions that they vanish at the end points. Another argument is that with these boundary conditions, the expansion into modes defines a well-defined functional space, at least as well defined as for the x. Of course, any choice of functional space is a priori equally acceptable, since the role of the ghosts is to remove ambiguities in the $\dot{x}\dot{x}$ propagators: one might even prefer to use cosines instead of sines in the expansion of the ghosts, but one would obtain the same answers. To conclude, we expand all ghosts into a series with sines.[2]

Hence

$$\phi^i(\tau) = \sum_{m=1}^{\infty} \phi^i_m \sin(\pi m \tau) \tag{3.11}$$

where ϕ^i denotes all the quantum fields q^i, a^i, b^i, c^i. The functional space of paths is now concretely defined by the space of all Fourier coefficients $\phi^i_m = (q^i_m, a^i_m, b^i_m, c^i_m)$. Similarly, the path-integral measure is properly defined in terms of integration over the Fourier coefficients ϕ^i_m as

[2] A suggestive way to interpret this is by considering a background/quantum split, where the background carries the boundary conditions implied by the classical equation of motion, while the quantum part is required to vanish at the time boundaries so as not to modify the boundary conditions of the background. In our case the classical solutions of the ghost field equations are $a^i = b^i = c^i = 0$.

follows:

$$\mathcal{D}x = \prod_{t_i<t<t_f} \sqrt{\det g_{ij}(x(t))}\, Dx = Dq \int Da\, Db\, Dc\; e^{-\frac{1}{\beta}S_{gh}}$$

$$Dq\, Da\, Db\, Dc = \lim_{M\to\infty} A \prod_{m=1}^{M}\prod_{i=1}^{n} m\, dq_m^i\, da_m^i\, db_m^i\, dc_m^i, \tag{3.12}$$

where A is a constant and the ghosts have been rescaled $(a,b,c) \to \frac{1}{\beta}(a,b,c)$ to normalize the ghost action as $S_{gh} = \int_{-1}^{0} d\tau\, \frac{1}{2} g_{ij}(x)(a^i a^j + b^i c^j)$. Note that we have used $Dx \equiv \prod_\tau d^n x(\tau) = \prod_\tau d^n q(\tau) \equiv Dq$ which is formally justified by the translational invariance of these free measures. In any case the second line in (3.12) defines precisely what we mean by path integration. Note also that with this definition the path-integral for a free particle in Cartesian coordinates reduces to

$$\int \mathcal{D}x\, \exp\Big(-\frac{1}{\beta} S_{free}\Big) = A\, \exp\Big(-\frac{1}{2\beta}\delta_{ij}\xi^i\xi^j\Big) \tag{3.13}$$

where

$$S_{free} = \int_{-1}^{0} d\tau\, \frac{1}{2}\delta_{ij}(\dot{x}^i\dot{x}^j + a^i a^j + b^i c^j). \tag{3.14}$$

(We used the fact that the set $\sqrt{2}\sin(\pi m\tau)$ is orthonormal and that Grassmann integration yields $\int db\, dc\, b\, c = -1$.) It is well known that $A = (2\pi\beta)^{-n/2}$; however, this value can also be deduced from the consistency requirement of satisfying the "renormalization conditions", as will be shown later on. Note that any other constant metric in (3.13) and (3.14), such as for example the choice $g_{ij}(x_f)$ we are going to use, does not change the normalization of the measure in (3.12); the Jacobian for the change of variables of the commuting fields q^i, a^i is exactly canceled by the corresponding Jacobian for the anticommuting fields b^i, c^i (i.e. this linear change of variables has a unit super-Jacobian).

The way to implement mode regularization is now quite clear and already suggested by (3.12): limiting the integration for each field up to a finite mode number M gives a natural regularization of the path-integral. One computes all quantities of interest at finite M. This necessarily gives a finite and unambiguous result. Then one sends $M \to \infty$ to reach the continuum limit. This regularization is enough to resolve all ambiguities in the product of distributions, as we shall see.

We now start to describe in detail the perturbative expansion and give the formulas for the propagators in mode regularization. The perturbative expansion is generated by splitting the action into a quadratic part S_2, which defines the propagators, and an interacting part S_{int}, which gives the vertices.[3] We do this splitting by expanding the action about the final point $x_{\rm f}^i$. Recalling that

$$\begin{aligned} x^i(\tau) &= x_{\rm f}^i - \xi^i \tau + q^i(\tau), \quad (\xi^i \equiv x_{\rm i}^i - x_{\rm f}^i) \\ \dot{x}^i(\tau) &= \dot{q}^i(\tau) - \xi^i \end{aligned} \tag{3.15}$$

we obtain

$$S = S_2 + S_{int} \tag{3.16}$$

where

$$S_2 = \int_{-1}^{0} d\tau \, \frac{1}{2} \, g_{ij}(x_{\rm f}) \, (\xi^i \xi^j + \dot{q}^i \dot{q}^j + a^i a^j + b^i c^j) \tag{3.17}$$

$$\begin{aligned} S_{int} = \int_{-1}^{0} d\tau \bigg\{ & \frac{1}{2} \left[g_{ij}(x) - g_{ij}(x_{\rm f})\right] (\dot{x}^i \dot{x}^j + a^i a^j + b^i c^j) \\ & + i\,\beta A_i(x) \dot{x}^i + \beta^2 [V(x) + V_{MR}(x)] \bigg\}. \end{aligned} \tag{3.18}$$

Note that a term linear in $\dot{q}^i$ also appears in S_2, but due to the boundary conditions on q^i its integral vanishes, and thus has been dropped. Inserting the mode expansions (3.11) into S_2 one obtains

$$S_2 = \frac{1}{2} g_{ij}\,(x_{\rm f}) \xi^i \xi^j + \frac{1}{4} g_{ij}(x_{\rm f}) \sum_{m=1}^{M} (\pi^2 m^2 q_m^i q_m^j + a_m^i a_m^j + b_m^i c_m^j). \tag{3.19}$$

The propagators are easily obtained by using this S_2 in the path-integral, adding sources and completing squares as usual. As an example, let us see in detail the derivation for the mode regulated propagator $\langle q^i(\tau) q^j(\sigma) \rangle$. Using the mode expansion (3.11) we obtain

$$\begin{aligned} \langle q^i(\tau) q^j(\sigma) \rangle &= \left\langle \sum_{m=1}^{M} q_m^i \sin(\pi m \tau) \sum_{n=1}^{M} q_n^j \sin(\pi n \sigma) \right\rangle \\ &= \sum_{m=1}^{M} \sum_{n=1}^{M} \langle q_m^i q_n^j \rangle \sin(\pi m \tau) \sin(\pi n \sigma). \end{aligned} \tag{3.20}$$

Adding sources for the q_m^i modes, completing squares and shifting integration variables produces the correlator

$$\langle q_m^i q_n^j \rangle = \beta g^{ij}(x_{\rm f}) \delta_{mn} \frac{2}{\pi^2 m^2} \tag{3.21}$$

[3] In the previous chapter the quadratic part was called the free part and was denoted by $S^{(0)}$.

which is just the inverse of the quadratic form Q appearing in the exponent $(\exp[-\frac{1}{2}\phi Q \phi])$. Using (3.21) into (3.20) one obtains

$$\langle q^i(\tau)q^j(\sigma)\rangle = \beta g^{ij}(x_{\rm f}) \sum_{m=1}^{M} \frac{2}{\pi^2 m^2} \sin(\pi m\tau)\sin(\pi m\sigma). \tag{3.22}$$

To check the normalization, note that acting with the field operator $-\frac{1}{\beta} g_{ki}(x_{\rm f})\partial^2/\partial\tau^2$ produces the Dirac delta function in the space of functions that vanish at the boundaries, $\delta(\tau,\sigma) = \sum 2\sin(\pi m\tau)\sin(\pi m\sigma)$. Similarly one obtains the ghost propagators. Thus we obtain the following list of propagators:

$$\begin{aligned} \langle q^i(\tau)q^j(\sigma)\rangle &= -\beta g^{ij}(x_{\rm f})\Delta(\tau,\sigma) \\ \langle a^i(\tau)a^j(\sigma)\rangle &= \beta g^{ij}(x_{\rm f})\Delta_{gh}(\tau,\sigma) \\ \langle b^i(\tau)c^j(\sigma)\rangle &= -2\beta g^{ij}(x_{\rm f})\Delta_{gh}(\tau,\sigma) \end{aligned} \tag{3.23}$$

where Δ and Δ_{gh} are regulated by the mode cut-off

$$\Delta(\tau,\sigma) = \sum_{m=1}^{M} \left[-\frac{2}{\pi^2 m^2} \sin(\pi m\tau)\sin(\pi m\sigma) \right] \tag{3.24}$$

$$\Delta_{gh}(\tau,\sigma) = \sum_{m=1}^{M} 2\sin(\pi m\tau)\sin(\pi m\sigma). \tag{3.25}$$

Note that at the regulated level (M big, but fixed) one has the relation $\Delta_{gh}(\tau,\sigma) = {}^{\bullet\bullet}\!\Delta(\tau,\sigma) = \Delta^{\bullet\bullet}(\tau,\sigma)$, where as usual left and right dots indicate derivatives with respect to left and right variables. These functions have the following limiting value for $M \to \infty$:

$$\Delta(\tau,\sigma) \quad \to \quad \tau(\sigma+1)\theta(\tau-\sigma) + \sigma(\tau+1)\theta(\sigma-\tau) \tag{3.26}$$

$$\Delta_{gh}(\tau,\sigma) \quad \to \quad \delta(\tau-\sigma). \tag{3.27}$$

Conversely, the Fourier transform of these relations again yields (3.24) and (3.25).

More generally, in loop computations one also needs the propagators for $\langle \dot q^i(\tau)q^j(\sigma)\rangle$, $\langle q^i(\tau)\dot q^j(\sigma)\rangle$ and $\langle \dot q^i(\tau)\dot q^j(\sigma)\rangle$, so that it is useful to explicitly record the corresponding formulas:

$${}^\bullet\!\Delta(\tau,\sigma) = \sum_{m=1}^{M} \left[-\frac{2}{\pi m} \cos(\pi m\tau)\sin(\pi m\sigma) \right] \tag{3.28}$$

$$\Delta^\bullet(\tau,\sigma) = \sum_{m=1}^{M} \left[-\frac{2}{\pi m} \sin(\pi m\tau)\cos(\pi m\sigma) \right] \tag{3.29}$$

$${}^\bullet\!\Delta^\bullet(\tau,\sigma) = \sum_{m=1}^{M} \left[-2\cos(\pi m\tau)\cos(\pi m\sigma) \right] \tag{3.30}$$

for which the limiting values for $M \to \infty$ can be computed as

$$
\begin{aligned}
{}^\bullet\!\Delta(\tau,\sigma) &\rightarrow \sigma + \theta(\tau-\sigma) &(3.31)\\
\Delta^{\bullet}(\tau,\sigma) &\rightarrow \tau + \theta(\sigma-\tau) &(3.32)\\
{}^\bullet\!\Delta^{\bullet}(\tau,\sigma) &\rightarrow 1 - \delta(\tau-\sigma). &(3.33)
\end{aligned}
$$

In addition, at coinciding times $\sigma = \tau$, one has

$$
\begin{aligned}
\Delta(\tau,\tau) &\rightarrow \tau(\tau+1) &(3.34)\\
{}^\bullet\!\Delta(\tau,\tau) &\rightarrow \tau + \frac{1}{2} &(3.35)\\
\Delta^{\bullet}(\tau,\tau) &\rightarrow \tau + \frac{1}{2}. &(3.36)
\end{aligned}
$$

These limiting values, and in fact all formal expressions in the limit $M \to \infty$, are the same as in time slicing. However, at finite M these regularized propagators have different properties from the propagators which are regularized by time slicing. Consider as an example the expression

$$I = \int_{-1}^{0}\int_{-1}^{0} d\tau\, d\sigma\; {}^\bullet\!\Delta(\tau,\sigma)\, {}^\bullet\!\Delta^{\bullet}(\tau,\sigma)\, \Delta^{\bullet}(\tau,\sigma).$$

With time slicing the result is $I(TS) = -\frac{1}{6}$. However, with mode number regularization one obtains a different answer, $I(MR) = -\frac{1}{12}$. To derive this result we use the fact that at the regulated level boundary terms in partial integration are well defined,

$$
\begin{aligned}
&\int_{-1}^{0}\int_{-1}^{0} d\tau\, d\sigma\; {}^\bullet\!\Delta(\tau,\sigma)\, {}^\bullet\!\Delta^{\bullet}(\tau,\sigma)\, \Delta^{\bullet}(\tau,\sigma)\\
&= \int_{-1}^{0}\int_{-1}^{0} d\tau\, d\sigma\; \frac{1}{2}\partial_\sigma({}^\bullet\!\Delta(\tau,\sigma))^2\, \Delta^{\bullet}(\tau,\sigma).
\end{aligned}
$$

We can partially integrate with ∂_σ without encountering boundary terms because ${}^\bullet\!\Delta(\tau,\sigma)$ vanish at the boundary points $\sigma = 0$ and $\sigma = -1$. We then obtain $-\frac{1}{2}[{}^\bullet\!\Delta(\tau,\sigma)]^2 \Delta^{\bullet\bullet}(\tau,\sigma)$ in the integrand. Next, we may replace $\Delta^{\bullet\bullet}(\tau,\sigma)$ by ${}^{\bullet\bullet}\!\Delta(\tau,\sigma)$ because this relation is clearly satisfied at the regulated level. Finally, we combine $-\frac{1}{2}[{}^\bullet\!\Delta(\tau,\sigma)]^{2\,\bullet\bullet}\!\Delta(\tau,\sigma) = -\frac{1}{6}\partial_\tau[{}^\bullet\!\Delta(\tau,\sigma)]^3$. The integration over τ can be performed, and at this point one may take the continuum limit because the integrand is finite and well behaved. This yields

$$I(MR) = -\frac{1}{6}\int_{-1}^{0} d\sigma\; ({}^\bullet\!\Delta)^3\Big|_{\tau=-1}^{\tau=0} = -\frac{1}{6}\int_{-1}^{0} d\sigma\; [(\sigma+1)^3 - \sigma^3] = -\frac{1}{12}.$$

3.2 The two-loop amplitude and the counterterm V_{MR}

We now compute the transition amplitude at the two-loop level, using mode regularization. We count the coordinate displacement $\xi^i = x_{\rm i}^i - x_{\rm f}^i$ as being of order $\sqrt{\beta}$. More precisely, we evaluate all graphs which contribute to order β; these are not only the two-loop graphs but also one-loop graphs with vertices of order β and tree graphs with vertices of order β^2. We take ξ^i to be of order $\sqrt{\beta}$ because at the end we will use the resulting transition amplitude to evolve wave functions, and a Gaussian integral over the displacements ξ^i will make them effectively of order $\sqrt{\beta}$. Taking this into account we can Taylor expand the interaction potentials in S_{int} given in (3.18) around the final point $x^i_{\rm f}$. We classify the vertices as

$$S_{int} = S_3 + S_4 + \cdots \tag{3.37}$$

with

$$\begin{aligned} S_3 &= \int_{-1}^{0} d\tau \left[\frac{1}{2}\partial_k g_{ij}(q^k - \xi^k\tau)(\xi^i\xi^j - 2\,\xi^i\dot{q}^j + \dot{q}^i\dot{q}^j + a^i a^j + b^i c^j)\right] \\ &\quad - i\beta A_i\xi^i \qquad (3.38) \\ S_4 &= \int_{-1}^{0} d\tau \left[\frac{1}{4}\partial_k\partial_l g_{ij}(q^k q^l + \xi^k\xi^l\tau^2 - 2\,q^k\xi^l\tau) \right. \\ &\quad \times (\xi^i\xi^j - 2\,\xi^i\dot{q}^j + \dot{q}^i\dot{q}^j + a^i a^j + b^i c^j) \\ &\quad \left. + i\beta\partial_j A_i(q^j - \xi^j\tau)(\dot{q}^i - \xi^i)\right] + \beta^2(V + V_{MR}). \qquad (3.39) \end{aligned}$$

In this expansion all geometrical quantities, such as g_{ij} and $\partial_k g_{ij}$, as well as A_i, V, V_{MR} and derivatives thereof, are constants since they are evaluated at the final point $x^i_{\rm f}$, but for notational simplicity we do not exhibit this dependence explicitly, as no confusion can arise. Each term $\frac{1}{\beta}S_n$ contributes effectively as $\beta^{n/2-1}$. For example, S_3 is of order $\beta^{3/2}$ because ξ is of order $\beta^{1/2}$ and each q is also of order $\beta^{1/2}$ because the q propagator is of order β. Similarly for the ghost fields and their propagators. Note also that a term originating from the expansion of the velocity $\dot{x}^i = \dot{q}^i - \xi^i$ in the A_i term of (3.38) integrates to zero and has been canceled. To obtain all corrections to the amplitude of a free particle to order β^2 we need at most the vertex S_4. (Two-loops come from terms with β^{L-1} with $L = 2$.)

Thus, the perturbative expansion reads as

$$\begin{aligned} \langle x^k_{\rm f}, t_{\rm f}|x^k_{\rm i}, t_{\rm i}\rangle &= \int_{\rm BC} \mathcal{D}x \exp\left[-\frac{1}{\beta}S\right] = A\,{\rm e}^{-\frac{1}{2\beta}g_{ij}\xi^i\xi^j}\langle {\rm e}^{-\frac{1}{\beta}S_{int}}\rangle \\ &= A\,{\rm e}^{-\frac{1}{2\beta}g_{ij}\xi^i\xi^j}\left(\left\langle 1 - \frac{1}{\beta}S_3 - \frac{1}{\beta}S_4 + \frac{1}{2\beta^2}S_3^2\right\rangle + O(\beta^{3/2})\right) \\ &= A\,{\rm e}^{-\frac{1}{2\beta}g_{ij}\xi^i\xi^j}\exp\left(-\frac{1}{\beta}\langle S_3\rangle - \frac{1}{\beta}\langle S_4\rangle + \frac{1}{2\beta^2}\langle S_3^2\rangle_c + O(\beta^{3/2})\right) \end{aligned} \tag{3.40}$$

where the brackets $\langle\cdots\rangle$ denote the averaging with the free action S_2, and amount to using the propagators given in (3.23). In fact, we have extracted the coefficient A together with the exponential of the quadratic action S_2 evaluated on the background trajectory so that the normalization of the remaining path-integral is such that $\langle 1 \rangle = 1$. In the last line only connected graphs appear in the exponent; this is indicated by the subscript c where it is needed.

Using standard Wick contractions one obtains

$$\begin{aligned} -\frac{1}{\beta}\langle S_3 \rangle &= -\frac{1}{\beta}\frac{1}{2}\partial_k g_{ij}\left(\beta\xi^k g^{ij}\,\mathbf{I}_1 + 2\beta\xi^i g^{jk}\mathbf{I}_2 + \frac{1}{2}\xi^i\xi^j\xi^k\right) + iA_i\xi^i \\ &= -\frac{1}{4\beta}\partial_k g_{ij}\xi^i\xi^j\xi^k + iA_i\xi^i. \end{aligned} \tag{3.41}$$

On the right-hand side there are terms without quantum fields and terms due to the contraction of two quantum fields. The latter contributions are denoted by $\mathbf{I}_1$ and $\mathbf{I}_2$, and correspond to the Feynman diagrams in (3.44) and (3.45). For example, $\mathbf{I}_1$ is due to $\int_{-1}^{0} d\tau\,\tau\,\langle \dot{q}\dot{q} + aa + bc\rangle$. Similarly,

$$\begin{aligned} -\frac{1}{\beta}\langle S_4 \rangle &= -\frac{1}{\beta}\frac{1}{4}\partial_k\partial_l g_{ij}\Big[\beta^2(g^{ij}g^{kl}\,\mathbf{I}_3 + 2g^{ik}g^{jl}\mathbf{I}_4) \\ &\quad - \beta(g^{ij}\xi^k\xi^l\,\mathbf{I}_5 + g^{kl}\xi^i\xi^j\mathbf{I}_6 + 4g^{jk}\xi^i\xi^l\,\mathbf{I}_7) + \frac{1}{3}\xi^i\xi^j\xi^k\xi^l\Big] \\ &\quad - i\partial_j A_i\left(-\beta g^{ij}\,\mathbf{I}_8 - \frac{1}{2}\xi^i\xi^j\right) - \beta(V + V_{MR}) \\ &= \partial_k\partial_l g_{ij}\Big[\frac{\beta}{24}(g^{ij}g^{kl} - g^{ik}g^{jl}) + \frac{1}{24}(2g^{jk}\xi^i\xi^l - g^{ij}\xi^k\xi^l - g^{kl}\xi^i\xi^j) \\ &\quad - \frac{1}{12\beta}\xi^i\xi^j\xi^k\xi^l\Big] + \frac{i}{2}\partial_j A_i\xi^i\xi^j - \beta(V + V_{MR}). \end{aligned} \tag{3.42}$$

Now in the expression for the connected graphs with two S_3 vertices one does not need terms corresponding to ξ^6 because they could only come from squaring the classical contributions in S_3, and would correspond to disconnected graphs. Thus, we find

$$\begin{aligned} \frac{1}{2\beta^2}\langle S_3^2 \rangle_c = \frac{1}{2\beta^2}\frac{1}{4}\partial_k g_{ij}\partial_l g_{mn}\{&-\beta^3(2g^{kl}g^{im}g^{jn}\,\mathbf{I}_9 + 4g^{km}g^{il}g^{jn}\,\mathbf{I}_{10} \\ &+ g^{kl}g^{ij}g^{mn}\,\mathbf{I}_{11} + 4g^{ki}g^{jl}g^{mn}\,\mathbf{I}_{12} + 4g^{ki}g^{lm}g^{jn}\,\mathbf{I}_{13}) \\ &+ \beta^2[4\xi^i\xi^m(g^{kl}g^{jn}\,\mathbf{I}_{14} + g^{kn}g^{jl}\,\mathbf{I}_{15}) + 2\xi^k\xi^l g^{im}g^{jn}\,\mathbf{I}_{16} \\ &+ 8\xi^k\xi^m g^{il}g^{jn}\,\mathbf{I}_{17} + 2\xi^i\xi^j(g^{kl}g^{mn}\,\mathbf{I}_{18} + 2g^{km}g^{ln}\,\mathbf{I}_{19}) \\ &+ 4\xi^k\xi^i(g^{jl}g^{mn}\,\mathbf{I}_{20} + 2g^{jm}g^{ln}\,\mathbf{I}_{21})] - \beta(\xi^i\xi^j\xi^m\xi^n g^{kl}\,\mathbf{I}_{22} \\ &+ 4\xi^k\xi^i\xi^m\xi^n g^{jl}\,\mathbf{I}_{23} + 4\xi^k\xi^i\xi^l\xi^m g^{jn}\,\mathbf{I}_{24})\} \end{aligned}$$

$$
\begin{aligned}
&= \partial_k g_{ij} \partial_l g_{mn} \Big\{ -\frac{\beta}{96}(6g^{kl}g^{im}g^{jn} - 4g^{km}g^{il}g^{jn} - g^{kl}g^{ij}g^{mn} \\
&\quad + 4g^{ki}g^{jl}g^{mn} - 4g^{ki}g^{lm}g^{jn}) + \frac{1}{48}[2\xi^i\xi^m(g^{kl}g^{jn} - g^{kn}g^{jl}) \\
&\quad + \xi^k\xi^l g^{im}g^{jn} - \xi^i\xi^j(g^{kl}g^{mn} - 2g^{km}g^{ln}) \\
&\quad + 2\xi^k\xi^i(g^{jl}g^{mn} - 2g^{jm}g^{ln})] \\
&\quad + \frac{1}{96\beta}(\xi^i\xi^j\xi^m\xi^n g^{kl} - 4\xi^k\xi^i\xi^m\xi^n g^{jl} + 4\xi^k\xi^i\xi^l\xi^m g^{jn})\Big\}.
\end{aligned}
\tag{3.43}
$$

These results inserted into (3.40) give the transition amplitude at the two-loop approximation. The integrals needed for computing the various Feynman diagrams are evaluated using mode regularization, namely first they are computed at finite M (and thus without ambiguities) and then the $M \to \infty$ limit is taken. We first list them here, and then explain in the next section how the computations in mode regularization are most easily performed:

$\mathbf{I}_1 =$ 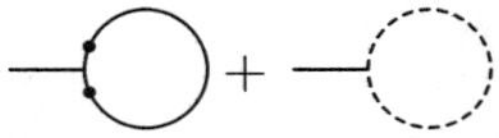 $= \int_{-1}^{0} d\tau\, \tau\, ({}^{\bullet}\Delta^{\bullet} + {}^{\bullet\bullet}\Delta)|_\tau = 0$ (3.44)

$\mathbf{I}_2 =$ 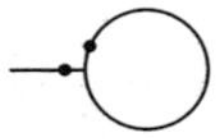 $= \int_{-1}^{0} d\tau\, {}^{\bullet}\Delta|_\tau = 0$ (3.45)

$\mathbf{I}_3 =$ 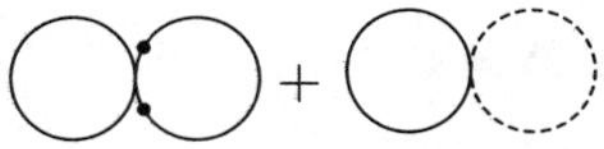 $= \int_{-1}^{0} d\tau\, \Delta|_\tau\, ({}^{\bullet}\Delta^{\bullet} + {}^{\bullet\bullet}\Delta)|_\tau = -\frac{1}{6}$ (3.46)

$\mathbf{I}_4 =$ 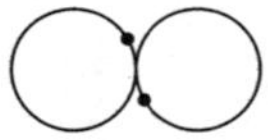 $= \int_{-1}^{0} d\tau\, {}^{\bullet}\Delta^2|_\tau = \frac{1}{12}$ (3.47)

$\mathbf{I}_5 =$ 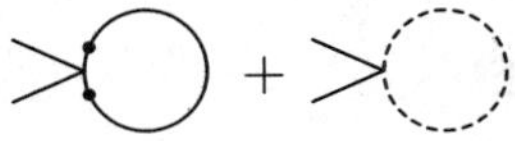 $= \int_{-1}^{0} d\tau\, \tau^2\, ({}^{\bullet}\Delta^{\bullet} + {}^{\bullet\bullet}\Delta)|_\tau = -\frac{1}{6}$ (3.48)

$\mathbf{I}_6 =$ 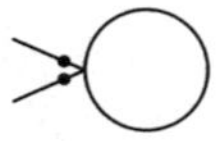 $= \int_{-1}^{0} d\tau\, \Delta|_\tau = -\frac{1}{6}$ (3.49)

$\mathbf{I}_7 =$ 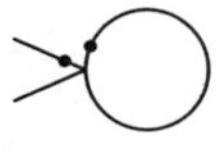 $= \int_{-1}^{0} d\tau\, \tau\, {}^{\bullet}\Delta|_\tau = \frac{1}{12}$ (3.50)

$\mathbf{I}_8 =$  $= \int_{-1}^{0} d\tau\, {}^{\bullet}\Delta|_\tau = 0$ (3.51)

$\mathbf{I}_9 =$ 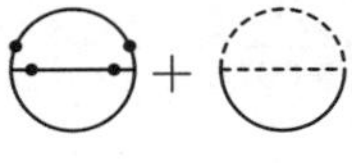 $= \int_{-1}^{0}\int_{-1}^{0} d\tau\, d\sigma\, \Delta\, ({}^{\bullet}\Delta^{\bullet 2} - {}^{\bullet\bullet}\Delta^2) = \frac{1}{4}$ (3.52)

$\mathbf{I}_{10} =$ 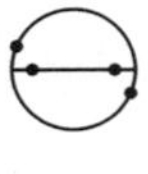 $= \int_{-1}^{0}\int_{-1}^{0} d\tau\, d\sigma\, {}^{\bullet}\Delta\, {}^{\bullet}\Delta^{\bullet}\, \Delta^{\bullet} = -\frac{1}{12}$ (3.53)

$\mathbf{I}_{11} =$ 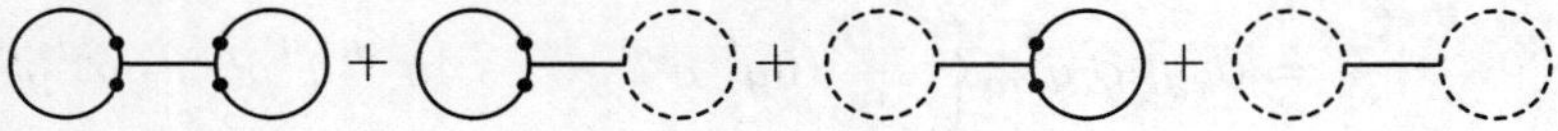

$$= \int_{-1}^{0}\int_{-1}^{0} d\tau\, d\sigma\, ({}^{\bullet}\Delta^{\bullet} + {}^{\bullet\bullet}\Delta)|_{\tau}\, \Delta\, ({}^{\bullet}\Delta^{\bullet} + {}^{\bullet\bullet}\Delta)|_{\sigma} = -\frac{1}{12} \tag{3.54}$$

$\mathbf{I}_{12} =$ 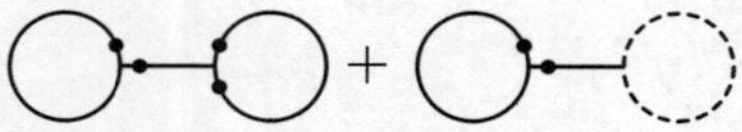

$$= \int_{-1}^{0}\int_{-1}^{0} d\tau\, d\sigma\, {}^{\bullet}\Delta|_{\tau}\, {}^{\bullet}\Delta\, ({}^{\bullet}\Delta^{\bullet} + {}^{\bullet\bullet}\Delta)|_{\sigma} = \frac{1}{12} \tag{3.55}$$

$\mathbf{I}_{13} =$ 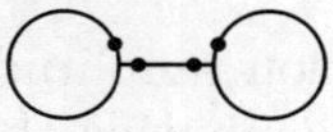 $= \int_{-1}^{0}\int_{-1}^{0} d\tau\, d\sigma\, {}^{\bullet}\Delta|_{\tau}\, {}^{\bullet}\Delta^{\bullet}\, \Delta^{\bullet}|_{\sigma} = -\frac{1}{12}$ (3.56)

$\mathbf{I}_{14} =$ 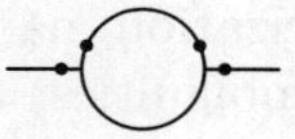 $= \int_{-1}^{0}\int_{-1}^{0} d\tau\, d\sigma\, {}^{\bullet}\Delta^{\bullet}\, \Delta = \frac{1}{12}$ (3.57)

$\mathbf{I}_{15} =$ 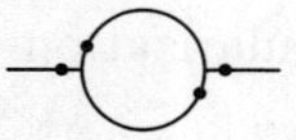 $= \int_{-1}^{0}\int_{-1}^{0} d\tau\, d\sigma\, {}^{\bullet}\Delta\, \Delta^{\bullet} = -\frac{1}{12}$ (3.58)

$\mathbf{I}_{16} =$ 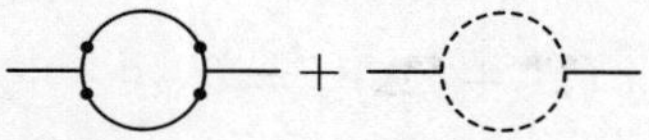

$$= \int_{-1}^{0}\int_{-1}^{0} d\tau\, d\sigma\, \tau\, ({}^{\bullet}\Delta^{\bullet 2} - {}^{\bullet\bullet}\Delta^{2})\, \sigma = \frac{1}{12} \tag{3.59}$$

$\mathbf{I}_{17} =$ 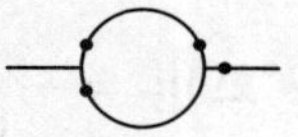 $= \int_{-1}^{0}\int_{-1}^{0} d\tau\, d\sigma\, \tau\, {}^{\bullet}\Delta^{\bullet}\, {}^{\bullet}\Delta = 0$ (3.60)

$\mathbf{I}_{18} =$ 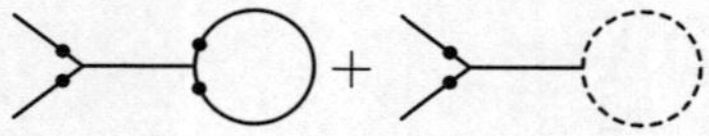

$$= \int_{-1}^{0}\int_{-1}^{0} d\tau\, d\sigma\, \Delta\, ({}^{\bullet}\Delta^{\bullet} + {}^{\bullet\bullet}\Delta)|_{\sigma} = -\frac{1}{12} \tag{3.61}$$

$\mathbf{I}_{19} =$ 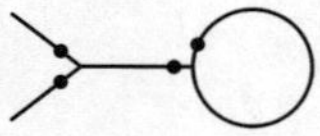 $= \int_{-1}^{0}\int_{-1}^{0} d\tau\, d\sigma\, \Delta^{\bullet}\, \Delta^{\bullet}|_{\sigma} = \frac{1}{12}$ (3.62)

$\mathbf{I}_{20} =$ 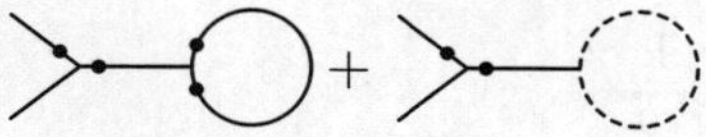

$$= \int_{-1}^{0}\int_{-1}^{0} d\tau\, d\sigma\, \tau\, {}^{\bullet}\Delta\, ({}^{\bullet}\Delta^{\bullet} + {}^{\bullet\bullet}\Delta)|_{\sigma} = \frac{1}{12} \tag{3.63}$$

$\mathbf{I}_{21} =$ 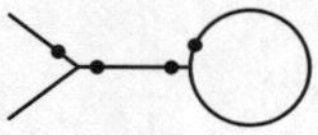 $= \int_{-1}^{0}\int_{-1}^{0} d\tau\, d\sigma\, \tau\, {}^{\bullet}\Delta^{\bullet}\, \Delta^{\bullet}|_{\sigma} = -\frac{1}{12}$ (3.64)

$\mathbf{I}_{22} =$ $= \int_{-1}^{0}\int_{-1}^{0} d\tau\, d\sigma\, \Delta = -\frac{1}{12}$ (3.65)

$\mathbf{I}_{23} =$  $= \int_{-1}^{0}\int_{-1}^{0} d\tau\, d\sigma\, \tau\, {}^{\bullet}\!\Delta = \frac{1}{12}$ (3.66)

$\mathbf{I}_{24} =$  $= \int_{-1}^{0}\int_{-1}^{0} d\tau\, d\sigma\, \tau\, {}^{\bullet}\!\Delta^{\bullet}\, \sigma = -\frac{1}{12}.$ (3.67)

These are the tree, one- and two-loop graphs which contribute to the transition amplitude to order β or less. Dots denote derivatives and the cross on $\mathbf{I}_8$ denotes $\partial_j A_i$. Dotted lines denote ghosts, solid internal lines denote q-propagators and external lines denote factors of ξ. Note how ghost graphs combine with divergent graphs without ghosts to yield finite results. This aspect will be discussed at length in the next section.

Now we come to the task of imposing the "renormalization conditions" which fix the overall normalization of the path-integrals as well as the counterterm V_{MR}. We **require** that the transition amplitude (3.40) should yield the correct time evolution of an arbitrary wave function $\Psi(x,t)$,

$$\Psi(x_\mathrm{f}, t_\mathrm{f}) = \int d^n x_\mathrm{i} \sqrt{g(x_\mathrm{i})} \langle x_\mathrm{f}^i, t_\mathrm{f} | x_\mathrm{i}^i, t_\mathrm{i} \rangle \Psi(x_\mathrm{i}, t_\mathrm{i}) \tag{3.68}$$

and **impose** that $\Psi(x_\mathrm{f}, t_\mathrm{f})$ solves the Schrödinger equation with an a priori given Hamiltonian. We choose as the Hamiltonian the one with the covariant Laplacian $\nabla_A^2 = g^{ij}(\nabla_i + A_i)(\partial_j + A_j)$ and without any coupling to the scalar curvature

$$H = -\frac{1}{2}\nabla_A^2 + V. \tag{3.69}$$

This Hamiltonian can arise as a possible quantization of the classical model in (3.1) and thus is a consistent requirement. It is the x-space representation of the abstract operator in (2.1) with A_i and V terms added.

Since the transition amplitude is given in terms of an expansion around the final point $(x_\mathrm{f}, t_\mathrm{f})$, we Taylor expand the wave function $\Psi(x_\mathrm{i}, t_\mathrm{i})$ and the measure $\sqrt{g(x_\mathrm{i})}$ in (3.68) about that point, perform the integration over $d^n x_\mathrm{i}$ and match the various terms. Thus, we insert

$$\begin{aligned}
\Psi(x_\mathrm{i}, t_\mathrm{i}) &= \Psi(x_\mathrm{f}, t_\mathrm{f}) - \beta\partial_t \Psi(x_\mathrm{f}, t_\mathrm{f}) + \xi^i \partial_i \Psi(x_\mathrm{f}, t_\mathrm{f}) \\
&\quad + \frac{1}{2}\xi^i\xi^j \partial_i\partial_j \Psi(x_\mathrm{f}, t_\mathrm{f}) + O(\beta^{3/2}) \\
\sqrt{g(x_\mathrm{i})} &= \sqrt{g(x_\mathrm{f})}\left[1 + \xi^i \Gamma_{ik}^k + \frac{1}{2}\xi^i\xi^j(\partial_i \Gamma_{jk}^k + \Gamma_{ik}^k \Gamma_{jl}^l) + O(\xi^3)\right]\Bigg|_{x_\mathrm{f}}
\end{aligned} \tag{3.70}$$

as well as (3.40) into (3.68). In the last expansion we have used the fact that $\frac{1}{\sqrt{g}}\partial_i\sqrt{g} = \frac{1}{2}g^{mn}\partial_i g_{mn} = \Gamma_{ik}^k$. All quantities are now evaluated at the point $(x_{\rm f}, t_{\rm f})$. For notational simplicity we do not indicate this dependence from now on, as no confusion can arise. The integrals over $d^n x_{\rm i} = d^n \xi$ give Gaussian averages since the transition amplitude (3.40) contains the exponential factor $e^{-\frac{1}{2\beta} g_{ij}\xi^i\xi^j}$. These averages are easily carried out using "Wick contractions" with the basic "propagator" $\langle \xi^i \xi^j \rangle = \beta g^{ij}$. This also explains why we counted $\xi^i \sim \sqrt{\beta}$ in the expansion of the wave functions in (3.70).

From the various terms in the expansion of (3.68) we find the following. The leading term (order β^0) fixes A

$$\Psi = A(2\pi\beta)^{n/2}\Psi \quad \rightarrow \quad A = (2\pi\beta)^{-n/2}. \tag{3.71}$$

This yields the Feynman measure, as expected.

The terms of order β involve the counterterm V_{MR}. We fix it by requiring that (3.68) yields the prescribed Schrödinger equation for Ψ. At order β one finds

$$\beta\left[-\partial_t\Psi + \frac{1}{2}\nabla_A^2\Psi - \left(V + V_{MR} - \frac{1}{8}R + \frac{1}{24}g^{ij}g^{mn}g_{kl}\Gamma^k_{im}\Gamma^l_{jn}\right)\Psi\right] = 0. \tag{3.72}$$

For example, $V + V_{MR}$ comes from (3.42) and the A^2 term in $\frac{1}{2}\nabla_A^2\Psi$ comes from expanding $\exp(-\frac{1}{\beta}\langle S_3\rangle)$ in (3.40). Thus, fixing

$$V_{MR} = \frac{1}{8}R - \frac{1}{24}g^{ij}g^{mn}g_{kl}\Gamma^k_{im}\Gamma^l_{jn} \tag{3.73}$$

gives the correct Schrödinger equation with the Hamiltonian in (3.69).

Higher-order terms in β yield equations which must be automatically satisfied, since we have completely fixed all the "free" parameters entering mode regularization. This can be explicitly checked. A related check is obtained by applying MR to evaluate trace anomalies in four dimensions (a three-loop calculation). This produces the correct results, see Chapter 7.

We see here a difference with the TS method: in TS we first determined the counterterm from Weyl ordering, and then we performed loop calculations. In MR we needed first to perform loop calculations to order β to fix the counterterm, but then one can go ahead and perform further loop calculations without any ambiguity.

To summarize, we have described the mode regularization scheme for computing the path-integral and have derived the corresponding counterterm V_{MR}. With precisely this counterterm the path-integrals will produce a solution of the Schrödinger equation with Hamiltonian $H = -\frac{1}{2}\nabla_A^2 + V$.

Any Hamiltonian for the Schrödinger equation can always be cast in the form (3.69) with suitable A_i and V. In particular, the mode-regulated path-integral with V_{MR} gives a general coordinate-invariant results for the transition element. We stress that given an arbitrary but fixed Hamiltonian $\hat{H}$ we obtain always the same V_{MR} in the action for the path-integral, but of course the action which corresponds to $\hat{H}$ will look different for different $\hat{H}$. The total action for the path-integral is the sum of:

(i) the sigma model action in (3.1);

(ii) the counterterm V_{MR} ((i) + (ii) now produce the covariant Hamiltonian in (3.69)); and

(iii) the extra terms present when the Hamiltonian is noncovariant (or contains an additional coupling to the scalar curvature); these are given by the extra terms in V and A_i by which the noncovariant Hamiltonian differs from the covariant Hamiltonian.

Thus, the mode regularization method can handle any Hamiltonian which is at most quadratic in the momenta.

3.3 Calculation of Feynman graphs in mode regularization

In this section we analyze in detail mode regularization, and explain how to efficiently evaluate Feynman diagrams.

First of all, all possible divergences are canceled by the ghost contributions. This is seen in diagrams like those in (3.44) or (3.46). Let us consider, for example, the case of I_3 in (3.46)

$I_3 =$ 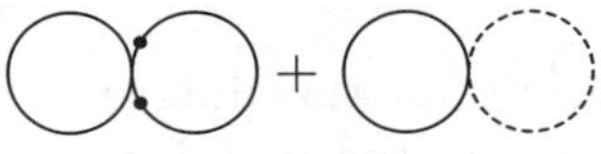 $= \int_{-1}^{0} d\tau\, \Delta|_\tau \, ({}^{\bullet}\Delta^{\bullet} + {}^{\bullet\bullet}\Delta)|_\tau$ (3.74)

where we must insert for each Δ on the right-hand side the discretized propagator as given in (3.24). At finite M each of the two diagrams produces a finite result since each one corresponds to a finite sum of finite integrals. However, only the sum of the two diagrams has a finite limit for $M \to \infty$. To compute this final value one can evaluate both terms at finite M, combine them, and then take the limit $M \to \infty$. This way of proceeding, though correct, is extremely laborious.

An easier way to proceed is to use partial integration to cast the integral into a form which can be computed directly and without ambiguities by taking the $M \to \infty$ limit inside the integral. Along the way one may use simple identities valid at the regulated level, such as the following one:

$$
\begin{aligned}
({}^\bullet\!\Delta^\bullet + {}^{\bullet\bullet}\!\Delta)|_\tau &= \sum_{m=1}^{M} \Big[-2\cos^2(\pi m\tau) + 2\sin^2(\pi m\tau) \Big] \\
&= \partial_\tau \sum_{m=1}^{M} \Big[-\frac{2}{\pi m} \sin(\pi m\tau)\cos(\pi m\tau) \Big] \\
&= \partial_\tau ({}^\bullet\!\Delta|_\tau) .
\end{aligned}
\tag{3.75}
$$

Thus, we compute

$$
\begin{aligned}
I_3 &= \int_{-1}^{0} d\tau\, \Delta|_\tau\, ({}^\bullet\!\Delta^\bullet + {}^{\bullet\bullet}\!\Delta)|_\tau = \int_{-1}^{0} d\tau\, \Delta|_\tau\, \partial_\tau({}^\bullet\!\Delta|_\tau) \\
&= -\int_{-1}^{0} d\tau\, \partial_\tau(\Delta|_\tau)\, {}^\bullet\!\Delta|_\tau .
\end{aligned}
\tag{3.76}
$$

In the partial integration no boundary terms are picked up since both $\Delta|_\tau$ and ${}^\bullet\!\Delta|_\tau$ vanish at $\tau = -1, 0$. In fact, notice that at the regulated level ${}^\bullet\!\Delta|_\tau$ always vanishes at those boundaries, even though its limit for $M \to \infty$ is discontinuous at those points, see (3.35) (continued along the whole line $-\infty < \tau < \infty$, the function ${}^\bullet\!\Delta|_\tau$ limits to the periodic triangular "saw-tooth"). Finally, the last integral in (3.76) can be computed directly in the continuum limit, since only step functions and no delta functions arise in single derivatives acting on Δ. Thus for $M \to \infty$ one can use the limits (3.34)–(3.36) directly inside the integral to obtain

$$
I_3 = -\int_{-1}^{0} d\tau\, [\partial_\tau(\tau^2 + \tau)] \left(\tau + \frac{1}{2}\right) = -2 \int_{-1}^{0} d\tau \left(\tau + \frac{1}{2}\right)^2 = -\frac{1}{6} .
\tag{3.77}
$$

Next, let us discuss the computations of the diagrams in (3.52) and (3.53), the values of which differ in all three different regularization schemes discussed in this book. First, we look at

$I_9 =$ 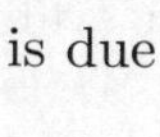 $+$ 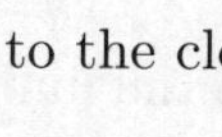 $= \int_{-1}^{0}\int_{-1}^{0} d\tau\, d\sigma\, \Delta\, ({}^\bullet\!\Delta^{\bullet 2} - {}^{\bullet\bullet}\!\Delta^2) .$ (3.78)

The minus sign is due to the closed ghost loop. Using partial integration we compute

$$
\begin{aligned}
I_9 &= \iint \Delta\, ({}^\bullet\!\Delta^\bullet\, {}^\bullet\!\Delta^\bullet - {}^{\bullet\bullet}\!\Delta\, {}^{\bullet\bullet}\!\Delta) \\
&= \iint (-{}^\bullet\!\Delta\, \Delta^\bullet\, {}^\bullet\!\Delta^\bullet - \Delta\, \Delta^\bullet\, {}^{\bullet\bullet}\!\Delta^\bullet + {}^\bullet\!\Delta\, {}^\bullet\!\Delta\, {}^{\bullet\bullet}\!\Delta + \Delta\, {}^\bullet\!\Delta\, {}^{\bullet\bullet\bullet}\!\Delta) .
\end{aligned}
\tag{3.79}
$$

There are no boundary contributions because Δ vanishes at the boundaries. Now we notice that the second and fourth term cancel because at the regulated level we can exchange two left derivatives with two right ones (i.e. ${}^{\bullet\bullet}\!\Delta = \Delta^{\bullet\bullet}$, as seen by inspecting (3.24)). Once again we see how the ghosts cancel a potential divergence. The first term in (3.79) equals $-I_{10}$ while the remaining third term gives

$$\int\!\!\int {}^{\bullet}\!\Delta\, {}^{\bullet}\!\Delta\, {}^{\bullet\bullet}\!\Delta = \int\!\!\int {}^{\bullet}\!\Delta\, {}^{\bullet}\!\Delta\, \Delta^{\bullet\bullet} = -2\int\!\!\int {}^{\bullet}\!\Delta^{\bullet}\, {}^{\bullet}\!\Delta\, \Delta^{\bullet} = -2I_{10}. \tag{3.80}$$

The boundary terms cancel because ${}^{\bullet}\!\Delta(\tau,\sigma)$ vanish at $\sigma = 0, -1$. Thus $I_9 = -3I_{10}$. So let us look at

$I_{10} =$ 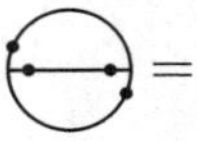

$$= \int_{-1}^{0}\int_{-1}^{0} d\tau\, d\sigma\, {}^{\bullet}\!\Delta\, {}^{\bullet}\!\Delta^{\bullet}\, \Delta^{\bullet}. \tag{3.81}$$

By using partial integration we obtain

$$\begin{aligned} I_{10} &= \int\!\!\int {}^{\bullet}\!\Delta\, {}^{\bullet}\!\Delta^{\bullet}\, \Delta^{\bullet} = \frac{1}{2}\int\!\!\int \Delta^{\bullet}\, ({}^{\bullet}\!\Delta^2)^{\bullet} = -\frac{1}{2}\int\!\!\int \Delta^{\bullet\bullet}\, ({}^{\bullet}\!\Delta^2) \\ &= -\frac{1}{2}\int\!\!\int {}^{\bullet\bullet}\!\Delta\, {}^{\bullet}\!\Delta^2 = -\frac{1}{6}\int_{-1}^{0}\int_{-1}^{0} d\tau\, d\sigma\, {}^{\bullet}({}^{\bullet}\!\Delta^3) = -\frac{1}{6}\int_{-1}^{0} d\sigma\, {}^{\bullet}\!\Delta^3\Big|_{\tau=-1}^{\tau=0} \\ &= -\frac{1}{6}\int_{-1}^{0} d\sigma\, [(\sigma+1)^3 - \sigma^3] = -\frac{1}{12}. \end{aligned} \tag{3.82}$$

Again we first used ${}^{\bullet\bullet}\!\Delta = \Delta^{\bullet\bullet}$ and then used the fact that ${}^{\bullet}\!\Delta(0,\sigma) = \sigma + 1$ and ${}^{\bullet}\!\Delta(-1,\sigma) = \sigma$, see (3.31). In this last step one should be careful in checking that the discretized functions really limit one to the above values (up to sets of points of zero measure). Indeed, one can verify that

$${}^{\bullet}\!\Delta(0,\sigma) = \sum_{m=1}^{M}\left[-\frac{2}{\pi m}\sin(\pi m\sigma)\right] \quad\to\quad \sigma + 1 \tag{3.83}$$

$${}^{\bullet}\!\Delta(-1,\sigma) = \sum_{m=1}^{M}\left[-\frac{2}{\pi m}(-1)^m\sin(\pi m\sigma)\right] \quad\to\quad \sigma. \tag{3.84}$$

Thus, we obtained $I_{10} = -\frac{1}{12}$ and $I_9 = \frac{1}{4}$.

To summarize, in computing mode-regulated integrals it is convenient to use partial integration together with the following identities valid at finite M:

$${}^{\bullet\bullet}\!\Delta(\tau,\sigma) = \Delta^{\bullet\bullet}(\tau,\sigma) \tag{3.85}$$

$${}^{\bullet}\!\Delta^{\bullet}(\tau,\tau) + {}^{\bullet\bullet}\!\Delta(\tau,\tau) = \partial_\tau({}^{\bullet}\!\Delta(\tau,\tau)) \tag{3.86}$$

$$\partial_\tau(\Delta(\tau,\tau)) = 2\Delta^{\bullet}(\tau,\tau) \tag{3.87}$$

$$\Delta^{\bullet}(\tau,\tau) = 0 \quad \text{at } \tau = -1, 0 \tag{3.88}$$

and the following limits for $M \to \infty$:

$$\Delta(\tau,\sigma) \;\to\; \tau(\sigma+1)\theta(\tau-\sigma)+\sigma(\tau+1)\theta(\sigma-\tau) \tag{3.89}$$

$$ {}^{\bullet}\!\Delta(\tau,\sigma) \;\to\; \sigma+\theta(\tau-\sigma) \tag{3.90}$$

$$\Delta^{\bullet}(\tau,\sigma) \;\to\; \tau+\theta(\sigma-\tau) \tag{3.91}$$

$$\Delta(\tau,\tau) \;\to\; \tau^2+\tau \tag{3.92}$$

$$\Delta^{\bullet}(\tau,\tau) = {}^{\bullet}\!\Delta(\tau,\tau) \;\to\; \tau+\frac{1}{2}. \tag{3.93}$$

4
Dimensional regularization

In this chapter we discuss path integrals defined by dimensional regularization (DR). In contrast to the previous time slicing (TS) and mode regularization (MR) schemes, this type of regularization seems to have no meaning outside perturbation theory. However, it leads to the simplest set up for perturbative calculations. In fact, the associated counterterm V_{DR} turns out to be covariant, and the additional vertices obtained by expanding V_{DR}, needed at higher loops, can be obtained with relative ease (using, for example, Riemann normal coordinates).

Dimensional regularization is based on the analytic continuation in the number of dimensions of the momentum integrals corresponding to Feynman graphs ($1 \to D+1$ with arbitrary complex D, in our case). At complex D we assume that the regularization of ultraviolet (UV) divergences is achieved by the analytic continuation as usual. The limit $D \to 0$ is taken at the end. Again one does not expect divergences to arise in quantum mechanics when the regulator is removed ($D \to 0$), and thus no infinite counterterms are necessary to renormalize the theory: potential divergences are canceled by the ghosts.

To derive the dimensional regularization scheme, one can employ a set up quite similar to the one described in the previous chapter for mode regularization. The only difference will be the prescriptions of how to regulate ambiguous diagrams.

One novelty of the dimensional regularization described in this chapter is that it addresses UV regularization on a compact space, namely on a one-dimensional segment corresponding to the finite time $\beta = t_{\mathrm{f}} - t_{\mathrm{i}}$. On such a space there cannot be infrared (IR) divergences, so that occasional mixing between IR and UV divergences (which sometimes occurs in infinite space) does not arise. On a compact space the momenta are discrete, and the Feynman graphs contain discrete sums $\sum_{k_n} [\cdots]$ rather then continuous integrals $\int dk\, [\cdots]$. The latter are easily extended to arbitrary D

and computed, but the former are more difficult to treat. In general, we have not been able to compute explicitly the combined sum and integrals in complex $D+1$ dimensions, and test if poles arise only at some integer value of D. However, assuming that to be the case, we will show how one can compute the regulated graphs directly at $D\to 0$ and with relative ease.

Dimensional regularization for bosonic nonlinear sigma models with a finite propagation time and with the correct counterterm V_{DR} which we present in the following section was developed in [85]. It was extended to fermions and to supersymmetric nonlinear sigma models in [29]. In the infinite propagation time limit, dimensional regularization was previously employed in [11] and the corresponding covariant counterterm was identified in [86]. An extended use of DR for computing trace anomalies in six dimensions is described in [87, 88]. Moreover, DR has been employed in [28–30, 89] to describe quantum field theories in a gravitational background within the worldline formalism. Additional discussions have been presented in [90, 91].

For pedagogical purposes it may be useful to first read the chapter on mode regularization, but the expert reader interested in learning the DR scheme directly can start here.

4.1 Dimensional regularization in configuration space

We start from the classical action in Euclidean time for the fields x^i with $i=1,\ldots,n$

$$S=\int_{t_{\rm i}}^{t_{\rm f}} dt\left[\frac{1}{2}g_{ij}(x)\frac{dx^i}{dt}\frac{dx^j}{dt}+iA_i(x)\frac{dx^i}{dt}+V(x)\right] \tag{4.1}$$

and aim to quantize the theory by defining the transition amplitude directly as a path integral

$$\langle x_{\rm f}^k,t_{\rm f}|x_{\rm i}^k,t_{\rm i}\rangle=\int_{\rm BC}\mathcal{D}x\,{\rm e}^{-S} \tag{4.2}$$

$$\mathcal{D}x=\prod_{t_{\rm i}<t<t_{\rm f}}\sqrt{\det g_{ij}(x(t))}\,d^nx(t) \tag{4.3}$$

where BC indicates the boundary conditions at the initial and final times, $x^k(t_{\rm i})=x_{\rm i}^k$ and $x^k(t_{\rm f})=x_{\rm f}^k$. Since this quantum theory is superrenormalizable, we first proceed formally and derive the Feynman graphs, then we introduce the dimensional regularization procedure to give a meaning to the ambiguous integrals and compute them, and finally we calculate the transition amplitude at two-loops. Imposing the same "renormalization conditions" as used in the TS and MR schemes will determine the counterterm V_{DR} and the overall normalization of the path integral.

The measure $\mathcal{D}x$ in (4.3) is formally a scalar under general coordinate transformations, but the factor $\prod_t \sqrt{\det g_{ij}(x(t))}$ is field-dependent and makes the measure unsuitable for generating the perturbative expansion. Thus, it is useful to introduce ghost fields a^i, b^i, c^i (with a^i commuting and b^i, c^i anticommuting) to exponentiate this field-dependent factor. No boundary conditions should be imposed on the path integral for the ghosts as they are auxiliary algebraic fields (by this we mean that no initial or final point value for these fields can be specified in the transition amplitude (4.2)).

It is also convenient to shift and rescale the time parameter t, so that the total propagation time β can be extracted from the action $S \to \frac{1}{\beta}S$. Defining $t = t_{\rm f} + \beta\tau$ with $\beta = t_{\rm f} - t_{\rm i}$, so that $-1 \le \tau \le 0$, and rescaling the ghost fields suitably one obtains the following complete action:

$$S = \int_{-1}^{0} d\tau \bigg\{ \frac{1}{2} g_{ij}(x)(\dot{x}^i\dot{x}^j + a^i a^j + b^i c^j) \\ + i\beta A_i(x)\dot{x}^i + \beta^2[V(x) + V_{DR}(x)] \bigg\} \tag{4.4}$$

where $\dot{x}^i = dx^i/d\tau$. We have denoted by V_{DR} the counterterm which is needed for dimensional regularization. Note that since $\exp(-\frac{1}{\beta}S)$ is the weight factor for the sum over paths, the time β plays a role analogous to the Planck constant $\hbar$ (which is set to one in this chapter) and can be used to count the number of loops.

For an arbitrary metric $g_{ij}(x)$ one can calculate the path integral in a perturbative expansion in β and the coordinate displacements $\xi^i \equiv x^i_{\rm i} - x^i_{\rm f}$. We perform a background/quantum split and parametrize

$$x^i(\tau) = x^i_{bg}(\tau) + q^i(\tau) \tag{4.5}$$

where $x^i_{bg}(\tau)$ is a background trajectory and $q^i(\tau)$ denotes the quantum fluctuations. After choosing the coordinate system to be employed for carrying out the computations, the background trajectory is taken to satisfy the free equations of motion $g_{ij}(x_{\rm f})\partial^2_\tau x^j(\tau) = 0$. It incorporates the correct boundary conditions

$$x^i_{bg}(\tau) = x^i_{\rm f} - \xi^i\tau, \quad \text{with } \xi^i \equiv x^i_{\rm i} - x^i_{\rm f} \;. \tag{4.6}$$

Note that by free equations of motion we mean the ones arising from (4.4) by neglecting the potentials $V + V_{DR}$ (which are explicitly of order β^2) and A_i (which is explicitly of order β), and by keeping the constant leading term in the expansion of the metric $g_{ij}(x)$ around the final point $x^i_{\rm f}$ (thus making the space effectively flat). Of course one could as well take any other point to linearize the metric. For the ghost fields one can also perform a background/quantum split. However, the background ghost fields vanish as a consequence of their algebraic equation of motion.

The quantum fields are all taken to vanish at the time boundaries since the boundary conditions are already included in the background configurations. Therefore they can be expanded in a Fourier sine series

$$\phi^i(\tau) = \sum_{m=1}^{\infty} \phi^i_m \sin(\pi m \tau) \tag{4.7}$$

where ϕ^i denotes all the quantum fields q^i, a^i, b^i, c^i. The functional space of paths is now defined as the space of all Fourier coefficients $\phi^i_m = (q^i_m, a^i_m, b^i_m, c^i_m)$. Similarly, the path integral measure is defined in terms of integration over the Fourier coefficients ϕ^i_m. Thus, we obtain the following path integral:

$$\langle x^k_{\rm f}, t_{\rm f} | x^k_{\rm i}, t_{\rm i} \rangle = \int_{\rm BC} Dq\, Da\, Db\, Dc\ {\rm e}^{-\frac{1}{\beta} S} \tag{4.8}$$

$$S = \int_{-1}^{0} d\tau \Big\{ \frac{1}{2} g_{ij}(x)(\dot{x}^i \dot{x}^j + a^i a^j + b^i c^j) + i\beta A_i(x)\dot{x}^i + \beta^2 [V(x) + V_{DR}(x)] \Big\} \tag{4.9}$$

$$x^i(\tau) = x^i_{bg}(\tau) + q^i(\tau) \tag{4.10}$$

$$Dq\, Da\, Db\, Dc = A \prod_{m=1}^{\infty} \prod_{i=1}^{n} m\, dq^i_m\, da^i_m\, db^i_m\, dc^i_m, \tag{4.11}$$

where A is a constant which will fixed later on (we will find again the Feynman measure $A = (2\pi\beta)^{-n/2}$).

The perturbative expansion is generated by splitting the action into a quadratic part S_2 which defines the propagators, and an interacting part S_{int} which gives the vertices. If the theory is free and S_{int} vanishes, there would not be any real reason to introduce a regularization. However, when the theory is interacting with a nontrivial field-dependent metric, one must regulate the ambiguous Feynman graphs. In dimensional regularization these graphs are extended to $D+1$ dimensions. To recognize how to perform this extension uniquely in each Feynman graph, we introduce D extra infinite regulating dimensions $\mathbf{t} = (t^1, \ldots, t^D)$ and extend the action directly. After having obtained from this action the corresponding Feynman diagrams, and in principle computed them at arbitrary D, one takes the limit $D \to 0$. Introducing $t^\mu \equiv (\tau, \mathbf{t})$ with $\mu = 0, 1, \ldots, D$ and

$d^{D+1}t = d\tau\, d^D\mathbf{t}$, the action in $D+1$ dimensions reads

$$S = \int_\Omega d^{D+1}t \left[\frac{1}{2}\, g_{ij} \left(\partial_\mu x^i \partial_\mu x^j + a^i a^j + b^i c^j\right) + i\beta A_i \partial_0 x^i + \beta^2 (V + V_{DR})\right] \tag{4.12}$$

where $\Omega = I \times R^D$ is the region of integration containing the finite interval $I = [-1, 0]$. Note that the contraction of the indices μ in the term quadratic in derivatives tells us how momenta become contracted in higher dimensions. Note also that the coupling to the abelian gauge field A_i is not modified in higher dimensions ($\partial_0 x^i \equiv \partial_\tau x^i = \dot{x}^i$). In addition, in $D+1$ dimensions the background solution (4.6) is left unchanged, so that the split $S = S_2 + S_{int}$ is given by

$$S_2 = \frac{1}{2} g_{ij}\,(x_{\rm f})\xi^i \xi^j + \int_\Omega d^{D+1}t\ \frac{1}{2}\, g_{ij}(x_{\rm f})\,(\partial_\mu q^i \partial_\mu q^j + a^i a^j + b^i c^j) \tag{4.13}$$

$$S_{int} = \int_\Omega d^{D+1}t \left\{\frac{1}{2}\,[g_{ij}(x) - g_{ij}(x_{\rm f})]\,(\partial_\mu x^i \partial_\mu x^j + a^i a^j + b^i c^j) + i\beta A_i(x)\partial_0 x^i + \beta^2 [V(x) + V_{MR}(x)]\right\}. \tag{4.14}$$

A term linear in $\partial_0 q^i$ appearing in S_2 integrates to zero and thus has been dropped.

The regulated propagators are given by

$$\langle q^i(t) q^j(s)\rangle = -\beta g^{ij}(x_f)\Delta(t,s) \tag{4.15}$$

$$\langle a^i(t) a^j(s)\rangle = \beta g^{ij}(x_f)\Delta_{gh}(t,s) \tag{4.16}$$

$$\langle b^i(t) c^j(s)\rangle = -2\beta g^{ij}(x_f)\Delta_{gh}(t,s) \tag{4.17}$$

where

$$\Delta(t,s) = \int \frac{d^D\mathbf{k}}{(2\pi)^D} \sum_{m=1}^\infty \frac{-2}{(\pi m)^2 + \mathbf{k}^2} \sin(\pi m\tau)\sin(\pi m\sigma)\, \mathrm{e}^{i\mathbf{k}\cdot(\mathbf{t}-\mathbf{s})} \tag{4.18}$$

$$\Delta_{gh}(t,s) = \int \frac{d^D\mathbf{k}}{(2\pi)^D} \sum_{m=1}^\infty 2\sin(\pi m\tau)\sin(\pi m\sigma)\, \mathrm{e}^{i\mathbf{k}\cdot(\mathbf{t}-\mathbf{s})} = \delta(\tau,\sigma)\,\delta^D(\mathbf{t}-\mathbf{s}) = \delta^{D+1}(t,s). \tag{4.19}$$

Here

$$\delta(\tau,\sigma) = \sum_{m=1}^\infty 2\sin(\pi m\tau)\sin(\pi m\sigma) \tag{4.20}$$

is the Dirac delta function on the space of functions vanishing at $\tau, \sigma = -1, 0$. Note that the function $\Delta(t,s)$ satisfies the relation (Green equation)

$$\partial_\mu^2 \Delta(t,s) = \Delta_{gh}(t,s) = \delta^{D+1}(s,t). \tag{4.21}$$

Formally, the $D \to 0$ limits of these propagators are the usual ones,

$$\Delta(\tau,\sigma) = \tau(\sigma+1)\theta(\tau-\sigma) + \sigma(\tau+1)\theta(\sigma-\tau) \tag{4.22}$$

$$\Delta_{gh}(\tau,\sigma) = {}^{\bullet\bullet}\Delta(\tau,\sigma) = \delta(\tau,\sigma) \tag{4.23}$$

where dots on the left-/right-hand side of $\Delta(\tau,\sigma)$ denote derivatives with respect to the first/second variable, respectively. However, such limits can be used only after one has defined the integrands in an unambiguous form by making use of the manipulations allowed by the regularization scheme. It is difficult to compute the integrals in Feynman graphs for arbitrary D. However, this is not strictly necessary. We can use various manipulations which are identities at the regulated level (and thus can be safely performed) to cast the integrals in alternative forms. In this way one tries to reach a form which can be unambiguously computed by removing the regulator $D \to 0$. This is the same strategy we used in MR to compute quickly the various regulated integrals.

In particular, in DR one can often use partial integration without the need to include boundary terms: this is always allowed in the extra D dimensions because of momentum conservation, while it can be achieved along the direction of the original finite time interval whenever there is an explicit function vanishing at the boundaries $\tau = -1, 0$ (for example, the propagator of the coordinates $\Delta(t,s)$). Along the way one may find terms of the form $\partial_\mu^2 \Delta(t,s)$ which according to (4.21) give Dirac delta functions. The latter can be safely used only at the regulated level, i.e. in $D+1$ dimensions. By performing such partial integrations one tries to arrive at forms of the integrals which are unambiguous even in the limit $D \to 0$. At this point they can be safely and easily calculated in such a limit.

We will give more details on how to compute integrals in DR in Section 4.3. For the moment an explicit example will suffice to describe how the above rules are used concretely:

$$\begin{aligned}
I_{10} &= \int_{-1}^{0} d\tau \int_{-1}^{0} d\sigma \ ({}^\bullet\Delta)\,(\Delta^\bullet)\,({}^\bullet\Delta^\bullet) \\
&\to \int d^{D+1}t \int d^{D+1}s \ ({}_\mu\Delta)\,(\Delta_\nu)\,({}_\mu\Delta_\nu) \\
&= \int d^{D+1}t \int d^{D+1}s \ ({}_\mu\Delta)\ {}_\mu\!\left[\frac{1}{2}(\Delta_\nu)^2\right] \\
&= -\frac{1}{2}\int d^{D+1}t \int d^{D+1}s \ ({}_{\mu\mu}\Delta)\,(\Delta_\nu)^2
\end{aligned}$$

$$= -\frac{1}{2}\int d^{D+1}t \int d^{D+1}s\ \delta^{D+1}(t,s)\ (\Delta_\nu)^2$$
$$= -\frac{1}{2}\int d^{D+1}t\ (\Delta_\nu)^2|_t$$
$$\to -\frac{1}{2}\int_{-1}^{0} d\tau ({}^\bullet\!\Delta^\bullet)^2|_\tau = -\frac{1}{24} \tag{4.24}$$

where the symbol $|_\tau$ means that one should set $\sigma = \tau$. We have introduced the notation ${}_\mu\Delta$ and Δ_μ to indicate derivatives with respect to the first or second variable. Thus $I_{10}(\mathrm{DR}) = -\frac{1}{24}$. In MR this integral was equal to $I_{10}(\mathrm{MR}) = -\frac{1}{12}$. The difference between both schemes occurred when we obtained $({}_{\mu\mu}\Delta)\ (\Delta_\nu)^2$. In both schemes we can still replace ${}_{\mu\mu}\Delta$ by $\Delta_{\mu\mu}$, but in DR we then set $\Delta_{\mu\mu}$ equal to the delta function (a distribution), while in MR $\Delta_{\mu\mu} = \Delta^{\bullet\bullet}$ is still mode regulated and yields $\Delta^{\bullet\bullet}({}^\bullet\!\Delta^\bullet)^2 = \frac{1}{3}\partial_\tau({}^\bullet\!\Delta^\bullet)^3$.

Thus, we see that the rules of computing in DR are quite similar to those used in MR, except for the different options allowed for partial integrations. In DR the rule for contracting one index with another follows directly from the extended action in (4.12). Thus, only certain partial integrations will lead to the Green equation $\partial_\mu^2\Delta(t,s) = \delta^{D+1}(s,t)$ in $D+1$ dimensions. At the same time the complex number D is the regulator, so that the discrete mode sums in (4.7) and (4.11) are really summed up to infinity. In MR one regulates instead by cutting off all mode sums at a large mode number M and then performs partial integrations: now all derivatives are of the same nature and different options of partial integrations arise. This explains the origin of the differences between these two regularizations.

4.2 Two-loop transition amplitude and the counterterm V_{DR}

We now compute the transition amplitude in dimensional regularization at the two-loop level (to order β), treating the coordinate displacement ξ^i as being of order $\beta^{1/2}$. The perturbative expansions is precisely of the same form as given in Section 3.2 to which we refer. The only difference is in the calculation of the integrals $I_1, \ldots, I_{24}$, which must be evaluated with the rules of dimensional regularization just described.

Proceeding to this task, one notices that all these integrals computed in DR acquire the same value as in MR, except for I_9 and I_{10}. We already described in (4.24) how to compute $I_{10}(\mathrm{DR}) = -\frac{1}{24}$. As for I_9 we obtain $I_9(\mathrm{DR}) = -3I_{10}(\mathrm{DR}) = \frac{1}{8}$. We will discuss these integrals in the next section.

Thus, it is straightforward to compute the difference between the DR and MR transition amplitude without counterterms

$$
\begin{aligned}
\Delta\langle x_{\rm f}^k, t_{\rm f}|x_{\rm i}^k, t_{\rm i}\rangle &\equiv \langle x_{\rm f}^k, t_{\rm f}|x_{\rm i}^k, t_{\rm i}\rangle(\mathrm{DR}) - \langle x_{\rm f}^k, t_{\rm f}|x_{\rm i}^k, t_{\rm i}\rangle(\mathrm{MR}) \\
&= A\, \mathrm{e}^{-\frac{1}{2\beta}g_{ij}\xi^i\xi^j} \frac{1}{2\beta^2}\frac{1}{4}\partial_k g_{ij}\partial_l g_{mn}(-\beta^3) \\
&\quad \times\Big\{2g^{kl}g^{im}g^{jn}\left[\mathbf{I}_9(\mathrm{DR}) - \mathbf{I}_9(\mathrm{MR})\right] \\
&\quad + 4g^{km}g^{il}g^{jn}\left[\mathbf{I}_{10}(\mathrm{DR}) - \mathbf{I}_{10}(\mathrm{MR})\right]\Big\}. \qquad (4.25)
\end{aligned}
$$

After integration over ξ^i one finds

$$
\begin{aligned}
&\int d^n\xi\, \Delta\langle x_{\rm f}^k, t_{\rm f}|x_{\rm i}^k, t_{\rm i}\rangle \\
&= \beta\left(\frac{1}{32}g^{kl}g^{im}g^{jn}\partial_k g_{ij}\partial_l g_{mn} - \frac{1}{48}g^{km}g^{il}g^{jn}\partial_k g_{ij}\partial_l g_{mn}\right) \\
&= \frac{\beta}{24}g^{ij}g^{mn}g_{kl}\Gamma^k_{im}\Gamma^l_{jn}. \qquad (4.26)
\end{aligned}
$$

This result implies (recalling (3.72)) that

$$
V_{DR} - V_{MR} = \frac{1}{24}g^{ij}g^{mn}g_{kl}\Gamma^k_{im}\Gamma^l_{jn}. \qquad (4.27)
$$

Then, using V_{MR} given in (3.73), one recognizes that the complete counterterm necessary to satisfy the renormalization conditions in dimensional regularization is covariant and equals

$$
V_{DR} = \frac{1}{8}R \qquad (4.28)
$$

while the value of the constant A is again fixed to be $A = (2\pi\beta)^{-n/2}$. This result shows that general covariance as well as gauge invariance are automatically preserved by dimensional regularization on the finite interval.

4.3 Calculation of Feynman graphs in dimensional regularization

In this section we analyze in some detail the principles to be followed in the application of dimensional regularization and explain through examples how to evaluate efficiently all Feynman diagrams.

First of all, all possible divergences are canceled by ghost contributions. This is seen in diagrams like those in (3.44) or (3.46). Let us consider, for example, the case of I_3 in (3.46) which is regulated in DR as follows:

$$I_3(\mathrm{DR}) =$$ 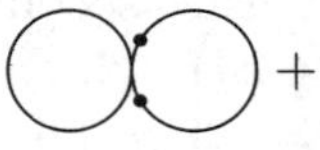 $$+$$ 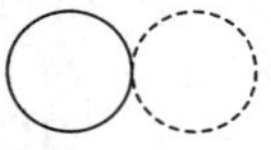 $$= \int d^{D+1}t\, \Delta|_t\, ({}_\mu\Delta_\mu + {}_{\mu\mu}\Delta)|_t. \tag{4.29}$$

Note that here we use $\Delta_{gh} = {}_{\mu\mu}\Delta$ for the ghost propagator, i.e. the Green equation (4.21). Because there is a Δ with two derivatives, we shall use various allowed manipulations to arrive at an expression where all Δ carry at most one derivative, and then we can take the limit $D \to 0$ without encountering ambiguities or divergences. Recall that we denote by a subscript 0 the derivative along the original compact time direction. By inspecting formulas (4.18) and (4.19), one obtains the following identity for $\Delta(t, s)$:

$$({}_\mu\Delta_\mu + {}_{\mu\mu}\Delta)|_t = {}_0({}_0\Delta|_t). \tag{4.30}$$

Inserting this identity into (4.29), one can partially integrate ∂_0 without picking up boundary terms and obtains

$$\begin{aligned} I_3(\mathrm{DR}) &= -\int d^{D+1}t\, \partial_0(\Delta|_t)\, {}_0\Delta|_t \to -\int_{-1}^{0} d\tau\, \partial_\tau(\Delta|_\tau)\, {}^\bullet\!\Delta|_\tau \\ &= -\int_{-1}^{0} d\tau\, [\partial_\tau(\tau^2+\tau)] \left(\tau + \frac{1}{2}\right) = -\frac{1}{6}. \end{aligned} \tag{4.31}$$

Let us now discuss the integral

$$I_9 =$$  $$+$$ 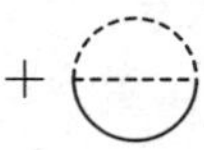 $$= \int_{-1}^{0}\int_{-1}^{0} d\tau\, d\sigma\, \Delta\, ({}^\bullet\!\Delta^{\bullet 2} - {}^{\bullet\bullet}\!\Delta^2). \tag{4.32}$$

In dimensional regularization

$$\begin{aligned} I_9(\mathrm{DR}) &= \int_{-1}^{0} d\tau \int_{-1}^{0} d\sigma\, \Delta\, ({}^\bullet\!\Delta^{\bullet 2} - \Delta_{gh}^2) \\ &\to \int d^{D+1}t \int d^{D+1}s\; \Delta({}_\mu\Delta_\nu\; {}_\mu\Delta_\nu - {}_{\mu\mu}\Delta\; {}_{\nu\nu}\Delta) \\ &= \int\!\!\int [-({}_\mu\Delta)\; (\Delta_\nu)\; ({}_\mu\Delta_\nu) - \Delta\; (\Delta_\nu)\; ({}_{\mu\mu}\Delta_\nu) \\ &\qquad + ({}_\mu\Delta)\; ({}_\mu\Delta)\; ({}_{\nu\nu}\Delta) + \Delta\; ({}_\mu\Delta)\; ({}_{\mu\nu\nu}\Delta)] \\ &= \int\!\!\int [-({}_\mu\Delta)\; (\Delta_\nu)\; ({}_\mu\Delta_\nu) + ({}_\mu\Delta)\; ({}_\mu\Delta)\; ({}_{\nu\nu}\Delta)] \\ &= -I_{10}(\mathrm{DR}) + \int d^{D+1}t \int d^{D+1}s\; ({}_\mu\Delta)^2 \delta^{D+1}(t,s) \\ &= -I_{10}(\mathrm{DR}) + \int d^{D+1}t\; ({}_\mu\Delta)^2|_t \\ &= -3I_{10}(\mathrm{DR}). \end{aligned} \tag{4.33}$$

We used the identity ${}_{\mu\mu}\Delta_\nu = \Delta_{\nu\mu\mu}$, obvious from (4.18), and recognized that the second and fourth term in the second line cancel. Finally, we used the last-but-one line of (4.24) which tells us that $I_{10}(\mathrm{DR}) = -\frac{1}{2}\int d^{D+1}t\ ({}_\mu\Delta)^2|_t$. Thus, $I_9(\mathrm{DR}) = -3I_{10}(\mathrm{DR})$, which is the same relation as in MR. The value of $I_{10}(\mathrm{DR}) = -\frac{1}{24}$ was already obtained in (4.24) and differs from $I_{10}(\mathrm{MR}) = -\frac{1}{12}$.

Finally, we study the integral I_8 which is related to gauge invariance

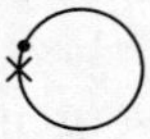

$$\mathbf{I}_8 = \text{(diagram)} = \int_{-1}^{0} d\tau\, {}^{\bullet}\!\Delta|_\tau. \tag{4.34}$$

where the cross indicates the location of the vertex and denotes the factor of $\partial_j A_i$ in (3.39) (there are no ξ^i factors, and hence no external lines according to our graphical notation). Computationally this diagram is rather simple. By power counting it is logarithmically divergent, and one could obtain any value for this diagram by using different prescriptions. Symmetric integration gives zero, but asymmetric integration schemes may easily produce a nonvanishing answer. All three schemes, DR, MR and TS, give the same answer which preserves gauge invariance. In DR one can write it as

$$I_8(\mathrm{DR}) = \int d^{D+1}t\ ({}_0\Delta)|_t \quad\rightarrow\quad \int_{-1}^{0} d\tau\, {}^{\bullet}\!\Delta|_\tau = 0 \tag{4.35}$$

which is directly computed in the $D = 0$ limit.

4.4 Path integrals for fermions

In this section we describe the dimensional regularization of fermionic path integrals. We shall discuss explicitly path integrals for Majorana fermions on a circle with periodic (PBC) or antiperiodic boundary conditions (ABC), as these are the only boundary conditions that will be needed directly in the applications to anomalies. First, we consider the path integral with ABC and describe how to extend dimensional regularization to fermions. The requirement that a two-loop computation with DR reproduces known results (namely those obtained by time slicing) fixes once for all the two-loop counterterms due to fermions. As we shall see this counterterm vanishes in DR. Since counterterms are due to ultraviolet effects, the infrared vacuum structure and the related boundary conditions on the fields should not matter in their evaluation. Therefore the same counterterm should apply to the fermionic path integral with PBC as well. No higher-loop contributions to the counterterm are expected as the model is super-renormalizable, just like the purely bosonic case. We end the section by presenting for completeness the essential formulas for the fermionic path integral with PBC.

Let us consider the $N = 1$ supersymmetric nonlinear sigma model written in terms of fermions with flat target-space indices

$$S = \int_{-1}^{0} d\tau \Big\{ \frac{1}{2} g_{ij}(x)(\dot{x}^i \dot{x}^j + a^i a^j + b^i c^j) + \frac{1}{2}\psi_a[\dot{\psi}^a + \dot{x}^i \omega_i{}^a{}_b(x)\psi^b] + \beta^2 \Big[V(x) + V_{CT}(x) + V'_{CT}(x)\Big]\Big\} \tag{4.36}$$

where $V'_{CT}(x)$ denotes the additional counterterm which may arise from the fermions ψ^a in the chosen regularization scheme. This $N=1$ model is described in detail in Appendix D. It is classically supersymmetric if all the potential terms which are multiplied by β^2 are set to zero (note that the ghosts can be set to zero by using their algebraic equations of motion). Supersymmetry may be broken by boundary conditions, e.g. periodic for the bosons and antiperiodic for the fermions. To start with we assume antiperiodic boundary conditions for the Majorana fermions $\psi^a(0) = -\psi^a(-1)$. Majorana fermions realize the Dirac gamma matrices in a path integral context, and ABC compute the trace over the Dirac matrices.[1]

Now we may explicitly compute by time slicing the transition amplitude for going from the background point x at time $t = 0$ back to the same point x at a later time $t = \beta$ using ABC for the Majorana fermions. In the two-loop approximation this computation gives

$$Z \equiv \mathrm{tr}\, \langle x | \mathrm{e}^{-\beta \hat{H}} | x \rangle = \frac{2^{n/2}}{(2\pi\beta)^{n/2}} \left[1 - \frac{\beta}{24} R + O(\beta^2) \right] \tag{4.37}$$

where the trace on the left-hand side is only over the Dirac matrices and where

$$\hat{H} = \hat{Q}^2 = -\frac{1}{2} \not{D}\not{D} = -\frac{1}{2}\left(D^i D_i + \frac{1}{4} R \right) \tag{4.38}$$

is the supersymmetric Hamiltonian of the $N=1$ model. This Hamiltonian is the square of the supercharge $\hat{Q}$, realized by the Dirac operator

$$\hat{Q} = \frac{i}{\sqrt{2}} \not{D} = \frac{i}{\sqrt{2}} \gamma^a e_a{}^i D_i, \quad D_i = \partial_i + \frac{1}{4}\omega_{iab}\gamma^a\gamma^b \tag{4.39}$$

with ω_{iab} being the spin connection (see Appendix A). Note that there is an explicit coupling to the scalar curvature arising in (4.38). Thus,

[1] One may avoid the path integral over Majorana fermions by explicitly using a matrix-valued action: one drops the kinetic term for fermions and replaces the potential term $\dot{x}^i \omega_{iab}\psi^a\psi^b$ by the matrix $\frac{1}{2}\dot{x}^i\omega_{iab}\gamma^a\gamma^b$. The path integral then requires an explicit time-ordering prescription to evaluate the exponential of the matrix-valued action and maintain gauge invariance ($T\mathrm{e}^{-\frac{1}{\beta}S}$) [40].

one needs to use $V=-\frac{1}{8}R$ in the action together with the time slicing counterterms $V_{TS}=\frac{1}{8}(R+g^{ij}\Gamma^k_{il}\Gamma^l_{jk})$ and $V'_{TS}=\frac{1}{16}g^{ij}\omega_i{}^{ab}\omega_{jab}$ (see (6.15) and (6.16)). For convenience we will later rederive this value of V'_{TS}.

Now we want to reproduce (4.37) with a path integral over Majorana fermions in dimensional regularization. This will unambiguously fix the additional counterterm $V'_{DR}(x)$ due to the fermions. Note that in dimensional regularization the potential $V=-\frac{1}{8}R$ cancels the counterterm $V_{DR}=\frac{1}{8}R$ exactly due to the bosons.

We focus directly on the regularization of the Feynman graphs arising in perturbation theory. To recognize how to dimensionally continue the various Feynman graphs we extend the action to D dimensions as follows:

$$S=\int_\Omega d^{D+1}t\Big[\frac{1}{2}\,g_{ij}(\partial_\mu x^i\partial_\mu x^j+a^ia^j+b^ic^j)$$
$$+\frac{1}{2}\bar{\psi}_a\gamma^\mu(\partial_\mu\psi^a+\partial_\mu x^i\omega_i{}^a{}_b\psi^b)+\beta^2V'_{DR}\Big] \tag{4.40}$$

where $\Omega=I\times R^D$ is the region of integration containing the finite interval $I=[-1,0]$ and γ^μ are the gamma matrices in $D+1$ dimensions ($\{\gamma^\mu,\gamma^\nu\}=2\delta^{\mu\nu}$). As before $t^\mu=(\tau,\mathbf{t})$ with $\mu=0,1,\ldots,D$. Here we assume that we can continue to those Euclidean integer dimensions where Majorana fermions can be defined. The Majorana conjugate is defined by $\bar{\psi}_a=\psi_a^TC_\pm$ with a suitable charge conjugation matrix $C_\pm$ such that $\bar{\psi}^a\gamma^\mu\psi^b=-\bar{\psi}^b\gamma^\mu\psi^a$. This can be achieved, for example, in two dimensions.[2] This requirement guarantees that the coupling $\omega_{iab}\psi^a\psi^b=-\omega_{iab}\psi^b\psi^a$ in (4.36) is nonvanishing when extended to $D+1$. The actual details of how to represent $C_\pm$ and the gamma matrices in $D+1$ dimensions are not important. These gamma matrices only serve as a book-keeping device to keep track of how derivatives are going to be contracted in higher dimensions. Apart from the above requirements, no additional Dirac algebra for γ^μ in $D+1$ dimensions is needed.

The bosonic and ghost propagators are as in the previous sections. The fermionic fields with ABC on the worldline, $\psi^a(0)=-\psi^a(-1)$, can be expanded in half-integer modes

$$\psi^a(\tau)=\sum_{r\in Z+\frac{1}{2}}\psi^a_r\,\mathrm{e}^{2i\pi r\tau} \tag{4.41}$$

and have the following unregulated propagator:

$$\langle\psi^a(\tau)\psi^a(\sigma)\rangle=\beta\delta^{ab}\Delta_{AF}(\tau-\sigma)$$
$$\Delta_{AF}(\tau-\sigma)=\sum_{r\in Z+\frac{1}{2}}\frac{1}{2\pi ri}\mathrm{e}^{2i\pi r(\tau-\sigma)} \tag{4.42}$$

---

[2] In Euclidean two dimensions one can choose $\gamma^1=\sigma^3$, $\gamma^2=\sigma^1$ and $C_+=1$. Recall that $C_\pm$ is defined by $C_\pm\gamma^\mu C_\pm^{-1}=\pm\,\gamma^{\mu\,T}$.

where the subscript AF denotes "antiperiodic fermions". Note that the Fourier sum defining Δ_{AF} is conditionally convergent for $\tau \neq \sigma$ and yields

$$\Delta_{AF}(\tau - \sigma) = \frac{1}{2}\epsilon(\tau - \sigma) \tag{4.43}$$

where $\epsilon(x) = \theta(x) - \theta(-x)$ is the sign function (with the value $\epsilon(0) = 0$ obtained by symmetrically summing the Fourier series). The function Δ_{AF} satisfies

$$\partial_\tau \Delta_{AF}(\tau - \sigma) = \delta_{AF}(\tau - \sigma) \tag{4.44}$$

where $\delta_{AF}(\tau - \sigma)$ is the Dirac delta function on functions with antiperiodic boundary conditions

$$\delta_{AF}(\tau - \sigma) = \sum_{r \in Z + \frac{1}{2}} \mathrm{e}^{2i\pi r(\tau - \sigma)}. \tag{4.45}$$

The dimensionally regulated propagator obtained by adding the extra coordinates reads

$$\langle \psi^a(t) \bar{\psi}^b(s) \rangle = \beta\, \delta^{ab} \Delta_{AF}(t, s) \tag{4.46}$$

where the function

$$\Delta_{AF}(t, s) = -i \int \frac{d^D\mathbf{k}}{(2\pi)^D} \sum_{r \in Z + \frac{1}{2}} \frac{2\pi r \gamma^0 + \boldsymbol{\gamma} \cdot \mathbf{k}}{(2\pi r)^2 + \mathbf{k}^2} \mathrm{e}^{2i\pi r(\tau - \sigma)} \mathrm{e}^{i\mathbf{k} \cdot (\mathbf{t} - \mathbf{s})} \tag{4.47}$$

satisfies

$$\gamma^\mu \frac{\partial}{\partial t^\mu} \Delta_{AF}(t, s) = -\frac{\partial}{\partial s^\nu} \Delta_{AF}(t, s) \gamma^\nu = \delta_{AF}(\tau - \sigma) \delta^D(\mathbf{t} - \mathbf{s}). \tag{4.48}$$

These are the essential relations needed to extend DR to fermions. They keep track of which derivative is contracted to which vertex to produce the $D+1$ delta function. The delta function is only to be used in $D+1$ dimensions, as we assume that regularization due to the extra dimensions is taking place.[3] By using partial integration one casts the various loop integrals in a form which can be computed by sending first $D \to 0$. Then one can use $\gamma^0 = 1$ and no extra factors arise from the Dirac algebra in $D+1$ dimensions.

We are now ready to perform the two-loop calculation in the $N = 1$ nonlinear sigma model using DR. The bosonic vertices together with the ghosts, V and V_{DR} give the same contribution calculated in Section 3.2.

[3] Again, we are not able to show this in full generality, and at this stage this rule is taken as an assumption which has turned out to be consistent in all the examples we have been dealing with so far. One way to prove it explicitly would be to compute all integrals arising in perturbation theory at arbitrary D and then to check the location of the poles.

The overall normalization of the fermionic path integral gives the extra factor of $2^{n/2}$ which equals the number of components of a Dirac fermion in n (even) dimensions. This already produces the full expected result in (4.37).

Thus, the sum of the additional fermion graphs arising from the cubic vertex contained in $\Delta S = \int_{-1}^{0} d\tau\, \frac{1}{2}\, \dot{x}^i \omega_{iab} \psi^a \psi^b$ and the extra counterterm V'_{DR} must vanish at two-loops. The cubic vertex arise by evaluating the spin connection at the background point x and reads as $\Delta S_3 = \frac{1}{2}\omega_{iab} \int_{-1}^{0} d\tau\, \dot{q}^i \psi^a \psi^b$. Using Wick contractions we identify the following contribution to $\langle \mathrm{e}^{-\frac{1}{\beta}S^{int}} \rangle$:

$$\frac{1}{2\beta^2}\langle (\Delta S_3)^2 \rangle = \frac{1}{2\beta^2}(-2)\left(\frac{1}{2}\omega_{iab}\right)^2 (-\beta^3) \times \int_{-1}^{0}\int_{-1}^{0} d\tau\, d\sigma\, {}^\bullet\!\Delta^\bullet(\tau,\sigma)[\Delta_{AF}(\tau,\sigma)]^2. \tag{4.49}$$

Using DR this graph is regulated by

$$\int_{-1}^{0}\int_{-1}^{0} d\tau\, d\sigma\, {}^\bullet\!\Delta^\bullet(\tau,\sigma)[\Delta_{AF}(\tau,\sigma)]^2 \to -\iint {}_\mu\Delta_\nu(t,s)\, \mathrm{tr}\,[\gamma^\mu \Delta_{AF}(t,s)\gamma^\nu \Delta_{AF}(s,t)] \tag{4.50}$$

(note the minus sign obtained in exchanging t and s in the last propagator; it is the usual minus sign arising for fermionic loops). We can partially integrate ∂_μ without picking boundary terms and obtain

$$\begin{aligned} 2\iint \Delta_\nu(t,s)\, \mathrm{tr}\,[(\gamma^\mu \partial_\mu \Delta_{AF}(t,s))\gamma^\nu \Delta_{AF}(s,t)] &= 2\iint \Delta_\nu(t,s)\, \mathrm{tr}\,[\delta^{D+1}(t,s)\gamma^\nu \Delta_{AF}(s,t)] \\ &= 2\int \Delta_\nu(t,t)\, \mathrm{tr}\,[\gamma^\nu \Delta_{AF}(t,t)] \\ &\to 2\int_{-1}^{0} d\tau\, {}^\bullet\!\Delta(\tau,\tau)\Delta_{AF}(0) = 0 \end{aligned} \tag{4.51}$$

because $\Delta_{AF}(0) = \frac{1}{2}\epsilon(0) = 0$ (and $\gamma^0 = 1$ at $D = 0$). As this example shows, the Dirac gamma matrices in $D+1$ are just a book-keeping device to keep track of where one can use the Green equation (4.48).

Thus, no contribution arises from the fermions at this order, and this implies that the extra counterterm must vanish

$$V'_{DR} = 0. \tag{4.52}$$

This is what one expects to preserve supersymmetry: the counterterm V_{DR} is exactly canceled by the tree-level potential $V = -\frac{1}{8}R$ needed to have the correct coupling to the scalar curvature in the Hamiltonian (4.38), while no extra contribution to the counterterm arises from fermions. Thus, dimensional regularization **without any counterterm** and **without** an **extra order β^2 tree-level potential** preserves the supersymmetry of the classical $N=1$ action

$$S = \int_{-1}^{0} d\tau \left\{ \frac{1}{2} g_{ij}(x) \dot{x}^i \dot{x}^j + \frac{1}{2} \psi_a [\dot{\psi}^a + \dot{x}^i \omega_i{}^a{}_b(x) \psi^b] \right\}. \tag{4.53}$$

One may say that the amount of curvature coupling in the Hamiltonian H brought in by DR is of the exact amount to render it supersymmetric at the quantum level.

To compare with TS, we can compute the graph (4.49) using the TS rules. Now we must use the fact that ${}^{\bullet}\Delta^{\bullet}(\tau, \sigma) = 1 - \delta(\tau, \sigma)$. The Dirac delta function is ineffective as $\epsilon(0) = 0$, but the rest gives

$$\begin{aligned} \frac{1}{2\beta^2} \langle (\Delta S_3)^2 \rangle (TS) &= \frac{1}{2\beta^2} (-2) \left(\frac{1}{2} \omega_{iab} \right)^2 (-\beta^3) \int_{-1}^{0} \int_{-1}^{0} d\tau \, d\sigma \, \frac{1}{4} \\ &= \frac{\beta}{16} (\omega_{iab})^2. \end{aligned} \tag{4.54}$$

This is canceled by using an extra counterterm $V'_{TS} = \frac{1}{16}\beta^2(\omega_{iab})^2$ which at this order contributes with a term $-\frac{1}{\beta}V'_{TS}$ (evaluated at the background point x). This is precisely the term that was derived using Weyl ordering to deduce the TS prescriptions.

Let us conclude this section by considering briefly the case of Majorana fermions with PBC. Now the mode expansion of $\psi^a(\tau)$ requires integer modes,

$$\psi^a(\tau) = \sum_{n \in Z} \psi^a_n \, e^{2i\pi n \tau}. \tag{4.55}$$

The zero modes ψ^a_0 of the kinetic operator (∂_τ) are treated separately, and the unregulated propagator in the sector of periodic functions orthogonal to the zero modes, $\psi'^a(\tau) = \psi^a(\tau) - \psi^a_0$, reads as

$$\langle \psi'^a(\tau) \psi'^b(\sigma) \rangle = \beta \delta^{ab} \Delta_{PF}(\tau - \sigma) \tag{4.56}$$

$$\Delta_{PF}(\tau - \sigma) = \sum_{n \neq 0} \frac{1}{2\pi n i} e^{2i\pi n(\tau - \sigma)} \tag{4.57}$$

where the function Δ_{PF} satisfies

$$\partial_\tau \Delta_{PF}(\tau - \sigma) = \delta_{PF}(\tau - \sigma) - 1 \tag{4.58}$$

with $\delta_{PF}(\tau-\sigma)$ being the Dirac delta function on periodic functions. Its continuum limit can be obtained by summing up the series and reads (for $(\tau-\sigma)\in[-1,1]$) as

$$\Delta_{PF}(\tau-\sigma)=\frac{1}{2}\epsilon(\tau-\sigma)-(\tau-\sigma). \tag{4.59}$$

Curved indices

It is interesting to consider the case of fermions with curved target-space indices. This is equivalent to the case of fermions with flat target-space indices: it is just a change of integration variables in the path integral. However, it is a useful exercise to work out since some formulas will become simpler. The classical $N=1$ supersymmetric sigma model is written as

$$S=\int_{-1}^{0}d\tau\,\frac{1}{2}g_{ij}(x)\Big\{\dot{x}^i\dot{x}^j+\psi^i[\dot{\psi}^j+\dot{x}^l\Gamma^j_{lk}(x)\psi^k]\Big\}. \tag{4.60}$$

The fermionic term could also be written more compactly in terms of the covariant derivative $\frac{D}{d\tau}\psi^j=\dot{\psi}^j+\dot{x}^l\Gamma^j_{lk}(x)\psi^k$. Note that the action is written in terms of the metric and Christoffel connection and there is no need to introduce the vielbein and spin connection. Writing out the Christoffel connection in terms of the metric also shows that the coupling to the metric g_{ij} is linear (see Appendix D, eq. (D.7)).

The bosonic part of the path integral has already been described and goes unchanged. For the fermionic part we can now derive the correct path integral measure by taking into account the Jacobian from the change of variable from the free measure with flat indices

$$\begin{aligned}D\psi^a&=D(e^a{}_i(x)\psi^i)=\mathrm{Det}^{-1}(e^a{}_i(x))\,D\psi^i\\&=\left(\prod_{-1\le\tau<0}\frac{1}{\sqrt{\det g_{ij}(x(\tau))}}\right)D\psi^i.\end{aligned} \tag{4.61}$$

Note the inverse determinant which is due to the Grassmann nature of the integration variables. We have denoted the determinant by 'Det' instead of 'det' to indicate that it is a functional determinant. The extra factor appearing in the measure can be exponentiated using bosonic ghosts $\alpha^i(\tau)$ with the same boundary condition on the fermions (ABC or PBC) and it leads to the extra term in the ghost action

$$S_{gh}^{extra}=\int_{-1}^{0}d\tau\,\frac{1}{2}g_{ij}(x)\alpha^i\alpha^j. \tag{4.62}$$

One can check that the counterterms of dimensional regularization are left unchanged. The full quantum action for the $N=1$ supersymmetric sigma model reads as

$$S = \int_{-1}^{0} d\tau \, \frac{1}{2} g_{ij}(x) \Big\{ \dot{x}^i \dot{x}^j + a^i a^j + b^i c^j + \psi^i [\dot{\psi}^j + \dot{x}^l \Gamma^j_{lk}(x) \psi^k] + \alpha^i \alpha^j \Big\} \tag{4.63}$$

and appears in the path integral as

$$Z = \int Dx \, Da \, Db \, Dc \, D\psi \, D\alpha \; \mathrm{e}^{-\frac{1}{\beta} S}. \tag{4.64}$$

Supersymmetry is not broken by boundary conditions if one uses periodic boundary conditions for both bosons and fermions. Then the effect of the ghosts cancels (they have the same boundary conditions) and can be eliminated altogether

$$\left[\prod_{-1 \le \tau < 0} \sqrt{\det g_{ij}(x(\tau))} \right] \left[\prod_{-1 \le \tau < 0} \frac{1}{\sqrt{\det g_{ij}(x(\tau))}} \right] = 1. \tag{4.65}$$

One can now recognize that the potential divergence arising in the bosonic $\dot{x}\dot{x}$ contractions are canceled by the fermionic $\psi\dot{\psi}$ contractions. This is an example of the improved ultraviolet behavior of supersymmetric quantum field theories. The remaining UV ambiguities are treated by dimensional regularization as usual. This scheme seems to be the best one to test, for example, that the Witten index (i.e. the gravitational contribution to the abelian chiral anomaly for a spin-$\frac{1}{2}$ field) does not get higher-order corrections in worldline loops, and is thus β-independent [89].

If one used ABC, the ghosts have different boundary conditions, their cancellation is not complete and they should be kept.

Part II
Applications to anomalies

5
Introduction to anomalies

We now start the second part of this book, namely the computation of anomalies in higher-dimensional quantum field theories using quantum mechanical (QM) path integrals. Anomalies arise when the symmetries of a classical system cannot all be preserved by the quantization procedure. Those symmetries which turn out to be violated by the quantum corrections are called anomalous. The anomalous behavior is encoded in the quantum effective action which fails to be invariant: its nonvanishing variation is called the anomaly. As we shall see, the ordinary Dirac action for a chiral fermion in n dimensions has anomalies which can be computed by using an $N = 1$ supersymmetric (susy) nonlinear sigma model in one (timelike) dimension. Although this relation between a nonsusy quantum field theory (QFT) and a susy QM system may seem surprising at first sight, it becomes plausible if one notices that the Dirac operator $\gamma^\mu D_\mu$ contains hermitian Dirac matrices γ^m (where $\gamma^\mu = \gamma^m e_m{}^\mu$, with $e_m{}^\mu$ being the inverse vielbein field) satisfying the same Clifford algebra $\{\gamma^l, \gamma^m\} = 2\delta^{lm}$ (with $l, m = 1, \ldots, n$ flat indices) as the equal-time anticommutation rules of a real (Majorana) fermionic quantum mechanical point particle $\psi^a(t)$ with $a = 1, \ldots, n$, namely[1] $\{\psi^a(t), \psi^b(t)\} = \hbar\delta^{ab}$. This suggests a representation of operators which appear in the QFT (γ^m) in terms of QM operators ($\psi^a(t)$), namely

$$\gamma^m \leftrightarrow \sqrt{\frac{2}{\hbar}}\psi^a(t). \tag{5.1}$$

[1] To distinguish objects in quantum field theory from objects in quantum mechanics, we use vector indices μ (curved) and m (flat) in field theory, and vector indices i (curved) and a (flat) for the point particle in quantum mechanics. We are always in Euclidean space with metric $\delta_{mn} = (1, \ldots, 1)$ in tangent space, unless stated otherwise.

It is also natural to represent the coordinates x^μ in the QFT by a corresponding point particle $x^i(t)$ in QM. Hence, one is led to suspect that the expression for the anomaly in terms of the operators $\frac{\partial}{\partial x^\mu}$, γ^m, etc. of the quantum field theory can be rewritten as an expression in terms of the operators of a corresponding QM model with bosonic $x^i(t)$ and fermionic $\psi^a(t)$. These QM models are often supersymmetric. Of course, it had been known long before the 1980s that many calculations in field theory can be simplified by just using first quantization (with point particles) instead of second quantization [92, 93, 93a, 18]. For a review see [27]. Thus one might also expect that the calculation of anomalies is drastically simplified if one uses quantum mechanics. This is indeed the case.

The anomalies we shall compute are chiral and gravitational anomalies for n-dimensional chiral fermions and self-dual antisymmetric tensor fields (AT) coupled to external gravitational and gauge fields, and trace anomalies for various fields coupled to gravity in two and four dimensions. These anomalies are anomalies in the axial $U(1)$ symmetry, local Lorentz symmetry and scale symmetry. As we shall discuss, we only use regularization schemes that maintain Einstein (general coordinate) invariance so there are no separate Einstein anomalies. Before analyzing the formalism of quantum mechanics to calculate anomalies, it is useful to demonstrate that such anomalies really exist. We therefore start this chapter in Section 5.1 with an explicit computation of anomalies in the simplest case: two-dimensional quantum field theories (one space and one time dimension). In this case we use a regularization scheme that is special for two dimensions: analytic regularization. Analytic regularization uses complex variables, just like conformal field theory, but in our application of analytic regularization we start in Minkowski space with real variables and then continue them into the complex plane in order to use contour integration, whereas in conformal field theory the basic variable $z = x + it$ is itself already made complex by a Wick rotation. So in this subsection we are for once in Minkowski space. First, we calculate the chiral anomaly for a complex one-component fermion coupled to an external Maxwell or Yang–Mills field. Then we compute the gravitational anomaly for a real one-component chiral fermion coupled to an external gravitational field. Finally, we compute the trace anomaly for a real nonchiral (two-component) fermion coupled to external gravity. These calculations will confirm the existence of chiral, gravitational and trace anomalies in two dimensions, and in the rest of the book we calculate similar anomalies in higher (but always Euclidean) dimensions, using quantum mechanics.

In Section 5.2 we discuss general aspects of the approach of calculating anomalies in field theories using quantum mechanics. Field theories coupled to external gravitational fields will lead to quantum mechanical nonlinear sigma models, while field theories coupled to gauge fields will lead to linear sigma models. To describe the anomalies in terms of quantum mechanical operators, we shall use Fujikawa's approach [12]. In this approach the anomaly for a QFT is given by the trace of the regulated Jacobian associated with a given symmetry, and this Jacobian is an expression which depends on $\partial/\partial x^\mu$, external fields $A_\mu^\alpha(x)$ and $e_\mu^m(x)$, Lie-algebra matrices T_α and Dirac matrices γ^m. First, we shall construct an explicit expression for the anomaly in terms of quantum mechanical operators. For example, as we already observed, Dirac matrices correspond to fermionic point particle operators $\psi^a(t)$. Then we rewrite the trace of the regulated Jacobians as quantum mechanical path integrals. At that point we take over all results of the first part of the book on the construction and properties of path integrals for linear and nonlinear sigma models, and compute the anomalies.

In Section 5.3 we give a brief history of anomalies.

5.1 The simplest case: anomalies in two dimensions

In this section we present an explicit calculation of chiral, gravitational and trace anomalies in a toy model: massless fermions coupled to external gauge and gravitational fields in two dimensions. In this model the full effective action can be computed explicitly, and its response to gauge and gravitational transformations can be easily studied.

5.1.1 The chiral anomaly in flat space

We shall start by discussing the classical Lagrangian of Dirac and Weyl fermions coupled to an external gauge field. Then we proceed to analyze three typical cases.

(i) We first present the calculation of the chiral anomaly due to a Weyl fermion. This is an example of a gauge anomaly, i.e. an anomaly in a current which is coupled to a gauge field. The corresponding effective action is not gauge invariant, and there is no local counterterm that can be added to the effective action to restore gauge invariance. The anomaly is thus a genuine anomaly. It satisfies certain consistency conditions, and thus it is called a "consistent anomaly". One particular consequence of these consistency conditions is that this anomaly itself cannot be gauge invariant. Nevertheless, it can be related to a "covariant anomaly", which is gauge invariant but cannot be interpreted as the gauge variation of an effective action.

(ii) Then we add the contribution of another Weyl fermion, but with opposite chirality. The two Weyl fermions with opposite chiralities make up a massless Dirac fermion. The vector current of the Dirac fermion is coupled to a $U(1)$ gauge field A_μ. The classical action has a local vector symmetry $U_V(1)$ and a rigid chiral symmetry

$U_A(1)$.[2] The total anomaly cancels in the $U_V(1)$ symmetry and the full effective action is $U_V(1)$ gauge invariant.

(iii) However, the $U_A(1)$ symmetry is anomalous and we compute the corresponding anomaly. It is invariant under transformations of the gauge group $U_V(1)$ and it is an example of a rigid anomaly since the corresponding current is not coupled to gauge fields (although we could couple it to an external axial vector field, see footnote 2). It is again a genuine anomaly and again it satisfies consistency conditions.

Let us start by describing the classical Lagrangian of a massless Dirac field λ coupled to a $U_V(1)$ gauge field A_μ,

$$\mathcal{L} = -\bar{\lambda}\gamma^\mu(\partial_\mu - iA_\mu)\lambda, \qquad \bar{\lambda} \equiv \lambda^\dagger i\gamma^0 \tag{5.2}$$

where the Dirac matrices satisfy $\{\gamma^\mu, \gamma^\nu\} = 2\eta^{\mu\nu}$, with $(\gamma^0)^2 = -1$. The classical symmetries are the $U_V(1)$ gauge transformations with an infinitesimal local parameter $\alpha(x)$,

$$\begin{aligned} \delta\lambda(x) &= i\alpha(x)\lambda(x) \\ \delta\bar{\lambda}(x) &= -i\alpha(x)\bar{\lambda}(x) \\ \delta A_\mu(x) &= \partial_\mu\alpha(x) \end{aligned} \tag{5.3}$$

and the axial $U_A(1)$ transformations with an infinitesimal constant parameter β,

$$\begin{aligned} \delta\lambda(x) &= i\beta\gamma_5\lambda(x) \\ \delta\bar{\lambda}(x) &= i\beta\bar{\lambda}(x)\gamma_5 \\ \delta A_\mu(x) &= 0. \end{aligned} \tag{5.4}$$

The chiral matrix γ_5 is chosen to satisfy $\gamma_5^2 = 1$ and $\{\gamma_5, \gamma^\mu\} = 0$. This model exists in any even spacetime dimension n, since then one can construct a matrix γ_5 with the required properties. In this section we restrict our attention to $n = 2$ by taking $\eta^{11} = -\eta^{00} = 1$ and $\gamma_5 \equiv \gamma^1\gamma^0$. We will also use the antisymmetric tensor $\epsilon_{\mu\nu}$, which we normalize to $\epsilon^{01} = -\epsilon_{01} = 1$. With this normalization one can verify that the following identity is satisfied by the gamma matrices: $\gamma^\mu\gamma_5 = \epsilon^{\mu\nu}\gamma_\nu$. We choose the following representation of the gamma matrices:

$$\gamma^0 = -i\sigma^2 = \begin{pmatrix} 0 & -1 \\ 1 & 0 \end{pmatrix}, \quad \gamma^1 = \sigma^1 = \begin{pmatrix} 0 & 1 \\ 1 & 0 \end{pmatrix}, \quad \gamma_5 = \begin{pmatrix} 1 & 0 \\ 0 & -1 \end{pmatrix} \tag{5.5}$$

---

[2] Actually, in two dimensions, the gauge field A_μ can be decomposed into light-cone components A_+ and A_- which only couple to left-moving and right-moving massless fermions, respectively. This shows that this model has even a local $U_A(1)$ symmetry, but in our discussion we consider the $U_A(1)$ symmetry only as a rigid symmetry.

and decompose the Dirac spinor as follows:

$$\lambda = 2^{-1/4} \begin{pmatrix} \lambda_L \\ \lambda_R \end{pmatrix} \tag{5.6}$$

where the Weyl components λ_L and λ_R are eigenstates of γ_5 with eigenvalues $+1$ and -1, respectively. It is useful to eliminate the gamma matrices completely from the Lagrangian (5.2). Using light-cone coordinates $x^\pm = \frac{1}{\sqrt{2}}(x^0 \pm x^1)$ one obtains

$$\mathcal{L} = i\lambda_L^\dagger(\partial_+ - iA_+)\lambda_L + i\lambda_R^\dagger(\partial_- - iA_-)\lambda_R \tag{5.7}$$

with $\partial_\pm = \frac{1}{\sqrt{2}}(\partial_0 \pm \partial_1)$. It is evident that one can also consider a model with a single Weyl fermion; for example, the left-moving fermion λ_L, by setting the other chirality to zero. In this case the two classical symmetries discussed above are not independent.

(i) Now let us consider the path integral quantization of the left-moving fermion λ_L, setting $\hbar = 1$ for notational simplicity,

$$S[\lambda_L, \lambda_L^\dagger, A] = \int d^2x \, i\lambda_L^\dagger(\partial_+ - iA_+)\lambda_L$$

$$\int \mathcal{D}\lambda_L \, \mathcal{D}\lambda_L^\dagger \, \mathrm{e}^{iS[\lambda_L,\lambda_L^\dagger,A]} = \mathrm{e}^{iW_L[A]}. \tag{5.8}$$

The one-loop effective action $W_L[A]$ is a functional of the gauge field A_+, and is formally given by the logarithm of a functional determinant

$$W_L[A] = -i \log \mathrm{Det}\,(i\partial_+ + A_+). \tag{5.9}$$

For our purposes it is simpler to view the effective action perturbatively, namely as the sum of all one-loop graphs with external gauge fields,

$W_L[A] =$ 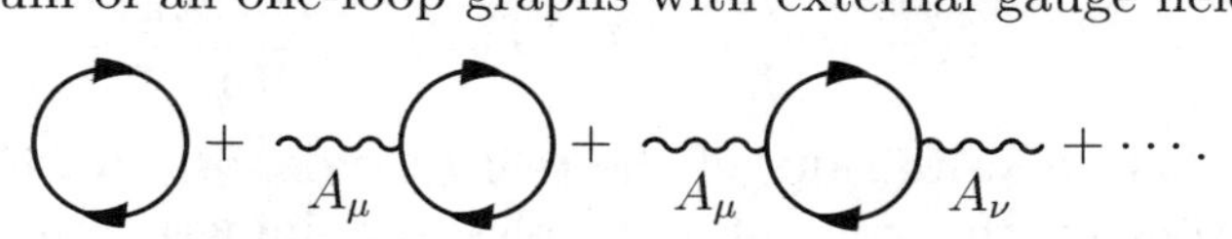 (5.10)

The first graph is a constant. It can be removed by a suitable normalization of the path integral. The second graph vanishes by symmetric integration. Thus, let us take a closer look at the third graph. Expanding (5.8) to second order in A_μ we obtain

$$\begin{aligned} iW_L^{(2)}[A] &= \frac{1}{2}\langle (iS_{int})^2 \rangle \\ &= -\frac{1}{2}\int d^2x \, d^2y \, A_+(x) \Big\langle \lambda_L^\dagger(x)\lambda_L(x) \, \lambda_L^\dagger(y)\lambda_L(y) \Big\rangle A_+(y) \end{aligned} \tag{5.11}$$

where we have split $S = S_0 + S_{int}$, with $S_0 = \int d^2x \; i\lambda_L^\dagger \partial_+ \lambda_L$ being the free action which yields the propagator, and $S_{int} = \int d^2x \; A_+ \lambda_L^\dagger \lambda_L$ which describes the interaction with the gauge field A_μ.

The propagator is readily obtained by coupling the fermions to external sources and completing squares,

$$\langle \lambda_L(x) \lambda_L^\dagger(y) \rangle = \frac{1}{\partial_+} \delta^2(x-y) = \int \frac{d^2p}{(2\pi)^2} \, \mathrm{e}^{ip\cdot(x-y)} \, \frac{2ip_-}{p^2 - i\epsilon} \tag{5.12}$$

where $p \cdot x = p_+ x^+ + p_- x^-$, $p^2 \equiv p_\mu p^\mu = -2p_+p_-$ with $p_\pm = \frac{1}{\sqrt{2}}(p_0 \pm p_1)$ and $-i\epsilon$ is the Feynman prescription that enforces the correct boundary conditions. It is easier to Fourier transform to momentum space by setting $A_+(x) = \int \frac{d^2p}{(2\pi)^2} \, \mathrm{e}^{-ip\cdot x} A_+(p)$. We obtain[3]

$$W_L^{(2)}[A] = \frac{i}{2} \int \frac{d^2p}{(2\pi)^2} A_+(p) U(p) A_+(-p) \tag{5.13}$$

with

$$U(p) \equiv \int d^2x \, \mathrm{e}^{-ip\cdot x} \, \langle \lambda_L^\dagger(x) \lambda_L(x) \, \lambda_L^\dagger(0) \lambda_L(0) \rangle$$

$$=$$ 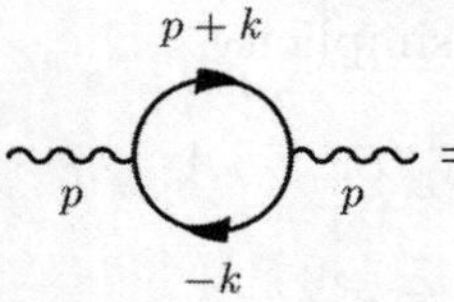

$$= \int \frac{d^2k}{(2\pi)^2} \, \frac{2(p_- + k_-)}{(p+k)^2 - i\epsilon} \, \frac{2k_-}{k^2 - i\epsilon}$$

$$= \int \frac{dk_- \, dk_+}{(2\pi)^2} \, \frac{1}{p_+ + k_+ + i\epsilon/2(p_- + k_-)} \, \frac{1}{k_+ + i\epsilon/2k_-}. \tag{5.14}$$

We now perform analytic regularization [94]. This scheme is suitable for a chiral theory. One can first perform the integral over k_+ by using a contour in the complex k_+-plane. To obtain a nonvanishing result, the two poles at

$$k_+ = -\frac{i\epsilon}{2k_-}, \quad k_+ = -p_+ - \frac{i\epsilon}{2(p_- + k_-)} \tag{5.15}$$

must be on opposite sides of the real k_+-axis, otherwise one could close the contour on the side without poles, obtaining a vanishing result. Let us first assume $p_- > 0$. Then the two poles are on opposite sides if $k_- < 0$ and $k_- + p_- > 0$, and

$$U(p) = \frac{1}{(2\pi)^2} \int_{-p_-}^{0} dk_- \, 2\pi i \, \frac{1}{p_+} = \frac{i}{2\pi} \frac{p_-}{p_+}. \tag{5.16}$$

Similarly, when $p_- < 0$, the two poles are on opposite sides if $k_- > 0$ and $k_- + p_- < 0$, and the same final result is obtained

$$U(p) = \frac{1}{(2\pi)^2} \int_{0}^{-p_-} dk_- \, (-2\pi i) \, \frac{1}{p_+} = \frac{i}{2\pi} \frac{p_-}{p_+}. \tag{5.17}$$

---

[3] The notation $A_+(x)$ for the function and $A_+(p)$ for its Fourier transform should not cause confusion, as we indicate the arguments explicitly.

The effective action to this order is thus

$$W_L^{(2)}[A] = -\frac{1}{4\pi}\int \frac{d^2p}{(2\pi)^2} A_+(p)\frac{p_-}{p_+}A_+(-p)$$
$$= -\frac{1}{4\pi}\int d^2x\, A_+(x)\frac{\partial_-}{\partial_+}A_+(x). \qquad (5.18)$$

One may check that for abelian gauge fields higher-order contributions to the effective action vanish, so this result is exact. For nonabelian gauge fields one finds nonlocal terms with three, four, five, and more external A_μ, but only the term with two A_μ fields contributes to the anomaly.[4]

Let us now analyze the gauge invariance. Under a gauge transformation

$$\delta A_\mu(x) = \partial_\mu \alpha(x) \quad \rightarrow \quad \delta A_\mu(p) = -ip_\mu \alpha(p) \qquad (5.19)$$

the effective action is not gauge invariant

$$\delta W_L[A] = \frac{i}{2\pi}\int \frac{d^2p}{(2\pi)^2}\, \alpha(p)\, p_- A_+(-p)$$
$$= \frac{1}{2\pi}\int d^2x\, \alpha(x)\, \partial_- A_+(x). \qquad (5.20)$$

Thus, it seems that the gauge symmetry has an anomaly. However, before deciding that this is a true anomaly, one must make sure that there does not exist a local counterterm for which the variation cancels the anomaly. Since the anomaly is Lorentz invariant, we consider the most general Lorentz-invariant local counterterm with the correct dimension

$$W_{loc}[A] = \beta\int d^2x\, A_\mu(x)A^\mu(x) = -2\beta\int d^2x\, A_-(x)A_+(x)$$
$$= -2\beta\int \frac{d^2p}{(2\pi)^2} A_-(p)A_+(-p) \qquad (5.21)$$

where β is an arbitrary parameter. Its gauge variation is easily computed

$$\delta W_{loc}[A] = 2i\beta\int \frac{d^2p}{(2\pi)^2}\, \alpha(p)\, [p_- A_+(-p) + p_+ A_-(-p)]. \qquad (5.22)$$

Clearly, no value of β can make the effective action $W_L[A]+W_{loc}[A]$ gauge invariant. Thus, the final conclusion is that there is an anomaly in the gauge symmetry.

Let us pause for a moment and make various comments.

---

[4]For nonabelian gauge fields one can parametrize A_+ as $\partial_+ g g^{-1}$ and the effective action $\Gamma(g)$ becomes a Wess–Zumino–Novikov–Witten (WZNW) model [95]. This WZNW [96] model can formally be written as an infinite sum of multiple commutators, and this explains why in the abelian case only the quadratic term remains [97]. A simpler explanation is that in conformal field theory the operator product of two abelian currents, $\lambda^\dagger\lambda$ contains only an anomaly term.

- To compute the effective action we have used analytic regularization which regulates the logarithmic divergent graph in (5.14) and automatically removes the divergence. This is one possible renormalization condition. For other renormalization conditions we may need to add the local counterterm in (5.21) with a particular value of β.

- One-loop effective actions which are computed with different regularization schemes can only differ by local counterterms. The addition of the most general local counterterm allows one to scan all possible regularizations at once. If the anomaly does not vanish for any possible counterterm, it means that the anomaly is not an artefact of the chosen regularization scheme. It is a genuine effect appearing in the quantum theory.

- One can also write the gauge variation of the effective action as follows:

$$\delta W[A] = \int d^2x \, \delta A_\mu(x) \frac{\delta W[A]}{\delta A_\mu(x)} = -\int d^2x \, \alpha(x) \partial_\mu J^\mu(x) \quad (5.23)$$

where $J^\mu(x) \equiv \delta W[A]/\delta A_\mu(x)$ is sometimes called the induced current. (It corresponds to the expectation value of the current coupled to the gauge field, $j^\mu = i\bar{\lambda}\gamma^\mu\lambda$, namely $J^\mu = \langle j^\mu \rangle$.) The gauge anomaly then reads as

$$\partial_\mu J^\mu(x) = -\frac{1}{2\pi}\partial_- A_+(x). \quad (5.24)$$

One may note that this expression is not gauge invariant.

- The consistency conditions are integrability conditions which follow from applying the commutator algebra of the symmetries to the effective action. The algebra of the gauge symmetry in (5.3) is abelian and reads as

$$[\delta(\alpha_1), \delta(\alpha_2)] = 0 \quad (5.25)$$

while the anomaly can be denoted by

$$\delta(\alpha) W[A] \equiv An[\alpha, A]. \quad (5.26)$$

Combining these two equations gives the consistency condition for the anomaly of a chiral fermion

$$\delta(\alpha_1) An[\alpha_2, A] = \delta(\alpha_2) An[\alpha_1, A]. \quad (5.27)$$

The anomaly is given by

$$An[\alpha, A] = \frac{1}{2\pi} \int d^2x \, \alpha(x) \partial_- A_+(x) \tag{5.28}$$

and clearly satisfies (5.27).

- A consequence of the consistency conditions is that the consistent gauge anomaly cannot be gauge invariant. This is immediately obvious by inspection of (5.24). More generally one can prove this property as follows. Let us introduce the shift transformation

$$\delta_s A_\mu = s_\mu \tag{5.29}$$

as a trick to study the gauge current J^μ, since then $\delta_s W[A] = \int d^2x \, s_\mu J^\mu$ (recall (5.23)). It is clear that

$$[\delta_s, \delta(\alpha)] A_\mu = 0 \tag{5.30}$$

if one defines $\delta(\alpha) s_\mu = 0$ [2]. However, evaluating this commutator on the effective action produces

$$[\delta_s, \delta(\alpha)] W[A] = \delta_s \mathcal{A}[\alpha, A] - \int d^2x \, (\delta(\alpha) J^\mu) s_\mu = 0 \tag{5.31}$$

which shows that the current J^μ must transform nontrivially under a gauge transformation and thus cannot be gauge invariant (unless the anomaly $\mathcal{A}$ vanishes or is A_μ independent).

- One can introduce another anomaly, called the covariant anomaly, which is not obtained by varying the effective action, but which is gauge covariant (or rather gauge invariant in our case). It is by definition the divergence of a "covariant current" obtained by adding a suitable local (in general, noncovariant) term $\tilde{J}^\mu$ to the consistent current J^μ. For $J_- = \frac{1}{2\pi} \frac{\partial_-}{\partial_+} A_+$ (recall that $J^+ = -J_-$) the addition of $\tilde{J}_- = -\frac{1}{2\pi} A_-$ yields a gauge-invariant nonlocal current

$$J_- + \tilde{J}_- = \frac{1}{2\pi} \frac{1}{\partial_+} (\partial_- A_+ - \partial_+ A_-) \tag{5.32}$$

for which the anomaly is covariant and local,

$$\partial_+ (J_- + \tilde{J}_-) = \frac{1}{2\pi} (\partial_- A_+ - \partial_+ A_-). \tag{5.33}$$

We stress that this "covariant anomaly" cannot be obtained as the gauge variation of the effective action, as it does not satisfy the consistency conditions.

(ii) Let us now add the contribution of a right-handed fermion to the previous model. The total action is the sum of the chiral actions and the path integral factorizes. So we only need to consider the extra terms arising from

$$S[\lambda_R, \lambda_R^\dagger, A] = \int d^2x\, i\lambda_R^\dagger(\partial_- - iA_-)\lambda_R$$
$$\int \mathcal{D}\lambda_R\, \mathcal{D}\lambda_R^\dagger\; \mathrm{e}^{iS[\lambda_R,\lambda_R^\dagger,A]} = \mathrm{e}^{iW_R[A]}. \tag{5.34}$$

The calculation is quite similar to the one described above and produces the following contribution to the effective action:

$$W_R[A] = -\frac{1}{4\pi}\int \frac{d^2p}{(2\pi)^2} A_-(p)\frac{p_+}{p_-}A_-(-p). \tag{5.35}$$

The sum $W_L[A] + W_R[A]$ is still not gauge invariant, but adding the local counterterm W_{loc} in (5.21) with $\beta = -1/4\pi$ makes the final effective action gauge invariant,

$$\begin{aligned} W[A] &= W_L[A] + W_R[A] + W_{loc}[A] \\ &= -\frac{1}{4\pi}\int \frac{d^2p}{(2\pi)^2}\,[p_+A_-(p) - p_-A_+(p)]\,\frac{1}{p_+p_-} \\ &\quad \times [p_+A_-(-p) - p_-A_+(-p)] \\ &= \frac{1}{4\pi}\int d^2x\, F_{+-}\frac{1}{\partial_+\partial_-}F_{+-} \\ &= \frac{1}{4\pi}\int d^2x\, F_{\mu\nu}\frac{1}{\Box}F^{\mu\nu}. \end{aligned} \tag{5.36}$$

We have denoted the d'Alembertian by $\Box = \partial^\mu\partial_\mu = -2\partial_+\partial_-$. Equivalently, the induced gauge current is conserved: $\partial_\mu J^\mu(x) = 0$. This is an example of anomaly cancellation. One may notice that our simple model turned out to have just vectorial couplings. A manifestly gauge-invariant regularization can be used in similar models, and this is enough to guarantee absence of anomalies. For example, in the case above a Pauli–Villars regularization maintains gauge invariance and would have produced automatically the correct local counterterm in the effective action [98]. In models with chiral couplings, where a manifestly gauge-invariant regularization is lacking, cancellation of anomalies must be checked by hand, as in the Standard Model of particle physics.

(iii) Finally, let us discuss the anomaly in a global current. The model above with a Dirac fermion enjoys at the classical level the axial symmetry given in (5.4). It is quite simple to see that this symmetry is anomalous. The Noether current associated with this symmetry is $j_5^\mu = i\bar{\lambda}\gamma^\mu\gamma_5\lambda$.

Thanks to the reducibility of the Lorentz group in two dimensions, we can actually use a trick to reinterpret A_μ as the gauge field (or source) coupled to j_5^μ. In fact, we can substitute $A_\mu = B^\nu \epsilon_{\nu\mu}$ (so $A_+ = B_+$ but $A_- = -B_-$) and use $\epsilon_{\nu\mu}\gamma^\mu = \gamma_\nu\gamma_5$

$$iA_\mu \bar{\lambda}\gamma^\mu\lambda = iB^\nu \epsilon_{\nu\mu}\bar{\lambda}\gamma^\mu\lambda = iB_\mu\bar{\lambda}\gamma^\mu\gamma_5\lambda. \tag{5.37}$$

The transformation

$$\delta(\beta)B_\mu(x) = \partial_\mu\beta(x) \tag{5.38}$$

gauges the symmetry in (5.4) and can be used to test the conservation of the axial current $j_5^\mu = i\bar{\lambda}\gamma^\mu\gamma_5\lambda$. On one hand, we can compute the variation of the effective action $W[A]$ as

$$\begin{aligned}\delta(\beta)W[A] &= \int d^2x \left[\delta(\beta)B_\mu(x)\right]\frac{\delta W[A(B)]}{\delta B_\mu(x)} \\ &= -\int d^2x\, \beta(x)\partial_\mu\langle j_5^\mu(x)\rangle. \end{aligned} \tag{5.39}$$

On the other hand, the explicit form of $W[A]$ computed in (5.36) yields, using $\delta(\beta)A_+ = \partial_+\beta$ and $\delta(\beta)A_- = -\partial_-\beta$,

$$\delta(\beta)W[A] = -\frac{1}{\pi}\int d^2x\, \beta(x)F_{+-}(x). \tag{5.40}$$

Thus,

$$\partial_\mu\langle j_5^\mu\rangle = \frac{1}{\pi}F_{+-}(x) = -\frac{1}{2\pi}\epsilon^{\mu\nu}F_{\mu\nu}(x). \tag{5.41}$$

This is the anomaly in the global axial $U_A(1)$ current. It is manifestly invariant under the $U_V(1)$ gauge group, as expected since the $U_V(1)$ gauge symmetry is not anomalous. Since the only local counterterm which depends on A_+ and A_-, and is Lorentz invariant, is $\Delta\mathcal{L} = \beta A_+ A_-$, and since its coefficient was already fixed by requiring cancellation of the vector anomaly, one cannot remove the chiral (or better the axial-vector) anomaly by a local counterterm: the axial anomaly is a genuine anomaly. The four-dimensional analog of this anomaly is the original Adler–Bell–Jackiw anomaly which is gauge invariant and related to pion decay.

5.1.2 The gravitational anomaly

Next, we construct the gravitational anomaly for a real chiral spin-$\frac{1}{2}$ field coupled to gravity in 1+1 dimensions. In [1] a similar calculation was performed.

The classical Lagrangian reads

$$\mathcal{L} = -\frac{1}{2} e\bar{\lambda}\gamma^m e_m{}^\mu \partial_\mu \lambda, \qquad \bar{\lambda} \equiv \lambda^T i\gamma^0 \tag{5.42}$$

where $e = \det e_\mu{}^m$ and $e_m{}^\mu$ is the inverse of the vielbein $e_\mu{}^m$. The Lagrangian is invariant under general coordinate transformations given by

$$\begin{aligned} \delta\lambda &= \xi^\mu \partial_\mu \lambda \\ \delta e_\mu{}^m &= \xi^\nu \partial_\nu e_\mu{}^m + \partial_\mu \xi^\nu e_\nu{}^m \\ \delta e &= \partial_\mu(\xi^\mu e). \end{aligned} \tag{5.43}$$

It is also invariant under local Lorentz transformations given by

$$\begin{aligned} \delta\lambda &= \frac{1}{4}\lambda^{mn}\gamma_{mn}\lambda, \quad \gamma_{mn} \equiv \frac{1}{2}[\gamma_m, \gamma_n] \\ \delta e_\mu{}^m &= \lambda^m{}_n e_\mu{}^n \\ \delta e &= 0. \end{aligned} \tag{5.44}$$

To prove the last statement, note that the Lorentz variation

$$\delta\mathcal{L} = -\frac{1}{2} e\bar{\lambda}\gamma^m e_m{}^\mu \frac{1}{4}(\partial_\mu \lambda^{pq})\gamma_{pq}\lambda \tag{5.45}$$

vanishes since a Majorana spinor satisfies the identity $\bar{\lambda}\gamma_m\lambda = 0$ while $\gamma^m\gamma_{pq} = \delta_p^m \gamma_q - \delta_q^m \gamma_p$. In higher dimensions, or for a complex (i.e. Dirac) spinor in two dimensions, one needs a term with the spin connection

$$\mathcal{L} = -e\bar{\lambda}\gamma^m e_m{}^\mu \left(\partial_\mu + \frac{1}{4}\omega_\mu{}^{pq}\gamma_{pq}\right)\lambda, \qquad \bar{\lambda} \equiv \lambda^\dagger i\gamma^0 \tag{5.46}$$

but in two dimensions this term vanishes for real λ, as we explained previously.

Consider now a chiral fermion satisfying $(1 - \gamma_5)\lambda = 0$. Since $\gamma_5 = \gamma^1\gamma^0 = \left(\begin{smallmatrix} 1 & 0 \\ 0 & -1 \end{smallmatrix}\right)$, this field has only an upper component which we denote by λ_-, so $\lambda = \left(\begin{smallmatrix} \lambda_- \\ 0 \end{smallmatrix}\right)$. The action for $\lambda_- = \lambda_L$ reduces to

$$\begin{aligned} \mathcal{L} &= -\frac{1}{2} e\lambda^T i\gamma^0(\gamma^m e_m{}^\mu)\partial_\mu\lambda \\ &= \frac{i}{2} e\lambda^T (e_0{}^\mu + \gamma_5 e_1{}^\mu)\partial_\mu\lambda \\ &= \frac{i}{2} e\lambda_L^T (e_0{}^\mu + e_1{}^\mu)\partial_\mu\lambda_L \\ &= \frac{i}{\sqrt{2}} e\lambda_-(e_+{}^\mu\partial_\mu)\lambda_-, \qquad e_\pm{}^\mu = \frac{1}{\sqrt{2}}(e_0{}^\mu \pm e_1{}^\mu). \end{aligned} \tag{5.47}$$

There follow now a few typical two-dimensional manipulations which allow us to write the action such that it only depends on one component of the gravitational field [98]. First, we write $e_+{}^\mu\partial_\mu$ as $e_+{}^{\tilde{+}}\partial_+ + e_+{}^{\tilde{-}}\partial_-$, where $\partial_\pm = \frac{1}{\sqrt{2}}(\partial_0 \pm \partial_1)$ and hence $e_m{}^{\tilde{\pm}} = \frac{1}{\sqrt{2}}(e_m{}^0 \pm e_m{}^1)$. The twiddle above the indices + and − indicates that these are curved indices. Then we extract the field $e_+{}^{\tilde{+}}$ and redefine λ_- such that it absorbs $e_+{}^{\tilde{+}}$,

$$\mathcal{L} = \frac{i}{\sqrt{2}}(\sqrt{e\,e_+{}^{\tilde{+}}}\lambda_-)(\partial_+ + h_{++}\partial_-)(\lambda_-\sqrt{e\,e_+{}^{\tilde{+}}}) \tag{5.48}$$

where

$$h_{++} = \frac{e_+{}^{\tilde{-}}}{e_+{}^{\tilde{+}}}. \tag{5.49}$$

Clearly, h_{++} is Lorentz invariant, and also

$$\tilde{\lambda}_- = \sqrt{e\,e_+{}^{\tilde{+}}}\lambda_- \tag{5.50}$$

is Lorentz invariant because vielbeins rotate twice as fast as spinors under Lorentz rotations.[5]

One can find a simpler expressions for the spinor $\tilde{\lambda}_-$ by using the explicit form of the inverse vielbein density

$$e\,e_m{}^\mu = \begin{pmatrix} e_{\tilde{-}}{}^- & -e_{\tilde{+}}{}^- \\ -e_{\tilde{-}}{}^+ & e_{\tilde{+}}{}^+ \end{pmatrix}, \qquad e_\mu{}^m \equiv \begin{pmatrix} e_{\tilde{+}}{}^+ & e_{\tilde{+}}{}^- \\ e_{\tilde{-}}{}^+ & e_{\tilde{-}}{}^- \end{pmatrix}. \tag{5.51}$$

Then,

$$\tilde{\lambda}_- = \sqrt{e_{\tilde{-}}{}^-}\lambda_-, \qquad h_{++} = -\frac{e_{\tilde{+}}{}^-}{e_{\tilde{-}}{}^-} \tag{5.52}$$

and the Lagrangian becomes

$$\mathcal{L} = \frac{i}{\sqrt{2}}\tilde{\lambda}_-\partial_+\tilde{\lambda}_- + \frac{1}{\sqrt{2}}h_{++}T_{--}, \qquad T_{--} = i\tilde{\lambda}_-\partial_-\tilde{\lambda}_-. \tag{5.53}$$

The fields $\tilde{\lambda}_-$ are left-moving fields that transforms as follows under general coordinate transformations:

$$\begin{aligned} \delta\tilde{\lambda}_- &= \xi^\mu\partial_\mu\tilde{\lambda}_- + \frac{1}{2}\frac{1}{\sqrt{e_{\tilde{-}}{}^-}}(\partial_-\xi^\alpha e_\alpha{}^-)\lambda_- \\ &= \xi^\mu\partial_\mu\tilde{\lambda}_- + \frac{1}{2}(\partial_-\xi^- - \partial_-\xi^+ h_{++})\tilde{\lambda}_-. \end{aligned} \tag{5.54}$$

[5] The Lorentz invariance is manifest from the matching of + and − indices. Some authors write ++ and = for the indices of vector fields, and others write $\sqrt{+}$ and $\sqrt{-}$ for the indices of spinor fields. We use only the indices + and − but the reader should keep in mind that vectors and spinors transforms differently.

In particular, under transformations with ξ^-, $\tilde{\lambda}_-$ transforms as a "half-vector" (due to the factor of $\frac{1}{2}$), and this forms the starting point for conformal field theory where $\tilde{\lambda}_-$ has a conformal spin of $\frac{1}{2}$.

The field h_{++} transforms as follows under general coordinate transformations:

$$\begin{aligned}\delta h_{++} &= \xi^\alpha \partial_\alpha h_{++} - \frac{1}{e_{\tilde{-}}{}^{-}}(\partial_+ \xi^\alpha e_\alpha{}^-) + \frac{e_{\tilde{+}}{}^{-}}{(e_{\tilde{-}}{}^{-})^2}(\partial_- \xi^\alpha e_\alpha{}^-) \\ &= \xi^\alpha \partial_\alpha h_{++} + (\partial_+ \xi^+) h_{++} - \partial_+ \xi^- + (\partial_- \xi^+) h_{++}^2 - (\partial_- \xi^-) h_{++}.\end{aligned} \tag{5.55}$$

So, to lowest order in h_{++}, we have $\delta h_{++} = -\partial_+ \xi^- = \partial_+ \xi_+$ (since $\eta^{+-} = -1$).

A similar treatment of right-moving fermions satisfying $(1+\gamma_5)\lambda = 0$ shows that they couple only to

$$h_{--} = -\frac{e_{\tilde{-}}{}^{+}}{e_{\tilde{+}}{}^{+}}. \tag{5.56}$$

As the third field which parametrizes the space of symmetric vielbeins we take

$$\begin{aligned}h_{+-} = h_{-+} &= \frac{1}{2} e_{\tilde{+}}{}^{m} e_{\tilde{-}}{}^{n} \eta_{mn} = -\frac{1}{2}(e_{\tilde{+}}{}^{+} e_{\tilde{-}}{}^{-} + e_{\tilde{+}}{}^{-} e_{\tilde{-}}{}^{+}) \\ \delta h_{+-} &= \partial_+ \xi_- + \partial_- \xi_+ + \cdots .\end{aligned} \tag{5.57}$$

All three fields h_{++}, h_{--} and h_{+-} are Lorentz invariant.

Let us now compute the one-loop effective action for the theory with $\tilde{\lambda}_-$. It is given by the sums of graphs

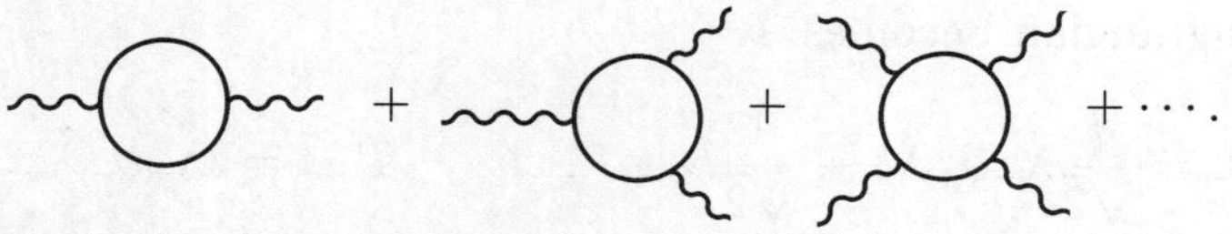
(5.58)

The anomaly resides only in the first graph, so we only evaluate this graph. Afterwards we will comment on the graphs with three and more h-fields. The $\tilde{\lambda}$ propagator is of course unchanged and is given by (5.12),

$$\frac{1}{\partial_+}\delta^2(x-y) = \frac{\partial_-}{\partial_+\partial_-}\delta^2(x-y) = \int \frac{d^2k}{(2\pi)^2}\frac{2ik_-}{k^2 - i\epsilon} e^{ik(x-y)}. \tag{5.59}$$

Hence,

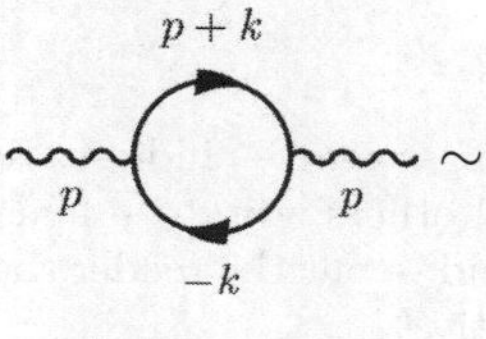

$$\sim \iint d^2x\, d^2y\, h_{++}(x) \langle T_{--}(x) T_{--}(y) \rangle h_{++}(y)$$

$$\sim \iint d^2x\, d^2y\, h_{++}(x)h_{++}(y)\langle\tilde{\lambda}_-(x)\partial_-\tilde{\lambda}_-(x)\,\lambda_-(y)\partial_-\tilde{\lambda}_-(y)\rangle$$

$$\sim \int d^2p\, h_{++}(p)h_{++}(-p)\int \frac{d^2k}{(2\pi)^2}(2k_- + p_-)^2\frac{k_- + p_-}{(k+p)^2 - i\epsilon}\frac{k_-}{k^2 - i\epsilon}$$

$$\sim \int d^2p\, h_{++}(p)h_{++}(-p)$$

$$\times \int \frac{dk_+ dk_-}{(2\pi)^2}(2k_- + p_-)^2\frac{1}{k_+ + p_+ + i\epsilon/2(k_- + p_-)}\frac{1}{k_+ + i\epsilon/2k_-}. \qquad (5.60)$$

We use again contour integration for the integral over k_+, which is non-vanishing (for $p_- > 0$) when $-p_- < k_- < 0$. We are then left with an integral of the form

$$\int_{-p_-}^{0} dk_- \frac{(2k_- + p_-)^2}{p_+} \sim \frac{p_-^3}{p_+}. \qquad (5.61)$$

Hence the effective action is proportional to

$$S_{\mathit{eff}} \sim \int d^2p\, h_{++}(p)\frac{p_-^3}{p_+}h_{++}(-p). \qquad (5.62)$$

We are now in a position to check whether the effective action is gauge (Einstein) invariant. Using the linearized transformation rules

$$\delta h_{++}(p) = \partial_+\xi_+ + \cdots = -ip_+\xi_+(p) + \cdots \qquad (5.63)$$

we find a candidate for an anomaly

$$\delta S_{\mathit{eff}} \sim \int d^2p\, \xi_+(p)p_-^3 h_{++}(-p). \qquad (5.64)$$

The result is rigidly Lorentz invariant (+ and − indices match; **rigid** Lorentz transformations also act on curved indices).

So the effective action is not gauge invariant, but one should still check that its variation cannot be canceled by the variation of a suitable local counterterm in the action. The most general counterterm for which the variation has the same number of fields as S_{eff} and is Lorentz invariant reads as

$$\begin{aligned}\Delta S = \int d^2p\, [&Ah_{++}(p)p_-p_-h_{+-}(-p)\\ &+ Bh_{++}(p)p_+p_-h_{--}(-p)\\ &+ Ch_{+-}(p)p_+p_-h_{+-}(-p)\\ &+ Dh_{+-}(p)p_+p_+h_{--}(-p)].\end{aligned} \qquad (5.65)$$

One may check that for no value of the constants A, B, C, D can the variation of ΔS cancel δS_{eff}.[6] Hence there exists a genuine gravitational anomaly in 1+1 dimensions for chiral spinors.

One expects that for nonchiral spinors there is no genuine gravitational anomaly. Consider the sum of the actions for λ_L and λ_R. The effective action is now a sum of an effective action for $\tilde{\lambda}_-$ depending on h_{++}, and another effective action for $\tilde{\lambda}_+$ depending on h_{--}. The variation of this sum of effective actions is proportional to

$$\begin{aligned}\delta S_{\mathit{eff}} &= \delta(S^L_{\mathit{eff}} + S^R_{\mathit{eff}}) \\ &\sim 2\int d^2p\,[\xi_+(p)p_-^3 h_{++}(-p) + \xi_-(p)p_+^3 h_{--}(-p)]. \end{aligned} \tag{5.66}$$

Now, however, there is a local counterterm for which the variation cancels δS_{eff}, namely

$$\begin{aligned}\Delta S \sim \int d^2p\,[&-4h_{++}(p)p_-p_-h_{+-}(-p) \\ &+ 2h_{++}(p)p_+p_-h_{--}(-p) \\ &+ 4h_{+-}(p)p_+p_-h_{+-}(-p) \\ &- 4h_{+-}(p)p_+p_+h_{--}(-p)]. \end{aligned} \tag{5.67}$$

In fact, using Appendix A, one may show that the effective action $S_{\mathit{eff}} + \Delta S$ is given by

$$S_{\mathit{eff}} + \Delta S \sim \int d^2p\,\frac{R(p)R(-p)}{p_+p_-} \sim \int d^2x\, eR(x)\frac{1}{\Box}R(x) \tag{5.68}$$

where $R(p) \sim p_+^2 h_{--}(p) + p_-^2 h_{++}(p) - 2p_+p_-h_{+-}(p)$ is the linearized form of the scalar curvature.

5.1.3 The trace anomaly

Finally, we also demonstrate the presence of a trace anomaly in two dimensions. We consider a real nonchiral fermion. The classical action in (5.53) is independent of h_{+-}, so that the classical stress tensor is traceless: $T_{+-} = 0$. (We obtain T_{+-} by varying the classical action with respect to h_{+-}.) At the quantum level, S_{eff} is still independent of h_{-+}, so T_{+-} still vanishes at the one-loop level. However, if we make the effective action Einstein invariant by adding the local counterterm ΔS, the effective action starts depending on h_{-+}, and thus there is a trace anomaly.

[6] The variation proportional to $D\,\xi_-\,p_-^3 h_{--}$ requires $D = 0$, and similarly one also finds $A = B = C = 0$.

From the expression for ΔS given in (5.67) and $\delta R \sim 2p_+p_-\delta h_{+-} + \cdots$ we see that the trace anomaly is given by

$$T_{+-} \sim T_\mu{}^\mu \sim R. \tag{5.69}$$

We have seen in concrete two-dimensional models the various aspects of anomalies: anomalies in gauge and gravitational currents, anomalies in rigid currents and cancellation of anomalies. We have used analytic regularization, but it is a nice exercise to repeat these calculations using other regularization schemes, for example Pauli–Villars regularization, ordinary ('t Hooft–Veltman) dimensional regularization or regularization by dimensional reduction. In higher dimensions the full effective action cannot be calculated explicitly in closed form, but one can still study the anomalous behavior of the various Feynman graphs. In the remaining part of the book we will use the general method based on (susy) quantum mechanics to compute the anomalies.

5.2 The Fujikawa method

Instead of computing anomalies by using Feynman diagrams, one may use a path integral approach. This allows for a conceptually more satisfactory way of obtaining anomalies because the anomalies are present from the beginning, instead of being discovered afterwards as the nonvanishing of the divergence of the quantum currents. The fundamental discovery, on which all subsequent work on anomalies in the path integral context is based, was the observation of K. Fujikawa in 1979 (ten years after anomalies in Feynman graphs were discovered) that anomalies are located in the path integral Jacobian [12, 99]. Namely, if one makes a change of integration variables in the path integral (which should not change the value of the path integral), the change in the quantum action (the action that appears in the exponent of the path integral) is compensated by minus the Jacobian. If this change of integration variables is part of the transformation rules of a symmetry of the quantum action, the change of the quantum action is equal to the divergence of the quantum Noether current. In this way the path integral expresses the anomaly in terms of the Jacobian.

The Fujikawa method is usually applied to the path integral of a quantum field theory itself. In later sections we shall evaluate the result for the Jacobian using quantum mechanics, but in this section we use quantum field theory methods. To appreciate the origin of the method, we give here a discussion of the Fujikawa approach applied to the quantum field theory with a rigid $U(1)$ chiral anomaly in curved space for a Dirac spinor in four Euclidean dimensions. At the end of this section we shall comment on extensions of this method to other models and other symmetries.

Consider a complex massless spin-$\frac{1}{2}$ fermion λ in a gravitational background described by the vielbein $e_\mu{}^m$. The partition function reads as

$$Z[e] = \int D\lambda\, D\bar{\lambda}\; \mathrm{e}^{-\int d^4x\, e\, \bar{\lambda}\, e_m{}^\mu \gamma^m D_\mu \lambda}. \tag{5.70}$$

Under an infinitesimal local chiral transformation of integration variables

$$\lambda \to [1 + i\alpha(x)\gamma_5]\lambda, \qquad \bar{\lambda} \to \bar{\lambda}[1 + i\alpha(x)\gamma_5] \tag{5.71}$$

the path integral does not change. The action is only invariant for constant α, hence for local $\alpha(x)$ it varies into the Noether current

$$\delta S = \int d^4x e\, \bar{\lambda} e_m{}^\mu \gamma^m i(\partial_\mu \alpha)\gamma_5 \lambda. \tag{5.72}$$

Hence, the Jacobian cancels the divergence of the Noether current

$$-\mathrm{Tr}(2i\alpha\gamma_5) - i\int d^4x\, e\, \langle \bar{\lambda}\gamma^\mu\gamma_5\lambda\rangle \partial_\mu \alpha = 0. \tag{5.73}$$

The minus sign is due to the superJacobian. If the trace "Tr" is nonvanishing, the Noether current $\bar{\lambda}\gamma^\mu\gamma_5\lambda$ is no longer conserved at the quantum level, and then there is an anomaly.

To regularize the trace Tr over spacetime points and spinor indices in a gauge covariant way, Fujikawa showed that any function $f(\not{D}^2)$ can be used, provided $f(0) = 1$, $f(\infty) = 0$ and $x^p(\frac{d}{dx})^p f(x)$ vanishes at $x = 0$ and $x = \infty$ for $p > 0$.

A convenient choice is $f = \mathrm{e}^{\not{D}^2/M^2}$. Hence, using plane waves

$$\begin{aligned}\mathrm{Tr}(2i\alpha\gamma_5) &= \sum \langle k|\mathrm{e}^{\not{D}^2/M^2} 2i\alpha\gamma_5|k\rangle \\ &\equiv \int \frac{d^4k}{(2\pi)^4} \mathrm{e}^{-ikx} \mathrm{tr}(\mathrm{e}^{\not{D}^2/M^2} 2i\alpha\gamma_5)\mathrm{e}^{ikx}\end{aligned} \tag{5.74}$$

where "tr" denotes the trace over spinor indices. By pulling the plane waves e^{ikx} to the left, the operator $\not{D}$ is replaced by $\not{D} + i\not{k}$, and one obtains

$$\begin{aligned}\mathrm{Tr}(2i\alpha\gamma_5) &= \int \frac{d^4k}{(2\pi)^4} \mathrm{tr}(2i\alpha\gamma_5 \mathrm{e}^{-k^2/M^2} \mathrm{e}^{B/M^2}) \\ B &= 2ik \cdot D + D^2 + \frac{1}{4}R.\end{aligned} \tag{5.75}$$

In deriving this result, we have used the cyclic identity for the Riemann tensor

$$\frac{1}{4}[\gamma^\mu, \gamma^\nu][D_\mu, D_\nu] = \frac{1}{8}\gamma^{\mu\nu} R_{\mu\nu}{}^{ab}(\omega)\gamma_{ab} = \frac{1}{4}R. \tag{5.76}$$

The evaluation of the integral is performed by expanding e^{B/M^2}, and only retaining terms which do not vanish when M^2 tends to infinity. The relevant terms are

$$\begin{aligned}\mathrm{Tr}(2i\alpha\gamma_5) = \int \frac{d^4k}{(2\pi)^4}\mathrm{tr}(2i\alpha\gamma_5)\, e^{-k^2/M^2}\Big\{1 + \frac{1}{1!}\left(D^2 + \frac{1}{4}R\right)M^{-2} \\ + \frac{1}{2!}\left[(2ik\cdot D)(2ik\cdot D) + \left(D^2 + \frac{1}{4}R\right)\left(D^2 + \frac{1}{4}R\right)\right]M^{-4} \\ + \frac{1}{3!}\Big[(2ik\cdot D)(2ik\cdot D)\left(D^2 + \frac{1}{4}R\right) \\ + (2ik\cdot D)\left(D^2 + \frac{1}{4}R\right)(2ik\cdot D) \\ + \left(D^2 + \frac{1}{4}R\right)(2ik\cdot D)(2ik\cdot D)\Big]M^{-6} \\ + \frac{1}{4!}[(2ik\cdot D)^4]M^{-8}\Big\}. \end{aligned} \tag{5.77}$$

Using

$$\int d^4k = M^4\pi^2 \int d(k^2/M^2)(k^2/M^2) \tag{5.78}$$

and

$$\int_0^\infty dy\, e^{-y} y^n = \Gamma(n+1) \tag{5.79}$$

one obtains

$$\begin{aligned}\mathrm{Tr}(2i\alpha\gamma_5) = \frac{2i\alpha}{16\pi^2}\,\mathrm{tr}\,\gamma_5\Big\{M^2D^2 \\ + \frac{1}{2!}\left[(-2)M^2D^2 + \left(D^2 + \frac{1}{4}R\right)\left(D^2 + \frac{1}{4}R\right)\right] \\ + \frac{1}{3!}(-2)\Big[D^2\left(D^2 + \frac{1}{4}R\right) + D_\mu\left(D^2 + \frac{1}{4}R\right)D^\mu \\ + \left(D^2 + \frac{1}{4}R\right)\Big]D^2\Big] \\ + \frac{1}{4!}\frac{6\times 16}{24}\left(D^2D^2 + D_\mu D^2 D^\mu + D_\mu D_\nu D^\mu D^\nu\right)\Big\}. \end{aligned} \tag{5.80}$$

The terms proportional to M^2 contain D^2, which contains four Dirac matrices since $D_\mu = \partial_\mu + \frac{1}{4}\omega_\mu{}^{mn}\gamma_m\gamma_n$, but these terms are seen to cancel. The terms containing the scalar curvature R cancel too, since they are given by

$$\frac{1}{2}\left(\frac{1}{4}D^2R + \frac{1}{4}RD^2\right) - \frac{1}{3}\left(\frac{1}{4}D^2R + \frac{1}{4}D_\mu R D^\mu + \frac{1}{4}RD^2\right) = \frac{1}{24}(D^2R) \tag{5.81}$$

which vanish since $(D_\mu R) = (\partial_\mu R)$ so that the trace over $\gamma_5(D^2 R)$ vanishes. In the remaining terms, those proportional to $(D^2)(D^2)$ cancel too, and one is left with

$$\begin{aligned}
\mathrm{Tr}(2i\alpha\gamma_5) &= \frac{i\alpha}{8\pi^2}\mathrm{tr}\,\gamma_5\left[-\frac{1}{3}D_\mu D^2 D^\mu + \frac{1}{6}(D_\mu D^2 D^\mu + D_\mu D_\nu D^\mu D^\nu)\right] \\
&= \frac{i\alpha}{96\pi^2}\mathrm{tr}\,\gamma_5[D_\mu, D_\nu][D^\mu, D^\nu] \\
&= \frac{i\alpha}{96\pi^2}\mathrm{tr}\,\gamma_5\left(\frac{1}{4}R_{\mu\nu}{}^{ab}\gamma_a\gamma_b\right)\left(\frac{1}{4}R^{\mu\nu cd}\gamma_c\gamma_d\right) \\
&= \frac{i\alpha}{96\pi^2}\frac{1}{4}(\epsilon^{abcd}R_{\mu\nu ab}R^{\mu\nu}{}_{cd}). \qquad (5.82)
\end{aligned}$$

This is the correct gravitational contribution to the spin-$\frac{1}{2}$ chiral anomaly.

We conclude with some comments.

(i) The flat-space counterpart of a Dirac fermion coupled to gauge fields instead of gravity uses $B = 2ik \cdot D + \frac{1}{2}\gamma^\mu\gamma^\nu F_{\mu\nu}$ in (5.75). Only the square of the last term contributes for $n = 4$, and it yields an anomaly proportional to $\epsilon^{\mu\nu\lambda\rho}F_{\mu\nu}F_{\lambda\rho}$ [12, 99].

(ii) For the spin-$\frac{3}{2}$ case [100] with field operator $(\not{D})^{\mu\sigma} \equiv (\Gamma^\rho)^{\mu\sigma}D_\rho = \gamma^\sigma\gamma^\rho\gamma^\mu D_\rho$, the expression for $\not{D}\not{D}$ is the same as for the spin-$\frac{1}{2}$ case, except that the term $\frac{1}{4}R$ is replaced by a new term $\frac{1}{2}R_{\lambda\rho}{}^{ab}(\omega)\gamma_a\gamma_b$. In the spin-$\frac{1}{2}$ case the term $\frac{1}{4}R$ did not contribute but in the spin-$\frac{3}{2}$ case there is a contribution from the square of the new term. The final result is four times $An\left(\frac{1}{2}\right)$ (the result for the spin-$\frac{1}{2}$ case) because one must now also trace over the vector index of the gravitino, minus $24An\left(\frac{1}{2}\right)$ due to the new term, minus $An\left(\frac{1}{2}\right)$ due to the ghosts, yielding in total $An\left(\frac{3}{2}\right) = -21An\left(\frac{1}{2}\right)$.

(iii) One can also calculate the axial anomaly in $N = 1$ conformal supergravity [100]. The various Faddeev–Popov ghosts and Nielsen–Kallosh ghosts for ordinary and conformal supergravity contribute, but the sum of all contributions vanishes. For the spin-$\frac{3}{2}$ field, the field equation is $(\not{D})^3$ but instead of $(\not{D})^6$ as a regulator one may use $(\not{D})^2$, since the result for the axial anomaly does not depend on the precise form of the regulator [1]. Hence, one finds the result of (i) except for the ghost contribution: $An\left(\frac{3}{2}\right) = -20An\left(\frac{1}{2}\right)$.

(iv) For the $N = 4$ conformal supergravity the sum of all anomalous contributions to the local chiral $U(1)$ symmetry cancels [101].

5.3 How to calculate anomalies using quantum mechanics

Anomalies arise when a classical action has a symmetry but the corresponding effective action is no longer invariant under this symmetry. The anomaly is then by definition the variation of the effective action under the symmetry. At the one-loop level the anomaly is a local polynomial in the fields and derivatives of the fields with finite coefficients. The anomaly can only occur if there are an infinite number of degrees of freedom in the theory, but the anomaly itself is finite. Qualitatively one can understand the finiteness of anomalies as follows. The conservation of the classical current yields zero, but the loop graphs yield an infinity, and the product of this zero and this infinity yields a finite result for the anomaly.

In the path integral for quantum field theories the anomaly appears if one makes an infinitesimal change of integration variables which amounts to a symmetry transformation [12]. The action in the path integral is invariant under this change of variables, but if there is an anomaly the Jacobian $1 + \mathrm{Tr}\, J$ for an infinitesimal change of integration variables is not unity. This is a true anomaly only if it cannot be removed by adding local counterterms to the action without spoiling other symmetries. The infinitesimal part of the Jacobian, $\mathrm{Tr}\, J$, is the trace of an operator summed over all points in spacetime (i.e. the trace of an infinite-dimensional matrix) which must be properly defined by regularization. Thus, the expression for the anomaly in terms of the differential operators of the QFT is

$$An = \lim_{\beta \to 0} \mathrm{Tr}\, J \mathrm{e}^{-\beta \mathcal{R}} \tag{5.83}$$

where $\mathcal{R}$ is the regulator (also a differential operator). The trace is over a complete set of states, for which we can take, for example, the set of plane waves or the set of eigenfunctions of the regulator $\mathcal{R}$. (One would prefer to use a positive-definite self-adjoint regulator $\mathcal{R}$ because then the eigenfunctions form a complete orthonormal set with positive eigenvalues.)

Consider first the case where the integration variable in the path integral which we use to describe the spinor field in the quantum field theory is an Einstein scalar $\lambda(x)$ (and, of course, a Lorentz spinor). We shall actually use $\tilde{\lambda}(x) = g^{1/4}\lambda(x)$ as a basis, but it is useful to compare both choices. A natural choice as a regulator $\mathcal{R}$ for the Jacobian of the n-dimensional Dirac action is the square of the Dirac operator, $\mathcal{R} \sim \not{D}\not{D}$. Using standard manipulations with Dirac matrices, one can simplify $\not{D}\not{D}$ to $\not{D}\not{D} = D^\mu D_\mu + \frac{1}{4}R$, where the second D_μ on the right-hand side is given by $D_\mu = \partial_\mu + \frac{1}{4}\omega_\mu{}^{mn}\gamma_m\gamma_n$ while the first D^μ on the right-hand side contains an extra term with a Christoffel connection which acts on the index μ of the second D_μ. One can remove this Christoffel term by rewriting

$D^\mu D_\mu$ as $\frac{1}{\sqrt{g}} D_\mu \sqrt{g} g^{\mu\nu} D_\nu$, where both D_μ and D_ν now only contain the spin connection,[7]

$$\mathcal{R} \sim \frac{1}{\sqrt{g}} D_\mu \sqrt{g} g^{\mu\nu} D_\nu + \frac{1}{4} R, \quad D_\mu = \partial_\mu + \frac{1}{4} \omega_{\mu mn} \gamma^m \gamma^n. \tag{5.84}$$

Replacing γ^m by $\sqrt{\frac{2}{\hbar}}\, \psi^a$ and also making the usual identification[8]

$$\frac{\hbar}{i} \frac{\partial}{\partial x^j} \longleftrightarrow g^{1/4}\, p_j\, g^{-1/4} \tag{5.85}$$

we interpret the regulator $\mathcal{R}$ as the Hamiltonian of a system with bosonic point particles $x^i(t)$ and fermionic point particles $\psi^a(t)$. We multiply $\mathcal{R}$ in (5.84) by $(\hbar/i)^2$ to replace ∂_μ by p_i, and by a factor of $\frac{1}{2}$ to obtain the conventional normalization $H = \frac{1}{2}p^2 + \cdots$. This leads to the following Hamiltonian for the bosonic point particle $x^i(t)$ and the fermionic point particle $\psi^a(t)$:

$$\hat{H} = \frac{1}{2} g^{-1/4} \Big(p_i - \frac{i}{2} \omega_i{}^{ab} \psi_a \psi_b \Big) g^{1/2} g^{ij} \Big(p_j - \frac{i}{2} \omega_j{}^{cd} \psi_c \psi_d \Big) g^{-1/4} - \frac{\hbar^2}{8} R. \tag{5.86}$$

Note that x, p and ψ are all operators in this expression. To avoid confusion we mention that we shall later Weyl-order this Hamiltonian, which will produce another term with R.

Let us now consider what one obtains if one uses $\tilde{\lambda}(x)$ as a basis. One might expect that by choosing a different basis in the quantum field theory one would obtain a different Hamiltonian in the quantum mechanical model. Thus, consider $\tilde{\lambda} = g^{1/4}\lambda$ as the integration variable for the path

---

[7] The expansion proceeds as follows: $\not{D}\not{D} = \frac{1}{2}\{\gamma^\mu, \gamma^\nu\} D_\mu D_\nu + \frac{1}{2}[\gamma^\mu, \gamma^\nu] D_\mu D_\nu = D^\mu D_\mu + \frac{1}{4}[\gamma^\mu, \gamma^\nu][D_\mu, D_\nu]$, where all derivatives D_μ are fully gravitationally covariant, so for example $[D_\mu, \gamma^\nu] = 0$ and $D_\mu e_\nu{}^m = 0$. The second term yields a curvature $\frac{1}{8}\gamma^\mu\gamma^\nu R_{\mu\nu}{}^{mn}(\omega)\gamma_{mn}$. Since we always take for $\omega_\mu{}^{mn}$ the usual spin connection $\omega_\mu{}^{mn}(e)$ which corresponds to the Christoffel connection via the vielbein postulate, see Appendix A, the curvature $R_{\mu\nu}{}^{mn}(\omega)$ satisfies the cyclic identity $R_{\mu[\nu mn]} = 0$ and the Ricci tensor is symmetric. It follows that $\gamma^\nu \gamma_{mn} R_{\mu\nu}{}^{mn} = 2 e^\nu_m \gamma_n R_{\mu\nu}{}^{mn}$ since $\gamma^\nu{}_{mn} R_{\mu\nu}{}^{mn} = 0$ due to the cyclic identity. Furthermore, $\gamma^\mu\, [e^\nu_m \gamma_n R_{\mu\nu}{}^{mn}(\omega)] = e^\nu_m e^\mu_n R_{\mu\nu}{}^{mn}(\omega) = R$ because the Ricci tensor is symmetric. The final result is the term $\frac{1}{4}R$ in $\not{D}\not{D}$. Moreover in general relativity one proves that the covariant divergence of a contravariant vector density $\sqrt{g}v^\mu$ is equal to the ordinary derivative, i.e. $D_\mu(\sqrt{g}v^\mu) = \partial_\mu(\sqrt{g}v^\mu)$. This yields (5.84).

[8] We derived this relation in (2.231); we recall that it follows from the hermiticity of p_j and the hermiticity of $g^{-1/4}\frac{\hbar}{i}\frac{\partial}{\partial x^j} g^{1/4}$ with the inner product $\langle\psi|\varphi\rangle = \int \sqrt{g}\, \psi^*(x)\varphi(x)\, d^n x$.

integral, and take for $\mathcal{R}$ the square of the field operator for $\tilde{\lambda}$ instead of λ. Then,

$$\mathcal{R} \sim g^{-1/4} D_\mu \sqrt{g} g^{\mu\nu} D_\nu g^{-1/4} + \frac{1}{4} R \tag{5.87}$$

but since the inner product is now $\langle \tilde{\psi} | \tilde{\varphi} \rangle = \int \tilde{\psi}^*(x) \tilde{\varphi}(x)\, d^n x$, the relation between $\frac{\hbar}{i} \frac{\partial}{\partial x^j}$ and p_j is now

$$\frac{\hbar}{i} \frac{\partial}{\partial x^j} = p_j. \tag{5.88}$$

Replacing $\frac{\partial}{\partial x^j}$ by $\frac{i}{\hbar} p_j$ in (5.87) yields the same expression for $\mathcal{R}$ as in (5.86). Thus, the regulator in terms of hermitian quantum mechanical operators $\hat{x}^i$ and $\hat{p}_i$ is the same, whether one starts from λ or $\tilde{\lambda}$ as basic field variables.

We now ask the crucial question: which quantum mechanical nonlinear sigma model leads to the Hamiltonian in (5.86)? The answer is the $N = 1$ supersymmetric nonlinear sigma model which in Minkowskian time is given by

$$L = \frac{1}{2} g_{ij} \dot{x}^i \dot{x}^j + \frac{i}{2} \psi_a (\dot{\psi}^a + \dot{x}^i \omega_i{}^{ab} \psi_b). \tag{5.89}$$

In this expression x^i and ψ^a are of course classical functions of t, not operators. We discuss this model in Appendix D, where it is shown that the classical action $S = \int L\, dt$ is supersymmetric. For this model the conjugate momenta are given by $p_i(x) = \frac{\partial}{\partial \dot{x}^i} L = g_{ij} \dot{x}^j + \frac{1}{2} i \omega_i{}^{ab} \psi_a \psi_b$ and $\pi_a(\psi) = \frac{\partial}{\partial \dot{\psi}^a} L = -\frac{1}{2} i \psi_a$. The classical Hamiltonian is given by

$$H_{cl} = \dot{x}^i p_i + \dot{\psi}^a \pi_a - L. \tag{5.90}$$

The terms with $\dot{\psi}^a$ cancel in this expression and elimination of $\dot{x}^i$ yields the following result:

$$H_{cl} = \frac{1}{2} \Big(p_i - \frac{i}{2} \omega_i{}^{ab} \psi_a \psi_b \Big) g^{ij} \Big(p_j - \frac{i}{2} \omega_j{}^{cd} \psi_c \psi_d \Big). \tag{5.91}$$

The term $-\frac{1}{8} \hbar^2 R$ is absent in this expression because it is a quantum effect, as is clear from the $\hbar^2$ term in (5.86).

To write this classical Hamiltonian at the quantum level as an operator which is general coordinate and local Lorentz invariant (meaning it should commute with the operators which generate infinitesimal general coordinate and local Lorentz transformations), one must add the factors with $g^{-1/4}$ and $g^{1/2}$ as in (5.86). We discussed this in the beginning of

Section 2.5. The scalar curvature is Einstein and locally Lorentz invariant by itself, so its coefficient is not fixed by requiring Einstein and local Lorentz invariance only. However, the term $-\frac{1}{8}\hbar^2 R$ in (5.86) with this precise coefficient is fixed by rigid susy (on the worldline). The argument goes as follows. The quantum susy generator[9] is given by

$$\begin{aligned}\hat{Q} &= \psi^c e^i_c \Big(g^{1/4} p_i g^{-1/4} - \frac{i}{2}\omega_i{}^{ab}\psi_a\psi_b \Big) \\ &= \Big(g^{-1/4} p_i g^{1/4} - \frac{i}{2}\omega_i{}^{ab}\psi_a\psi_b \Big)\psi^c e^i_c . \end{aligned} \tag{5.92}$$

It commutes with the generator of general coordinate transformations in (2.191) and (2.196) [19],

$$G_E = \frac{1}{2i\hbar}[p_k \xi^k(x) + \xi^k(x) p_k] + \text{terms acting on } e^i_a \text{ and } \omega_{iab}. \tag{5.93}$$

Requiring that they commute fixes the factors $g^{1/4}$ and $g^{-1/4}$ in Q. It also commutes with the generator of local Lorentz rotations, $J = [\frac{1}{2}\lambda_{ab}(x)\psi^a\psi^b + \text{terms acting on } e^i_a \text{ and } \omega_{iab}]$ because $\psi^c e^i_c$ and $[g^{1/4} p_i g^{-1/4} - \frac{1}{2} i\omega_i{}^{ab}\psi_a\psi_b]$ are separately locally Lorentz invariant, see below (2.198). Defining $\hat{H} = \frac{1}{2}\{\hat{Q}, \hat{Q}\}$ one finds (5.86) including the term $\frac{1}{8}\hbar^2 R$. Thus, whereas Einstein and local Lorentz symmetry do not fix the coefficient of the R term in the quantum Hamiltonian, the rigid susy of the QM model does.

Although we are going to use covariant regulators, let us make a few remarks concerning consistent regulators. Once the action for the quantum field theory in n dimensions is given, the consistent regulator for the Jacobian can be constructed. By a consistent regulator we mean a regulator which produces consistent anomalies, namely anomalies that satisfy the consistency conditions. These consistency conditions follow from the fact that the anomalies are the gauge variation of the one-loop effective action Γ. If there are no anomalies, the effective action (due to fermion loops with external Yang–Mills or gravitational fields) is gauge invariant, but if there are anomalies, the gauge variation with the parameter $\lambda^a(x)$ of the effective action Γ leads to the consistent anomaly $\delta_{gauge}(\lambda^a(x))\,\Gamma = \int d^n x \lambda^a(x) An_a(x)$. In [13] consistent regulators for quantum field theories are constructed using Pauli–Villars regularization of the action.

[9] In general, a generator of a symmetry is the space integral of the time component of the corresponding Noether current. For quantum mechanics there are no space coordinates, so the charge is the current. The susy Noether current is most easily obtained by making a susy variation with a local (time-dependent) susy parameter $\epsilon(t)$, and collecting all terms proportional to $\dot{\epsilon}$ [75]. This yields the expression for Q in the text. The order of the factors with $g^{\pm 1/4}$ in $\hat{Q}$ is determined by the transformation rule $(g^{1/4} p_i g^{-1/4})' = \frac{\partial x^j}{\partial x'^i}(g^{1/4} p_j g^{-1/4})$, see (2.196) and (2.197).

However, it is not necessary to use consistent regulators for the purpose of calculating anomalies; one may also use covariant regulators. When anomalies cancel with one regulator, they also cancel for another regulator (possibly after adding local counterterms to the action) and working with covariant regulators has the advantage that calculations are simpler. For this reason we shall use covariant regulators (such as (5.86) for spin $\frac{1}{2}$) to evaluate the anomalies in the local Lorentz symmetry and gauge symmetry of loops with chiral fermions and self-dual antisymmetric tensors.

The anomaly in the field theory is proportional to $\mathrm{Tr}\, J e^{-\beta\mathcal{R}}$ where $1 + J$ is the infinitesimal Jacobian for the symmetry with the anomaly we want to compute. For example, for the rigid chiral symmetry $\delta\psi \sim \gamma_5\psi$ of massless Dirac actions the Jacobian is proportional to γ_5 and we shall construct a quantum mechanical representation for γ_5, denoted by $\hat{\gamma}_5$. One is then led to the expression

$$An = \lim_{\beta\to 0} \mathrm{Tr}\, \hat{\gamma}_5 \mathrm{e}^{-\frac{\beta}{\hbar}\hat{H}} \tag{5.94}$$

as an operator expression in the quantum mechanical model. This expression can now be rewritten as a quantum mechanical path integral by inserting complete sets of states as explained in the first part of this book. In this way we see how the problem of evaluating a functional trace in n dimensions gets mapped into a problem in susy quantum mechanics. We shall systematically calculate the following anomalies.

(i) The usual abelian γ_5 anomaly for complex (Dirac) spin-$\frac{1}{2}$ fields coupled to external gravity (the gravitational contribution to the chiral anomaly). The transformation rules for this rigid symmetry multiply fermions by $i\alpha\gamma_5$. The Feynman graphs which yield this anomaly are fermion loops with external gravitons at all vertices, except at one vertex where the axial vector current is present. Taking the divergence of this axial vector current (contracting the vertex with the momentum which flows in or out at this vertex) produces the anomaly. This anomaly will be shown to be present only in $4k$ dimensions. The calculation of this anomaly is very simple, but for didactical reasons we shall spell out each step in detail. As a curious technical point we already mention that on the worldline only loops with scalar point particles contribute, but no loops with fermions or ghosts.

(ii) Next, we compute the same γ_5 anomaly for spin-$\frac{1}{2}$ fields, but now coupled to external Yang–Mills fields instead of external gravitons fields. We call this the abelian chiral anomaly, to distinguish it from the gauge anomaly for chiral fermions coupled to Yang–Mills fields which corresponds to $\mathrm{Tr}\,\gamma_5 T_\alpha$ in Fujikawa's approach. The latter is called the nonabelian chiral anomaly. In that case the gauge fields are also transformed under symmetry transformations. To deal with the internal symmetry generators T_α in the quantum mechanical model we introduce new ghosts c^* and c. The corresponding nonlinear sigma model is discussed in Appendix E. A few technical problems are encountered and solved: one must take traces only over one-particle states, and to achieve this we construct a suitable projection operator [58]. We can then give a full path integral treatment of these anomalies. (In the work by Alvarez-Gaumé and Witten [1], and also earlier work by us [40], the internal sector was still treated with operatorial methods.)

(iii) Then we consider "Einstein–Lorentz anomalies" for chiral spin-$\frac{1}{2}$ fields. These are also called gravitational anomalies and are anomalies in the local Lorentz symmetry. The Feynman graphs which yield the effective action are polygons with fermions in the loop and gravitons sticking out. Just as with vector and axial vector symmetry in gauge theories, one can push the anomaly from the Einstein to the Lorentz sector or back [2]. Following [1], we shall find it advantageous to consider a suitable linear combination of these two local symmetries to compute its anomaly. The Jacobian then becomes covariant: it can be written as follows in terms of covariant derivatives $J = \frac{1}{2}(D_\mu \xi^\mu + \xi^\mu D_\mu)$ in which, as we shall see, the Christoffel connections cancel. It will turn out that gravitational anomalies only exist in $4k + 2$ dimensions.

(iv) Next, we consider mixed gravitational and nonabelian chiral Yang–Mills anomalies, corresponding to loops with chiral spin-$\frac{1}{2}$ fields coupled to both external gravitational and gauge fields. The Jacobians are proportional to $\frac{1}{2}(D_\mu \xi^\mu + \xi^\mu D_\mu)$ and $i\eta^\alpha T_\alpha \gamma_5$, respectively. As a particular case they contain the purely gravitational anomaly as well as the purely nonabelian chiral anomaly. These anomalies correspond to a breakdown of the reparametrization and gauge invariances of the effective action for chiral fermions coupled to gravity and nonabelian gauge fields. These anomalies are fatal if they occur in pure Yang–Mills theories: in four dimensions they imply a breakdown of renormalizability and unitarity of the QFT, and one should try to find a collection of fields for which the anomalies cancel each other. In higher dimensions both gauge and gravitational quantum field theories are not renormalizable, but it is believed that anomalies should still cancel in order for the theory to make sense after dimensional reduction to four dimensions.

(v) After these studies of anomalies for spin-$\frac{1}{2}$ fields we turn to spin-$\frac{3}{2}$ fields. For readers unfamiliar with supergravity we give a short self-contained discussion of the quantization and the ghost structure of supergravity. We then compute the gravitational contribution to the abelian γ_5 anomaly for spin-$\frac{3}{2}$ in $4k$ dimensions. The corresponding Feynman graphs consist of a spin-$\frac{3}{2}$ loop with gravitons sticking out, and at one vertex the axial current is present. From a technical point of view this calculation is amusing because it combines the results in (1) and (2). (The reason is that we shall treat the spin connection when it acts on the vector index of the gravitino in the same way as a Yang–Mills field.)

(vi) Next, we calculate the gravitational anomaly for spin-$\frac{3}{2}$ fields, corresponding to loops of chiral $\frac{3}{2}$ fields coupled to external gravity. Now we are dealing with spin-$\frac{3}{2}$ loops with gravitons at all vertices. So this section is the spin-$\frac{3}{2}$ counterpart of the discussion in Section 5.3. Again there is an anomaly only in $4k+2$ dimensions. Spin-$\frac{3}{2}$ fields do not couple in supergravity to Yang–Mills fields, hence there is no discussion of mixed anomalies for loops with spin-$\frac{3}{2}$ fields.

(vii) Finally, we discuss gravitational anomalies due to loops with self-dual antisymmetric tensor fields coupled to external gravity. Again couplings to an arbitrary Yang–Mills group do not exist. The problem of there being no covariantly gauge-fixed action for self-dual antisymmetric tensor fields was circumvented by Alvarez-Gaumé and Witten [1] by using ordinary (unconstrained, namely non-self-dual) fields in loops, and coupling only one vertex to the self-dual part of the stress tensor. It seems not well known that there exist **local** actions for self-dual antisymmetric tensor fields in even dimensions. These actions can be found in [102–104], and one can use them to calculate the self-dual tensor anomalies in $4k+2$ dimensions in exactly the same way as for the other anomalies.[10] One obtains the same result as the one by Alvarez-Gaumé and Witten [104].

After these chiral anomalies we turn to trace anomalies. Here the situation is much more delicate: one needs to evaluate higher-loop graphs on the worldline and the calculations depend very much on the precise definition of the measure, Hamiltonian and Feynman rules. Let us once again state that by Feynman rules we mean not only certain formal expressions for the propagators and vertices, but also the precise rules of how to compute integrals over products of these. The precise rules were derived in

[10] Covariant actions for self-dual antisymmetric fields can be formulated at the classical level [105], but their covariant quantization remains problematic. When quantized noncovariantly they reduce to the actions in [102–104] and thus lead to the same anomaly computation [106].

great detail in Part I, and we shall now reap the fruits of that labor. We consider

(i) trace anomalies for scalar and spin-$\frac{1}{2}$ fields in two dimensions,

(ii) trace anomalies for spin-0, spin-$\frac{1}{2}$ and spin-1 fields in four dimensions. For spin-1 fields we need to include the contributions to the spin-1 trace anomaly which come from the Faddeev–Popov ghosts.

Before turning to the calculation of these anomalies, we want to test the QM approach in a case where we know beforehand that there should be no anomalies. This case is Einstein symmetry. Consider the Einstein transformation of a scalar field φ, given by $\delta_E \varphi = \xi^\mu(x)\partial_\mu \varphi(x)$. It simplifies the analysis if one takes instead the variable $\tilde{\varphi} = g^{1/4}\varphi$ as a fundamental variable. The inner product in the space of variables $\tilde{\varphi}$ is $\langle \tilde{\varphi}_1 | \tilde{\varphi}_2 \rangle = \int \tilde{\varphi}_1^*(x)\tilde{\varphi}_2(x)\, d^n x$ without extra factors of $\sqrt{g}$ (the usual factor of $\sqrt{g}$ has been absorbed into the definition of $\tilde{\varphi}$). Then $\delta_E \tilde{\varphi} = \xi^\mu \partial_\mu \tilde{\varphi} + \frac{1}{2}(\partial_\mu \xi^\mu)\tilde{\varphi}$ because $\tilde{\varphi}$ is a scalar half-density, and we can write this as $\delta_E \tilde{\varphi} = \frac{1}{2}(\xi^\mu \partial_\mu + \partial_\mu \xi^\mu)\tilde{\varphi}$, where also the second derivative ∂_μ can act on $\tilde{\varphi}$. The Jacobian is now

$$J = \frac{1}{2}(\xi^\mu \partial_\mu + \partial_\mu \xi^\mu) \tag{5.95}$$

where the derivative ∂_μ can act past ξ^μ. The anomaly is then

$$An = \lim_{\beta \to 0} \operatorname{Tr} J e^{-\beta \mathcal{R}}. \tag{5.96}$$

We now show that this symmetry has no anomaly. As regulator we consider an arbitrary operator $\mathcal{R}$ with complete set of eigenfunctions $\tilde{\varphi}_N$ with eigenvalues λ_N^2. One then finds

$$\begin{aligned} An(E) &= \frac{1}{2}\operatorname{Tr}(\xi^\mu \partial_\mu + \partial_\mu \xi^\mu) e^{-\beta \mathcal{R}} \\ &= \frac{1}{2}\int \sum_N \tilde{\varphi}_N^* (\xi^\mu \partial_\mu + \partial_\mu \xi^\mu) e^{-\beta \lambda_N^2} \tilde{\varphi}_N \, d^n x \\ &= \frac{1}{2}\int \partial_\mu \Big(\sum_N \tilde{\varphi}_N^* \xi^\mu \tilde{\varphi}_N e^{-\beta \lambda_N^2} \Big) d^n x \end{aligned} \tag{5.97}$$

as long as the complete set contains both $\tilde{\varphi}_N$ and $\tilde{\varphi}_N^*$ (plane waves are an example). In general, the λ_N^2 increase fast enough with increasing N so that the sum over N converges, and assuming that $\xi^\mu(x)$ vanishes for large x one finds that Einstein symmetry indeed has no anomaly.

In practice one can calculate anomalies by using a complete set of plane waves, as shown by Fujikawa for chiral anomalies [12]. For Einstein symmetries a two-parameter class of regulators has been considered in [107],

$$\mathcal{R} = -g^{-\alpha}\partial_\mu g^{\mu\nu} g^\beta \partial_\nu g^{-\alpha}. \tag{5.98}$$

This operator is hermitian in the space of fields $\tilde{\varphi}$ with inner product $\langle\tilde{\varphi}|\tilde{\psi}\rangle = \int \tilde{\varphi}^*(x)\tilde{\psi}(x)\, d^n x$. Hence, it can be diagonalized and $e^{-\beta\mathcal{R}}$ then becomes $e^{-\beta\lambda_N^2}$ when acting on $\tilde{\psi}_N$. The explicit calculation of the Einstein anomaly with this regulator using a complete set of plane waves is tedious but straightforward (one has to use the Baker–Campbell–Hausdorff formula), and the result is that the Einstein anomaly given by (5.96) indeed vanishes for arbitrary α and β.

So let us now repeat this calculation using the QM model, and check that the Einstein anomaly still vanishes. This provides a test of the method. We start from the field $\tilde{\varphi}$ and represent[11] ∂_μ by $\frac{i}{\hbar}p_i$ (we recall our notation that $\mu, \nu, \ldots$ denote indices in the QFT and $i, j, \ldots$ correspond to indices in the QM model). The operator $\frac{1}{2}(\xi^\mu\partial_\mu + \partial_\mu\xi^\mu)$ turns into

$$J = \frac{i}{2\hbar}[\xi^i(x)p_i + p_i\xi^i(x)] \tag{5.99}$$

which is Weyl-ordered. We can then rewrite the trace as a path integral, as explained in Part I of this book. We recall that we replace x by $x_0 + q$ where the quantum fluctuations q vanish at the endpoints $t = -\beta$ and $t = 0$. It further simplifies the analysis if we write J as $\exp(J)$ and later take the term linear in ξ. It is then convenient to integrate out the momenta and obtain a path integral in configuration space.[12] The Einstein anomaly (if nonvanishing) is then obtained by expanding

$$An = \lim_{\beta\to 0} \int dx_0 \sqrt{g(x_0)} \frac{1}{(2\pi\beta\hbar)^{n/2}} \left\langle e^{-\frac{1}{\hbar}S^{int} - \frac{1}{\beta}\int_{-\beta}^{0} \frac{1}{\hbar}\xi^i(x) g_{ij}(x)\frac{dx^j}{dt}dt} \right\rangle \tag{5.100}$$

and keeping the terms linear in ξ^i. The factor of $1/\beta$ in front of the second term in the exponent is important, so let us explain its origin in detail. We write J as N times $(\epsilon/\beta)J$, where $\beta = N\epsilon$. Then we exponentiate. The sum over the N terms with ϵ turns into an integral $\frac{1}{\beta}\int_{-\beta}^{0} dt$, and this yields the result.

The interactions were discussed in Part I of this book and read

$$\begin{aligned} -\frac{1}{\hbar}S^{int} &= -\frac{1}{\beta\hbar}\int_{-1}^{0} \frac{1}{2}[g_{ij}(x_0+q) - g_{ij}(x_0)](\dot{q}^i\dot{q}^j + b^i c^j + a^i a^j)\, d\tau \\ &\quad - \frac{\beta\hbar}{8}\int_{-1}^{0} (R + g^{ij}\Gamma^l_{ik}\Gamma^k_{jl})\, d\tau. \end{aligned} \tag{5.101}$$

---

[11] With the Fujikawa variables $\tilde{\varphi}$ the scalar product is given by $\langle\tilde{\varphi}|\tilde{\psi}\rangle = \int d^n x\, \tilde{\varphi}^*(x)\tilde{\psi}(x)$ and the hermitian operator p_i is simply represented by $(\hbar/i)\partial_i$.

[12] Details are as follows. The phase-space action contains $(1/\hbar)\int_{-\beta}^{0}(ip_i\dot{x}^i - \frac{1}{2}g^{ij}p_ip_j)\, dt + (i/\beta\hbar)\int_{-\beta}^{0} p_i\xi^i\, dt$. Completing squares and integrating over p_i yields (5.100).

(Because in the trace the initial and final points coincide, the classical trajectory $x_{cl}(t)$ is simply x_0.) Expanding $g_{ij}(x_0+q)-g_{ij}(x_0)$, we find in normal coordinates (in which $\partial_k g_{ij}(x_0)=0$) terms of the form $\frac{1}{\beta\hbar}q^k q^l R_{iklj}(\dot{q}^i\dot{q}^j+b^i c^j+a^i a^j)$ and higher-order terms. The term with ξ can be rewritten as

$$
\begin{aligned}
-V_\xi &= -\frac{1}{\beta\hbar}\int_{-1}^{0}\xi^i(x_0+q)g_{ij}(x_0+q)\dot{q}^j\,d\tau \\
&= -\frac{1}{\beta\hbar}\int_{-1}^{0}q^k\dot{q}^j[\partial_k\xi^i(x_0)g_{ij}(x_0)+\xi^i(x_0)\partial_k g_{ij}(x_0)]\,d\tau+\cdots \\
&= -\frac{1}{\beta\hbar}\Big(\int_{-1}^{0}q^k\dot{q}^j\,d\tau\Big)g_{ij}(x_0)D_k\xi^i(x_0)+\cdots \\
&= -\frac{1}{\beta\hbar}\Big(\int_{-1}^{0}q^k\dot{q}^j\,d\tau\Big)D_k\xi_j(x_0)+\cdots
\end{aligned}
\tag{5.102}
$$

where the terms denoted by $\cdots$ contain more q fields. We rescaled $t=\beta\tau$, but this did not change the prefactor $1/\beta$ because $\dot{q}^j(t)\,dt=\dot{q}^j(\tau)\,d\tau$. Because q vanishes at the endpoints ($\tau=0$ and $\tau=-1$) the integral $\int_{-1}^{0}q^k\dot{q}^j\,d\tau$ is antisymmetric in k and j.

The Einstein anomaly is given by the β-independent terms, hence the factor of $(\beta\hbar)^{-n/2}$ in the Feynman measure in (5.100) should be compensated by factors of $\beta\hbar$ produced by loops. Since we expect no anomaly in the Einstein transformations, the terms in the final expression of order $(\beta\hbar)^0$ should vanish. We need Feynman graphs with precisely one vertex V_ξ and any number of other vertices. All vertices are proportional to $1/\beta\hbar$ (or $\beta\hbar$, see the last term in S^{int}) and the q propagators are proportional to $\beta\hbar$.

In $n=2$ dimensions there is a factor of $(\beta\hbar)^{-1}$ in the measure, hence the sum of all Feynman graphs with one factor of $\beta\hbar$ should vanish.[13] There is one graph which could possibly contribute

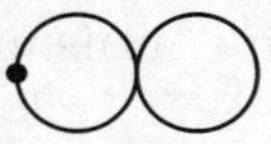

where the dot denotes the vertex V_ξ. The vertex in the middle contains R_{iklj}. The whole graph is of order $\beta\hbar$, and if it were nonvanishing, there would be an Einstein anomaly. However, the result must be of the form of $D_m\xi_n$ times a curvature, and this product always vanishes since the curvature can have at most two indices, hence it is either Ricci curvature R_{mn} or $g_{mn}R$, and in both cases the contraction with the antisymmetric $D_m\xi_n$ in (5.102) vanishes.

[13]There is even a graph of zeroth order in $\beta\hbar$, namely . If nonvanishing this would yield a divergent contribution proportional to $1/\beta$ in the anomaly. Fortunately its contribution vanishes due to $\int {}^\bullet\!\Delta^{\!\bullet}(\tau,\tau)\,d\tau=0$, see Section 2.5.

We could go on to also check that in $n = 4$ dimensions the QM approach yields a vanishing Einstein anomaly. Now the sum of all graphs proportional to $(\beta\hbar)^2$ should vanish. The graphs to be analyzed are the irreducible graphs

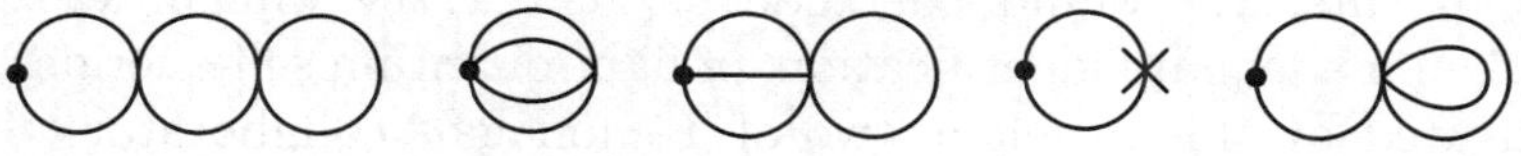

and the product of graphs

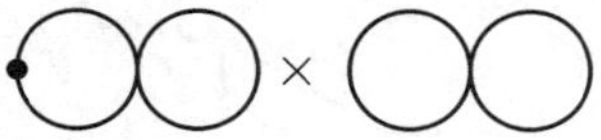

The cross in the fourth graph indicates the counterterm, and all vertices with a dot come from V_ξ. We shall encounter similar graphs in the chiral and trace anomalies, and since we are mostly interested in nonvanishing anomalies we leave the analysis that there is no Einstein anomaly in $d = 4$ as an exercise.

5.4 A brief history of anomalies

When physicists tried to compute radiative corrections to processes in QED in the 1930s, they of course stumbled on divergences and other inconsistencies. Even the simplest loop diagrams presented enormous difficulties, and some physicists (Heisenberg and Pauli at one time or another, and also Dirac and Oppenheimer) blamed QED itself for these difficulties. In the 1940s the problems became more focused. A diagram which exhibited some difficulties very clearly was the photon self-energy diagram due to an electron loop (we of course use modern terminology)

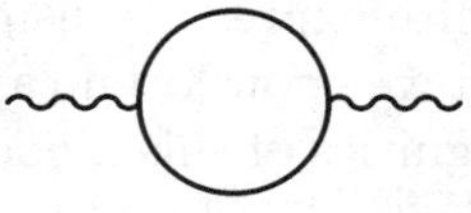

$$\partial_\mu \langle 0 \mid T j^\mu_{em}(x) j^\nu_{em}(y) \mid 0 \rangle = 0? \qquad (5.103)$$

Gauge invariance required that this diagram should be transversal, and on-shell it should vanish because the photon should remain massless, but Tomonaga and collaborators found it to be infinite, as well as not gauge invariant [108]. They studied the e^2 corrections to the Klein–Nishina formula for Compton scattering and reported that "there is an infinity containing [the] electromagnetic potential bilinearly . . . in . . . the vacuum polarization effect. [It] cannot be subtracted by amalgamation [removal by renormalization] as in the case of mass-type and charge-type infinities". This divergence could be identified as a photon mass, but unlike the mass divergence of the electron which could be "amalgamated" into an already existing electron mass, the photon mass divergence could not be dealt with in the same way because there is no photon mass in Maxwell's equations [109]. Oppenheimer commented in a note attached to this article: "As ... Schwinger and others have shown, the very greatest care must be taken in evaluating such self-energies lest, instead of the zero value they should have, they give non-gauge covariant, noncovariant, in general infinite results. ... I would conclude ... [that] ... the difficulties ... result from ... an inadequate identification, of light quantum self-energies." [109]

Motivated by this problem, two of Tomonaga's collaborators, Fukuda and Miyamoto [110], examined the next simplest diagram, namely the triangle diagram

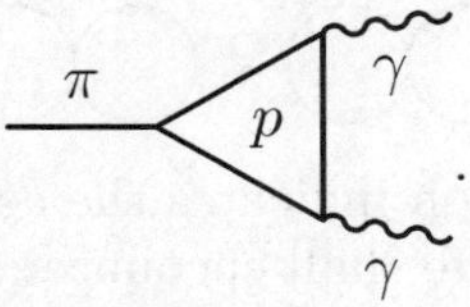

It was supposed to describe the decay $\pi \rightarrow p\bar{p} \rightarrow \gamma + \gamma$. They considered the cases where the neutral meson (π^0, Yukawa's U particle) was a scalar, pseudoscalar or pseudovector, with couplings $fU\bar{\psi}\psi$, $fU\bar{\psi}\gamma_5\psi$ and $(f/2m)\bar{\psi}\gamma_5\gamma^\mu\psi U_\mu$, respectively, where m is the proton mass. They found two problems:

(i) the results were not gauge covariant since bare gauge fields A_μ appeared in the result,

(ii) the results for the decay into two photons of a pseudovector U_μ and a pseudoscalar U particle were not the same if they set $U_\mu = \partial_\mu U$, even though the interactions were the same after partial integration and using the Dirac equation of motion.

They concluded: "Evidently these inconsistent results arise from the mathematical difficulty of obtaining [a] definite expression using the singular function of Jordan and Pauli. At present we know [of] no appropriate prescription which makes one free from ambiguities of this kind". The singular function in question was $D(x) = \int \frac{d^3k}{(2\pi)^3} \frac{\sin(kx-\omega t)}{2\omega}$ which appears in the equal-time canonical commutation relations.

Steinberger [111], then a theorist at Princeton, heard from Yukawa (who was visiting Princeton) about the work of Fukuda and Miyamoto (see footnote 11 of his article) and he applied the brand new Pauli–Villars regularization scheme [112] to the triangle graph and an array of other problems. Tomonaga was of course also quite interested in these consistency problems, and with coworkers he also applied the Pauli–Villars regularization scheme to the calculation of the triangle graph [113]. The conclusion of these studies was a partial success: the scheme did maintain gauge invariance and Lorentz invariance, and it led to a finite result for the triangle graph, but the actual value for this finite result seemed to depend on how the calculations were performed, and the relation between pseudovector and pseudoscalar couplings was still not satisfied [111, 113]. In modern terms: there was a chiral anomaly! However, this was not yet fully understood at that time. Rather, it seemed to lead to the perplexing conclusion that the lifetime of the neutral pion was ambiguous: "We see that there remains still some ambiguity [in] how to use the regulator, and this ambiguity would be solved only by some experiment which could detect the γ-decay of [the] neutretto" [113]. (Neutretto was another name for π^0.)

In 1951 Schwinger made a fresh attack on the problem of gauge invariance of the photon self-energy and the triangle diagrams. He introduced a regularization scheme (point splitting) which preserves gauge invariance at all intermediate stages. As he wrote in "On gauge invariance and vacuum polarization" [93]: "This paper is based on the elementary remark that the extraction of gauge invariant results from a formally gauge invariant theory is ensured if one employs methods of solution that involve only gauge covariant quantities". He then proceeded to solve the equations of motion of an electron in an electromagnetic field

$$\frac{dx^\mu}{ds} = 2\pi_\mu; \qquad \frac{d\pi_\mu}{ds} = e(F_{\mu\nu}\pi^\nu + \pi^\nu F_{\mu\nu}) + \frac{1}{2}e\sigma^{\lambda\nu}\frac{\partial F_{\lambda\nu}}{\partial x^\mu} \tag{5.104}$$

where $\pi_\mu = p_\mu - eA_\mu$ and s is the proper time. He found that the photon self-energy did vanish on-shell, so gauge invariance was preserved. However, he also concluded that the pseudovector coupling gave the same result for the triangle graph describing π^0 decay as the pseudoscalar coupling, namely

$$\mathcal{L}_{\mathit{eff}} = \frac{\alpha}{\pi}\frac{f}{m}\pi^0\vec{E}\cdot\vec{H}. \tag{5.105}$$

Although this was the result which seemed to solve the earlier problems, we now know that the pseudovector and pseudoscalar couplings should not be the same: there is an axial anomaly! It has been argued that he moved the anomaly from the right-hand side of the anomaly equation to the left-hand side [114][14].

---

[14] In section 5 of [93] he used "point splitting", a regularization scheme in x-space that is completely equivalent to the Pauli–Villars scheme in momentum space. According to this scheme the axial current is written as

$$\bar\psi\Big(x + \frac{1}{2}\epsilon\Big)\gamma_5\gamma_\mu\Big(\exp ie\int_{x-\frac{1}{2}\epsilon}^{x+\frac{1}{2}\epsilon} A_\mu\, dx^\mu\Big)\psi\Big(x - \frac{1}{2}\epsilon\Big) \tag{5.106}$$

and the exponential factor (later called a Wilson line) is added to keep electromagnetic gauge invariance. Schwinger defined $\partial_\mu[\mathrm{tr}\,\gamma_5\gamma^\mu G(x,x)]$ by $\lim_{x',\,x''\to x}[(\partial'_\mu - ieA_\mu(x')) + (\partial''_\mu + ieA_\mu(x''))]\mathrm{tr}\,\gamma_5\gamma^\mu G(x',x'')$ because "[this] structure is dictated by the requirement that only gauge covariant quantities be employed". However if one

In the 1950s and 1960s field theory fell from favor, and alternative physical theories took the limelight: Regge theory, the S-matrix program of Chew and current algebra. Although the first two alternatives were meant to replace field theory, it was natural to try to build field theoretical models which gave a representation of current algebra and in which the consistency of current algebra could be tested. In fact, many of the physicists who worked on current algebra in those days later helped to create modern quantum gauge field theory.

One such attempt was a beautiful little article in 1960 by Gell-Mann and Levy on the linear sigma model [115], in which PCAC (the partially conserved axial-vector current relation) was satisfied: $\partial_\mu j_5^\mu = f_\pi m_\pi^2 \pi(x)$, where f_π is the π-decay constant (93 MeV). The model contained, in addition to the nucleons, the three pions $\pi^\pm, \pi^0$ and a scalar meson σ, with an $SO(4)$ symmetry which was spontaneously broken, giving the nucleons a mass. If a term linear in σ was added to the action, this explicit symmetry breaking also gave the pions a mass. This model became obligatory reading for graduate students at Utrecht University (where one of us obtained his PhD). In Stony Brook B. Lee started studying the renormalization program of spontaneously broken field theories and wrote an influential small book [116] on the renormalization of this model.[15] G. 't Hooft heard B. Lee at the Cargèse summer school lecture on this topic, and upon returning to Utrecht, he decided to start applying these ideas to gauge theories, with well-known consequences.

In 1969 two important articles were submitted for publication within two weeks of each other, one by Bell and Jackiw [118], and the other by Adler [119]. Bell and Jackiw noted that the amplitude for $\pi^0 \to \gamma\gamma$ could be parametrized as follows:

$$T^{\mu\nu}(p,q) = \epsilon^{\mu\nu\alpha\beta} p_\alpha q_\beta T(k^2) \tag{5.107}$$

where p and q were the on-shell photon momenta, and $k = p + q$ was the pion momentum. They used the linear sigma model and considered both the case with k^2 off-shell as well as the case with $k^2 + m_\pi^2 = 0$ for an on-shell pion. Their amplitude satisfied gauge invariance ($p_\mu T^{\mu\nu}(p,q) = q_\nu T^{\mu\nu}(p,q) = 0$) as well as Bose symmetry ($T^{\mu\nu}(p,q) = T^{\nu\mu}(q,p)$). They noted that Steinberger had calculated $T(k^2)$ using the same graphs that

adds a Wilson line, one should use ordinary instead of covariant derivatives. If one had required in the Pauli–Villars scheme that the $U(1)$ vector gauge invariance is maintained, the ambiguities in this scheme would also have been fixed, and one would have obtained the same result as point splitting.

[15] Because there were no direct axial-vector couplings in this model, no problems with the chiral anomaly were encountered. (However, the chiral symmetry between pions and the σ meson allowed one to define an axial vector current, and its renormalization was also studied [117].)

occur in the linear sigma model and had found a nonzero result ($T(0) = 4\pi^2 g/m$). On the other hand, Veltman and Sutherland [120] had found that $T(0) = 0$ if one used an off-mass-shell pion field that was equal to the divergence of the axial current (PCAC). The puzzle that $T(0)$ should on the one hand be nonvanishing and on the other hand be vanishing was the problem Bell and Jackiw decided to tackle. They noted that the problem was "in the same tradition as that of the photon mass, noncanonical terms in commutators – Schwinger terms – and violations of the Jacobi identity". They claimed that this "demonstrates in a very simple example [the linear σ model] the unreliability of the formal manipulations common to current-algebra calculations", but then they went on to "develop a variation which respects PCAC, as well as Lorentz and gauge invariance, and find that indeed the explicit perturbation calculation also then yields $T(0) = 0$".[16] In their appendix they noted the hallmark of an anomaly: "Since the integral is linearly divergent a shift of variable picks up a surface term". (The procedure which yielded $T(0) = 0$ amounted to adding a nonlocal counterterm to the action [121], but this violates renormalizability.)

Adler just studied the axial-vector-vector (AVV) triangle graph in spinor QED, and took the results as they came: "... we demonstrate the uniqueness of the triangle diagrams [by imposing vector gauge invariance] ... and discuss a possible connection between our results and the $\pi^0 \to 2\gamma$ and $\eta \to 2\gamma$ decays ... [The] partial conservation of the axial-vector current ... must be modified in a well-defined manner, which completely alters the PCAC predictions for the π^0 and the η two-photon decays". Here is the axial anomaly in all its glory: it could not be clearer. He used an explicit expression for the triangle graph which Rosenberg had obtained already in 1963 [122]. Rosenberg considered electromagnetic properties of neutrinos in the V-A theory, and expanded the amplitude for the triangle graph coupled to two photons and a neutrino current in form factors, some of which were divergent and others of which were finite. Then he imposed vector gauge invariance, and this expressed the divergent form factors in terms of convergent ones. However, Rosenberg did not study whether the (naive) axial vector Ward identity failed in the case of the triangle graphs; that was done by Bell and Jackiw, and Adler.

With the demonstration of 't Hooft in 1971 that nonabelian pure gauge theories are renormalizable, it was realized that anomalies would spoil renormalizability and unitarity [123]. Thus, one had to make sure that anomalies (more precisely anomalies in the gauge transformations of chiral spin-$\frac{1}{2}$ fields, the quarks and leptons) would cancel. In the Standard

[16] This variation was the old Pauli–Villars regularization scheme, applied to the Steinberger calculation, but with *mass-dependent* coupling constants for the extra regulator fermions.

Model the gauge group $SU(3)$ has no anomalies because it does not couple to chiral quarks, while $SU(2)$ has no anomalies because all of its representations are pseudoreal. Only the $U(1)$ hypercharge gauge symmetry is potentially anomalous, but its anomalies cancel because the sum of electric charges of all quarks and leptons in a given family cancels.[17] Thus, the threat of anomalies in the Standard Model was averted.

Having settled the issue of the chiral anomalies in nongravitational theories, it was realized first by Kimura, and later by Delbourgo and Salam, and then by Eguchi and Freund (who corrected a factor of 2 in the paper by Delbourgo and Salam) that one could also encounter anomalies if one couples fermions to external gravity instead of external electromagnetism [124]. These authors considered triangle graphs in four dimensions with nonchiral (Dirac) fermions in the loop, with one vertex given by the axial current $\bar{\psi}\gamma_5\gamma_\mu\psi$ and the other two vertices given by $h^{\mu\nu}T_{\mu\nu}$, where $T_{\mu\nu}$ is the stress tensor for fermions. They indeed found anomalies of the form $\epsilon^{\mu\nu\rho\sigma}R_{\mu\nu}{}^{mn}R_{\rho\sigma mn}$ if one sets the metric $g_{\mu\nu}$ equal to $\eta_{\mu\nu}+\kappa h_{\mu\nu}$ and retains the terms quadratic in $h_{\mu\nu}$. A generalization of the gravitational γ_5 anomaly for spin $\frac{3}{2}$ was given in [125, 126].

This, in turn, led to a related problem: if one couples chiral fermions to external gravity, are there anomalies in the conservation of the stress tensor which are the counterpart of the anomalies in the gauge invariance of chiral gauge theories? It was soon realized that the nonconservation of the stress tensor is closely related to the presence of local Lorentz anomalies and the symmetry of the stress tensor; in fact, we discuss the precise relation in Section 6.3. It was then found that gravitational contributions to the chiral anomaly do cancel in the Standard Model,[18] while local Lorentz anomalies can only occur in $4k+2$ dimensions, and thus yield no potential problems for the Standard Model. Also, in the minimally supersymmetric Standard Model all nongravitational and gravitational contributions to the chiral anomalies cancel, because the two Higgsinos

[17] Triangle graphs with one $U(1)$ gauge field and two $SU(2)$ gauge fields are proportional to the sum of the hypercharges of the left-handed doublets. This sum vanishes for each family: $\frac{1}{6}\times 3\times 2+(-\frac{1}{2})\times 2=0$. Furthermore, triangle graphs with three $U(1)$ gauge fields are proportional to the sum of the cubes of the hypercharges of all fermions (rewriting right-handed fermions as charge conjugates of left-handed fermions), which also vanishes for each family: $(\frac{1}{6})^3\times 6+(-\frac{2}{3})^3\times 3+(\frac{1}{3})^3\times 3+(-\frac{1}{2})^3\times 2+(1)^3=0$. Triangle graphs coupled to three $SU(2)$ gauge fields yield no anomaly because the d-symbol $\mathrm{Tr}\,\sigma^a\{\sigma^b,\sigma^c\}$ vanishes.

[18] These triangle graphs with one $U(1)$ gauge field and two gravitons are proportional to the sum of the hypercharges of **all** fermions (rewriting right-handed fermions as charge conjugates of left-handed fermions), which is also the sum of their electric charges because the hypercharge is the average electric charge for each multiplet. Again this sum vanishes.

have opposite electric charge. However, one can also write down models in which the nongravitational anomalies cancel, but the gravitational anomalies do not cancel.[19] Thus (external) gravity fits remarkably well with the Standard Model and its minimal supersymmetric extension. All of these anomalies were treated in a uniform way, and for all dimensions at once, in the fundamental paper by Alvarez-Gaumé and Witten [1], on which part of this book is based.

In addition to anomalies in chiral models, there are also trace anomalies which occur when (rigid or local) scale invariance of the classical action is broken at the quantum level. For rigid scale transformations this was first shown by Coleman and Jackiw in 1971 [127], while the breakdown of local (Weyl) scale invariance for massless vectors and spinors in four dimensions coupled to gravity was first observed by Capper and Duff [128]. In the latter case the most general form of the trace anomaly was found to be given by [129]

$$\begin{aligned} T_\mu{}^\mu &= aR \qquad (d=2) \\ T_\mu{}^\mu &= aR^2 + bR_{\mu\nu}^2 + cR_{\mu\nu\rho\sigma}^2 + d\Box R + e(F_{\mu\nu}^a)^2 \qquad (d=4). \end{aligned} \tag{5.108}$$

The term $\Box R$ could be removed by a local counterterm $\Delta\mathcal{L} \sim R^2$, but the other terms were genuine anomalies. The coefficients in the $d=4$ trace anomaly are not all independent, but rather, as required by the consistency conditions, they combine as follows [130]:

$$T_\mu{}^\mu = \alpha\Big(C_{\mu\nu\rho\sigma}^2 + \frac{2}{3}\Box R\Big) + \beta\Big(\epsilon\epsilon RR\Big) + \gamma(F_{\mu\nu}^a)^2 \tag{5.109}$$

where $C_{\mu\nu\rho\sigma}^2 = R_{\mu\nu\rho\sigma}^2 - 2R_{\mu\nu}^2 + \frac{1}{3}R^2$ is the square of the Weyl tensor and $(\epsilon\epsilon RR) = R_{\mu\nu\rho\sigma}^2 - 4R_{\mu\nu}^2 + R^2$ yields the Euler invariant.

The constants a in $d=2$ and α, β, γ in $d=4$ also parametrize the one-loop divergences due to matter loops with external gravity [131].

For scalars an improvement term $\sim R\varphi^2$ can be added to the action which then becomes classically Weyl invariant, but a genuine trace anomaly develops at the quantum level. In a theory with N_S real scalars, N_F spin-$\frac{1}{2}$ Dirac fermions and N_V real vectors fields, the $d=4$ trace anomaly is given by

$$\begin{aligned} \alpha &= \frac{1}{120(4\pi)^2}\Big(N_S + 6N_V + 12N_F\Big) \\ \beta &= -\frac{1}{360(4\pi)^2}\Big(N_S + 11N_V + 62N_F\Big). \end{aligned} \tag{5.110}$$

[19] These models contain extra chiral fermions which couple to gravity and $U(1)$ but are $SU(3) \times SU(2)$ neutral.

It follows from unitarity that all coefficients in α must be positive, so that trace anomalies cannot cancel in rigidly susy $d = 4$ models. In models where the scale invariance is already explicitly broken at the classical level, one can nevertheless define a trace anomaly by

$$An(\text{Weyl}) = g^{\mu\nu}\langle T_{\mu\nu}\rangle_{reg} - \langle g^{\mu\nu}T_{\mu\nu}\rangle_{reg}. \tag{5.111}$$

For example, using dimensional regularization, one uses $\gamma^\mu\gamma_\mu = g^{\mu\nu}g_{\mu\nu} = n$ in the first term, but $\gamma^\mu\gamma_\mu = g^{\mu\nu}g_{\mu\nu} = 4$ in the second term. Moreover it was found in [132] that one can write a scalar in $d = 4$ either as φ or as a rank-two antisymmetric gauge field $\varphi_{\mu\nu}$, but the trace anomalies are different. A rank-three antisymmetric gauge field is dual to nothing, but it nevertheless yields a nonvanishing trace anomaly. It has been argued that one can make the effective action equal by suitable renormalization procedures [133]. If one reduces by a trivial Kaluza–Klein reduction $d = 10$ type IIB supergravity, one finds in $d = 4$ not the usual 70 scalars, but rather 63 scalars, seven fields $\varphi_{\mu\nu}$ and one field $\varphi_{\mu\nu\rho}$. One can use index theorems to compute the axial and conformal anomalies for arbitrary spin in gravity and supergravity [134]. By considering background fields with $R_{\mu\nu} = 0$ the trace anomaly becomes proportional to $(\alpha + \beta)R^2_{\mu\nu\rho\sigma}$. The combined trace anomaly for these spin $2, \frac{3}{2}, 1, \frac{1}{2}, 0$ fields then cancels in $N = 8$, $d = 4$ supergravity.

We end this section by giving some references to related aspects of anomalies. For textbooks on anomalies, the reader may turn to [3, 99, 135–137]. See also the reviews [138–140]. For an analysis of the Atiyah–Singer index theorem using supersymmetric quantum mechanics see [35, 36], while a simplified version without supersymmetry is given in [141]. The relations between Einstein anomalies and local Lorentz anomalies which we derive in Chapter 6 are found in [139–143].

6
Chiral anomalies from susy quantum mechanics

6.1 The abelian chiral anomaly for spin-$\frac{1}{2}$ fields coupled to gravity in $4k$ dimensions

As a first application of the formalism we have developed, we shall compute the anomaly in the chiral symmetry $\delta\lambda = i\alpha\gamma_5\lambda$ for a massless Dirac fermion λ in n dimensions coupled to external gravity (n is even) [124]. The real parameter α is an infinitesimal constant and γ_5 is proportional to the product $\gamma^1 \cdots \gamma^n$ and hermitian (hence $(\gamma_5)^2 = 1$). This anomaly is sometimes called the gravitational chiral anomaly, although a more precise name would be the gravitational contribution to the abelian chiral anomaly. If there is an anomaly, the axial vector current is no longer conserved at the quantum level (some matrix elements of the divergence of the axial current are nonvanishing).

The Lagrangian of the field theory in n Minkowskian or Euclidean dimensions is given by

$$\mathcal{L} = -e\,\bar{\lambda}\, e_m^{\mu}\gamma^m D_\mu\lambda, \qquad \bar{\lambda} = \lambda^\dagger i\gamma^0, \qquad (\gamma^0)^2 = -1 \tag{6.1}$$

where $e = (\det e_\mu^{m})$, e_m^{μ} is the inverse of the vielbein field e_μ^{m} and $D_\mu\lambda = \partial_\mu\lambda + \frac{1}{4}\omega_{\mu mn}(e)\gamma^m\gamma^n$, with $\omega_{\mu mn}(e)$ being the spin connection of Appendix A. In Minkowski space the chiral transformation law $\delta\lambda = i\alpha\gamma_5\lambda$ implies that $\delta\bar{\lambda} = i\alpha\bar{\lambda}\gamma_5$ because $\bar{\lambda} = \lambda^\dagger i\gamma^0$, but in Euclidean space λ and $\bar{\lambda}$ are independent complex spinors and then $\delta\bar{\lambda} = i\alpha\bar{\lambda}\gamma_5$ follows from requiring chiral invariance of the action.[1]

If in the path integral

$$Z[e_\mu^{m}] = \int \mathcal{D}\lambda\mathcal{D}\bar{\lambda}\, \mathrm{e}^{-\int d^n x\,\mathcal{L}} \tag{6.2}$$

[1] In the Euclidean case, α can even be complex, but since only α and not α^* appears in the transformation laws of λ and $\bar{\lambda}$, the fact that α may be complex does not enlarge the symmetry group.

one makes a chiral change of integration variables $\lambda' = (1 + i\alpha\gamma_5)\lambda$ and $\bar{\lambda}' = \bar{\lambda}(1 + i\alpha\gamma_5)$ with local $\alpha(x)$, one obtains the Jacobian we shall compute and a term $\int (\partial_\mu \alpha) j_5^\mu \, d^n x$ in the action, where the Noether current j_5^μ is the axial-vector current $-ie\bar{\lambda}\gamma^\mu\gamma_5\lambda$. Since the path integral of course does not change under a change of integration variables, the Jacobian yields the expectation value of the divergence of the axial-vector current. (More precisely, the connected n-point correlation functions with the Jacobian at one point are the same as the n-point correlation functions with $\partial_\mu j_5^\mu$ at that point. The one-point correlation function then gives the expectation values of these operators.) The corresponding Feynman graphs are single loops with λ in the loops, and gravitons sticking out from all vertices, except at one vertex where one has $(\partial_\mu \alpha) j_5^\mu$. One could have gauged the axial $U(1)$ symmetry by introducing a gauge field with coupling $-A_\mu j_5^\mu$ at this vertex. Then the chiral anomaly causes a breakdown of the gauge invariance of the effective action under $\delta A_\mu = \partial_\mu \alpha$.

The infinitesimal chiral transformations of λ and $\bar{\lambda}$ are equal: they are given for both λ and $\bar{\lambda}$ by the matrix

$$J = i\alpha\gamma_5. \tag{6.3}$$

Its trace yields the Jacobian, but one should regulate this trace. Of course, one can compute Feynman diagrams with the spin-$\frac{1}{2}$ field in the loop, with the axial-vector current at one vertex, while at the other vertices external gravitons couple to the fermion. There are vertices with one, two or more gravitons. Clearly, a background field formalism is called for, which takes the sum of all vertices into account simultaneously. Such a background field formalism has been developed by Fujikawa [12], who showed that in path integrals the anomaly is given by the regulated Jacobian

$$An = \lim_{\beta \to 0} \operatorname{Tr} J \, \mathrm{e}^{-\beta \mathcal{R}}. \tag{6.4}$$

As regulator we use a covariant regulator which is obtained as follows. As spin-$\frac{1}{2}$ fields we take $\tilde{\lambda} = g^{1/4}\lambda$ and $\tilde{\bar{\lambda}} = g^{1/4}\bar{\lambda}$ (recall that if one takes these fields as integration variables in the path integral, the Einstein anomalies are immediately seen to be absent for any self-adjoint regulator). The field operator for $\tilde{\lambda}$ and $\tilde{\bar{\lambda}}$ is $g^{1/4} \not{D} g^{-1/4}$. The covariant regulator is proportional to the square of this Dirac operator. Hence,

$$\mathcal{R} = -\frac{1}{2} g^{1/4} \not{D}^2 g^{-1/4} \tag{6.5}$$

and this regulator is the same for $\tilde{\lambda}$ and $\tilde{\bar{\lambda}}$ because the Dirac operator stands between $\tilde{\lambda}$ and $\tilde{\bar{\lambda}}$. (The factor of $\frac{1}{2}$ is conventional and could have been absorbed into β.) One could now directly calculate the trace in (6.4) using this regulator and a complete set of plane waves. This was done in curved space in [107]. However, for higher dimensions n, the calculations become progressively more complicated due to the algebra of the many Dirac matrices, and a simpler method than the Fujikawa method is needed. This is the method of supersymmetric quantum mechanics (susy QM). As we discussed in Chapter 5, one uses a representation of the operators x^μ, ∂_μ and γ^m (which are the only ingredients entering in J and $\mathcal{R}$) in terms of quantum mechanical operators $\hat{x}^i, \hat{p}_i$ and $\hat{\psi}^a$, with the same (anti)commutation relations in a Hilbert space with the same dimension and with the same hermiticity properties. In the QM model the regulator becomes the Hamiltonian of a simple susy QM model, the so-called $N = 1$ model. As we shall see, the presence of fermions in the QM model will remove all factors of β from the measure, and consequently, in the limit β tending to zero, only one-loop graphs need be computed. For this particular anomaly, loops with QM ghosts or QM fermions do not even contribute, but this is in general not the case.

Underlying this approach is the fact that all different representations of the canonical (anti)commutation relations which preserve the hermiticity of the operators are unitarily equivalent.[2] Since the anomaly we are going to calculate is proportional to a trace in a Hilbert space, and traces are invariant under similarity transformations, the anomaly does not depend on the representation chosen. The representation in terms of QM leads to particularly simple calculations, and this is the reason why we transform the quantum field theory problem into a problem in quantum mechanics.

We choose to work in Euclidean space because in this case the Gaussian integrals we need to evaluate are well defined. One could have started in Minkowski space, but then one would need at some point to make a Wick rotation to evaluate these Gaussian integrals, so it is easier to start from the beginning in Euclidean space.

The matrix γ_5 denotes the product of all Dirac matrices, and in order that $(\gamma_5)^2 = +1$ we normalize it as follows:

$$\gamma_5 = (-i)^{n/2}\gamma^1 \cdots \gamma^n \tag{6.6}$$

where $\{\gamma^m, \gamma^k\} = 2\delta^{mk}$ and all γ are hermitian (including γ_5). For $n = 2$, with $\gamma^1 = \sigma^1$ and $\gamma^2 = \sigma^2$, γ_5 equals the Pauli matrix σ^3, while in $n = 4$ we have $\gamma_5 = -\gamma^1\gamma^2\gamma^3\gamma^4$. The anomaly can be written as $\mathrm{Tr}\,\gamma_5 \exp(-\beta\mathcal{R})$, where the regulator $\mathcal{R}$ which preserves Einstein (general coordinate) and

[2] For the quantum mechanical variables p and q this is a theorem due to von Neumann [147]. For the fermionic extension with $\psi^a \sim \gamma^m$ one can use finite group theory to prove that there is only one faithful irreducible representation of the Clifford algebra in even dimensions [77, 148], hence the dimension of the fermionic part of the Hilbert space is fixed and is equal to $2^{n/2}$.

local Lorentz invariance is given by

$$\begin{aligned}\mathcal{R} &= -\frac{1}{2} g^{1/4} \not{D}\not{D} g^{-1/4} \\ &= -\frac{1}{2}\Big[g^{1/4} g^{\mu\nu} D_\mu^{(\omega,\Gamma)} D_\nu g^{-1/4} + \gamma^m\gamma^n \frac{1}{4} R_{mnpq}(\omega)\gamma^p\gamma^q\Big] \\ &= -\frac{1}{2} g^{-1/4} D_\mu \sqrt{g} g^{\mu\nu} D_\nu\, g^{-1/4} - \frac{1}{8} R \end{aligned} \tag{6.7}$$

with $\not{D} = \gamma^\mu D_\mu$, $\gamma^\mu = e^\mu_m \gamma^m$ and $D_\mu = \partial_\mu + \frac{1}{4}\omega_{\mu mn}(e)\gamma^m\gamma^n$. In the second line we have written $D_\mu^{(\omega,\Gamma)}$ for the first derivative because it contains (in addition to the spin connection) a Christoffel symbol that acts in the index ν of the derivative D_ν. To obtain the third line, we used the Bianchi identity $R_{[mnp]q} = 0$ and the fact that in general relativity the covariant derivative of a contravariant vector density equals the ordinary derivative. Hence, no Christoffel connections are present in D_μ, but D_μ of course contains terms with the spin connection.

We represent $\mathcal{R}$ in terms of quantum mechanical operators $\hat{x}^i$, $\hat{p}_i$ and $\hat{\psi}^a$, and denote the result by $\hat{H}$. Of course, $[\hat{p}_i, \hat{x}^j] = \frac{\hbar}{i}\delta_i^j$. However this does not fix the x-representation of $\hat{p}_j$ completely, namely $(p_x)_j = g^\alpha \frac{\hbar}{i}\frac{\partial}{\partial x^j} g^{-\alpha}$ is still possible for arbitrary α. Hermiticity of $(p_x)_j$ fixes the factors of g^α. The relation is then $\frac{\hbar}{i}\partial_\mu = p_\mu$ without extra factors of $g^{1/4}$ because we use $\tilde{\lambda}$ as basic fields, and $\frac{\hbar}{i}\frac{\partial}{\partial x^\mu}$ is hermitian if $\int \tilde{\bar{\lambda}}_1 \tilde{\lambda}_2\, d^n x$ is the inner product.

The Dirac matrices $\gamma^m (m = 1, \ldots, n)$ can be viewed as operators in a $2^{n/2}$-dimensional linear vector space, with anticommutation relations $\{\gamma^m, \gamma^k\} = 2\delta^{mk}$. Of course, we must take n to be an even number if we want to define a matrix γ_5. In the QM model, we introduce corresponding operators ψ_1^a $(a = 1, \ldots, n)$ satisfying $\{\psi_1^a, \psi_1^b\} = \delta^{ab}$. The reason for the subscript 1 will become clear shortly. Hence, $\gamma^m \leftrightarrow \sqrt{2}\psi_1^a$. Flat vector indices $m, n, \ldots$ in quantum field theory correspond to indices $a, b, \ldots$ in the QM model. In a given dimension of spacetime there may or may not exist a Majorana representation of the Dirac matrices, but we always use a hermitian representation of the Dirac matrices in Euclidean space, and hence ψ_1^a are hermitian.

In our formalism, we need operators $\psi_a^\dagger$ and ψ^b satisfying $\{\psi_a^\dagger, \psi^b\} = \delta_a{}^b$ (and $\{\psi^a, \psi^b\} = \{\psi_a^\dagger, \psi_b^\dagger\} = 0$). We therefore introduce **new** operators $\psi_2^a (a = 1, \ldots, n)$ which are **free** (i.e. the Hamiltonian $\hat{H}$ (to be constructed) will be independent of ψ_2^a) and satisfy $\{\psi_2^a, \psi_2^b\} = \delta^{ab}$ and $\{\psi_1^a, \psi_2^b\} = 0$. We then define

$$\psi^a \equiv (\psi_1^a + i\psi_2^a)/\sqrt{2}; \qquad \psi_a^\dagger = (\psi_1^a - i\psi_2^a)/\sqrt{2}. \tag{6.8}$$

In particular, $\gamma^m \leftrightarrow (\psi^a + \psi_a^\dagger)$. The operators ψ^a and $\psi_b^\dagger$ then indeed have the desired anticommutation relations. (At this point, ψ_1^a and ψ_2^a have been introduced without any considerations involving canonical quantization. Hence, in $\{\psi^a, \psi_b^\dagger\} = \delta^a{}_b$ there are no factors of $\hbar$. This simplifies the notation. One could rescale ψ_a and $\psi_a^\dagger$ with factors of $\sqrt{\hbar}$ to revert to the usual normalization of fermion fields.)

The space in which ψ_2^a acts also has dimension $2^{n/2}$, just like the space for ψ_1^a. So we take as the Hilbert space the direct product of the spaces for ψ_1^a an ψ_2^a. **In the Hamiltonian ψ_2^a is absent** but when we convert the operator expression to a path integral, we will find terms $-\int_{-1}^0 \bar{\psi}_a \dot{\psi}^a \, d\tau$ in the action, so that effectively terms with ψ_2^a are present in the action which appears in the path integrals. The linear vector space obtained by acting with $\psi_a^\dagger$ on the ψ-vacuum, has dimension 2^n. (The ψ-vacuum is defined by $\psi^a|0\rangle = 0$.) In traces over the direct product of both spaces we therefore divide by hand by $2^{n/2}$, since the original problem only involved the space of ψ_1^a.

There is a more minimal but also more cumbersome way of deriving the results without an extra ψ_2^a. We can begin with operators $\hat{\psi}_1^a$ satisfying $\{\hat{\psi}_1^a, \hat{\psi}_1^b\} = \delta^{ab}$, but then we can combine pairs of them into ψ and $\bar{\psi}^\dagger$. For example, $(\hat{\psi}_1^1 + i\hat{\psi}_1^2)/\sqrt{2} = \psi^{\mathrm{I}}$ and $(\hat{\psi}_1^1 - i\hat{\psi}_1^2)/\sqrt{2} = \psi_{\mathrm{I}}^\dagger$. The Hilbert space now has dimension $2^{n/2}$. The Feynman rules for the approach with extra ψ_2^a and the approach in which one combines Majorana spinors differ, but physical results (such as the transition element) are the same. We shall discuss and use both approaches. The approach in which one combines spinors is purely deductive and uses only the original Hilbert space, but the approach with ψ_2^a is algebraically somewhat simpler.

The γ_5 anomaly for the spinor field λ in the QFT can now be written in the QM model as the trace in (6.4) with $J = \gamma_5$ in (6.6) and $\mathcal{R}$ in (6.7) written in terms of the QM operators

$$
\begin{aligned}
An &= \lim_{\beta\to 0} \frac{1}{2^{n/2}} (-i)^{n/2} \operatorname{Tr} \prod_{a=1}^{n} (\hat{\psi}^a + \hat{\psi}_a^\dagger) \, \mathrm{e}^{-\frac{\beta}{\hbar}\hat{H}} \\
\hat{H} &= \frac{1}{2} g^{-1/4} \pi_i \sqrt{g} g^{ij} \pi_j g^{-1/4} - \frac{\hbar^2}{8} R \\
\pi_i &= \hat{p}_i - \frac{i\hbar}{2} \omega_{iab}(e) \hat{\psi}_1^a \hat{\psi}_1^b
\end{aligned}
\tag{6.9}
$$

where ψ_1^a and ψ_1^b in π_i are to be written in terms of ψ^a and $\psi_a^\dagger$ using (6.8). We have redefined $\beta \to \beta/\hbar$ such that it has the dimensions of a time. We shall first compute (6.9), which is proportional to the anomaly for the field λ in (6.1), but at the end we must add to this result the contribution from $\bar{\lambda}$. Since these two contributions are equal (because the Jacobians and the regulators are equal), we shall just multiply the final result by a factor of $-2i\alpha$ to obtain the correct normalization. (The minus sign is due to the fact that the traces over fermions acquire a minus sign. This, in turn, is due to the fact that the Jacobian for bosonic and fermionic fields in quantum field theory is a super-determinant, see the appendix of [77].)

Using the trace formula in (2.184)

$$\operatorname{Tr} A = \int \sqrt{g(x_0)} \Big(\prod_{i=1}^{n} dx_0^i \Big) \Big(\prod_{a=1}^{n} d\chi^a \, d\bar{\chi}_a \Big) \, e^{\bar{\chi}\chi} \langle \bar{\chi}, x_0 | A | \chi, x_0 \rangle \quad (6.10)$$

and the completeness relation in (2.153)

$$I = \int \Big(\prod_{a=1}^{n} d\bar{\eta}_a \, d\eta^a \Big) |\eta\rangle e^{-\bar{\eta}_a \eta^a} \langle \bar{\eta}| \quad (6.11)$$

we obtain for the chiral anomaly (omitting the overall factor of $-2i\alpha$ for the time being, and not yet taking the limit of vanishing β)

$$\begin{aligned} An = {} & \frac{(-i)^{n/2}}{2^{n/2}} \int \Big(\prod_{i=1}^{n} dx_0^i \Big) \sqrt{g(x_0)} \Big(\prod_{a=1}^{n} d\bar{\eta}_a \, d\eta^a \, d\chi^a \, d\bar{\chi}_a \Big) \\ & \times e^{\bar{\chi}\chi} \langle \bar{\chi} | \prod_{a=1}^{n} (\hat{\psi}^a + \hat{\psi}_a^\dagger) | \eta \rangle \, e^{-\bar{\eta}\eta} \langle \bar{\eta}, x_0 | e^{-\frac{\beta}{\hbar}\hat{H}} | \chi, x_0 \rangle. \end{aligned} \quad (6.12)$$

Since $\prod(\hat{\psi}^a + \hat{\psi}_a^\dagger)$ is already Weyl ordered (each factor is separately Weyl ordered, and different factors anticommute), we can at once evaluate the first matrix element. For the matrix element $\langle \bar{\eta}, x_0 | e^{-\frac{\beta}{\hbar}\hat{H}} | \chi, x_0 \rangle$ we substitute the result derived in Chapter 2. We recall that this involves a Dirac action with fields ψ and $\bar{\psi}$, together with an extra term $\bar{\eta}\chi$, see (2.178). We found that

$$\langle \bar{\eta}, x_0 | e^{-\frac{\beta}{\hbar}\hat{H}} | \chi, x_0 \rangle = \frac{1}{(2\pi\beta\hbar)^{n/2}} \Big\langle e^{-\frac{1}{\hbar}\int_{-\beta}^{0} S^{(int)} dt + \bar{\eta}_a \chi^a} \Big\rangle \quad (6.13)$$

where, in general, $S^{(int)}$ depends on both $\bar{\eta}_a + \bar{\psi}_a(t)$ and $\chi^a + \psi^a(t)$, with propagator $\langle \psi^a(t) \bar{\psi}_b(t') \rangle = \theta(t - t') \delta^a{}_b$.

In Chapter 2 we discussed all aspects of the path integral at the discretized level, but we shall now use a continuum notation, and only go back to discretized expressions when this is necessary to resolve ambiguities. This simplifies the notation, but it should be stressed that all of our continuum expressions stand for more complicated but well-defined discretized expressions.

Since in our case the Hamiltonian depends only on ψ_1^a, the expectation value of $\exp\big(-\frac{1}{\hbar} S^{int}\big)$ will only depend on x_0 and $(\chi^a + \bar{\eta}_a)/\sqrt{2}$, but not on

$\chi^a - \bar{\eta}_a$ (the boundary term $\exp(\bar{\eta}\chi)$ depends of course on both variables). Hence, three of the four Grassmann integrations over η, $\bar{\eta}$, χ and $\bar{\chi}$ will be very simple. Finally, the measure factor $[g(z)/g(y)]^{1/4}$ in the transition element in (2.81) becomes unity since $z = y = x_0$. After integrating out p, we obtain

$$An = \frac{1}{2^{n/2}} \int dx_0 \sqrt{g(x_0)}\, d\bar{\eta}\, d\eta\, d\chi\, d\bar{\chi}\, \mathrm{e}^{\bar{\chi}\chi}\, \mathrm{e}^{-\bar{\eta}\eta}\, \mathrm{e}^{\bar{\eta}\chi} \times \left[(-i)^{n/2} \mathrm{e}^{\bar{\chi}\eta} \prod_{a=1}^{n} (\eta^a + \bar{\chi}_a) \right] \frac{1}{(2\pi\beta\hbar)^{n/2}} \left\langle \mathrm{e}^{-\frac{1}{\hbar} S^{int}(x_0, (\bar{\eta}+\chi)/\sqrt{2})} \right\rangle . \tag{6.14}$$

The factor in square brackets comes from the matrix element of γ_5 and the rest of the second line comes from the transition element $\langle \bar{\eta}, x_0 | \mathrm{e}^{-\frac{\beta}{\hbar}\hat{H}} | \chi, x_0 \rangle$.

We take a closer look at the action. Rewriting the quantum Hamiltonian in (6.9) in Weyl-ordered form yields

$$H_W = \left(\frac{1}{2} g^{ij} \pi_i \pi_j \right)_S + \frac{\hbar^2}{8} g^{ij} \left(\Gamma^l_{ik} \Gamma^k_{jl} + \frac{1}{2} \omega_i{}^{ab} \omega_{jab} \right) \tag{6.15}$$

because the scalar curvature R from Weyl ordering the bosonic sector, see (B.25), cancels the scalar curvature R in (6.9). The letter S denotes that in $\frac{1}{2} g^{ij} \pi_i \pi_j$ all operators appear in (anti)symmetrized form. In Appendix C we show that Weyl ordering of the fermions in the $N = 2$ Hamiltonian gives a contribution $\frac{1}{8}\omega\omega$. The result in (6.15) refers to an $N = 1$ model, see Appendix D, and for that reason a factor of $\frac{1}{2}$ appears in front of the $\omega\omega$ term.

In order to derive the explicit form of S^{int} in (6.14) we go back to the complete action S before we added external sources to obtain the propagators. We use a continuum notation. Having integrated over p_i, one finds in the path integral the following configuration-space action:

$$\begin{aligned} -\frac{1}{\hbar} S = &-\frac{1}{\beta\hbar} \int_{-1}^{0} \frac{1}{2} g_{ij}(x_0 + q)(\dot{q}^i \dot{q}^j + b^i c^j + a^i a^j)\, d\tau + \bar{\eta}\psi(0) \\ &- \int_{-1}^{0} \bar{\psi}_a \dot{\psi}^a\, d\tau - \int_{-1}^{0} \frac{1}{2} \psi_1^a \dot{q}^i \omega_{iab}(x_0 + q) \psi_1^b\, d\tau \\ &- \frac{\beta\hbar}{8} \int_{-1}^{0} (\Gamma\Gamma + \frac{1}{2}\omega\omega)\, d\tau \end{aligned} \tag{6.16}$$

where $\psi_1 = (\psi + \bar{\psi})/\sqrt{2}$. This is the $N = 1$ model in Appendix D, (D.7), but in Euclidean space, and with the order $\hbar^2$ counterterms and the extra term $\bar{\eta}\psi(0)$, the presence of which we derived in (2.164) and the role of which is to cancel boundary terms in the ψ field equation. Substituting $\bar{\psi}_a = \bar{\eta}_a + \bar{\psi}_{qu,a}$ and $\psi^a = \chi^a + \psi^a_{qu}$ with constant background fermions χ^a and $\bar{\eta}_a$, one finds

$$
\begin{aligned}
-\frac{1}{\hbar}S = &-\frac{1}{\beta\hbar}\int_{-1}^{0}\frac{1}{2}g_{ij}(x_0+q)(\dot{q}^i\dot{q}^j + b^ic^j + a^ia^j)\,d\tau \\
&+\bar{\eta}\chi - \int_{-1}^{0}\bar{\psi}_{qu,a}\dot{\psi}^a_{qu}\,d\tau \\
&-\frac{1}{2}\int_{-1}^{0}\dot{q}^i\omega_{iab}(x_0+q)(\psi^a_{1,bg}+\psi^a_{1,qu})(\psi^b_{1,bg}+\psi^b_{1,qu})\,d\tau \\
&-\frac{\beta\hbar}{8}\int_{-1}^{0}\left(\Gamma\Gamma+\frac{1}{2}\omega\omega\right)\,d\tau
\end{aligned}
\tag{6.17}
$$

where $\psi^a_{1,bg} = (\bar{\eta}_a + \chi^a)/\sqrt{2}$ and $\psi^a_{1,qu} = (\psi^a_{qu}) + \bar{\psi}^a_{qu})/\sqrt{2}$.[3] The last two lines and the interactions in the first line yield S^{int} in (6.13),

$$
\begin{aligned}
-\frac{1}{\hbar}S^{int} = &-\frac{1}{\beta\hbar}\int_{-1}^{0}\frac{1}{2}[g_{ij}(x_0+q)-g_{ij}(x_0)](\dot{q}^i\dot{q}^j + b^ic^j + a^ia^j)\,d\tau \\
&-\frac{1}{2}\int_{-1}^{0}\dot{q}^i\omega_{iab}(x_0+q)(\psi^a_{1,bg}+\psi^a_{1,qu})(\psi^b_{1,bg}+\psi^b_{1,qu})\,d\tau \\
&-\frac{\beta\hbar}{8}\int_{-1}^{0}\left(\Gamma\Gamma+\frac{1}{2}\omega\omega\right)\,d\tau.
\end{aligned}
\tag{6.18}
$$

Note that the term $\bar{\eta}\psi_{qu}(0)$ (with undetermined $\psi_{qu}(0)$) has canceled. The terms $-(1/\beta\hbar)\int_{-1}^{0}\frac{1}{2}g_{ij}(x_0)(\dot{q}^i\dot{q}^j + b^ic^j + a^ia^j)\,d\tau - \int_{-1}^{0}\bar{\psi}_{qu,a}\dot{\psi}^a_{qu}\,d\tau$ yield the propagators, and the rest yields the vertices. These results were derived in Chapter 2, and the reader may look there for more details on the derivation.

After performing the loop integrations, the result for $\langle\exp(-\frac{1}{\hbar}S^{int})\rangle$ will only depend on $\psi^a_{1,bg} = (\chi^a + \bar{\eta}_a)/\sqrt{2}$. Hence, we can first do the $\bar{\chi}$ and η integrals, while the $\chi - \bar{\eta}$ integral will effectively remove ψ^a_2 from the trace. We shall then be left with an integral over $\psi^a_{1,bg} = (\chi^a + \bar{\eta}_a)/\sqrt{2}$. There now follows an orgy of Grassmann integrations. Readers who are only interested in the final result may jump to three lines below (6.24).

In the $\bar{\chi}, \eta$ sector we find the following integral (use $\int d\bar{\eta}\,d\eta\,d\chi\,d\bar{\chi} = \int d\bar{\eta}\,d\chi\,d\bar{\chi}\,d\eta$):

$$
\int d\bar{\chi}\,d\eta\,\mathrm{e}^{-\bar{\eta}\eta}\,\mathrm{e}^{\bar{\chi}\chi}\,\mathrm{e}^{\bar{\chi}\eta}\prod_{a=1}^{n}(\eta^a+\bar{\chi}_a).
\tag{6.19}
$$

[3] More precisely, in the discretized approach the interactions depend on $\psi^a_{k-1/2} = (\psi^a_k + \psi^a_{k-1})/2$ and $\bar{\psi}_{k,a}$, and then this leads to $\psi^a_{1,k} = (\psi^a_{k-1/2} + \bar{\psi}_{k,a})/\sqrt{2}$ and $\psi^a_{1,qu} = (\psi^a_{qu,k-1/2} + \bar{\psi}_{k,a})/\sqrt{2}$. From these discretized results we derived the propagators in Chapter 2.

The last factor is a fermionic delta function $\delta(\eta + \bar{\chi})$, hence $\exp(\bar{\chi}\eta)$ can be replaced by unity. For the same reason we can rewrite the exponent in the following way:

$$-\bar{\eta}\eta + \bar{\chi}\chi = -\frac{1}{2}(\eta - \bar{\chi})(\chi - \bar{\eta}). \tag{6.20}$$

Using

$$\begin{aligned} d\bar{\chi}\, d\eta &= d\bar{\chi}^n \cdots d\bar{\chi}^1\, d\eta^1 \cdots d\eta^n \\ &= 2^n d(\bar{\chi}^n + \eta^n) \cdots d(\bar{\chi}^1 + \eta^1)\, d(\eta^1 - \bar{\chi}^1) \cdots d(\eta^n - \bar{\chi}^n) \end{aligned} \tag{6.21}$$

(*not* with a factor of 2^{-n} because we need the super-Jacobian) and pulling $\prod(\eta + \bar{\chi})$ to the left past $d(\eta - \bar{\chi})$, we obtain a factor of $(-)^n$ times $\int d(\bar{\chi} + \eta) \prod(\eta + \bar{\chi}) = 1$. Then, we obtain

$$2^n \int d(\eta - \bar{\chi})(-)^n \mathrm{e}^{-\frac{1}{2}(\eta - \bar{\chi})(\chi - \bar{\eta})} = \prod_{a=1}^{n} (\chi^a - \bar{\eta}_a). \tag{6.22}$$

Hence we end up with another fermionic delta function, which again will make the corresponding Grassmann integral trivial.

At this point we have obtained

$$\begin{aligned} An = {} & \frac{(-i)^{n/2}}{(4\pi\beta\hbar)^{n/2}} \int dx_0 \sqrt{g(x_0)}\, d\bar{\eta}\, d\chi \prod(\chi - \bar{\eta})\, \mathrm{e}^{\bar{\eta}\chi} \\ & \times \Bigg\langle \exp\Bigg[-\frac{1}{\beta\hbar} \int_{-1}^{0} \frac{1}{2} \left[g_{ij}(x_0 + q) - g_{ij}(x_0) \right] (\dot{q}^i \dot{q}^j + b^i c^j + a^i a^j)\, d\tau \\ & - \frac{1}{2} \int_{-1}^{0} \dot{q}^i \omega_{iab}(x_0 + q)(\psi^a_{1,bg} + \psi^a_{1,qu})(\psi^b_{1,bg} + \psi^b_{1,qu})\, d\tau \\ & - \frac{\beta\hbar}{8} \int_{-1}^{0} \Gamma\Gamma + \frac{1}{2}\omega\omega\, d\tau \Bigg] \Bigg\rangle. \end{aligned} \tag{6.23}$$

We combined the bosonic measure $(2\pi\beta\hbar)^{-n/2}$ with the factor of $2^{-n/2}$ which accounted for the dimension of the space in which ψ_2^a acts. **The "extra term" $\mathrm{e}^{\bar{\eta}\chi}$ in the action is annihilated by the fermionic delta function** $\prod(\chi^a - \bar{\eta}_a)$, and we proceed to perform the $(\chi - \bar{\eta})$ integrals.

We once again perform the transition from the variables $\bar{\eta}$ and χ to $\chi + \bar{\eta}$ and $\bar{\eta} - \chi$. For any function F of $(\chi + \bar{\eta})/\sqrt{2}$ one has

$$\begin{aligned} & \int d\bar{\eta}\, d\chi \prod(\chi - \bar{\eta}) F\Big(\frac{\chi + \bar{\eta}}{\sqrt{2}}\Big) \\ & = \int 2^n d(\bar{\eta} + \chi)\, d(\chi - \bar{\eta}) \prod(\chi - \bar{\eta}) F\Big(\frac{\chi + \bar{\eta}}{\sqrt{2}}\Big) \end{aligned}$$

$$= \int 2^n\, d(\chi+\bar\eta) F\Big(\frac{\chi+\bar\eta}{\sqrt{2}}\Big) = \int 2^n\, d(\sqrt{2}\psi^a_{1,bg}) F(\psi^a_{1,bg})$$

$$= \int 2^{n/2} \prod_{a=1}^{n} d\psi^a_{1,bg} F(\psi^a_{1,bg}). \tag{6.24}$$

We used the fact that $d(\bar\eta+\chi)\, d(\chi-\bar\eta) \equiv d(\bar\eta+\chi)^n \cdots d(\bar\eta+\chi)^1\, d(\chi-\bar\eta)^1 \cdots d(\chi-\bar\eta)^n$ equals $d(\chi+\bar\eta)\, d(-\bar\eta+\chi) \equiv d(\chi+\bar\eta)^1 \cdots d(\chi+\bar\eta)^n\, d(-\bar\eta+\chi)^n \ldots d(-\bar\eta+\chi)^1$ to do the integral over $\prod(\chi-\bar\eta) = (\chi-\bar\eta)^1 \cdots (\chi-\bar\eta)^n$.

Next, we rescale $\psi_{1,qu}$ and $\psi_{1,bg}$ by a factor of $(\sqrt{\beta\hbar})^{-1}$. **The rescaling of $\psi^a_{1,bg}$ removes the $\beta\hbar$ dependence of the measure** (use $d\psi = d(\psi'/\sqrt{\beta\hbar}) = \sqrt{\beta\hbar}\, d\psi'$), and the fact that the measure is $\beta\hbar$ independent will have enormous consequences. The rescaling of ψ_1 adds a factor of $1/\beta\hbar$ to the vertices with fermions and a factor of $\beta\hbar$ to the propagators of the fermions. Dropping the primes on ψ' we arrive at

$$An = \frac{(-i)^{n/2}}{(2\pi)^{n/2}} \int \prod_{i=1}^{n} dx_0^i \sqrt{g(x_0)} \prod_{a=1}^{n} d\psi^a_{1,bg} \Big\langle \exp\Big(-\frac{1}{\hbar} S^{int}\Big)\Big\rangle$$

$$-\frac{1}{\hbar} S^{int} = -\frac{1}{\beta\hbar} \int_{-1}^{0} \frac{1}{2} [g_{ij}(x_0+q) - g_{ij}(x_0)](\dot q^i \dot q^j + b^i c^j + a^i a^j)\, d\tau$$
$$- \frac{1}{\beta\hbar} \int_{-1}^{0} \frac{1}{2} \dot q^i \omega_{iab}(x_0+q)(\psi^a_{1,bg} + \psi^a_{1,qu})(\psi^b_{1,bg} + \psi^b_{1,qu})\, d\tau$$
$$- \frac{\beta\hbar}{8} \int_{-1}^{0} \Big(\Gamma\Gamma + \frac{1}{2}\omega\omega\Big)\, d\tau. \tag{6.25}$$

The expectation value $\langle\cdots\rangle$ indicates that all quantum fields ($q^i, \psi^a_{1,qu}$ and a, b, c ghosts) must be contracted using the propagators of Chapter 2. However, in the end we must take the limit $\beta \to 0$, and since all propagators are proportional to $\beta\hbar$ and all vertices are proportional to $1/\beta\hbar$ (or even $\beta\hbar$ for the $\Gamma\Gamma + \frac{1}{2}\omega\omega$ term), we conclude:

(1) only one-loop graphs survive the $\beta\hbar \to 0$ limit (at higher loops there are more propagators than vertices);

(2) the a, b, c ghosts do not contribute at the one-loop level because their vertices involve at least three quantum fields;

(3) the $\Gamma\Gamma + \frac{1}{2}\omega\omega$ term can be discarded as it is of higher order in β;

(4) there are no terms linear in quantum fields, and hence no tadpoles, because the integral of $\dot q^i \omega_{iab}(x_0)\psi^a_{1,bg}\psi^b_{1,bg}$ vanishes due to the boundary conditions on q^i;

(5) we can, for convenience, choose a frame with $\omega_{iab}(x_0) = 0$. Then $\omega_{iab}(x_0+q)$ is at least linear in quantum fields. Expanding

$\omega_{iab}(x_0+q)$ to first order, one can set $\psi^a_{1,qu}=0$ in the last-but-one line of (6.25); and

(6) the only remaining vertex is

$$-\frac{1}{\hbar}S^{int} = -\frac{1}{\beta\hbar}\int_{-1}^{0}\frac{1}{2}\dot{q}^i q^j \partial_j\omega_{iab}(x_0)\psi^a_{1,bg}\psi^b_{1,bg}\,d\tau$$
$$= -\frac{1}{\beta\hbar}\frac{1}{4}\int_{-1}^{0} q^i\dot{q}^j R_{ijab}(\omega(x_0))\psi^a_{1,bg}\psi^b_{1,bg}\,d\tau \qquad (6.26)$$

where $R_{ijab}(\omega)(x_0) = \partial_i\omega_{jab}(e(x_0)) + \omega_{iac}\omega_j{}^c{}_b - (i\leftrightarrow j)$ and we have used the fact that $\int q^i\dot{q}^j d\tau$ is antisymmetric in i and j. (Since q^i vanishes at the endpoints we are allowed to partially integrate.)

Hence, **we need only compute closed** q**-loops**, with q-propagators and $R_{ij}(x_0)\equiv R_{ijab}(\omega(x_0))\psi^a_{1,bg}\psi^b_{1,bg}$ sticking out of each vertex. Then,

$$An = \frac{(-i)^{n/2}}{(2\pi)^{n/2}}\int dx_0\sqrt{g(x_0)}\,d\psi_{1,bg}\left\langle \mathrm{e}^{-\frac{1}{\beta\hbar}\frac{1}{4}R_{ij}(x_0)\int_{-1}^{0}q^i\dot{q}^j d\tau}\right\rangle. \qquad (6.27)$$

This formula contains the chiral anomaly for any dimension, and the explicit evaluation of the expression $\langle\cdots\rangle$ is far simpler than the corresponding Feynman graph calculation. Expanding the exponent, one obtains disconnected graphs: sums of **products** of closed q loops are found, which yield terms like $(\mathrm{tr}\,R^2)^2$ and $\mathrm{tr}\,R^4$, for example. If one writes the result for $\langle\cdots\rangle$ as $\exp[-\frac{1}{\hbar}W(\text{loops})]$, then $W(\text{loops})$ contains only single closed loops (because $\exp[-\frac{1}{\hbar}W(\text{loops})]$ is the generating functional for connected graphs). To obtain the final formula for the anomaly one must expand the exponent, and then one again finds the products of closed loops.

The evaluation of the sum of connected closed loops in $W(\text{loops})$ yields a sum of graphs, each with k vertices and k propagators. The propagators read

$$\langle q^i(\sigma)q^j(\tau)\rangle = -\beta\hbar g^{ij}(x_0)\Delta(\sigma,\tau) \qquad (6.28)$$

where $\Delta(\sigma,\tau)$ was defined in (1.23) The $g^{ij}(x_0)$ contract the first two indices of the curvatures to a trace over k curvature tensors. Hence,

$$-\frac{1}{\hbar}W(\text{loops}) = \sum_{k=2}^{\infty}\frac{1}{k!}\frac{1}{4^k}(\mathrm{tr}\,R^k)(k-1)!\,2^{k-1}$$
$$\times\int_{-1}^{0}d\tau_1\cdots\int_{-1}^{0}d\tau_k {}^{\bullet}\!\Delta(\tau_1,\tau_2)\,{}^{\bullet}\!\Delta(\tau_2,\tau_3)\ldots{}^{\bullet}\!\Delta(\tau_k,\tau_1)$$
$$(6.29)$$

where $\operatorname{tr} R^k = R_{i_1 i_2} R_{i_2 i_3} \cdots R_{i_{2k-1} i_1}$. The factor of $(k-1)!$ states that one can contract the k vertices in $(k-1)!$ ways, while the symmetry of each vertex in both q yields a factor of 2^{k-1} (partial integration is allowed since $q^i(\sigma) = 0$ at the end points). If we replace R by $R/2$, there remains only an overall factor of $\frac{1}{2}$. Because $\operatorname{tr} R = 0$ we started the summation at $k = 2$.

The integrals

$$I_k = \int_{-1}^{0} d\tau_1 \cdots \int_{-1}^{0} d\tau_k \, [\tau_2 + \theta(\tau_1 - \tau_2)] \times [\tau_3 + \theta(\tau_2 - \tau_3)] \cdots [\tau_1 + \theta(\tau_k - \tau_1)] \tag{6.30}$$

are most easily evaluated by first computing the generating function $\sum_{k=1}^{\infty} (y^k/k) I_k$. In fact, the expression for the anomaly has precisely this structure, with $y = R/2$. The first few I_k are easily evaluated. One finds $I_1 = 0$ and

$$\begin{aligned} I_2 &= \int_{-1}^{0} d\tau_1 \int_{-1}^{0} d\tau_2 \, [\tau_2 + \theta(\tau_1 - \tau_2)][\tau_1 + \theta(\tau_2 - \tau_1)] \\ &= \frac{1}{4} + 2 \int_{-1}^{0} d\tau_1 \int_{\tau_1}^{0} \tau_2 \, d\tau_2 = -\frac{1}{12} \end{aligned} \tag{6.31}$$

etc. Using induction (see appendix A.4 of [58]), the general result is found,

$$\sum_{k=2}^{\infty} \frac{y^k}{k} I_k = \ln \frac{y/2}{\sinh y/2} = -\frac{1}{3!} \left(\frac{y}{2}\right)^2 + \cdots . \tag{6.32}$$

Using this result, we find for **the gravitational contribution to the chiral anomaly of a Dirac fermion** in n dimensions due to the transformation law $\delta\lambda = i\alpha\gamma^5\lambda$ and $\delta\bar{\lambda} = i\alpha\bar{\lambda}\gamma^5$,

$$\begin{aligned} An &= (-2i\alpha) \frac{(-i)^{n/2}}{(2\pi)^{n/2}} \int dx_0^i \sqrt{g(x_0)} \, d\psi_{1,bg}^a \, \exp\left[\frac{1}{2} \operatorname{tr} \ln \left(\frac{R/4}{\sinh R/4}\right)\right] \\ R &= R_{ij} = R_{ijab} \psi_{1,bg}^a \psi_{1,bg}^b . \end{aligned} \tag{6.33}$$

The factor of $-2i\alpha$ is of course due to the Jacobian in (6.3), and the factor of $\frac{1}{2}$ in the exponent is the overall factor of $\frac{1}{2}$ mentioned below (6.29). Furthermore, the factor of $(-i)^{n/2}$ is due to the definition of γ_5, and the factor of $(2\pi)^{-n/2}$ is due to the Feynman measure. Sometimes in the literature one defines $\gamma_5 = (+i)^{n/2} \gamma^1 \cdots \gamma^n$ instead of (6.6), and then one finds a factor of $(+i)^{n/2}$ instead of $(-i)^{n/2}$ in the anomaly. The integration over the n Grassmann variables $\psi_{1,bg}^a$ yields an ϵ-tensor $\epsilon^{a_1 \cdots a_n}$. Since only the term with precisely n factors $\psi_{1,bg}$ can contribute to the

Grassmann integral (hence only the terms proportional to $R^{n/2}$), we can absorb the overall normalization factor into the trace,

$$An = (-2i\alpha)\left(\int \prod_{i=1}^{n} dx_0^i \sqrt{g(x_0)}\right)$$
$$\times \left(\int \prod_{a=1}^{n} d\psi_{1,bg}^a\right) \exp \frac{1}{2} \operatorname{tr} \ln \left[\frac{-iR/8\pi}{\sinh(-iR/8\pi)}\right]. \tag{6.34}$$

This is our final result. The matrix $-iR_{ijab}\psi_{1,bg}^a\psi_{1,bg}^b$ is real, so the anomaly is purely imaginary. In the path integral in Euclidean space, we find $\langle(\partial_\mu\alpha)j_5^\mu\rangle - \langle 2i\alpha\gamma_5\rangle = 0$, but j_5^μ has no definite reality properties in Euclidean space. In Minkowski space one has i times the action in the path integral: now $i\langle(\partial_\mu\alpha)j_5^\mu\rangle$ is antihermitian. The anomaly we have computed is imaginary, both in Euclidean and in Minkowski space. (Making a Wick rotation to Minkowski space, the factor of i from the ϵ symbol cancels the factor of i from d^nx.) Only traces with an even number of Riemann tensors are present (because $x^{-1}\sinh x$ is even in x; consequently, no factors of i survive in the expansion of the exponent). Since each R_{ij} contains two ψ, this means that **there is a gravitational contribution to the chiral anomaly only in $n = 4k$ dimensions.** In particular, there is a gravitational contribution to the chiral anomaly in $d = 4$ but not in $d = 2$ or $d = 10$. For $n = 4$, the terms with four ψ_1^a yields $\epsilon^{abcd}R_{ijab}R_{ijcd}$ times a factor of $\frac{1}{3!}\frac{1}{(8\pi)^2}\frac{1}{2} = \frac{1}{192\,(2\pi)^2}$, times $(-2i\alpha)$, which is the correct result for Dirac spinors [124].

This result contains the complete dependence on all gravitational fields (not only the leading term) since we used the spin connections $\omega_{iab}(e)$ as external fields. Although we put ω_{iab} at the point x_0 equal to zero, the complete γ_5 anomaly does not contain further terms with bare ω. This follows either from a direct calculation, or from the fact that chiral transformations and local Lorentz transformations commute. Since we used a regulator which preserves Einstein and local Lorentz symmetry, the anomaly must be locally Lorentz invariant:

$$\delta_{lL}\, An(\text{chiral}) = 0. \tag{6.35}$$

The same holds true for the Einstein symmetry. Hence, (6.34) is the complete answer.

Note also that one obtains in the exponent a sum of terms, so that the anomaly corresponds to sums of products of traces over Riemann tensors. For example, in $d = 8$ one finds two terms with four curvatures, proportional to $\operatorname{tr} R^4$ and $(\operatorname{tr} R^2)^2$. This is different from the Yang–Mills contribution to the abelian chiral anomaly which always has the form

$\mathrm{Tr}(F^{n/2})$.[4] In Feynman diagram language this means that disconnected graphs contribute to the gravitational chiral anomaly. The group theoretical reason is that the Lorentz generators do not commute with the Dirac matrices, whereas the Yang–Mills generators of course commute with the Dirac matrices.

One might worry that our procedure of introducing free ψ_2^a at the beginning violates local Lorentz invariance. Since we started with a regulator which is locally Lorentz invariant, and all other steps were mathematical identities, local Lorentz invariance cannot be lost. As a check one might repeat the calculations with $\omega_{iab}(x_0)$ not vanishing. One should find that terms with bare ω cancel. In fact, if one defines the ψ_2^a to be inert under local Lorentz transformations, the action preserves local Lorentz symmetry because the ψ_2^a do not couple, and then local Lorentz invariance should remain preserved at all stages.

For the next anomaly we use the alternative approach in which one does not add free ψ_2^a ("doubling") but combines pairs of spinors into ψ and $\bar{\psi}$ ("halving"). One could repeat the calculations of this section using halving instead of doubling; the answer should be the same.

A comment on supersymmetry. Starting from the non-susy Dirac action in n-dimensional space, we found an action for the QM path integral which turns out to be the $N=1$ susy QM model for spinors ψ_1^a plus terms of order $\hbar^2$ due to Weyl ordering. This classical $N=1$ action (with Euclidean time) is given by

$$L = \frac{1}{2} g_{ij} \dot{x}^i \dot{x}^j + \frac{1}{2} \psi_1^a [\dot{\psi}_1^a + \dot{x}^i \omega_{iab}(e) \psi_1^b] \tag{6.36}$$

and is invariant under

$$\delta x^i = -i\epsilon\, \psi_1^a e_a{}^i, \qquad \delta \psi_1^a = i e_i{}^a \dot{x}^i \epsilon - \delta x^i \omega_i{}^a{}_b \psi_1^b. \tag{6.37}$$

(In one-dimensional worldspace there are Euclidean Majorana spinors.) In flat space $L = \frac{1}{2}(\dot{x}^i)^2 + \frac{1}{2}\psi_1^i \dot{\psi}_1^i$ is clearly invariant under $\delta x^i = -i\epsilon\psi_1^i$ and $\delta\psi_1^i = i\dot{x}^i\epsilon$ (because $\delta L = -\dot{x}^i(i\epsilon\dot{\psi}_1^i) + (i\dot{x}^i\epsilon)\dot{\psi}_1^i = 0$), while in curved space one just covariantizes these rules ($\delta\psi_1^a + \delta x^j \omega_j{}^a{}_b \psi_1^b$ is covariant under local Lorentz transformations, see (D.56)). We refer to Appendix D for more details. The $\Gamma\Gamma + \omega\omega/2$ terms we found in the regulator were

[4] Jumping ahead, we shall see that for spin $\frac{1}{2}$ the Yang–Mills contributions to the abelian chiral anomaly are due to the factor of $\mathrm{Tr}\, e^{\frac{1}{2}F}$, whereas the gravitational contributions are due to the factor of $\exp[\frac{1}{2}\mathrm{tr}\, \ln(\frac{R/4}{\sinh R/4})]$. Because in the latter case the trace occurs in the exponent, one obtains an expression of the form $\exp[\, a\, \mathrm{tr}\, R^2 + b\, \mathrm{tr}\, R^4 + \cdots]$. Expanding one obtains products of traces. From $\mathrm{Tr}\, e^{\frac{1}{2}F}$ one obtains of course only a single trace.

quantum effects due to regularization by time slicing. They did not contribute to the gravitational γ_5 anomaly, but they do contribute to the trace anomaly. Using dimensional regularization these noncovariant terms are absent. Another way of obtaining the regulator $\mathcal{R}$ would have been to first construct the supersymmetry generator Q at the quantum level, for which operator ordering is fixed by requiring that it be Einstein and local Lorentz invariant. We discussed this in Chapter 5. Then $\hat{H} = \hat{Q}\hat{Q}$ is clearly a supersymmetric, Einstein and locally Lorentz invariant regulator. This reproduces the regulator in (6.9); in particular, the coefficient of the curvature in that expression is fixed by supersymmetry. We did not impose supersymmetry from the beginning, but rather we chose a regulator which was the square of the field operator of the fermions. So we took the action of the quantum field theory as our starting point, and this fixed the R term in the quantum action.

Let us now discuss the relation between our calculations and those of [1]. One can either evaluate the Feynman graphs with l external sources R_{ij} and then sum over l, as we have done, or directly evaluate the propagator in a gravitational background and use the fact that the sum of one-loop graphs corresponds to the determinant of the field operator. The former approach is the most natural at this point, since we have already determined the propagators and interaction vertices. However, it needs some special tricks [58]. The latter approach is used in [1] and we sketch the connection. If one goes back to the discretized path integral with $z = y = x_0$ and adds the final integration with dx_0 to the $N-1$ integrations $dx_1 \cdots dx_{N-1}$, one obtains for $N \to \infty$ a continuous path integral over $Dx(\tau)$ with periodic boundary conditions. The part quadratic in quantum fields q then yields the one-loop determinant,

$$\left\langle \exp\left(-\frac{1}{\beta\hbar}\frac{1}{4}R_{ij}(x_0)\int_{-1}^{0} q^i\dot{q}^j\, d\tau\right)\right\rangle = \left\{\frac{\det[-g_{ij}(x_0)\partial_\tau^2 + \frac{1}{2}R_{ij}(x_0)\partial_\tau]}{\det[-g_{ij}(x_0)\partial_\tau^2]}\right\}^{-1/2}. \tag{6.38}$$

Transforming $g_{ij}(x_0)$ to δ_{ij} and diagonalizing the hermitian $n \times n$ matrix $\frac{1}{2}R_{ijab}\psi_1^a\psi_1^b$ with eigenvalues $(y_1, \ldots, y_n)$, one obtains for the ratio of the two determinants

$$\prod_{k=1}^{n}\prod_{n=-\infty}^{\infty}{}'\left(1+\frac{iy_k}{2\pi n}\right) = \prod_{k=1}^{n}\prod_{n=1}^{\infty}\left(1+\frac{y_k^2}{4\pi^2 n^2}\right) = \prod_{k=1}^{n}\frac{\sinh(y_k/2)}{y_k/2}. \tag{6.39}$$

The factors $2\pi n$ come from the plane wave $\exp 2\pi i n\tau$ which form a complete set with periodic boundary conditions. The prime indicates that one should omit the zero mode with $n=0$. (It is sometimes stated that one can regularize the infrared divergence corresponding to $n = 0$ by giving the quantum field a small mass. Then for $n = 0$ there is a nonvanishing contribution to both determinants which cancels in the ratio (6.38).

However, there are many more subtle points to be worked out, and since we do not follow this approach we do not discuss this different infrared regularization scheme further.) This can be rewritten as $\exp \operatorname{tr} \ln \left(\frac{\sinh R/4}{R/4} \right)$. Raising this to a power of $-\frac{1}{2}$ indeed yields the same exponent as in (6.33) This exponent takes care of the nonzero modes, whereas the integration over x_0^i and ψ_1^a takes care of the bosonic and fermionic zero modes. The factor of $(2\pi)^{-n/2}$ is of course the remnant of the Feynman measure after the factor of $\beta\hbar$ has been scaled away.

6.2 The abelian chiral anomaly for spin-$\frac{1}{2}$ fields coupled to the Yang–Mills fields in $2k$ dimensions

The next anomaly we consider is the abelian chiral (γ_5) anomaly in loops with Dirac fermions, coupled to external Yang–Mills fields instead of external gravitational fields. This is the same calculation as in the previous section, but with gravity replaced by Yang–Mills fields. We can again make a local chiral transformation of the integration variables of the path integral, and then find that the anomaly is equal to the divergence of the abelian axial-vector current $\bar{\lambda}\gamma_5\gamma^\mu\lambda$. Since there is now no metric $g_{ij}(x)$, the QM approach we are now dealing with contains **linear** sigma models, and no a, b, c ghosts will be present (equivalently the a, b, c ghosts are now free fields). The regulator for the quantum field theory will contain new objects, namely the matrices for the generators in the representation of the gauge group for the fermions. These matrices will be denoted by $(T_\alpha)^M{}_N$, and in the QM model new internal ghosts must be introduced. These ghosts will be denoted by $\hat{c}^M$ and $\hat{c}^*{}_M$ and satisfy the anticommutation relations $\{\hat{c}^M, \hat{c}^*{}_N\} = \delta^M{}_N$. We shall omit the hats most of the time.

As we shall explain, we are interested only in the one-particle subspace of the whole Fock space obtained by acting with one c^* on the c-vacuum, and the matrices T_α are represented in this subspace by $c^*{}_M(T_\alpha)^M{}_N c^N$. (It is also possible to represent T_α with ghosts satisfying commutation relations, such as $d^*{}_M(T_\alpha)^M{}_N d^N$ where $[d^M, d^*{}_N] = \delta^M{}_N$, but this has the disadvantage that the Fock space becomes infinite dimensional. Incidentally, note that one cannot represent the Dirac matrices $(\gamma^m)^\alpha{}_\beta$ with either commuting or anticommuting ghosts as $d^\dagger\gamma^m d$ or $c^\dagger\gamma^m c$, since they satisfy *anti*commutation relations.) In their pioneering article, Alvarez-Gaumé and Witten [1] used an operator (Hamiltonian) approach for the internal ghost sector, but we shall treat the ghosts on equal footing with the nonghost sector, namely by a path integral approach.

The action of the quantum field theory we consider is given by

$$\mathcal{L} = -\bar{\lambda}\gamma^\mu D_\mu\lambda; \quad D_\mu\lambda^M = \partial_\mu\lambda^M + gA_\mu{}^\alpha(T_\alpha)^M{}_N\lambda^N \tag{6.40}$$

with $[T_\alpha, T_\beta] = f_{\alpha\beta}{}^\gamma T_\gamma$ and $[D_\mu, D_\nu] = gF_{\mu\nu}{}^\alpha T_\alpha$. The regulator in the quantum field theory, as obtained in the previous section, or from the construction of [13], is $\mathcal{R} \sim \not{D}\not{D} = D_\mu D^\mu + \frac{1}{2}\gamma^\mu\gamma^\nu(gF_{\mu\nu})$. The massless Dirac action has the rigid chiral symmetry

$$\delta\lambda = i\alpha\gamma_5\lambda, \quad \delta\bar{\lambda} = i\alpha\bar{\lambda}\gamma_5 \tag{6.41}$$

and in quantum field theory the anomaly is given by

$$An = -2i\alpha \lim_{\beta\to 0} \mathrm{Tr}\, \gamma_5\, \mathrm{e}^{-\beta\mathcal{R}}. \tag{6.42}$$

We shall omit for the time being the overall prefactor of $-2i\alpha$. The operators γ_5 and $\mathcal{R}$ depend on $x^\mu, \partial_\mu, \gamma^\mu$ (in flat space we do not distinguish between flat and curved indices) and $(T_\alpha)^M{}_N$, and the trace is over the internal indices M, N as well as over the spinor indices, and, of course, over all points in spacetime.

To construct the corresponding QM model, we again represent x^μ by $\hat{x}^i$, the hermitian operator $(\hbar/i)\partial_\mu$ by $\hat{p}_i$, and γ^m by the hermitian operators $\sqrt{2}\hat{\psi}_1^a$, with $[\hat{x}^i, \hat{p}_j] = i\hbar\delta^i{}_j$ and $\{\hat{\psi}_1^a, \hat{\psi}_1^b\} = \delta^{ab}$. This again yields the regulator in (6.7), but now in flat space and coupled to Yang–Mills fields. The matrices $(T_\alpha)^M{}_N$ are represented by operators

$$\hat{T}_\alpha = \hat{c}^*{}_M(T_\alpha)^M{}_N\hat{c}^N \tag{6.43}$$

which act in the sector of the one-particle states $|N\rangle$ (omitting hats)

$$|N\rangle = c^*{}_N|0\rangle, \quad c^M|0\rangle = 0. \tag{6.44}$$

There are, of course, also two-particle, etc. states $c^*{}_M c^*{}_N|0\rangle$, etc., but if we start with a one-particle state, the operators $\hat{T}_\alpha$ will never bring us outside this subspace. The action of $\hat{T}_\alpha$ on $|N\rangle$ is just like the matrix $(T_\alpha)^M{}_N$ acting on vectors in the carrier space of the representation R, and products of the operators $\hat{T}^\alpha$ lead to matrix multiplication.

To construct operators $\psi_a^\dagger$ and ψ^a from ψ_1^a, we could again add a set of free fermions ψ_2^a and proceed as in the previous section. The reader may follow this approach as an exercise; the answer for the anomaly will be the same. Here we follow an alternative approach: we combine pairs of hermitian fermions into Dirac spinors. Namely, we define

$$\begin{aligned}
\chi^A &= \frac{1}{\sqrt{2}}(\psi_1^{2A-1} + i\psi_1^{2A}); \quad \chi_A^\dagger = \frac{1}{\sqrt{2}}(\psi_1^{2A-1} - i\psi_1^{2A}) \\
\psi_1^a &= \frac{1}{\sqrt{2}}[\chi^{(a+1)/2} + \chi^\dagger_{(a+1)/2}] \qquad \text{for } a \text{ odd} \\
\psi_1^a &= \frac{-i}{\sqrt{2}}(\chi^{a/2} - \chi^\dagger_{a/2}) \qquad \text{for } a \text{ even.}
\end{aligned} \tag{6.45}$$

In order to define a matrix γ_5, we need an even number of dimensions $n = 2k$. The indices a run from 1 to n as always, but $A = 1, \ldots, n/2$, and

$$\{\chi^A, \chi^\dagger_B\} = \delta^A{}_B; \quad \{\chi^A, \chi^B\} = \{\chi^\dagger_A, \chi^\dagger_B\} = 0. \tag{6.46}$$

Weyl ordering will be defined with respect to the operators χ and $\chi^\dagger$.

We first must write γ_5 as an operator constructed from χ and $\chi^\dagger$. Recalling the definition $\gamma_5 = (-i)^{n/2}\gamma^1\gamma^2 \cdots \gamma^n$ and $\gamma^m \sim \sqrt{2}\psi^a$ we obtain

$$\begin{aligned}
\gamma_5 &= (-i)^{n/2}(\chi^1 + \chi^\dagger_1)(-i)(\chi^1 - \chi^\dagger_1) \\
&\quad \cdots (\chi^{n/2} + \chi^\dagger_{n/2})(-i)(\chi^{n/2} - \chi^\dagger_{n/2}) \\
&= (-1)^{n/2} \prod_{A=1}^{n/2} (\chi^A + \chi^\dagger_A)(\chi^A - \chi^\dagger_A) \\
&= \prod_{A=1}^{n/2} (\chi^A \chi^\dagger_A - \chi^\dagger_A \chi^A) = \prod_{A=1}^{n/2} (1 - 2\chi^\dagger_A \chi^A)
\end{aligned} \tag{6.47}$$

where we have used $\chi^A\chi^\dagger_A = 1 - \chi^\dagger_A\chi^A$. Since $\chi^\dagger_A\chi^A$ for fixed A is a projection operator, we can also write this as

$$\gamma_5 = \prod_{A=1}^{n/2} \mathrm{e}^{-i\pi\chi^\dagger_A\chi^A} = \mathrm{e}^{-i\pi\sum_{A=1}^{n/2}\chi^\dagger_A\chi^A} = (-)^F \tag{6.48}$$

where $F = \sum_{A=1}^{n/2} \chi^\dagger_A\chi^A$ is the fermion number operator. Indeed, the operator $\exp(-i\pi\chi^\dagger_A\chi^A)$ for fixed A is equal to

$$1 + \left[-i\pi + \frac{(-i\pi)^2}{2!} \cdots \right] \chi^\dagger_A\chi^A = 1 + (\mathrm{e}^{-i\pi} - 1)\chi^\dagger_A\chi^A = 1 - 2\chi^\dagger_A\chi^A. \tag{6.49}$$

Hence, the Jacobian can be written in two ways

$$J = \prod_{A=1}^{n/2} (1 - 2\chi^\dagger_A\chi^A) = (-)^F. \tag{6.50}$$

In [1] $(-)^F$ is used; we shall also use this expression.[5]

Next, we consider the regulator $\mathcal{R}$. The Hamiltonian of the QM model is obtained from the Euclidean regulator $\mathcal{R} \sim D_\mu D^\mu + \frac{1}{2}\gamma^\mu\gamma^\nu(gF_{\mu\nu})$ of the quantum field theory by replacing all operators of the quantum field

[5] In string theory, the operator $(-)^F$ is part of the so-called GSO projection operator introduced by Gliozzi, Scherk and Olive [149].

theory by corresponding operators in the QM model, and reads, after multiplication by $\frac{1}{2}(\hbar/i)^2$, as

$$\hat{H} = \frac{1}{2}\hat{\pi}_i\hat{\pi}_j\delta^{ij} - \frac{\hbar^2}{2}\hat{\psi}_1^a\hat{\psi}_1^b gF_{ab}{}^\alpha(A(\hat{x}))\hat{c}^*T_\alpha\hat{c}$$

$$D_\mu = \partial_\mu + gA_\mu{}^\alpha T_\alpha \quad \rightarrow \quad \hat{\pi}_i = \hat{p}_i - i\hbar gA_i{}^\alpha(\hat{x})\hat{c}^*T_\alpha\hat{c}. \tag{6.51}$$

The operator $\hat{H}$ is an operator constructed from the action of the quantum field theory in Euclidean space. The operators $\hat{\psi}_1^a$ are hermitian, and if one defines that $\hat{c}_M^*$ is the hermitian conjugate of $\hat{c}^M$, then $\hat{H}$ is formally hermitian. Since we are in flat (Euclidean) space, the indices of π_i and π_j are contracted with δ^{ij}, so we are dealing with linear sigma models, and there is no difference between curved indices i, j and flat indices a, b. The $\hat{\psi}_1^a$ in (6.51) are understood to be expressed in terms of $\hat{\chi}$ and $\hat{\chi}^\dagger$ by (6.45). Note that there are no counterterms generated if we rewrite the fermions $\hat{\psi}_1^a$ in $\hat{H}$ in Weyl-ordered form because F_{ab} is antisymmetric, whereas Weyl ordering only produces an extra term with δ^{ab}. The ghosts $\hat{c}$ and $\hat{c}^*$ should not be rewritten in Weyl-ordered form for reasons soon to be explained. We again drop hats.

The anomaly is now given by

$$An = \lim_{\beta\to 0} \mathrm{Tr}' \, \mathrm{e}^{-i\pi\chi^\dagger\chi}\, \mathrm{e}^{-\frac{\beta}{\hbar}H} \tag{6.52}$$

where the prime indicates that we are evaluating the trace only over the one-particle ghost sector (the states $|N\rangle = c^*{}_N|0\rangle$). To write the trace as an unconstrained trace, we introduce the one-particle ghost projection operator P_{gh}. We claim that [58]

$$P_{gh} = \,:x\mathrm{e}^{-x}:, \quad x \equiv c^*{}_M c^M \tag{6.53}$$

where : : indicates normal ordering with respect to $c^*{}_M$ and c^N. Indeed, on the vacuum $P_{gh}|0\rangle = 0$, while on the one-particle states

$$P_{gh}|N\rangle = \,: x - x^2 + \cdots : |N\rangle = \,: x : |N\rangle = |N\rangle. \tag{6.54}$$

Of course $: x^2 :$ vanishes on $|N\rangle$ because it contains two annihilation operators which stand to the right of the two creation operators. On two-particle states

$$|M, N\rangle \equiv c^*{}_M c^*{}_N|0\rangle \tag{6.55}$$

P_{gh} vanishes:

$$\begin{aligned} P_{gh}|M, N\rangle &= \,: x - x^2 + \frac{1}{2}x^3 + \cdots : |M, N\rangle \\ &= \,: x - x^2 : |M, N\rangle \\ &= (2- : x^2 :)|M, N\rangle = 0 \end{aligned} \tag{6.56}$$

since $: x^2 : |M, N\rangle = c^*{}_P c^*{}_Q c^Q c^P c^*{}_M c^*{}_N|0\rangle = 2|M, N\rangle$. The reader may check that P_{gh} also vanishes on three-particle states. (Only $: x - x^2 + \frac{1}{2}x^3 :$ contributes, and yields $3 - 3{\cdot}2 + \frac{1}{2}3! = 0$.) For a proof that P_{gh} vanishes on all ghost states, see section 12 of [58].

The anomaly can thus be written as the following unconstrained trace:

$$An = \lim_{\beta\to 0} \mathrm{Tr}\, \mathrm{e}^{-i\pi\chi_A^\dagger\chi^A} : c^*{}_M c^M \mathrm{e}^{-c^*{}_N c^N} : \mathrm{e}^{-\frac{\beta}{\hbar}H}. \tag{6.57}$$

To write out the trace on a basis of fermionic coherent states we introduce decompositions of unity in the internal (Yang–Mills ghost) space and in the fermionic $(\chi, \chi^\dagger)$ space

$$I_{gh} = \int d\bar{\eta}_{gh}\, d\eta_{gh}\, |\eta_{gh}\rangle \mathrm{e}^{-\bar{\eta}_{gh}\eta_{gh}} \langle\bar{\eta}_{gh}|$$
$$I_f = \int d\bar{\eta}_f\, d\eta_f\, |\eta_f\rangle \mathrm{e}^{-\bar{\eta}_f\eta_f} \langle\bar{\eta}_f| \tag{6.58}$$

as we discussed in Section 2.4. The coherent states denoted by a subscript gh (for ghosts) are constructed from the operators c^M and $c^*{}_N$, and the coherent states with a subscript f (for fermions) are constructed from the operators χ^A and $\chi_A^\dagger$. The trace over internal ghost states and fermionic states of a bosonic operator $\hat{A}$ is given by, respectively,

$$\mathrm{tr}_{gh}\, \hat{A} = \int d\chi_{gh}\, d\bar{\chi}_{gh}\, \mathrm{e}^{\bar{\chi}_{gh}\chi_{gh}} \langle\bar{\chi}_{gh}|\hat{A}|\chi_{gh}\rangle$$
$$\mathrm{tr}_f\, \hat{A} = \int d\chi_f\, d\bar{\chi}_f\, \mathrm{e}^{\bar{\chi}_f\chi_f} \langle\bar{\chi}_f|\hat{A}|\chi_f\rangle \tag{6.59}$$

as we discussed in the previous section. Whenever we write a multiple integral such as $d\eta$, it is ordered as $d\eta^1 \cdots d\eta^n$, while the integrals over barred fermions such as $d\bar{\eta}$ are ordered in the opposite order, $d\bar{\eta}^n \cdots d\bar{\eta}^1$. Recall that χ_f consists of χ_f^A with $A = 1, \ldots, n/2$, while χ_{gh} contains χ_{gh}^M with $M = 1, \ldots, \dim R$, where $\dim R$ is the dimension of the Yang–Mills representation of the fermions in the original QFT. Hence (omitting the symbol $\lim_{\beta\to 0}$ for the time being)

$$An = \mathrm{tr}_{x_0}\mathrm{tr}_f\, \mathrm{tr}_{gh}\, \langle x_0, \bar{\chi}_{gh}, \bar{\chi}_f|\mathrm{e}^{-i\pi\chi_f^\dagger\chi_f} P_{gh} I_{gh} I_f \mathrm{e}^{-\frac{\beta}{\hbar}H}|\chi_f, \chi_{gh}, x_0\rangle \tag{6.60}$$

where $\mathrm{tr}\, x_0 = \int \prod_{i=1}^n dx_0^i$. This trace factorizes into a ghost trace and a fermionic trace.

The ghost part yields

$$\int d\chi_{gh}\, d\bar{\chi}_{gh}\, \mathrm{e}^{\bar{\chi}_{gh}\chi_{gh}}\, d\bar{\eta}_{gh}\, d\eta_{gh}\, \mathrm{e}^{-\bar{\eta}_{gh}\eta_{gh}} \langle\bar{\chi}_{gh}|P_{gh}|\eta_{gh}\rangle\langle\bar{\eta}_{gh}|\mathrm{e}^{-\frac{\beta}{\hbar}H}|\chi_{gh}\rangle. \tag{6.61}$$

Since $P_{gh} = :x\mathrm{e}^{-x}:$ projects the coherent state $|\eta_{gh}\rangle$ onto its one-particle part, $P_{gh}|\eta_{gh}\rangle = c_M^* \eta_{gh}^M|0\rangle$, the first matrix element is easily computed and yields

$$\langle\bar{\chi}_{gh}|P_{gh}|\eta_{gh}\rangle = \bar{\chi}_{gh}\eta_{gh}. \tag{6.62}$$

Note that in this case we did not use Weyl ordering to evaluate this matrix element. We begin with the integral over $\bar{\chi}_{gh}$,

$$\int d\bar{\chi}_{gh}\, \mathrm{e}^{\bar{\chi}_{gh}\chi_{gh}} \bar{\chi}_{gh}\eta_{gh}$$
$$= \int d\bar{\chi}_{gh,\dim R} \cdots d\bar{\chi}_{gh,1} \Big(\sum_M \bar{\chi}_{gh,M}\eta_{gh}^M\Big) \mathrm{e}^{\bar{\chi}_{gh}\chi_{gh}}. \tag{6.63}$$

For a given M, each $\bar{\chi}_{gh,N}$ integral yields χ_{gh}^N, except for the integral which contains $\bar{\chi}_{gh,M}\eta_{gh}^M$ in the integrand, which yields η_{gh}^M. Hence the $\bar{\chi}_{gh}$ integrals yield

$$\sum_{M=1}^{\dim R} \chi_{gh}^{\dim R} \cdots \chi_{gh}^{M+1}\, \eta_{gh}^M\, \chi_{gh}^{M-1} \cdots \chi_{gh}^{1}. \tag{6.64}$$

Next, we perform the η_{gh} integrations

$$\int d\eta_{gh}^1 \cdots \int d\eta_{gh}^{\dim R} \mathrm{e}^{-\bar{\eta}_{gh}\eta_{gh}} \sum_M \Big(\prod_{N>M} \chi_{gh}^N\Big)\eta_{gh}^M \Big(\prod_{N<M} \chi_{gh}^N\Big). \tag{6.65}$$

Again, each $d\eta_{gh}$ integral yields a factor of $\bar{\eta}_{gh}$, except in one case where η_{gh}^M in the integrand yields unity. The result will be called $P_{\bar{\eta},\chi}^{gh}$ and reads as

$$P_{\bar{\eta},\chi}^{gh} = \sum_{M=1}^{\dim R} \left(\bar{\eta}_{gh,1}\chi_{gh}^1\right) \cdots \left(\bar{\eta}_{gh,M-1}\, \chi_{gh}^{M-1}\right)$$
$$\times \left(\bar{\eta}_{gh,M+1}\chi_{gh}^{M+1}\right) \cdots \left(\bar{\eta}_{gh,\dim R}\, \chi_{gh}^{\dim R}\right). \tag{6.66}$$

Clearly, in an arbitrary function of $\bar{\eta}_{gh}$ and χ_{gh}, this operator deletes all terms with two or more $\bar{\eta}_{gh}$ and χ_{gh}. It is thus a kind of projection operator onto terms which are linear in (or independent of) $\bar{\eta}$ and χ. We denote it by $P_{\bar{\eta},\chi}^{gh}$. We interrupt the discussion of the ghost sector at this point and first perform the trace in the fermionic sector.

In the fermionic sector we must first evaluate the matrix element of γ_5. We find

$$\langle\bar{\chi}_f|\mathrm{e}^{-i\pi\chi_A^\dagger\chi^A}|\eta_f\rangle = \langle\bar{\chi}_f|\prod_{A=1}^{n/2}\left(1 - 2\chi_A^\dagger\chi^A\right)|\eta_f\rangle$$
$$= \mathrm{e}^{\bar{\chi}_f\eta_f} \prod_{A=1}^{n/2}\left(1 - 2\bar{\chi}_{f,A}\eta_f^A\right) = \mathrm{e}^{-\bar{\chi}_f\eta_f}. \tag{6.67}$$

To obtain this result, we used the definition of coherent states and the identity

$$1 - 2\bar{\chi}_{f,A}\eta_f^A = \mathrm{e}^{-2\bar{\chi}_{f,A}\eta_f^A}. \tag{6.68}$$

The Grassmann integral over $\bar{\chi}_f$ and η_f can now be performed, yielding

$$\begin{aligned}
&\int d\chi_f\, d\bar{\chi}_f\, \mathrm{e}^{\bar{\chi}_f\chi_f}\, d\bar{\eta}_f\, d\eta_f\, \mathrm{e}^{-\bar{\eta}_f\eta_f}\, \mathrm{e}^{-\bar{\chi}_f\eta_f} \\
&= \int d\chi_f\, d\bar{\eta}_f \left(\int d\eta_f\, d\bar{\chi}_f\, \mathrm{e}^{\bar{\chi}_f(\chi_f-\eta_f)}\, \mathrm{e}^{-\bar{\eta}_f\eta_f}\right) \\
&= \int d\chi_f d\bar{\eta}_f \left[\int d\eta_f\, \left(\chi_f^{n/2} - \eta_f^{n/2}\right)\cdots\left(\chi_f^1 - \eta_f^1\right)\mathrm{e}^{-\bar{\eta}_f\eta_f}\right] \\
&= \int d\bar{\eta}_f\, d\chi_f\, \mathrm{e}^{-\bar{\eta}_f\chi_f}.
\end{aligned} \tag{6.69}$$

In the last line we used the fact that $\prod(\chi_f - \eta_f)$ is a fermionic delta function, and replaced $\exp(-\bar{\eta}_f\eta_f)$ by $\exp(-\bar{\eta}_f\chi_f)$. We also canceled a sign factor of $(-)^{n/2}$ by interchanging $d\chi_f$ and $d\bar{\eta}_f$. This yields

$$\begin{aligned}
An = &\int \Big(\prod_i dx_0^i\Big)\left(d\chi_{gh}\, d\bar{\eta}_{gh}\, P^{gh}_{\bar{\eta},\chi}\right)\left(d\bar{\eta}_f\, d\chi_f\, \mathrm{e}^{-\bar{\eta}_f\chi_f}\right) \\
&\times \langle x_0, \bar{\eta}_{gh}, \bar{\eta}_f|\mathrm{e}^{-\frac{\beta}{\hbar}H}|\chi_f, \chi_{gh}, x_0\rangle.
\end{aligned} \tag{6.70}$$

The regulated trace of γ^5 contains the transition element

$$\langle x_0, \bar{\eta}_{gh}, \bar{\eta}_f|\mathrm{e}^{-\frac{\beta}{\hbar}H}|\chi_f, \chi_{gh}, x_0\rangle = \frac{1}{(2\pi\beta\hbar)^{n/2}}\langle \mathrm{e}^{-\frac{1}{\hbar}S^{int}}\rangle \tag{6.71}$$

where the extra terms $\bar{\chi}_{gh}\psi_{gh}(0)$ and $\bar{\chi}_f\psi_f(0)$ are present in the exponent. The action S is obtained by inserting complete sets of states. For the fermions we use coherent states depending on $\psi^A_{f,k}$, $\bar{\psi}_{f,k,A}$ and $\psi^M_{gh,k}$, $\bar{\psi}_{gh,k,M}$. Then the operators $\hat{\psi}_1$ and $\hat{c}^*$ and $\hat{c}$ are replaced by $\psi_1 \equiv (\psi_f + \bar{\psi}_f)/\sqrt{2}$ and ψ_{gh} and $\bar{\psi}_{gh}$. The momenta are integrated out from

$$\begin{aligned}
\int_{-1}^{0} \Bigg\{ &\frac{i}{\hbar}p_i\dot{x}^i - \frac{\beta}{\hbar}\Big[\frac{1}{2}(p_i - i\hbar g A_i{}^\alpha\bar{\psi}_{gh}T_\alpha\psi_{gh})^2 \\
&-\frac{\hbar^2}{2}\psi_1^a\psi_1^b g F_{ab}{}^\alpha\bar{\psi}_{gh}T_\alpha\psi_{gh}\Big]\Bigg\}d\tau.
\end{aligned} \tag{6.72}$$

Then the terms quadratic in A_i^α cancel and only the familiar $\dot{x}^i A_i^\alpha$ interaction is left. In the path integral the integration variables are $\psi_f^A, \bar{\psi}_{f,A}$, $\psi^M_{gh}, \bar{\psi}_{gh,M}$ and q^i. We make a decomposition into background fields and quantum fields as follows:

$$\begin{aligned} \psi_f^A &= \chi_f^A + \psi^A, & \bar{\psi}_{f,A} &= \bar{\eta}_{f,A} + \bar{\psi}_A \\ \psi_{gh}^M &= \chi_{gh}^M + c^M, & \bar{\psi}_{gh,M} &= \bar{\eta}_{gh,M} + \bar{c}_M \end{aligned} \tag{6.73}$$

and find then along the same lines as in (6.16),[6]

$$\begin{aligned} -\frac{1}{\hbar} S = &-\frac{1}{\beta\hbar} \int_{-1}^{0} \frac{1}{2} \dot{q}^i \dot{q}^i \, d\tau + (\bar{\eta}_{gh}\chi_{gh} + \bar{\eta}_f \chi_f) \\ &- \int_{-1}^{0} \left(\bar{\psi}_A \dot{\psi}^A + \bar{c}_M \dot{c}^M \right) d\tau \\ &- \int_{-1}^{0} \dot{q}^i g A_i{}^{\alpha}(x_0 + q)(T_\alpha) \, d\tau \\ &+ \beta\hbar \int_{-1}^{0} \frac{1}{2} \psi_1^a \psi_1^b g F_{ab}{}^{\alpha}(x_0 + q)(T_\alpha) \, d\tau \end{aligned} \tag{6.74}$$

where

$$(T_\alpha) \equiv \bar{\psi}_{gh,M} (T_\alpha)^M{}_N \psi_{gh}^N = (\bar{\eta}_{gh,M} + \bar{c}_M)(T_\alpha)^M{}_N (\chi_{gh}^N + c^N) \tag{6.75}$$

and ψ_1^a is expressed in terms of $\psi_f^A, \bar{\psi}_{fA}$ as in (6.45), which themselves are further decomposed as in (6.73). Thus, $q^i, \psi^A, \bar{\psi}_A, c^M$ and $\bar{c}_M$ are the quantum variables, and x_0, χ_f^A, $\bar{\eta}_{f,A}$, χ_{gh}^M, $\bar{\eta}_{gh,M}$ are the background variables.

The coupling $-\dot{q}^i A_i{}^{\alpha}(T_\alpha)$ came from integrating out the momenta, and the $\bar{c}c$ terms in (T_α) combine with the kinetic term $\bar{c}_M \dot{c}^M$ to a covariant derivative

$$D_\tau c^M = \dot{c}^M + \dot{q}^i g A_i{}^{\alpha}(x_0 + q)(T_\alpha)^M{}_N c^N \tag{6.76}$$

but note that there are also background fields in the interaction term.

The anomaly now reduces to

$$\begin{aligned} An = &\frac{1}{(2\pi\beta\hbar)^{n/2}} \int dx_0 \, d\chi_{gh} \, d\bar{\eta}_{gh} \, P_{\bar{\eta},\chi}^{gh} \, e^{\bar{\eta}_{gh}\chi_{gh}} \, d\bar{\eta}_f \, d\chi_f \\ &\times \Big\langle \exp\left[-\int_{-1}^{0} \dot{q}^i g A_i{}^{\alpha}(x_0 + q)(\bar{\eta}_{gh} + \bar{c}) T_\alpha (\chi_{gh} + c) \, d\tau \right] \\ &\times \exp\left[\beta\hbar \int_{-1}^{0} \frac{1}{2} \psi_1^a \psi_1^b g F_{ab}{}^{\alpha} (\bar{\eta}_{gh} + \bar{c}) T_\alpha (\chi_{gh} + c) \, d\tau \right] \Big\rangle. \end{aligned} \tag{6.77}$$

The term $\exp(\bar{\eta}_{gh}\,\chi_{gh})$ in the ghost sector of this expression is the "extra term" in the action which remains after substitution of (6.73), but the

[6] Recall that except in the vertices, all terms linear in quantum fields cancel. In particular, $\exp[\bar{\eta}_f \psi_f(0) - \int \bar{\psi}_f \dot{\psi}_f \, d\tau]$ becomes equal to $\exp(\bar{\eta}_f \chi_f - \int \bar{\psi}_f \dot{\psi}_f \, d\tau)$.

corresponding term $\exp(\bar{\eta}_f \chi_f)$ in the fermionic sector has canceled with the factor of $\exp(-\bar{\eta}_f \chi_f)$ in (6.70). **We rescale the fermionic variables $\bar{\eta}_{f,A}$, χ_f^A, ψ^A and $\bar{\psi}_A$, but not the ghost variables**, by a factor of $(\beta\hbar)^{-1/2}$ (so one sets $\chi_f^A = \chi'^A_f/\sqrt{\hbar\beta}$ and then drops the prime). Then the measure becomes $\beta\hbar$ independent,

$$\begin{aligned} An = &\frac{1}{(2\pi)^{n/2}} \int dx_0\, d\bar{\eta}_f\, d\chi_f \int d\chi_{gh}\, d\bar{\eta}_{gh}\, P^{gh}_{\bar{\eta},\chi} e^{\bar{\eta}_{gh}\chi_{gh}} \\ &\times \Big\langle \exp\left[-\int_{-1}^0 \dot{q}^i g A_i{}^\alpha (x_0 + q)(\bar{\eta}_{gh} + \bar{c}) T_\alpha (\chi_{gh} + c)\, d\tau \right] \\ &\times \exp\left[\frac{1}{2} \int_{-1}^0 \psi_1^a \psi_1^b g F_{ab}{}^\alpha (x_0 + q)(\bar{\eta}_{gh} + \bar{c}) T_\alpha (\chi_{gh} + c)\, d\tau \right] \Big\rangle . \end{aligned} \tag{6.78}$$

After this rescaling the fermion and boson propagators $\langle \psi^A \bar{\psi}_B \rangle$ and $\langle q^i q^j \rangle$ are proportional to $\beta\hbar$, but the ghost propagators $\langle c^M \bar{c}_N \rangle$ are $\beta\hbar$ independent.[7] All vertices are $\hbar\beta$ independent. It follows that **in the limit of vanishing β, only graphs with ghost propagators or without any propagators contribute**. Hence we may set the quantum fields q, ψ^A and $\bar{\psi}_A$ equal to zero, and replace ψ_1^a and ψ_1^b by their background values, which we denote by ψ_{bg}^a and ψ_{bg}^b. Then only the vertex with $\frac{1}{2}\psi_{bg}^a \psi_{bg}^b g F_{ab}^\alpha (x_0)(\bar{\eta}_{gh} + \bar{c}) T_\alpha (\chi_{gh} + c)$ will contribute, but the vertex with with $\dot{q}^i g A_i^\alpha (x_0)(\bar{\eta}_{gh} + \bar{c}) T_\alpha (\chi_{gh} + c)$ does not contribute since we set $q = 0$.

The propagator due to $\int_{-1}^0 \bar{c}_M \dot{c}^M\, d\tau$ is the same as for $\int_{-1}^0 \bar{\psi}_{qu,A}\, \dot{\psi}_{qu}^A\, d\tau$ given by (2.174), namely

$$\langle c^M(\sigma) \bar{c}_N(\tau) \rangle = \delta^M{}_N \theta(\sigma - \tau). \tag{6.79}$$

It follows that closed ghost loops do not contribute (a closed loop always moves somewhere backwards in time). **Only tree graphs with ghosts can contribute**, with at one end a field $\bar{\eta}_{gh}$ and at the other end a field χ_{gh}. In fact, only terms with precisely one $\bar{\eta}_{gh}$ and one χ_{gh} contribute due to the operator $P^{gh}_{\bar{\eta},\chi}$.

The ghost tree graphs are obtained by expanding

$$\exp\left[\frac{1}{2} \psi_{bg}^a \psi_{bg}^b g F_{ab}{}^\alpha (x_0) \int_{-1}^0 (\bar{\eta}_{gh} + \bar{c}) T_\alpha (\chi_{gh} + c)\, d\tau \right] \tag{6.80}$$

---

[7] More precisely, having introduced external sources K and $\bar{K}$ which couple to the fermionic quantum integration variables $\bar{\psi}$ and ψ, completed squares, and integrated out the fermionic quantum variables, S^{int} depends on $\chi + \delta/\delta\bar{K}$ and $\eta - \delta/\delta K$. By rescaling $\delta/\delta K$ and $\delta/\delta\bar{K}$ in the same way as $\bar{\eta}$ and χ, we must also rescale K and $\bar{K}$ in the source term $\bar{K}AK$. This produces the $\beta\hbar$ dependence of the $\psi\bar{\psi}$ propagator, and the $\beta\hbar$ independence of the vertices.

and contracting the vertices with ghost propagators. If one has k vertices one obtains

$$\frac{1}{k!}k!\left(\frac{1}{2}\right)^k \int_{-1}^{0}\cdots\int_{-1}^{0} d\sigma_1\cdots d\sigma_k\, \bar{\eta}_{gh} F \underbrace{c\bar{c}} F \cdots \underbrace{c\bar{c}} F \chi_{gh} \tag{6.81}$$

where $F \equiv \psi^a_{bg}\psi^b_{bg} g F_{ab}{}^{\alpha}(x_0) T_\alpha$ is a matrix in the internal symmetry space and where the factor of $k!$ in the numerator is due to the fact that one can order the k vertices into a tree in precisely $k!$ ways. The hooks in (6.81) denote the propagator in (6.79). The integral over σ_k yields

$$\int_{-1}^{0} d\sigma_1 \cdots \int_{-1}^{0} d\sigma_k\, \theta(\sigma_1-\sigma_2)\cdots\theta(\sigma_{k-1}-\sigma_k) = \frac{1}{k!}. \tag{6.82}$$

Hence, the ghost trees yield

$$\bar{\eta}_{gh} \exp\left(\frac{1}{2}F\right)\chi_{gh}. \tag{6.83}$$

The term without any F vertices is provided by expanding the factor of $\exp(\bar{\eta}_{gh}\chi_{gh})$ in (6.78). Owing to $P^{gh}_{\bar{\eta},\chi}$, only terms with precisely one $\bar{\eta}_{gh,M}$ and χ^M_{gh} with the same M will contribute, and they yield a trace in the space of the representation of the fermions,

$$\int d\chi_{gh}\, d\bar{\eta}_{gh} \left(\sum_M \prod_{N\neq M} \bar{\eta}_{gh,N}\chi^N_{gh}\right) \bar{\eta}_{gh}\, e^{\frac{1}{2}F}\chi_{gh} = \mathrm{Tr}\, e^{\frac{1}{2}F}. \tag{6.84}$$

All ghost variables are now gone.

The anomaly becomes

$$An = \frac{1}{(2\pi)^{n/2}} \int dx_0\, d\bar{\eta}_f\, d\chi_f \,\mathrm{Tr}\exp\left[\frac{1}{2}\psi^a_{bg}\psi^b_{bg} g F_{ab}{}^{\alpha}(x_0)T_\alpha\right] \tag{6.85}$$

where ψ^a_{bg} is expressed in term of the constant background fields χ^A_f and $\bar{\eta}_{f,A}$ (introduced in (6.73)) as in (6.45). Finally, we transform integration variables from $d\bar{\eta}_{f,A}\, d\chi^A_f$ to $d\psi^a_{bg}$. We find[8]

$$d\bar{\eta}_{f,n/2}\cdots d\bar{\eta}_{f,1}\, d\chi^1_f \cdots \chi^{n/2}_f = (-i)^{n/2} d\psi^1_{bg}\cdots d\psi^n_{bg}. \tag{6.86}$$

The final result for **the abelian chiral anomaly for complex spin-$\frac{1}{2}$ fields coupled to external Yang–Mills gauge fields** in $n = 2k$

[8] As a check note that for $n = 2$ one finds

$$d\psi^1_{bg}\, d\psi^2_{bg} = d[(\chi+\bar{\eta})/\sqrt{2}]\, d[(-i)(\chi-\bar{\eta})/\sqrt{2}] = 2i\, d(\chi+\bar{\eta})\, d(\chi-\bar{\eta}).$$

Furthermore, $d(\chi+\bar{\eta})d(\chi-\bar{\eta}) = \frac{1}{2}d\bar{\eta}\, d\chi$ (and not $2\, d\bar{\eta}\, d\chi$).

dimensions is given by

$$An = (-2i\alpha)\frac{(-i)^{n/2}}{(2\pi)^{n/2}}\int dx_0\, d\psi^1_{bg}\cdots d\psi^n_{bg}\,\mathrm{Tr}\exp\left(\frac{1}{2}\psi^a_{bg}\psi^b_{bg}gF_{ab}\right)$$

$$= \frac{(-2i\alpha)}{(2\pi)^{n/2}}\left(\frac{ig}{2}\right)^{n/2}\epsilon^{a_1\cdots a_n}\int \mathrm{Tr}(F_{a_1a_2}\cdots F_{a_{n-1}a_n})\,dx_0 \qquad (6.87)$$

where we have reinstated the factor of $(-2i\alpha)$ mentioned before (6.42). A factor of $(-)^{n/2}$ has been canceled by another factor of $(-)^{n/2}$ arising from the formula

$$\int d\psi^1_{bg}\cdots d\psi^n_{bg}\,\psi^{a_1}_{bg}\cdots\psi^{a_n}_{bg} = (-)^{n/2}\epsilon^{a_1\cdots a_n}. \qquad (6.88)$$

Thus, we obtain the familiar result that the divergence of the axial-vector current is proportional to $\epsilon F\cdots F$, but the great advantage of this expression is that it yields the result for all dimensions n in a simple compact formula. The anomaly is proportional to the totally symmetrized trace of the generators of the gauge group in the representation of the spin-$\frac{1}{2}$ fields, and if for a particular group this Casimir invariant vanishes, there is no corresponding anomaly. Again the anomaly is imaginary. For example, in two dimensions, there is only a $U(1)$ anomaly because $\mathrm{tr}\,T_\alpha = 0$ for the representations of simple Lie algebras, and the factor of $T_\alpha = i$ of the $U(1)$ group cancels the factor of $(-i)^{n/2}$, leaving only the i in $-2i\alpha$.

On the other hand, when the symmetrized trace over a product of generators in a particular representation of the gauge group is nonvanishing, there is an abelian chiral Yang–Mills anomaly in that representation. For certain groups and representations the trace over an odd number of generators vanishes (for example, for a real representation of $SO(N)$) and in these cases there are only anomalies possible in $4k$ dimensions.

We obtained the Hamiltonian H in (6.51) from the regulator $\mathcal{R}\sim \not{D}\not{D}$ of the QFT by replacing the operators of the QFT by corresponding operators in the QM model. When we followed these steps for the case where $D_i = \partial_i + \frac{1}{4}\omega_i{}^{mn}\gamma_m\gamma_n$, we found the $N=1$ susy model. This suggest that also for $D_i = \partial_i + gA_i^\alpha T_\alpha$ there is a corresponding susy model. There is indeed such a model, given in (6.74), and this model is discussed further in Appendix E. The interaction term in (6.74) with the Yang–Mills curvature is needed to supersymmetrize the interaction with $\dot{x}^i A_i$. It is interesting to note that once again an ordinary nonsupersymmetric quantum gauge field theory has produced a supersymmetric QM model that yields its anomalies.

6.3 Lorentz anomalies for chiral spin-$\frac{1}{2}$ fields coupled to gravity in $4k+2$ dimensions

Another important anomaly concerns the violation of the conservation of the stress tensor at the quantum level. Actually, there are two local symmetries which can be violated at the quantum level: Einstein (general coordinate) invariance and local Lorentz symmetry. The anomalies in Einstein and local Lorentz symmetry can be moved from one to the other, just like anomalies in the vector or axial-vector gauge invariance [2]. In principle, one should consider at the quantum level the Noether current for the rigid BRST symmetry which receives contributions from all local symmetries in the classical action: Einstein symmetry and local Lorentz symmetry (and local supersymmetry if spin-$\frac{3}{2}$ fields are present). However, we consider external gravitational fields, and then the quantum actions still have classical Einstein and local Lorentz symmetries. Even when we consider spin-$\frac{3}{2}$ fields and add a gauge-fixing term for the local supersymmetry (which is needed to be able to construct propagators for the spin-$\frac{3}{2}$ fields), we still preserve Einstein and local Lorentz gauge symmetry in the quantum action if the gauge-fixing term for local supersymmetry preserves these spacetime symmetries.

The gravitational anomalies we shall obtain are covariant anomalies: they depend only on Riemann curvatures and do not contain terms with bare ω_{iab}. We achieve this by using regulators which are both Einstein and locally Lorentz invariant, and which are vector-like (treat left- and right-handed spinors in the same way), but in the Jacobian an extra factor of γ_5 takes into account the fact that we compute the anomalies for chiral complex Dirac fermions. (Chiral fermions satisfy $\tilde{\lambda} = \frac{1}{2}(1+\gamma_5)\tilde{\lambda}$, but the term with $\frac{1}{2}\tilde{\lambda}$ does not contribute, leaving only the term with $\frac{1}{2}\gamma_5\tilde{\lambda}$.) As a result, these anomalies will not satisfy the consistency conditions which are present if the anomaly is the response of the effective action under a gauge transformation. One could construct a consistent regulator to be sure that the anomalies satisfy the consistency conditions. However, proceeding in this way is extremely tedious, because the regulator is not manifestly Lorentz invariant. It is much simpler to instead use regulators which are also Lorentz invariant, since then the anomaly will also be Lorentz covariant. This can be done, but these regulators are not "consistent" and the anomaly will not satisfy the consistency conditions. This is not a problem because there is a well-defined procedure to obtain the consistent anomaly from the covariant one [2]. As to the Jacobian, we shall consider a particular combination of general coordinate and local Lorentz transformations which leads to covariant transformation laws, and then we shall prove that if one also uses our covariant regulator, one obtains in this way (twice) the local Lorentz anomaly.[9]

Now we shall first define the covariant transformation law. Then we shall determine the regulator by requiring that a certain identity involving the Jacobian and the regulator holds. These are issues which are not explicitly discussed in [1] and which have confused us for a very long time, but through the work of Endo [126, 151] we finally clarified these issues.

We begin with the concept of a covariant Einstein transformation, denoted by δ_{cov}. This is a combination of an ordinary Einstein (i.e. general coordinate) transformation with the usual parameter ξ^μ and a local Lorentz transformation with composite parameter $\lambda_{mn} = \xi^\mu \omega_{\mu mn}$,

$$\delta_{cov}(\xi) = \delta_E(\xi) + \delta_{lL}(\xi^\mu \omega_{\mu mn}) \tag{6.89}$$

where the ordinary Einstein and the local Lorentz transformation on the vielbein are given as usual by

$$\begin{aligned} \delta_E(\xi) e_\mu{}^m &= \xi^\nu \partial_\nu e_\mu{}^m + (\partial_\mu \xi^\nu) e_\nu{}^m \\ \delta_{lL}(\lambda_{mn}) e_\mu{}^m &= \lambda^m{}_n e_\mu{}^n . \end{aligned} \tag{6.90}$$

The covariant Einstein transformation on the vielbein is then

$$\begin{aligned} \delta_{cov}(\xi) e_\mu{}^m &= \xi^\nu \partial_\nu e_\mu{}^m + (\partial_\mu \xi^\nu) e_\nu{}^m + \xi^\nu \omega_\nu{}^m{}_n e_\mu{}^n \\ &= \xi^\nu D_\nu(\omega) e_\mu{}^m + (\partial_\mu \xi^\nu) e_\nu{}^m \\ &= \xi^\nu [D_\nu(\omega) e_\mu{}^m - D_\mu(\omega) e_\nu{}^m] + \xi^\nu D_\mu(\omega) e_\nu{}^m + (\partial_\mu \xi^\nu) e_\nu{}^m \\ &= D_\mu(\omega)(\xi^\nu e_\nu{}^m) = D_\mu(\omega) \xi^m . \end{aligned} \tag{6.91}$$

We used the vielbein postulate $D_\nu(\omega) e_\mu{}^m - D_\mu(\omega) e_\nu{}^m = 0$. The notation $D_\mu(\omega)$ indicates that this derivative contains spin connections but no Christoffel symbols, and $\xi^m = \xi^\nu e_\nu{}^m$.

Another symmetry which plays a role in the computation of the anomalies is a symmetrized version of δ_{cov}. This symmetrized covariant Einstein transformation of the vielbein field is defined by

$$\delta_{sym}(\xi) e_\mu{}^m = \frac{1}{2}[D_\mu(\omega)\xi^m + D^m(\Gamma)\xi_\mu]. \tag{6.92}$$

It is a combination of a covariant Einstein transformation and a local Lorentz transformation with parameter $\lambda_{mn} = \frac{1}{2}(D_m \xi_n - D_n \xi_m) \equiv D_{[m}\xi_{n]}$. Namely,

$$\begin{aligned} \delta_{sym}(\xi) e_\mu{}^m &= D_\mu \xi^m - \frac{1}{2}[D_\mu(\omega)\xi^m - D^m(\Gamma)\xi_\mu] \\ &= \delta_{cov}(\xi) e_\mu{}^m + \frac{1}{2}[D^m(\omega,\Gamma)\xi_n e_\mu^n - e_\mu^n D_n(\omega)\xi^m] \\ &= \delta_{cov}(\xi) e_\mu{}^m + \delta_{lL}(D_{[m}\xi_{n]}) e_\mu{}^m . \end{aligned} \tag{6.93}$$

[9] For an approach which only uses the transformation laws of local Lorentz symmetry, see [144, 150].

Its physical meaning is clear: if one begins with symmetric vielbein fields, then the particular combination of Einstein and local Lorentz transformations which constitutes δ_{sym} preserves the symmetry of the vielbein fields.

The anomaly due to δ_{cov} is the response of the effective action Γ under a covariant Einstein transformation. Using the chain rule we find

$$\begin{aligned} An_{cov}(\xi) &= \delta_{cov}(\xi)\Gamma[e_\mu{}^m] = \int dx\,[\delta_{cov}(\xi)e_\mu{}^m(x)]\frac{\delta\Gamma}{\delta e_\mu{}^m(x)} \\ &= \int dx\, e\, D_\mu\xi^m\, T_m{}^\mu = -\int dx\, e\, \xi^m(D_\mu T_m{}^\mu) \end{aligned} \tag{6.94}$$

where we have defined the effective stress tensor $T_m{}^\mu$ by

$$T_m{}^\mu = \frac{1}{e}\frac{\delta\Gamma}{\delta e_\mu{}^m(x)}. \tag{6.95}$$

For $\mathcal{L} = -\frac{1}{2}\sqrt{g}g^{\mu\nu}\partial_\mu\varphi\partial_\nu\varphi$ this definition yields the usual normalization $T_{\mu\nu} = \partial_\mu\varphi\partial_\nu\varphi + \cdots$. Hence **the covariant divergence of the effective stress tensor yields the anomaly in the covariant Einstein transformations**[10].

The local Lorentz anomaly is the response of the effective action under a local Lorentz transformation

$$\begin{aligned} An_{lL}(\lambda_{mn}) &= \delta_{lL}(\lambda_{mn})\Gamma[e_\mu{}^m] = \int dx\, \lambda^m{}_n e_\mu{}^n(x)\frac{\delta\Gamma}{\delta e_\mu{}^m(x)} \\ &= \int dx\, e\, \lambda^m{}_n e_\mu{}^n T_m{}^\mu = \int dx\, e\, \lambda_{mn}T^{mn} \\ &= \int dx\, e\, \lambda_{mn}T_A^{mn} \end{aligned} \tag{6.96}$$

where $T_A^{mn} = \frac{1}{2}(T^{mn} - T^{nm})$ indicates the antisymmetric part of T^{mn}. Hence **the antisymmetric part of the effective stress tensor yields the local Lorentz anomaly**.

We can also define an anomaly due to symmetric Einstein transformations,

$$\begin{aligned} An_{sym}(\xi) &= \delta_{sym}(\xi)\Gamma[e_\mu{}^m] = \int dx\,[\delta_{sym}(\xi)e_\mu{}^m(x)]\frac{\delta\Gamma}{\delta e_\mu{}^m(x)} \\ &= \int dx\, e\,\frac{1}{2}\Big[D_\mu(\omega)\xi^m + D^m(\Gamma)\xi_\mu\Big]T_m{}^\mu \\ &= -\int dx\, e\,\xi_\nu(D_\mu T_S^{\mu\nu}) \end{aligned} \tag{6.97}$$

[10] We do not call this anomaly the covariant Einstein anomaly because it is actually a consistent anomaly in the technical sense of Section (5.1.1).

where $T_S^{\mu\nu} = \frac{1}{2}(T^{\mu\nu} + T^{\nu\mu})$. Hence **the covariant divergence of the symmetrized effective stress tensor yields the anomaly in Einstein transformations which preserve the symmetric vielbein gauge**.

These three anomalies are not linearly independent. By applying (6.93) to the effective action we obtain the following relation between the anomalies:

$$An_{cov}(\xi) = An_{sym}(\xi) - An_{lL}(D_{[m}\xi_{n]}). \tag{6.98}$$

We can restate the results we have obtained for anomalies as two theorems.

Theorem I. The stress tensor is covariantly conserved if and only if the effective action is invariant under a covariant Einstein transformation (see (6.94).

Theorem II. The stress tensor is symmetric if and only if the effective action is locally Lorentz invariant (see (6.96)).

When there are matter fields φ on which Γ also depends, so $\Gamma = \Gamma[e_\mu^m, \varphi]$, these two theorems remain true "on-shell", namely when the matter field equations $\delta\Gamma/\delta\varphi = 0$ are satisfied. Again, anomalies in covariant Einstein transformations and local Lorentz anomalies break these symmetries. The anomaly (nonconservation) of $T_S^{\mu\nu} = \frac{1}{2}(T^{\mu\nu} + T^{\nu\mu})$ is sometimes called the "general coordinate anomaly" [152], but note that is really a combination of an anomaly in the covariant Einstein transformations (which itself is a combination of an ordinary Einstein anomaly and a local Lorentz anomaly) and the local Lorentz anomaly itself.

In the path integral formalism, the transformation of the fields in the measure yields the Jacobian, which in turn yields the anomaly. So we now study how spin-$\frac{1}{2}$ field transforms under these symmetries. For a spinor half-density $\tilde{\lambda} = g^{\frac{1}{4}}\lambda$, the covariant translation is given by

$$\delta_{cov}\tilde{\lambda} = \xi^\mu \partial_\mu \tilde{\lambda} + \frac{1}{2}(\partial_\mu \xi^\mu .)\tilde{\lambda} + \frac{1}{4}(\xi^\mu \omega_\mu{}^{mn})\gamma_{mn}\tilde{\lambda}. \tag{6.99}$$

The dot in $(\partial_\mu \xi^\mu .)$ indicates that the derivative ∂_μ does not act to the right of the dot. For later purposes we write the expression for δ_{cov} in terms of the following derivative:

$$\tilde{D}_\mu(\omega) \equiv g^{1/4} D_\mu(\omega) g^{-1/4}. \tag{6.100}$$

It is straightforward to verify that

$$\begin{aligned} \delta_{cov}\tilde{\lambda} &= \frac{1}{2}[\xi^\mu D_\mu(\omega) + D_\mu(\omega)\xi^\mu]\tilde{\lambda} \\ &= \frac{1}{2}[\xi^\mu D_\mu(\omega, \Gamma) + D_\mu(\omega, \Gamma, \Gamma)\xi^\mu]\tilde{\lambda} \\ &= \frac{1}{2}[\xi^\mu \tilde{D}_\mu(\omega) + \tilde{D}_\mu(\omega, \Gamma)\xi^\mu]\tilde{\lambda}. \end{aligned} \tag{6.101}$$

In the second line the notation $D_\mu(\omega, \Gamma)$ indicates that this derivative contains one Christoffel symbol (for the density character of $\tilde{\lambda}$), and $D_\mu(\omega, \Gamma, \Gamma)$ contains two Christoffel symbols (one for the density character of $\tilde{\lambda}$ and another for the index μ on ξ^μ). These Γ terms cancel as one easily verifies using $\frac{1}{2}\partial_\mu \ln g = \Gamma^\nu_{\mu\nu}$ and $D_\mu(\omega, \Gamma)\tilde{\lambda} = D_\mu(\omega)\tilde{\lambda} - \frac{1}{2}\Gamma^\nu_{\mu\nu}\tilde{\lambda}$. In the third line the Γ in $\tilde{D}_\mu(\omega, \Gamma)$ acts on ξ^μ and the tilde on $\tilde{D}_\mu$ produces the other two Γ terms and takes care of the density character of $\tilde{\lambda}$. This proves that the third line is equal to the second line.

We shall use covariant Einstein transformations to compute the gravitational anomalies. The reason for this choice is that the first line in (6.101) contains the same operator $D_\mu(\omega)$ in the first term and in the second term. The Jacobian is thus Weyl-ordered, and it may be replaced by a function in the path integral formalism according to Berezin's theorem. However, we use the third line in (6.101) to derive a property of the regulator we are going to use.

Consider the regulator

$$\mathcal{R} = \tilde{\not{D}}\tilde{\not{D}}. \tag{6.102}$$

The operator $\tilde{\not{D}}$ is the field operator for $\tilde{\lambda}$. This regulator satisfies a crucial identity.

Identity:
$$\operatorname{Tr} \gamma_5 \Big[\not{\xi}\tilde{\not{D}}(\omega) + \tilde{\not{D}}(\omega)\not{\xi}\Big]\, \mathrm{e}^{\mathcal{R}} = 0. \tag{6.103}$$

We stress that the four operators $\tilde{D}_\mu(\omega)$ in this expression are the same, and are given by

$$\tilde{\not{D}} = g^{1/4}\gamma^\mu\Big(\partial_\mu + \frac{1}{4}\omega_\mu{}^{mn}\gamma_{mn}\Big)\mathrm{g}^{-1/4}. \tag{6.104}$$

The proof of (6.103) is trivial: in the second term use $\gamma_5\tilde{\not{D}}(\omega) = -\tilde{\not{D}}(\omega)\gamma_5$, then use cyclicity to move $\tilde{\not{D}}(\omega)$ to the right and finally pull $\tilde{\not{D}}(\omega)$ past $\mathrm{e}^{\mathcal{R}}$ (which is possible because $\mathcal{R}$ only depends on $\tilde{\not{D}}(\omega)$). One then obtains minus the first term.

Next, we expand the identity in (6.103) as follows:

$$\begin{aligned}
0 &= \operatorname{Tr}\gamma_5\Big[\not{\xi}\tilde{\not{D}}(\omega) + \tilde{\not{D}}(\omega)\not{\xi}\Big]\,\mathrm{e}^{\mathcal{R}}\\
&= \operatorname{Tr}\gamma_5\gamma^\mu\gamma^\nu\Big[\xi_\mu\tilde{D}_\nu(\omega) + \tilde{D}_\mu(\omega,\Gamma)\xi_\nu)\Big]\,\mathrm{e}^{\mathcal{R}}\\
&= \operatorname{Tr}\gamma_5\Big\{\Big[\xi^\mu\tilde{D}_\mu(\omega) + \tilde{D}_\mu(\omega,\Gamma)\xi^\mu\Big] + \gamma^{\mu\nu}\Big[\xi_\mu\tilde{D}_\nu(\omega) - \tilde{D}_\nu(\omega,\Gamma)\xi_\mu\Big]\Big\}\,\mathrm{e}^{\mathcal{R}}\\
&= \operatorname{Tr}\gamma_5\Big\{\Big[\xi^\mu\tilde{D}_\mu(\omega) + \tilde{D}_\mu(\omega,\Gamma)\xi^\mu\Big] + \gamma^{\mu\nu}\Big[D_\mu(\Gamma)\xi_\nu.\Big]\Big\}\,\mathrm{e}^{\mathcal{R}}.
\end{aligned} \tag{6.105}$$

Again the dot in $[D_\mu(\Gamma)\xi_\nu.]$ indicates that $[D_\mu(\Gamma)\xi_\nu]$ is a function and that $D_\mu(\Gamma)$ does not act beyond ξ_ν. The Γ in $D_\mu(\Gamma)$ acts on the index of

ξ_ν. The first term is just twice the Jacobian for $\delta_{cov}\tilde{\lambda}$ (see the third line in (6.101)), while the second term is four times the Jacobian for a local Lorentz transformation with the parameter $D_{[m}\xi_{n]}$. Hence, we have found another relation between the gravitational anomalies of spin-$\frac{1}{2}$ fields,

$$An_{cov}(\xi) + 2An^{lL}(D_{[m}\xi_{n]}) = 0. \tag{6.106}$$

However, **this relation only holds if one uses $\tilde{\not{D}}\tilde{\not{D}}$ as regulator for the spin-$\frac{1}{2}$ field**.

Finally, we return to our original question: what anomaly are we going to calculate, and what regulator must we use to compute the covariant anomaly in the easiest way? The answer follows from (6.106). The regulator is $\mathcal{R} = \tilde{\not{D}}\tilde{\not{D}}$, and we compute the covariant Einstein anomaly. Note that for this particular regulator the covariant Einstein anomaly is equal to -2 times the local Lorentz anomaly.

We shall now construct the quantum mechanical model for this anomaly. On the basis with inner product $\langle \tilde{\lambda}_1, \tilde{\lambda}_2 \rangle = \int \tilde{\tilde{\lambda}}_1 \tilde{\lambda}_2 \, d^n x$ the operator $\frac{\hbar}{i}\frac{\partial}{\partial x^\mu}$ is hermitian. Hence, in the corresponding QM model $\frac{\hbar}{i}\frac{\partial}{\partial x^\mu}$ is replaced by p_i. The Jacobian $J = \frac{1}{2}[\xi^\mu D_\mu(\omega) + D_\mu(\omega)\xi^\mu]$ from the first line of (6.101) then corresponds to $\frac{i}{2\hbar}[\xi^i(x)\pi_i + \pi_i \xi^i(x)]$. The regulator $\mathcal{R} = \tilde{\not{D}}\tilde{\not{D}} = g^{-1/4} D_\mu(\omega)\sqrt{g} g^{\mu\nu} D_\nu(\omega) g^{-1/4} + \frac{1}{4}R$ then becomes the Hamiltonian derived previously,

$$\begin{aligned}
An(\text{grav}) \equiv An_{cov} &= -2 \,\text{Tr}\, \frac{1}{2}(D_\mu \xi^\mu + \xi^\mu D_\mu) \left(\frac{1+\gamma_5}{2}\right) \mathrm{e}^{\frac{\beta}{\hbar}\tilde{\mathcal{R}}} \\
&= -\frac{2i}{\hbar} \,\text{Tr}\, \frac{1}{2}(\pi_i \xi^i + \xi^i \pi_i) \left(\frac{1+\gamma_5}{2}\right) \mathrm{e}^{-\frac{\beta}{\hbar}H} \\
\pi_i &= p_i - \frac{i\hbar}{2}\omega_{iab}\psi_1^a \psi_1^b \\
H &= \frac{1}{2} g^{-1/4} \pi_i \sqrt{g} g^{ij} \pi_j g^{-1/4} - \frac{\hbar^2}{8} R.
\end{aligned} \tag{6.107}$$

The factor of -2 in front takes into account that $\tilde{\lambda}$ and $\tilde{\tilde{\lambda}}$ are independent fields in Euclidean space. The operator $\tilde{\mathcal{R}}$ should act in the space of chiral spinors, and in the space of nonchiral spinors one needs the projection operator $\frac{1}{2}(1+\gamma^5)$ to project on the chiral subspace. The term with $-\frac{1}{8}\hbar^2 R$ is due to expanding $-\frac{1}{2}\not{D}\not{D}$. Note that the operator $\tilde{\not{D}}$ does not map the space of chiral spinors into itself, rather it maps chiral spinor into antichiral spinors and vice versa. However, $\mathcal{R}$ maps chiral spinors into chiral spinors and antichiral spinors into antichiral spinors. Note also that $\pi_i\xi^i + \xi^i\pi_i$ is already Weyl ordered. Thus, in the path integral we will simply obtain the function $\pi_i\xi^i + \xi^i\pi_i = 2\pi_i\xi^i$. This is one reason why we chose $\tilde{\lambda}$ as a basic field variable. Another reason is that the Jacobian for Einstein transformations becomes a total derivative on a basis with $\tilde{\lambda}$, see Chapter 5. Only the term with γ_5 will contribute (for nonchiral spinors there is no gravitational anomaly). So we must evaluate

$$An(\text{grav}) = -\frac{i}{2\hbar}\mathrm{Tr}\,\gamma_5(\pi_i\xi^i + \xi^i\pi_i)\mathrm{e}^{-\frac{\beta}{\hbar}H}. \tag{6.108}$$

We would like to bring the operator $\pi_i\xi^i + \xi^i\pi_i$ into the exponent, as a term which is added to H. At the end we then expand in terms of ξ^i and take the term linear in ξ^i. To achieve this we decompose the operator $\frac{-i}{2\hbar}(\pi_i\xi^i+\xi^i\pi_i)$ into N times $\mathcal{O} = \frac{-i\epsilon}{2\hbar\beta}(\pi_i\xi^i+\xi^i\pi_i)$, where, $N\epsilon = \beta$, and making use of the cyclicity of the trace and the fact that γ_5 and H commute, we write the trace as

$$\begin{aligned}\mathrm{Tr}\,\gamma_5\Big(&\mathcal{O}\mathrm{e}^{-\frac{N\epsilon}{\hbar}H} + \mathrm{e}^{-\frac{\epsilon}{\hbar}H}\mathcal{O}\mathrm{e}^{-(N-1)\frac{\epsilon}{\hbar}H}\\ &+\cdots+\mathrm{e}^{-\frac{(N-1)\epsilon}{\hbar}H}\mathcal{O}\mathrm{e}^{-\frac{\epsilon}{\hbar}H}\Big); \qquad \epsilon = \frac{\beta}{N}.\end{aligned} \tag{6.109}$$

Instead of $\mathcal{O}$ we write $\exp\mathcal{O}$ and then obtain a path integral with modified Hamiltonian $H + \frac{i}{2\beta}(\pi_i\xi^i + \xi^i\pi_i)$. (Strictly speaking one should use the Baker–Campbell–Hausdorff formula to combine $\mathrm{e}^{-\frac{\epsilon}{\hbar}H}\mathrm{e}^{\mathcal{O}}$, but the terms involving commutators are of higher order in ϵ or ξ and can be neglected.) Inserting complete sets of x, p eigenstates and coherent states for the fermions ψ_1^a, ψ_2^a (with a free set ψ_2 added as explained in Section 6.1), one obtains the phase-space path integral.

The next step is to integrate out the momenta from

$$-\frac{\epsilon}{2\hbar}g^{ij}(\bar{x}_k)\pi_{k,i}\pi_{k,j} + \frac{i}{\hbar}p_{k,i}(x_k^i - x_{k-1}^i) - \frac{i\epsilon}{\beta\hbar}\pi_{k,i}\xi^i(\bar{x}_k) \tag{6.110}$$

where $\bar{x}_k = (x_k + x_{k-1})/2$. This yields, as before, the interaction term in the covariant derivative in $-\frac{1}{2}\int_{-1}^{0}\psi_1^a\frac{D}{D\tau}\psi_1^a\,d\tau$, while the gravitational Jacobian $\xi^i\pi_i$ is replaced by $\frac{1}{\beta\hbar}\int_{-1}^{0}\xi^i g_{ij}(x)\dot{x}^j\,d\tau$. Similarly to (6.25) before the rescaling of fermions, we arrive at

$$\begin{aligned}&An(\text{grav})\\ &= \frac{(-i)^{n/2}}{(2\pi\beta\hbar)^{n/2}}\int\prod_{i=1}^{n}dx_0^i\sqrt{g(x_0)}\prod_{a=1}^{n}d\psi_{1,bg}^a\\ &\times\Big\langle\exp\Big\{-\frac{1}{\beta\hbar}\frac{1}{2}\int_{-1}^{0}[g_{ij}(x_0+q)-g_{ij}(x_0)](\dot{q}^i\dot{q}^j + b^ic^j + a^ia^j)\,d\tau\Big\}\\ &\times\exp\Big[-\int_{-1}^{0}\frac{1}{2}\dot{q}^i\omega_{iab}(x_0+q)(\psi_{1,bg}^a+\psi_{qu}^a)(\psi_{1,bg}^b+\psi_{qu}^b)\,d\tau\Big]\end{aligned}$$

$$\times \exp\left[-\frac{\beta\hbar}{8}\int_{-1}^{0}\left(\Gamma\Gamma+\frac{1}{2}\omega\omega\right)d\tau\right]$$
$$\times \exp\left[\frac{1}{\beta\hbar}\int_{-1}^{0}\dot{q}^i\xi^j(x_0+q)g_{ij}(x_0+q)\,d\tau\right]\Bigg\rangle. \tag{6.111}$$

The last exponent does not contain a term linear in q since $\int \dot{q}\,d\tau = 0$. Expanding ξ^j and g_{ij} to first order in q, the last term can be written as $\frac{1}{\beta\hbar}\int_{-1}^{0}\dot{q}^i q^k D_k\xi_i(x_0)\,d\tau$, where $\int_{-1}^{0}\dot{q}^i q^k\,d\tau$ is antisymmetric in i and k. Rescaling ψ_1^a as before, the factors of $1/\hbar\beta$ in the measure cancel, while the vertex with $\dot{q}\omega\psi_1\psi_1$ acquires a factor of $1/\hbar\beta$. If we also choose a local Lorentz frame in which $\omega_{iab}(x_0)=0$, there is only the vertex with $q^i\dot{q}^j R_{ijab}\psi_1^a\psi_1^b$ which we encountered before in (6.26), and a new vertex of the form $(D_i\xi_j)q^i\dot{q}^j$. They combine into the vertex

$$-\frac{1}{\hbar}S^{(int)} = -\frac{1}{\beta\hbar}\left[\frac{1}{4}R_{ijab}(x_0)\psi_1^a\psi_1^b - D_i\xi_j\right]\int_{-1}^{0}q^i\dot{q}^j\,d\tau. \tag{6.112}$$

Hence, the gravitational anomaly is obtained from the abelian chiral anomaly in (6.34) by adding $-D_{[i}\xi_{j]}$ to $\frac{1}{4}R_{ijab}\psi_1^a\psi_1^b$, where $D_{[i}\xi_{j]} = \frac{1}{2}(D_i\xi_j - D_j\xi_i)$. As in Section 6.1 there are no contributions from the quantum parts in ψ_1^a, so we replace ψ_1^a by $\psi_{1,bg}^a$.

We conclude that **the gravitational anomaly of a complex chiral spinor** in n dimensions is given by

$$An(\text{grav, spin } \tfrac{1}{2})$$
$$= \int dx_0^i\sqrt{g(x_0)}\,d\psi_{1,bg}^a \exp\frac{1}{2}\text{tr}\ln\left[\frac{-i\tilde{R}/8\pi}{\sinh(-i\tilde{R}/8\pi)}\right]\Bigg|_{\text{linear in }\xi}$$
$$\tilde{R} \equiv R_{ijab}\psi_1^a\psi_1^b - 4D_i\xi_j. \tag{6.113}$$

This anomaly is (-2) times the local Lorentz anomaly with parameter $\frac{1}{2}(D_m\xi_n - D_n\xi_m)$ as we showed in (6.106). We absorbed the factors of $(-i)^{n/2}$ into the exponent because then $-i\tilde{R}$ is hermitian. One factor of i is left in front of the anomaly because one must expand the exponent to first order in ξ, hence, once again, the anomaly is purely imaginary. Expanding the terms in the exponent, and retaining the terms linear in ξ, one is now left with an odd number of R terms. Hence **there are only gravitational anomalies in $n = 4k+2$ dimensions** (there were only gravitational chiral anomalies in $n = 4k$ dimensions). So, when one discusses gravitational anomalies for chiral spinors in string models in 10 dimensions, one means local Lorentz anomalies (anomalies in the Lorentz-covariant conservation of the stress tensor) but not gravitational contributions to the rigid abelian chiral anomaly.

As a final comment, we discuss an alternative method of calculating the Lorentz anomaly, namely by straightforwardly evaluating the regulated trace of the Lorentz Jacobian $\frac{1}{4}\lambda_{mn}(x)\gamma^m\gamma^n\gamma_5$. This calculation combines aspects of the γ_5 anomaly and the gravitational anomaly. As in the case of the gravitational anomaly one exponentiates the Jacobian, and in the quantum mechanical model one finds the operator $\frac{1}{\beta\hbar}\lambda_{ab}(x)\psi^a\psi^b$, whereas γ_5 is treated as in the $U(1)$ case. However, when one rescales the fermions in order to remove the factors of $\beta\hbar$ in the measure, one finds a factor of $1/(\beta\hbar)^2$ in front of the $\lambda\psi\psi$ term. All other interactions are proportional to $1/(\beta\hbar)$ (or even $\beta\hbar$) and all propagators are proportional to $\beta\hbar$. Hence one-loop graphs are no longer enough to compute the Lorentz anomaly in this approach. For that reason the approach which starts from the covariant combination of Einstein and local Lorentz transformations is to be preferred.

6.4 Mixed Lorentz and non-abelian gauge anomalies for chiral spin-$\frac{1}{2}$ fields coupled to gravity and Yang–Mills fields in $2k$ dimensions

We consider a complex chiral (Weyl) spin-$\frac{1}{2}$ field in $n=2k$ dimensions, coupled to both external gravitational fields and to external Yang–Mills fields with gauge group G with antihermitian generators T_α. Mixed anomalies can occur, namely anomalies in the Einstein–Lorentz symmetry and in the Yang–Mills symmetry of one-loop graphs coupled both to gravitons and to gauge bosons. We determine them in this section. We shall again switch to nonchiral complex Dirac fermions on the basis $\tilde{\lambda} = g^{1/4}\lambda$, and use a projection operator $\frac{1}{2}(1+\gamma^5)$ in the Jacobian. As before only the term with $\frac{1}{2}\gamma^5$ contributes.

The first time anomalies can appear is in $(k+1)$-polygon graphs; for example, in triangle graphs in four dimensions. There can then be graphs with $(k+1)$ external gravitons, or with k gravitons and one external gauge field, or with $(k-1)$ gravitons and two gauge fields, up to $(k+1)$ external gauge fields. Given a graph with r external gauge fields, the trace over the generators T_α of the gauge fields yields a factor

$$\mathrm{Tr}_S(T_{\alpha_1}T_{\alpha_2}\cdots T_{\alpha_r}) \tag{6.114}$$

where the subscript S indicates that one should totally symmetrize with respect to the indices $\alpha_1, \alpha_2, \ldots, \alpha_r$. If and only if this trace does not vanish can there be an anomaly in this graph. However, even if the symmetrized trace is nonvanishing, there need not be an anomaly. For example, for $r=0$, we already saw that purely gravitational anomalies can only occur in $n=4k+2$ dimensions. Graphs with only external gauge fields (the case of $r=k+1$) are anomalous whenever the symmetrized trace with $r = n/2 + 1$ matrices is nonvanishing. In four dimensions this occurs whenever there is a cubic Casimir operator (the $d_{\alpha\beta\gamma}$ symbol) in the gauge group, and in n dimensions whenever there is a rank-$(k+1)$ Casimir operator. The general case is most easily explained if one has the explicit result in hand, so we first derive the general formula for mixed gravitational and gauge anomalies.

The Dirac operator in the n-dimensional field theory is given by

$$\not{D} = e_m^{\mu}\gamma^m D_\mu, \quad D_\mu = \partial_\mu + \frac{1}{4}\omega_{\mu mn}(e)\gamma^m\gamma^n + gA_\mu^\alpha T_\alpha. \tag{6.115}$$

The corresponding nonlinear sigma model is a combination of the $N = 1$ nonlinear sigma model of Section 6.1 and Appendix D, and the linear sigma model with extra ghosts of Section 6.2 and Appendix E. It reads in Minkowski time as

$$\begin{aligned} L_M = &\frac{1}{2}g_{ij}(x)\dot{x}^i\dot{x}^j + \frac{i}{2}\psi_a(\dot{\psi}^a + \dot{x}^k\omega_k{}^a{}_b\psi^b) \\ &+ ic_A^*[\dot{c}^A + \dot{x}^k A_k^\alpha(x)(T_\alpha)^A{}_B c^B] \\ &+ \frac{1}{2}\psi^i\psi^j F_{ij}{}^\alpha c_A^*(T_\alpha)^A{}_B c^B. \end{aligned} \tag{6.116}$$

The sum of the first three terms and the sum of the last three terms are separately supersymmetric. In Euclidean space we obtain

$$\begin{aligned} L_E = &\frac{1}{2}g_{ij}(x)\dot{x}^i\dot{x}^j + \frac{1}{2}\psi_a(\dot{\psi}^a + \dot{x}^k\omega_k{}^a{}_b\psi^b) \\ &+ c_A^*[\dot{c}^A + \dot{x}^k A_k^\alpha(x)(T_\alpha)^A{}_B c^B] \\ &- \frac{1}{2}\psi^i\psi^j F_{ij}{}^\alpha c_A^*(T_\alpha)^A{}_B c^B. \end{aligned} \tag{6.117}$$

The regulator for the n-dimensional spinor $\tilde{\lambda} = g^{1/4}\lambda$ of the quantum field theory is given by

$$\begin{aligned} \mathcal{R} &= -g^{1/4}\not{D}\not{D}g^{-1/4} \\ &= -g^{-1/4}D_\mu g^{1/2}g^{\mu\nu}D_\nu g^{-1/4} - \frac{1}{4}R - \frac{1}{2}\gamma^\mu\gamma^\nu F_{\mu\nu}^\alpha T_\alpha \end{aligned} \tag{6.118}$$

where D_μ and D_ν contain spin and gauge connections but no Christoffel symbol. We derived this regulator in Section 6.1, see (6.7).

The Jacobian for Einstein–Lorentz transformations and for separate Yang–Mills transformations of $\tilde{\lambda}$ and its conjugate field is given by combining (6.101) and a gauge transformation with the parameter η,

$$J = \left[\frac{1}{2}(\xi^\mu D_\mu + D_\mu\xi^\mu) - \eta^\alpha T_\alpha\right]\frac{1+\gamma_5}{2} \tag{6.119}$$

where we repeat that the derivative D_μ contains a spin connection and a gauge connection. The mixed anomalies in the QFT are then

$$An(\text{mixed}) = (-2) \lim_{\beta \to 0} \text{Tr}\, J \mathrm{e}^{-\frac{\beta}{\hbar}\mathcal{R}}. \qquad (6.120)$$

In the corresponding QM approach the anomalies are given by

$$An(\text{mixed}) = \text{Tr}\left\{\left[\gamma_5 \frac{-i}{2\hbar}(\pi_i \xi^i + \xi^i \pi_i) + \gamma_5 \eta^\alpha c^* T_\alpha c\right] \mathrm{e}^{-\frac{\beta}{\hbar}H}\right\} \qquad (6.121)$$

where π_i was defined in (6.107).

We exponentiate the Jacobian as in Section 6.3. After integrating out the momenta, we then find in the action the nonlinear sigma model of Appendix E, together with the term $\frac{1}{\beta\hbar}\int_{-1}^{0} \dot{q}^i \xi^j (x_0+q) g_{ij}(x_0+q)\, d\tau$ which we have already found in Section 6.3 and the term

$$\int_{-1}^{0} \eta^\alpha (x_0 + q)(\bar{\eta}_{gh} + \bar{c}_{gh}) T_\alpha (\chi_{gh} + c_{gh})\, d\tau \qquad (6.122)$$

which is new. Inspection of (6.78) shows then that only the combination

$$\frac{1}{2}\tilde{F} = \left(\frac{1}{2}\psi^a_{1,bg}\psi^b_{1,bg} g F^\alpha_{ab} + \eta^\alpha\right)(\bar{\eta}_{gh} + \bar{c}_{gh}) T_\alpha (\chi_{gh} + c_{gh}) \qquad (6.123)$$

appears. From (6.112) we also find that only the combination

$$\tilde{R}_{ij} = R_{ijab}\psi^a_{1,bg}\psi^b_{1,bg} - 2(D_i \xi_j - D_j \xi_i) \qquad (6.124)$$

occurs. The mixed anomaly is then given by the same formulas as derived in Sections 6.2 and 6.3, but with $\tilde{F}$ and $\tilde{R}$ instead of F and R, where now

$$\frac{1}{2}\tilde{F} = \frac{1}{2}\psi^a_{1,bg}\psi^b_{1,bg} g F^\alpha_{ab} T_\alpha + \eta^\alpha T_\alpha \qquad (6.125)$$

according to (6.87) and $\tilde{R}$ as given in (6.124). The **mixed (gravitational and gauge) anomaly of a complex chiral spinor** in n dimensions is given by

$$An(\text{mixed, spin } \tfrac{1}{2}) = \frac{(-i)^{n/2}}{(2\pi)^{n/2}} \int dx_0^i \sqrt{g(x_0)}\, d\psi^a_{1,bg} \times \left(\text{Tr}\, \mathrm{e}^{\frac{1}{2}\tilde{F}}\right) \exp \frac{1}{2} \text{tr} \ln \left(\frac{\tilde{R}/4}{\sinh \tilde{R}/4}\right). \qquad (6.126)$$

The terms linear in ξ and η yield the anomaly. The trace tr is over matrices

$\tilde{R}_{ij}$ but the trace Tr is in the space of the representation of the fermion (the space with matrices $(T_\alpha)_M{}^N$). The anomalies are obtained by expanding both factors and extracting the terms linear in ξ^m or η^α. A given anomaly depends in general both on F and R.

At this point we can make a consistency check between the abelian chiral anomaly of Section 6.2 and the gauge anomaly of this section. Consider a local chiral $U(1)$ gauge transformation (for example, the $U(1)$ of the Standard Model). We can view it as an abelian chiral transformation (with $T_\alpha = -i$) and evaluate $\mathrm{Tr}\, \mathrm{e}^{\frac{1}{2}F}$. We can also treat it as a gauge anomaly in which case we introduce ghosts to construct $c^* T_\alpha c$ with T_α a constant, and then we must evaluate $\mathrm{Tr}\, \mathrm{e}^{\frac{1}{2}\tilde{F}}$ by taking the term linear in η. The answer should be the same,[11] and it is the same because $\mathrm{Tr}\,[(F/2)^n/n!]$ equals the term linear in η in $\mathrm{Tr}\,[(\tilde{F}/2)^{n+1}/(n+1)!]$.

Now consider a one-loop graph with a complex chiral spin-$\frac{1}{2}$ field in the loop and with p external gravitons and q external gauge fields. The integration over $\psi^a_{1,bg}$ requires $p+q = \frac{1}{2}n+1$, so the first time anomalies are possible in n dimensions is in polygon graphs with $\frac{1}{2}n+1$ sides. The gauge variation of a graviton yields a factor of $D_{[m}\xi_{n]}$, while the gauge variation of a Yang–Mills field yields a factor of $\eta^\alpha T_\alpha$ in (6.126). Since the second factor in (6.126) is even in $\tilde{R}$, **there are only anomalies (both gravitational or gauge anomalies) if there are an even number of external gravitons (p even)**.

In 10 dimensions, the first time anomalies appear is in hexagon graphs. There is a purely gravitational anomaly with six external gravitons, because 10 is in the set $4k+2$. There is a mixed anomaly with four external gauge fields and two gravitons if the gauge group is such that $\mathrm{Tr}F^4$ is nonvanishing in the representation of the fermions. There is always a mixed anomaly with two external gauge fields and four gravitons, because every semisimple gauge group has a quadratic Casimir operator, but a purely gauge anomaly only exists if $\mathrm{Tr}F^6$ is nonvanishing. From (6.251) one can, for example, conclude that there are no mixed anomalies for E_8.

In four dimensions the first time anomalies appear is in triangle graphs. There is no purely gravitational anomaly, only a mixed anomaly if $G = U(1)$ (the abelian chiral gravitational anomaly of Section 6.1), and only a pure gauge anomaly if G has a nonvanishing $d_{\alpha\beta\gamma}$ symbol in the representation of the fermions.

[11] The abelian chiral anomaly we computed in Section 6.2 referred to a nonchiral fermion with Jacobian $-2i\alpha\gamma_5$. In this section we have used a chiral fermion by inserting the projection operator $\frac{1}{2}(1+\gamma_5)$ in the trace. Hence for comparison one should take $\frac{1}{2}$ times the abelian chiral anomaly of a nonchiral fermion.

6.5 The abelian chiral anomaly for spin-$\frac{3}{2}$ fields coupled to gravity in $4k$ dimensions

In this section we extend the calculation of the gravitational corrections to the abelian chiral anomaly ("the γ_5 anomaly") from the case of the spin-$\frac{1}{2}$ field to the case of spin $\frac{3}{2}$. This introduces supergravity, because the only consistent interactions for spin $\frac{3}{2}$ with other fields are the interactions of supergravity models. Earlier models, with spin $\frac{3}{2}$ only coupled to spin 1, turned out to be inconsistent or trivial, while the couplings of spin-$\frac{3}{2}$ fields to spin 2 are consistent if they are given by $N = 1$ supergravity. One can also couple spin-$\frac{3}{2}$ fields to spin-1 and spin-0 fields, but only if at the same time one couples the spin-$\frac{3}{2}$ field to gravity and these interactions are given by a supergravity model with $N > 1$ (N is the number of real spin-$\frac{3}{2}$ fields). In this section we consider only loops with spin-$\frac{3}{2}$ fields in the loop and external gravity fields. The spin-2 and spin-$\frac{3}{2}$ interactions may be part of a more complicated supergravity model, but that does not make a difference for the computation of the one-loop anomalies. Only the minimal gravitational couplings of gravity to spin $\frac{3}{2}$ contribute, so torsion due to gravitinos may be ignored. We assume that the reader has no knowledge of supergravity, and start from the beginning.

In $(3 + 1)$-dimensional Minkowski spacetime, the real spin-$\frac{3}{2}$ field $\psi_\mu{}^\alpha$ (with $\mu = 0, \cdots, 3$ the vector index and $\alpha = 1, \cdots, 4$ the corresponding four-component spinor index) is the gauge field for local supersymmetry (i.e. supergravity). The classical action for the spin-$\frac{3}{2}$ field in $(3+1)$-dimensional Minkowski spacetime coupled to external gravitational fields reads as

$$
\begin{aligned}
\mathcal{L}_{3/2} &= -\frac{1}{2} e \bar{\psi}_\mu \gamma^{\mu\rho\sigma} D_\rho \psi_\sigma \\
e &= \det e_\mu{}^m; \quad \gamma^\mu = e_m{}^\mu \gamma^m \\
D_\rho \psi_\sigma &= \partial_\rho \psi_\sigma + \frac{1}{4} \omega_{\rho mn}(e) \gamma^m \gamma^n \psi_\sigma \\
\bar{\psi} &= \psi^T C; \quad C\gamma^\mu C^{-1} = -(\gamma^\mu)^T
\end{aligned}
\tag{6.127}
$$

where $\gamma^{\mu\rho\sigma}$ equals the totally antisymmetrized product of γ^μ, γ^ρ and γ^σ (so $\gamma^{\mu\rho\sigma} = \frac{1}{6}(\gamma^\mu\gamma^\rho\gamma^\sigma +$ five other terms)), and γ^m are constant 4×4 matrices. The form of the action for the spin-$\frac{3}{2}$ gauge field is fixed by requiring invariance of the free spin-$\frac{3}{2}$ action in flat spacetime under $\delta\psi_\mu = \partial_\mu \epsilon$; in fact, it already follows from requiring that the residue of the free field propagator be without ghosts [77]. Thus, also for local supersymmetry, gauge invariance follows from unitarity. (When one does not treat gravity as external, one also, of course, needs the Einstein action and all the other paraphernalia of supergravity. In particular, the spin connection

$\omega_{\rho mn}$ then contains torsion terms bilinear in gravitinos. For the calculation of anomalies with spin-$\frac{3}{2}$ loops, we can restrict ourselves to external gravitational fields and use the spin connection $\omega_{\rho mn}(e)$ of Appendix A, which only depends on the external vielbein field $e_\mu{}^m$.) In principle, one should add a Christoffel connection in the definition of $D_\rho\psi_\sigma$, but it cancels in $\mathcal{L}_{3/2}$ because $D_\rho\psi_\sigma$ appears in the action only as antisymmetric in ρ and σ.

This classical spin-$\frac{3}{2}$ action is gauge invariant by itself provided the background fields are Ricci flat (Einstein spaces with $R_{\mu\nu}=0$): it is locally supersymmetric. Under $\delta\psi_\sigma = D_\sigma\epsilon, \delta e_\mu{}^m = 0$ one obtains, using the fact that the commutator of two covariant derivatives is a curvature,

$$\begin{aligned}\delta\mathcal{L}_{3/2} = &-\frac{1}{2}e\bar{\psi}_\mu\gamma^{\mu\rho\sigma}\frac{1}{2}\left(\frac{1}{4}R_{\rho\sigma mn}\gamma^m\gamma^n\epsilon\right)\\ &-\frac{1}{2}e(D_\mu\bar{\epsilon})\gamma^{\mu\rho\sigma}D_\rho\psi_\sigma. \end{aligned}\tag{6.128}$$

After partial integration (using the fact that D_μ commutes with e and γ^μ because we have omitted the torsion terms in the spin connection) one finds

$$\begin{aligned}\delta\mathcal{L}_{3/2} &= -\frac{1}{16}e\bar{\psi}_\mu\gamma^{\mu\rho\sigma}\gamma^m\gamma^n\epsilon R_{\rho\sigma mn} + \frac{1}{16}e\bar{\epsilon}\gamma^{\mu\rho\sigma}R_{\mu\rho mn}\gamma^m\gamma^n\psi_\sigma\\ &= -\frac{1}{16}eR_{\rho\sigma mn}\bar{\psi}_\mu(\gamma^{\mu\rho\sigma}\gamma^m\gamma^n + \gamma^m\gamma^n\gamma^{\mu\rho\sigma})\epsilon. \end{aligned}\tag{6.129}$$

(We used $\bar{\epsilon}\gamma^{\mu\rho\sigma}\gamma^m\gamma^n\psi_\sigma = -\bar{\psi}_\sigma\gamma^n\gamma^m\gamma^{\sigma\rho\mu}\epsilon$; this relation follows from the Majorana property of the spinor parameter, $\bar{\epsilon} = \epsilon^T C$, and the charge conjugation property $C\gamma^{m\,T} = -\gamma^m C$.) In the anticommutator $\{\gamma^{\mu\rho\sigma}, \gamma^m\gamma^n\}$ there are only terms with a totally antisymmetric product of five Dirac matrices or terms with one Dirac matrix. The terms with a totally antisymmetric product of three Dirac matrices cancel in the anticommutator. (They survive in the commutator.) Those with five Dirac matrices do not contribute in 3+1 dimensions because a tensor with five indices which is totally antisymmetric vanishes in fewer then five dimensions, and also because of the cyclic identity of the Riemann tensor. In higher dimensions, these terms still vanish due to the cyclic identity of the Riemann tensor. The variations with one Dirac matrix can only contract with a Ricci tensor $R_{\mu\nu}$. In fact, this could have been anticipated because there are not enough free indices to contract with a full Riemann tensor. One finds $\delta\mathcal{L}_{3/2} = \frac{1}{2}e(R_{\mu\nu} - \frac{1}{2}g_{\mu\nu}R)(\bar{\psi}^\mu\gamma^\nu\epsilon)$. When $R_{\mu\nu} = 0$, one calls the gravitational field Ricci flat. Hence, for Ricci-flat backgrounds the gravitino action is gauge invariant (locally supersymmetric).

If one treats the gravitational field dynamically, there is no longer a restriction to Ricci flatness, provided one also transforms the gravitational field (the vielbein field $e_\mu{}^m$) under local supersymmetry. It transforms as $\delta e_\mu{}^m = \bar{\epsilon}\gamma^m\psi_\mu$, and also this variation is multiplied by the Einstein tensor $G_{\mu\nu} = R_{\mu\nu} - \frac{1}{2}g_{\mu\nu}R$; in fact, the sum of all local supersymmetry variations cancels. From now on we continue with arbitrary gravitational fields; for spin-$\frac{3}{2}$ loops coupled to gravity the spacetime symmetries (general coordinate and local Lorentz transformations) of the fields (vielbein $e_\mu{}^m$ and gravitino ψ_μ) are as usual, and the classical action is also invariant under local supersymmetry if one transforms $e_\mu{}^m$ as we discussed. The previous derivation with external gravitational fields was given for readers who are not familiar with supergravity. Readers who want an introduction to supergravity may turn to [77].

Because there is a local supersymmetry one must add a gauge-fixing term and a ghost action. Without a gauge-fixing term the kinetic operator cannot be inverted, and no graphs with spin-$\frac{3}{2}$ in the loop can be constructed. With a gauge-fixing term the local supersymmetry gets broken, but we shall choose the local supersymmetry gauge-fixing term such that the classical spacetime symmetries remain unbroken. As a gauge-fixing term the most convenient choice is $\gamma^\mu\psi_\mu = 0$, leading to

$$\mathcal{L}(\text{fix}) = \frac{1}{4}e\bar{\psi}_\mu\gamma^\mu \not{D}\gamma^\nu\psi_\nu \qquad (6.130)$$

which preserves general coordinate invariance and local Lorentz symmetry but breaks local supersymmetry. The corresponding Faddeev–Popov ghost action then follows from applying $\delta\psi_\mu = D_\mu\epsilon$ to $\gamma^\mu\psi_\mu$ and contracting with the antighost of supersymmetry

$$\mathcal{L}(\text{FP ghost}) = -e\bar{b}_\alpha(\not{D}c)^\alpha \qquad (6.131)$$

where b_α and c^α are real commuting ghosts (the spinor index α will be dropped below). (One should also vary the vielbein in $\gamma^\mu = e^\mu_m\gamma^m$, but this yields a term $(\bar{b}\gamma^m\psi_\mu)(\bar{\psi}_m\gamma^\mu c)$ in the ghost action which does not contribute at the one-loop level. It is used, however, to fix the chiral symmetry transformation rules, see below.)

To obtain (6.130) in the exponent of the path integral, one starts from the gauge-fixing term $\delta[\gamma\cdot\psi - F]$ with F being an independent anticommuting Majorana spinor, and then one inserts unity into the path integral as follows: $I = \int[dF]\exp(\bar{F}\not{D}F)(\det\not{D})^{-1/2}$. Integration over F yields $(\det\not{D})^{1/2}$ which cancels the factor of $(\det\not{D})^{-1/2}$. The normalization factor $(\det\not{D})^{-1/2}$ can be exponentiated to give another ghost, the so-called Nielsen–Kallosh ghost. It is really a commuting complex ghost (a pair B, C of real ghosts[12]) and an anticommuting real ghost A, as we now explain, just like the a, b, c ghosts of the QM model but with opposite

[12] A pair of real ghosts can, of course, be replaced with one complex (Dirac) ghost.

statistics. Because the Dirac action for one real commuting ghost vanishes, one writes $(\det \not{D})^{-1/2}$ as $(\det \not{D})^{1/2}/(\det \not{D})$ and then exponentiation of $(\det \not{D})^{1/2}$ gives a real anticommuting A ghost while $\det \not{D}^{-1}$ yields commuting B, C ghosts. The total Nielsen–Kallosh ghost action reads as

$$\mathcal{L}(\text{NK ghosts}) = -\frac{e}{2}\bar{A}\not{D}A - \frac{e}{2}\bar{B}\not{D}C. \tag{6.132}$$

The classical action together with its gauge-fixing term can be written in a very simple form,

$$\mathcal{L}\left(\tfrac{3}{2}\right) = \frac{e}{4}\bar{\psi}_\mu\gamma^\sigma\not{D}\gamma^\mu\psi_\sigma. \tag{6.133}$$

This result follows from the identity $-\frac{1}{2}\gamma^{\mu\rho\sigma} + \frac{1}{4}\gamma^\mu\gamma^\rho\gamma^\sigma = \frac{1}{4}\gamma^\sigma\gamma^\rho\gamma^\mu$. This action is clearly invariant under rigid chiral transformations,

$$\delta\psi_\mu = i\alpha\gamma_5\psi_\mu, \quad \delta\bar{\psi}_\mu = i\alpha\bar{\psi}_\mu\gamma_5 \tag{6.134}$$

with real constant α, because the field operator contains an odd number of Dirac matrices. (These $\delta\psi_\mu$ are still Majorana spinors.) Each ghost action in (6.131) and (6.132) is by itself chirally invariant, for arbitrary chiral weights. One may fix these weights by considering the full quantum supergravity action (with Einstein action and various couplings between ghosts and gravitinos). Requiring that the whole ghost sector (b, c, A, B, C) should also be invariant under γ_5 transformations fixes the relative chiral weights of the various ghost fields, and the net result is that, as far as γ_5 transformations are concerned, **the ghost sector acts as if it contained only one anticommuting complex chiral spin-$\frac{1}{2}$ field with opposite chiral weight to the gravitino.**[13] So the final result for the anomaly will be the result of $\mathrm{Tr}\,\gamma_5$ for ψ_μ minus the result of $\mathrm{Tr}\,\gamma_5$ for one chiral Dirac fermion (a chiral Dirac fermion is equivalent to a nonchiral real (Majorana) fermion; one can rewrite one in terms of the other in four dimensions).

---

[13] The details are as follows. Variation of the vielbein in the gauge-fixing term $\gamma^\mu\psi_\mu$, and contracting the result with the antighost to obtain the ghost action as usual, produces a term $(\bar{b}\gamma^m\psi_\mu)(\bar{\psi}_m\gamma^\mu c)$ in the ghost action. Chiral invariance then requires that the Faddeev–Popov ghosts b and c have the same chiral weight as the gravitino. The operator $(\not{D})^{-1/2}$ is exponentiated by means of Nielsen–Kallosh ghosts; since it acts in the space which contains $\gamma \cdot \psi$, the spinors in this space have opposite chirality to the gravitino. We then find for the total effective chiral weight in ghost space (defining the chiral weight as the weight one chiral complex anticommuting spinor should have in order to reproduce the same anomaly): $-1(b) - 1(c) - 1(A) + 1(B) + 1(C) = -1$. Physically the role of the ghosts is as follows: the Faddeev–Popov ghosts b and c remove as usual the unphysical longitudinal and time components of the vector-spinor field ψ_μ^α. This corresponds to the gauge symmetry $\delta\psi_\mu^\alpha = \partial_\mu\epsilon^\alpha + \cdots$. On-shell a physical massless spin-$\frac{3}{2}$ particle should have two polarizations with helicity $\pm\frac{3}{2}$. This is indeed achieved because on-shell $\gamma^\mu\psi_\mu = 0$. The Nielsen–Kallosh ghosts remove the $\gamma \cdot \psi$ part from the gravitino.

In other dimensions we cannot always define real (Majorana) gravitinos. However, one can always consider complex nonchiral gravitinos. The action is then still given by (6.127) where $\bar{\psi}_\mu$ is now the Dirac conjugate, and the corresponding supergravity theory is then still locally supersymmetric if one defines the vielbein transformation laws such that they contain complex ϵ and ψ_μ. The gauge-fixing term is given by (6.140) and the ghost structure is as before. The anomalies one computes for these complex gravitinos are a factor of 2 larger than for real gravitinos (if the latter exist in a given dimension).

We now consider **the gravitational contribution to the abelian chiral (γ_5) anomaly.** Since the gravitino in $N=1$ supergravity has no Yang–Mills index, it can only couple to gravity. Hence we can only consider the chiral anomaly, due to a spin-$\frac{3}{2}$ loop with (infinitely many) external gravitons at the vertices, and the abelian chiral current as one of its vertices. In the path integral approach, the anomaly in the conservation of this current is, up to an overall constant $-2i\alpha$,

$$An = \mathrm{Tr}\,\gamma_5\,\mathrm{e}^{\beta\mathcal{R}}. \tag{6.135}$$

In addition there is a trace for the ghost which we add later. We must now discuss the regulator for the spin-$\frac{3}{2}$ field.

A mass term for the gravitino which preserves both Einstein invariance and local Lorentz symmetry is given by $\mathcal{L}(M) = -\frac{1}{2}eM\bar{\psi}_m\psi^m$, where $\psi_m = e_m{}^\mu\psi_\mu$ has flat indices. It follows that $\tilde{\psi}_m = e^{1/2}\psi_m$ is the field for which the mass term is proportional to the unit matrix. One could then use its field operator to yield the regulator. Namely, if $R_m{}^n$ is the full kinetic operator for $\tilde{\psi}_m$, then its square is the regulator which preserves general coordinate and local Lorentz symmetry,

$$\begin{aligned}
\mathcal{R} &\sim R_m{}^p R_p{}^n \\
R_m{}^n &= g^{1/4}\gamma^n\gamma^\rho\gamma_m D_\rho g^{-1/4} \\
D_\mu &= \delta_m^n\delta_\alpha^\beta\partial_\mu + \frac{1}{4}\omega_{\mu pq}(\gamma^p\gamma^q)^\alpha{}_\beta\delta_m^n + \omega_{\mu m}{}^n\delta_\alpha^\beta.
\end{aligned} \tag{6.136}$$

Actually the mass term which does not lead to tachyons is not $\bar{\psi}_m\psi^m$, but rather $\bar{\psi}_m\gamma^{mn}\psi_n$. The corresponding regulator would then be the square of $(T^{-1})_m{}^s R_s{}^n$, where T^{-1} is the inverse of γ^{mn}. These operators are all difficult to work with.

A much simpler regulator which yields the same chiral anomalies is the Dirac operator (see Alvarez-Gaumé and Witten [1])

$$\begin{aligned}
\mathcal{R}_{3/2}(D) &\sim g^{1/4}\not{D}\not{D}g^{-1/4} \\
&= \left(g^{-1/4}D_\mu\sqrt{g}g^{\mu\nu}D_\nu g^{-1/4}\right)^\alpha{}_\beta\,\delta_m{}^n
\end{aligned}$$

$$+\frac{1}{2}\left[\frac{1}{4}\gamma^\mu\gamma^\nu R_{\mu\nu pq}(\omega)\gamma^p\gamma^q\right]^\alpha{}_\beta\,\delta_m{}^n$$
$$+\frac{1}{2}(\gamma^\mu\gamma^\nu)^\alpha{}_\beta R_{\mu\nu m}{}^n(\omega). \tag{6.137}$$

The term with four Dirac matrices can be simplified to $\frac{1}{4}R$.[14] It still acts in the combined vector–spinor space, and D_μ again contains a spinor and a vector connection, as in (6.136). Hence, the anomaly becomes (up to a factor of $-2i\alpha$)

$$An = \mathrm{Tr}\,\gamma_5\,\mathrm{e}^{\mathcal{R}_{3/2}(D)} - \mathrm{tr}\,\gamma_5\,\mathrm{e}^{\mathcal{R}_{1/2}(D)} \tag{6.138}$$

where the first trace Tr is over vector and spinor indices of the gravitino, while the second trace tr is only over spinor indices of the ghost. We have added the subscripts 3/2 and 1/2 to $\mathcal{R}(D)$ to stress that $\mathcal{R}_{3/2}(D)$ has an extra term with respect to $\mathcal{R}_{1/2}(D)$.

Alvarez-Gaumé and Witten give a general argument that one may use $\not{D}\not{D}$ for the γ_5 anomaly of spin-$\frac{3}{2}$ fields. Here we present a direct proof, see also [126]. We do this for general dimensions. The spin-$\frac{3}{2}$ action in n dimensions reads as

$$\mathcal{L}_0 = -\frac{e}{2}\bar{\psi}_\mu\gamma^{[\mu}\gamma^\nu\gamma^{\rho]}\,D_\nu\psi_\rho. \tag{6.139}$$

Again, this results follows from the fact that in flat space this is the only action that is invariant under $\delta\psi_\mu = \partial_\mu\epsilon$. After adding a gauge-fixing term

$$\mathcal{L}_{fix} = \frac{n-2}{8}e\bar{\psi}_\mu\gamma^\mu\gamma^\nu\gamma^\rho D_\nu\psi_\rho \tag{6.140}$$

and choosing a new basis for the spin-$\frac{3}{2}$ fields

$$\chi_\mu = \psi_\mu - \frac{1}{2}\gamma_\mu\gamma\cdot\psi \quad\rightarrow\quad \psi_\mu = \chi_\mu - \frac{1}{n-2}\gamma_\mu\gamma\cdot\chi \tag{6.141}$$

the action becomes a sum of Dirac actions

$$\mathcal{L}_0 + \mathcal{L}_{fix} = -\frac{1}{2}e\bar{\chi}_\mu\not{D}\chi^\mu = -\frac{1}{2}e\bar{\chi}_m\not{D}\chi^m. \tag{6.142}$$

The field χ_m transforms, of course, in the same way as ψ_m under spacetime transformations and γ_5 transformations, and the regulator for the spin-$\frac{3}{2}$ field χ_m is thus $g^{1/4}\not{D}\not{D}g^{-1/4}$, for the same reasons as for the spin-$\frac{1}{2}$ field.[15]

[14] Write $\gamma^\nu\gamma^p\gamma^q$ as $\gamma^{\nu pq} + \eta^{\nu p}\gamma^q + \eta^{pq}\gamma^\nu - \eta^{\nu q}\gamma^p$. Then, $\gamma^{\nu pq}$ does not contribute due to the cyclic identity. There remain only Ricci tensors $R_{\mu\nu}$, and since these are symmetric (recall that we dropped the torsion terms), the remaining two Dirac matrices $\gamma^\mu\gamma^\nu$ can be replaced by $g^{\mu\nu}$.

[15] One can write down an Einstein and locally Lorentz invariant mass term for χ_m, namely $e\bar{\chi}_m\chi^m$. This is not the mass term which is without ghosts and tachyons, but it serves our purposes to construct a regulator [13].

We are now ready to compute the gravitational contribution to the γ^5 anomaly. We do this in n dimensions for one complex nonchiral gravitino. This is the procedure we have also followed for the spin-$\frac{1}{2}$ case. (In $3+1$ dimensions, a chiral gravitino is complex and can be rewritten as a real nonchiral gravitino, but in other dimensions this is not always true.) We repeat the steps taken in the case of the γ_5 anomalies for spin-$\frac{1}{2}$. We continue to write the term with the spin-$\frac{1}{2}$ Lorentz generator in terms of ψ_1^a, but the term with the spin-1 generator we treat like the internal generator of a Yang–Mills group.[16] Hence, the internal matrix $A_i{}^\alpha T_\alpha$ with T_α now being the Lorentz generators, is represented by $c^*{}_a\omega_i{}^a{}_b c^b$ (just like $A_i{}^\alpha T_\alpha$ was represented by $A_i{}^\alpha c^* T_\alpha c$). The spin-$\frac{1}{2}$ term $\frac{1}{4}\omega_{\mu mn}\gamma^m\gamma^n$ becomes $\frac{1}{2}\omega_{iab}\psi_1^a\psi_1^b$. Note that although in (6.136) the spin connection terms which act on the flat spinor and flat vector index of the gravitinos appear on an equal footing, we treat them differently in the QM model. Hence the QM treatment for spin $\frac{3}{2}$ combines the gravitational and Yang–Mills treatment for spin $\frac{1}{2}$.

The ghosts again require a one-particle projection operator, and this yields upon combining (6.25) and (6.78),

$$\begin{aligned}
An = {} & \frac{(-i)^{n/2}}{(2\pi)^{n/2}} \int dx_0 \sqrt{g(x_0)}\, d\psi_{1,bg} \int d\chi\, d\bar\eta\, P^{gh}_{\bar\eta,\chi}\, e^{\bar\eta\chi} \\
& \times \Big\langle \exp\Big\{-\frac{1}{\beta\hbar}\int_{-1}^{0} \frac{1}{2}[g_{ij}(x_0+q) - g_{ij}(x_0)](\dot q^i\dot q^j + b^i c^j + a^i a^j)\, d\tau\Big\} \\
& \times \exp\Big[-\frac{1}{\beta\hbar}\frac{1}{2}\int_{-1}^{0} \dot q^i \omega_{iab}(x_0+q)\psi_1^a\psi_1^b\, d\tau\Big] \\
& \times \exp\Big[-\int_{-1}^{0} \dot q^i \omega_i{}^a{}_b(x_0+q) c^*{}_a c^b\, d\tau\Big] \\
& \times \exp\Big[\frac{1}{2}\int_{-1}^{0} c_a^* R^a{}_{bcd}(\omega(x_0+q)) c^b \psi_1^c\psi_1^d\, d\tau\Big] \\
& \times \exp\Big[-\frac{\beta\hbar}{8}\int_{-1}^{0} \Big(\Gamma\Gamma + \frac{1}{2}\omega\omega\Big)\, d\tau\Big]\Big\rangle
\end{aligned} \tag{6.143}$$

where as in (6.25) ψ_1 denotes $\psi_{1,bg} + \psi_{1,qu}$, and as in (6.78) c (c^*) denotes $\chi + c_{qu}$ ($\bar\eta + c^*_{qu}$). Recall that we only rescaled ψ_1^a but not the ghosts; this removed the $\beta\hbar$ term from the measure. The term $\frac{1}{4}\hbar\gamma^\mu\gamma^\nu R_{\mu\nu m}{}^n$ in

[16] The deeper reason that these separate treatments of the vector part and the spinor part of the generators make sense is that the trace of a direct product is the product of the traces. In group theory one uses this simple fact to compute traces over products of generators in a given representation R (such as the $d^{(R)}_{abc}$ symbols) which are built from direct products of the fundamental representation F (yielding a relation between $d^{(R)}_{abc}$ and $d^{(F)}_{abc}$).

the regulator $\frac{1}{\hbar}(-\frac{1}{2}\hbar^2\mathcal{R})$ becomes a term $\beta\hbar\int_{-1}^{0}\frac{1}{2}c^* Rc\psi_1\psi_1$ in the action and becomes $\hbar\beta$-independent after the rescaling of ψ_1^a, see the last-but-one line. The term with $\Gamma\Gamma+\frac{1}{2}\omega\omega$ does not contribute, since it is proportional to $\hbar\beta$, and also the terms with $\dot{q}^i\dot{q}^j + b^i c^j + a^i a^j$ and $\dot{q}^i\omega_i{}^a{}_b c_a^* c^b$ do not contribute for the same reason as before. The vertex with $\frac{1}{\hbar\beta}\dot{q}^i\omega_{iab}\psi_1^a\psi_1^b$ yields the vertex $\int d\tau\, \dot{q}^i q^j R_{ijab}\psi_1^a\psi_1^b$ upon expanding ω_{iab}, but the vertex with $\int d\tau\, \dot{q}^i\omega_{iab}c^{*a}c^b$ does not contribute because the propagator for $\dot{q}^i$ brings in a factor of β, whereas this vertex is β-independent.

In fact, **only closed q-loops with $R\psi_1\psi_1$ at the vertices or ghost trees with $R\psi_1\psi_1$ at the vertices contribute**. As we have seen, the latter give a factor of $\operatorname{tr}\exp(\frac{1}{2}R_{..ab}\psi_{1,bg}^a\psi_{1,bg}^b)$, and the former give a factor with $\exp\frac{1}{2}\operatorname{tr}\ln[\frac{1}{4}R/\sinh(\frac{1}{4}R)]$. The final result for **the gravitational contribution to the chiral anomaly of a complex chiral gravitino** in n dimensions is

$$An(\gamma_5, \text{spin}\,\tfrac{3}{2}) = (-2i\alpha)\frac{(-i)^{n/2}}{(2\pi)^{n/2}}\int\left(\prod_{i=1}^{n} dx_0^i\sqrt{g(x_0)}\right)\left(\prod_{a=1}^{n} d\psi_{1,bg}^a\right)$$
$$\times\left[\left(\operatorname{tr} e^{\frac{1}{2}R}\right) - 1\right]\exp\frac{1}{2}\operatorname{tr}\ln\left(\frac{R/4}{\sinh(R/4)}\right)$$
$$R \equiv R_m{}^n{}_{ab}\psi_{1,bg}^a\psi_{1,bg}^b \tag{6.144}$$

where the factor of $-2i\alpha$ mentioned above (6.135) has been reinserted. The first trace corresponds to the result in (6.87) and contains the contributions due to the vector index m of the gravitino $\psi_m{}^\alpha$; the exponent with the second trace corresponds to the result in (6.33) and takes care of the contributions due to the spinor index α of the gravitino $\psi_m{}^\alpha$, and the factor of -1 accounts for the contributions of the supersymmetry ghosts. In the Yang–Mills case the trace could be over any representation of the gauge group, and we denoted this trace by Tr. Because both traces in (6.144) are over the indices m and n of R_{mn}, we denote both by the same symbol tr.

As an application we compute the γ_5 anomaly for spin-$\frac{3}{2}$ loops in four dimensions. The answer is known to be -21 times the same anomaly for a spin-$\frac{1}{2}$ loop [125]. Expanding the factor $\left(\operatorname{tr} e^{\frac{1}{2}R}\right)-1$ gives a contribution $\frac{1}{8}\operatorname{tr} R^2$. The second exponent gives the contribution of a spin-$\frac{1}{2}$ loop, which is $-\frac{1}{2}\operatorname{tr}\left(R/4\right)^2\frac{1}{3!} = -\frac{1}{192}\operatorname{tr} R^2$. This second contribution is multiplied by $(\operatorname{tr} I)-1=3$ (after gauge-fixing the field ψ_m represents four spin-$\frac{1}{2}$ spinors, but the ghosts remove one spin-$\frac{1}{2}$ spinor). Then one finds for $n=4$

$$
\begin{aligned}
An(\gamma_5, \text{spin } \tfrac{3}{2}, n=4) &= \tfrac{2i\alpha}{4\pi^2} \int \left(\tfrac{1}{8}\mathrm{tr}\, R^2 - 3\tfrac{1}{192}\mathrm{tr}\, R^2\right) \\
&= \frac{2i\alpha}{4\pi^2} \int \left(\frac{21}{192}\mathrm{tr}\, R^2\right) = -21 An(\gamma_5, \text{spin } \tfrac{1}{2}, n=4). \qquad (6.145)
\end{aligned}
$$

where $\mathrm{tr}\, R^2$ is equal to $\epsilon^{\mu\nu\rho\sigma} R_{\mu\nu}{}^{mn} R_{\rho\sigma mn}$. This is indeed the correct result.[16]

6.6 Lorentz anomalies for chiral spin-$\frac{3}{2}$ fields coupled to gravity in $4k+2$ dimensions

We now discuss the anomaly in the local Lorentz symmetries when a chiral spin-$\frac{3}{2}$ field in a loop couples to an external gravitational field. We take the spin-$\frac{3}{2}$ field to be complex. In certain dimensions, chiral spinors can be real (Majorana–Weyl spinors) and for these cases one must divide the result for a complex chiral gravitino by a factor of 2.

It is preferable to have a covariant expression for the spin-$\frac{3}{2}$ transformation rule under spacetime transformations, and a covariant expression for the corresponding Jacobian, because then the answer for the anomaly will be a relatively simple expression involving only curvatures. This can be achieved by taking certain linear combinations of Einstein transformations and local Lorentz transformations. For spin-$\frac{1}{2}$ fields, we have already explained this before. The case of spin $\frac{3}{2}$ is more complicated. To wet the appetite of the reader for this problem, we first quote the transformation rules[17] used by Alvarez-Gaumé and Witten [1],

$$
\begin{aligned}
\delta_{AGW}\psi_m &= \frac{1}{2}(\xi^\mu D_\mu + D_\mu \xi^\mu)\psi_m + [(D_m\xi^n) - (D^n\xi_m)]\psi_n \\
D_m\xi^n &= e_m{}^\mu(\partial_\mu \xi^n + \omega_\mu{}^n{}_p \xi^p), \qquad \xi^n = e_\mu{}^n \xi^\mu \\
D_\mu \psi^a_m &= \partial_\mu \psi^a_m + \omega_{\mu m}{}^n \psi^a_n + \frac{1}{4}\omega_m{}^{pq}(\gamma_p\gamma_q)^a{}_b \psi^b_m. \qquad (6.146)
\end{aligned}
$$

The spin-$\frac{3}{2}$ field ψ_m, the so-called gravitino, is a Lorentz vector–spinor. This explains the term $\omega_{\mu m}{}^n\psi_n$ in the last line. This spin-$\frac{3}{2}$ transformation rule contains both a covariant translation and **an extra covariant local**

[16] In [1] one finds the result $\prod_{i=1}^{n/2}[(x_i/2)/\sinh(x_i/2)]$ instead of $\exp\{\frac{1}{2}\mathrm{tr}\ln[(R/4)/\sinh(R/4)]\}$, where R denotes the matrix $R_m{}^n = R_m{}^n{}_{ab}\psi^a_{1,bg}\psi^b_{1,bg}$ which is made block diagonal with blocks of the type $\begin{pmatrix} 0 & x_i \\ -x_i & 0 \end{pmatrix}$ along the diagonal. These formulas agree and the factor of $\frac{1}{2}$ in the exponent is correct, as one may check. For example, $\sum_i (x_i/2)^2 = \frac{1}{2}\mathrm{tr}(R/4)^2$.

[17] It took us many years to find a clear rigorous derivation of these rules. We thank R. Endo, whose help was essential.

Lorentz transformation acting only on the vector index of the gravitino. Its meaning is at first sight rather mysterious.

We now proceed to derive this transformation rule. Several steps in the derivation are identical to the steps taken in the spin-$\frac{1}{2}$ case, but there are new aspects due to the spin-1 index of the gravitino, and in order not to have to refer back all the time to the spin-$\frac{3}{2}$ case, we give a complete derivation of the spin-$\frac{3}{2}$ case from scratch.

An Einstein transformation of a spin-$\frac{3}{2}$ gravitino with a flat index, $\psi_m = e_m{}^\mu \psi_\mu$, is given by

$$\delta_E(\xi)\psi_m = \xi^\mu \partial_\mu \psi_m. \tag{6.147}$$

We aim at covariant transformation rules and covariant regulators for reasons explained previously. Hence, we prefer to consider the following transformation rule:

$$\delta_{cov}(\xi)\psi_m = \xi^\mu D_\mu \psi_m = \xi^\mu \partial_\mu \psi_m + \frac{1}{4}\xi^\mu \omega_{\mu rs}\gamma^r\gamma^s\psi_m + \xi^\mu \omega_{\mu m}{}^n \psi_n. \tag{6.148}$$

This is a linear combination of an Einstein transformation $\xi^\mu \partial_\mu \psi_m$ and a local Lorentz transformation with parameter $\xi^\mu \omega_{\mu mn}$, namely $\frac{1}{4}\xi^\mu \omega_{\mu rs}\gamma^r\gamma^s\psi_m + \xi^\mu \omega_{\mu m}{}^n \psi_n$. We called this contribution a covariant Einstein transformation in Section 6.3. So $D_\mu \psi_m$ is completely covariant and includes the spin connection for both (vector and spinor) indices of ψ_m. When we use path integral techniques to convert the trace $\mathrm{Tr} J e^{\mathcal{R}}$ into a path integral, with the corresponding operator becomes a corresponding function provided it is Weyl-ordered. This is Berezin's theorem, which we discussed in Part I of this book. Thus, rather than $\delta\psi_m = \xi^\mu D_\mu \psi_m$ we would like to use the transformation rule $\delta\psi_m = \frac{1}{2}[\xi^\mu D_\mu(\omega) + D_\mu(\omega)\xi^\mu]\psi_m$. Note that the same operator $D_\mu(\omega)$ should appear in the term $\xi^\mu D_\mu(\omega)$ as in the term $D_\mu(\omega)\xi^\mu$. This is not a covariant expression, of course, because for it to be covariant we would also need a Christoffel symbol $\Gamma^\nu_{\nu\mu}$ to take care of the index of ξ^μ. **However, if we take $\tilde{\psi}_m = g^{1/4}\psi_m$ as a basic variable, the transformation rule $\delta\tilde{\psi}_m = \frac{1}{2}[\xi^\mu D_\mu(\omega) + D_\mu(\omega)\xi^\mu)]\tilde{\psi}_m$ is covariant**. Let us prove this.

Consider a covariant Einstein transformation, a combination of an Einstein transformation with parameter ξ^μ and a local Lorentz transformation with composite parameter $\lambda^{mn} = \xi^\mu \omega_\mu{}^{mn}$. Then the field $\tilde{\psi}_m$ transforms as follows:

$$\delta_{cov}(\xi)\tilde{\psi}_m = \xi^\mu \partial_\mu \tilde{\psi}_m + \frac{1}{2}(\partial_\mu \xi^\mu)\tilde{\psi}_m + \frac{1}{4}\xi^\mu \omega_{\mu rs}\gamma^r\gamma^s\tilde{\psi}_m + \xi^\mu \omega_{\mu m}{}^n \tilde{\psi}_n. \tag{6.149}$$

The term $\frac{1}{2}(\partial_\mu \xi^\mu)\tilde{\psi}_m$ is needed in the transformation rule of a half-density according to the rules of tensor calculus in general relativity. We can rewrite this result as follows:

$$\delta_{cov}(\xi)\tilde{\psi}_m = \frac{1}{2}[\xi^\mu D_\mu(\omega) + D_\mu(\omega)\xi^\mu]\tilde{\psi}_m \tag{6.150}$$

where both the derivatives can act on $\tilde{\psi}_m$. This is indeed a covariant transformation law in Weyl-ordered form and we used it in the spin-$\frac{1}{2}$ case, but it does not lead to the easiest way to compute the anomaly for spin $\frac{3}{2}$. The easiest way to derive the anomaly is to use the law used in [1], and we now proceed to derive it.

The crux to the derivation of the transformation law in (6.146) is an identity satisfied by the regulator $\mathcal{R}$ where

$$\mathcal{R} = \tilde{\not{D}}\tilde{\not{D}}, \qquad \tilde{\not{D}} = g^{1/4}\gamma^\mu D_\mu(\omega) g^{-1/4} \tag{6.151}$$

and $D_\mu(\omega)$ is the derivative which also appears in $\delta_{cov}\tilde{\psi}_m$. The operator $\tilde{\not{D}}$ is the field operator for $\tilde{\psi}_m$ in the Dirac action $\mathcal{L} = \bar{\psi}^m\sqrt{g}\not{D}\psi_m = \bar{\tilde{\psi}}^m\tilde{\not{D}}\tilde{\psi}_m$. One obtains the second form of $\mathcal{L}$ by changing variables from ψ_m to $\tilde{\psi}_m = g^{1/4}\psi_m$. We can write $\delta_{cov}\tilde{\psi}_m$ in terms of derivatives $\tilde{D}_\mu(\omega)$ as follows:

$$\delta_{cov}(\xi)\tilde{\psi}_m = \frac{1}{2}[\xi^\mu\tilde{D}_\mu(\omega) + \tilde{D}_\mu(\omega,\Gamma)\xi^\mu]\tilde{\psi}_m \tag{6.152}$$

where the Γ in $\tilde{D}_\mu(\omega,\Gamma)$ acts on the index ξ^μ as usual. It is easy to show that this expression is the same as (6.150), because the two terms $\frac{1}{2}\xi^\mu g^{1/4}\partial_\mu g^{-1/4}$ cancel the term $\frac{1}{2}\Gamma_{\mu\nu}{}^\mu\xi^\nu$. We showed this already for the spin-$\frac{1}{2}$ case in (6.101) We are now ready to derive the identity we need. It reads as

Lemma: $$An_{cov}(\xi^\mu) = -2An_{lL}{}^{(1/2)}(D_{[m}\xi_{n]}). \tag{6.153}$$

In other words, the anomaly $An = \mathrm{Tr}Je^{\mathcal{R}}$ with $\mathcal{R}$ given by (6.151) and where J is the Jacobian for δ_{cov} is the same as minus twice the anomaly with the same $\mathcal{R}$ but with J due to a local Lorentz transformation which **only acts on the spin-$\frac{1}{2}$ index** of the gravitino.

The proof of this lemma is the same as in the spin-$\frac{1}{2}$ case, see (6.103), but since there are now also terms acting on the vector index of the gravitino, we shall present the complete proof for the spin-$\frac{3}{2}$ case. Consider the expression

$$\mathrm{Tr}\,\gamma_5\left[\not{\xi}\tilde{\not{D}}^{(2\omega)} + \tilde{\not{D}}^{(2\omega)}\not{\xi}\right]\mathrm{e}^{\tilde{\not{D}}^{(2\omega)}\tilde{\not{D}}^{(2\omega)}} \tag{6.154}$$

where we repeat that $\tilde{\not{D}}^{(2\omega)} = \gamma^\mu\tilde{D}_\mu^{(2\omega)}$ has no Γ-term, but two ω-terms, one acting on the spin-$\frac{1}{2}$ index and the other acting on the spin-1 index. We pull the second $\tilde{\not{D}}^{(2\omega)}$ to the left past the matrix γ_5 (this yields a minus sign), and then, using cyclicity of the trace, and commuting $\tilde{\not{D}}^{(2\omega)}$ past $e^{\tilde{\not{D}}^{(2\omega)}\tilde{\not{D}}^{(2\omega)}}$ one obtains zero,

$$\begin{aligned} &\mathrm{Tr}\,\gamma_5\left[\not{\xi}\not{D}^{(2\omega)}+\not{D}^{(2\omega)}\not{\xi}\right]\mathrm{e}^{\tilde{\not{D}}^{(2\omega)}\tilde{\not{D}}^{(2\omega)}} \\ &\qquad=\mathrm{Tr}\,\gamma_5\left[\not{\xi}\not{D}^{(2\omega)}-\not{\xi}\not{D}^{(2\omega)}\right]\mathrm{e}^{\tilde{\not{D}}^{(2\omega)}\tilde{\not{D}}^{(2\omega)}}=0. \end{aligned} \tag{6.155}$$

Next, we rewrite (6.154). We pull the two Dirac matrices to the left, and taking symmetric and antisymmetric parts, we obtain inside the trace

$$\begin{aligned} 0 &= \left[\not{\xi}\tilde{\not{D}}^{(2\omega)}+\tilde{\not{D}}^{(2\omega)}\not{\xi}\right]=\gamma^\mu\gamma^\nu\left[\xi_\mu\tilde{D}_\nu^{(2\omega)}+\tilde{D}_\mu^{(2\omega,\Gamma)}\xi_\nu\right] \\ &= \left[\xi^\mu\tilde{D}_\mu^{(2\omega)}+\tilde{D}_\mu^{(2\omega,\Gamma)}\xi^\mu\right]+\gamma^{\mu\nu}\left[\xi_\mu\tilde{D}_\nu^{(2\omega)}-\tilde{D}_\nu^{(2\omega,\Gamma)}\xi_\mu\right]. \end{aligned} \tag{6.156}$$

The Γ-term in the last $\tilde{D}_\nu^{(2\omega,\Gamma)}$ derivative acts on the vector index of ξ_μ. Note now that in

$$\xi_\mu\tilde{D}_\nu^{(2\omega)}-\tilde{D}_\nu^{(2\omega,\Gamma)}\xi_\mu=-\left[D_\nu^{(\Gamma)}\xi_\mu\right] \tag{6.157}$$

the derivative no longer acts past ξ_μ. The Γ-term cancels after contracting with $\gamma^{\mu\nu}$. Thus, inside the regulated trace one has the identity

$$\left[\xi^\mu\tilde{D}_\mu^{(2\omega)}+\tilde{D}_\mu^{(2\omega,\Gamma)}\xi^\mu\right]-\gamma^{\mu\nu}(\partial_\nu\xi_\mu)=0. \tag{6.158}$$

The last term can be written with flat indices as follows:

$$\gamma^{\mu\nu}(\partial_\nu\xi_\mu)=\gamma^{mn}[D_n^{(\omega)}\xi_m] \tag{6.159}$$

where $D_n^{(\omega)}\xi_m=e_n{}^\mu(\partial_\mu\xi_m+\omega_{\mu m}{}^n\xi_n)$. In the first term we can replace $\tilde{D}_\mu$ by D_μ and drop the Γ-term in $\tilde{D}_\nu^{(2\omega,\Gamma)}$ because these three Γ-terms cancel each other. So, finally,

$$\left[\xi^\mu D_\mu^{(2\omega)}+D_\mu^{(2\omega)}\xi^\mu\right]-\gamma^{mn}[D_n^{(\omega)}\xi_m]=0. \tag{6.160}$$

Recalling the definition of δ_{cov} in (6.150), we have found

$$\mathrm{Tr}\left\{\delta_{cov}(\xi^\mu)+2\delta_{lL}^{(1/2)}\left[\frac{D_m^{(\omega)}\xi_n-D_n^{(\omega)}\xi_m}{2}\right]\right\}\mathrm{e}^{\mathcal{R}}=0. \tag{6.161}$$

This concludes the proof of the lemma in (6.153).

To calculate the gravitational anomalies, Alvarez-Gaumé and Witten did not use δ_{cov} to obtain the Jacobian, but rather $2\delta_{sym}$, where δ_{sym} is the same combinations of symmetries as in the spin-$\frac{1}{2}$ case

$$2\delta_{sym}(\xi)=2\delta_{cov}(\xi)+2\delta_{lL}\left(\frac{D_m\xi_n-D_n\xi_m}{2}\right). \tag{6.162}$$

For the spin-$\frac{3}{2}$ field a local Lorentz transformation contains a part which acts on the spin-$\frac{1}{2}$ index and also a part that acts on the vector index

$$2\delta_{sym}\tilde{\psi}_m = 2\delta_{cov}\tilde{\psi}_m + 2\delta_{lL}^{1/2}\tilde{\psi}_m + 2\delta_{lL}^{1}\tilde{\psi}_m. \qquad (6.163)$$

Using (6.161) this can also be written as

$$\delta_{AGW} = 2\delta_{sym} = \delta_{cov}(\xi^\mu) + 2\delta_{lL}^{(1)}(D_{[m}\xi_{n]}) \qquad (\text{spin } \tfrac{3}{2}). \qquad (6.164)$$

This is precisely the mysterious transformation law in (6.146)!

There is, of course, one question left. Are the transformation laws in the spin-$\frac{1}{2}$ and spin-$\frac{3}{2}$ cases the same combinations of Einstein and local Lorentz transformations? Obviously they should be the same if one wants to study cancellation of anomalies in theories with different spin contents. The spin-$\frac{1}{2}$ transformation law we used to compute the gravitational anomaly was given in (6.101)

$$\delta_{cov}(\xi)\tilde{\lambda} = \frac{1}{2}(\xi^\mu D_\mu + D_\mu \xi^\mu)\tilde{\lambda}. \qquad (6.165)$$

We proved the identity (see (6.106))

$$\delta_{cov}(\xi)\tilde{\lambda} + 2\delta_{lL}(D_{[m}\xi_{n]})\tilde{\lambda} = 0. \qquad (6.166)$$

We also encountered another combination of symmetries

$$\delta_{sym}(\xi)\tilde{\lambda} = \delta_{cov}(\xi)\tilde{\lambda} + \delta_{lL}(D_{[m}\xi_{n]})\tilde{\lambda}. \qquad (6.167)$$

Hence, for spin $\frac{1}{2}$, $\delta_{cov}\tilde{\lambda}$ is twice $\delta_{sym}\tilde{\lambda}$

$$\delta_{cov}\tilde{(\xi)}\lambda = 2\delta_{sym}(\xi)\tilde{\lambda}. \qquad (6.168)$$

For spin-$\frac{3}{2}$ fields, we have just derived that the Alvarez-Gaumé and Witten law is twice δ_{sym}

$$\delta_{AGW}\tilde{\psi}_m = 2\delta_{sym}(\xi)\tilde{\psi}_m. \qquad (6.169)$$

Hence, if one uses $\delta_{cov}\tilde{\lambda}$ to compute anomalies in the spin-$\frac{1}{2}$ case, we should use $\delta_{AGW}\tilde{\psi}_m$ to compute the same anomalies in the spin-$\frac{3}{2}$ case.

We now turn to the calculation of the gravitational anomaly for complex chiral spin-$\frac{3}{2}$ fields in n dimensions using

$$\delta_{AGW}\,\tilde{\psi}_m = \frac{1}{2}(\xi^\mu D_\mu + D_\mu \xi^\mu)\tilde{\psi}_m + [(D_m\xi^n) - (D^n\xi_m)]\tilde{\psi}_n. \qquad (6.170)$$

The calculation is similar to the calculation for the spin-$\frac{1}{2}$ case in Section 6.3, except that we treat the last term in (6.170) as a Yang–Mills symmetry, therefore with extra ghosts according to the methods of

Section 6.2. As in (6.119) the covariant derivative D_μ contains both a spin connection acting on the spinor index and a spin connection acting on the vector index of the gravitino; the whole D_μ becomes the covariant conjugate momentum π_i, see (6.121). After integrating out the momenta one obtains a term $\dot{q}^i\omega_{iab}\psi_1^a\psi_1^b$ which yields $\dot{q}^i q^j R_{ijab}\psi_1^a\psi_1^b$, as in (6.143).

The contribution from $\delta_{cov}(\xi)$ combines with a term $R_{ijkl}\psi^i\psi^j\dot{q}^k q^l$ in the action to give the combination

$$\Big(\frac{1}{4}R_{ijkl}\psi^i\psi^j - D_{[k}\xi_{l]}\Big)\dot{q}^k q^l. \tag{6.171}$$

We encountered this combination in the spin-$\frac{1}{2}$ case. The contribution from $2\delta^{(1)}_{lL}(D_{[m}\xi_{n]})$ combines with a term $\psi^i\psi^j R_{ijkl}c^{*k}c^l$ to give

$$\Big(\frac{1}{4}R_{ijkl}\psi^i\psi^j - D_{[k}\xi_{l]}\Big)c^{*k}c^l. \tag{6.172}$$

So, thanks to the extra spin-1 Lorentz transformation in (6.146), the final answer only depends on the combination $\left(\frac{1}{4}R_{ijkl}\psi^i\psi^j - D_{[k}\xi_{l]}\right)$, both in the spin-$\frac{1}{2}$ sector and in the spin-1 sector. This is closely related to the descent equations from two dimensions higher [2, 153].

We can then directly write down the result for **the gravitational anomaly for a complex chiral gravitino** in n dimensions,

$$An(\text{grav, spin } \tfrac{3}{2}) = \frac{(-i)^{n/2}}{(2\pi)^{n/2}}\int\left(d^n x_0^i\sqrt{g(x_0)}\right)\left(\prod_{a=1}^{n} d\psi^a_{1,bg}\right)$$
$$\times\left[\left(\operatorname{tr} e^{\frac{1}{2}\tilde{R}}\right)-1\right]\exp\frac{1}{2}\operatorname{tr}\ln\left[\frac{\tilde{R}/4}{\sinh(\tilde{R}/4)}\right] \tag{6.173}$$

where everywhere $\tilde{R}$ stands for $R_{abcd}\psi^a_{1,bg}\psi^b_{1,bg} - 2(D_a\xi_b - D_b\xi_a)$. Thus, the relative normalizations of the two terms in (6.170) are just so that the Einstein–Lorentz anomaly in $4k+2$ dimensions is obtained from the chiral anomaly (more precisely for the gravitational contribution to the chiral $U(1)$ anomaly in $4k+4$ dimensions) by the *uniform* shift $R \to R - \frac{1}{4}D\xi$.[18] The last factor comes from the spin-$\frac{1}{2}$ sector, see (6.113), and takes into account the transformation law $\delta\tilde{\psi}_m = \frac{1}{2}(\xi^n D_n + D_n\xi^n)\tilde{\psi}_m$. The first

[18] The D_μ in $\frac{1}{2}(\xi^\mu D_\mu + D_\mu\xi^\mu)$ lead again to a term with $\dot{q}^i\xi^j(x_0+q)g_{ij}(x_0+q)$ in the action which can be written in the form $\dot{q}^i q^k D_k\xi_i$, see the last term in (6.111). Together with the term with $R_{ijab}\psi_1^a\psi_1^b$ due to expanding the $\dot{q}^i\omega_{iab}$ term in the action, these two contributions yield the term with $\exp\frac{1}{2}\operatorname{tr}\ln\left[(\tilde{R}/4)/\sinh(\tilde{R}/4)\right]$. On the other hand, the term $c_a^*(R^a{}_{bcd}\psi_1^c\psi_1^d)c^b$ coming from the commutator $\gamma^\mu\gamma^\nu[D_\mu, D_\nu]$ in the regulator and the last term in (6.170) together produce $\operatorname{tr} e^{\frac{1}{2}\tilde{R}}$ as in (6.143) and (6.144).

factor takes into account the vector index of $\tilde{\psi}_m$, see (6.87), and obtains its contributions from $\delta\tilde{\psi}_m = (D_m\xi^n - D^n\xi_m)\tilde{\psi}_n$. The Yang–Mills curvature F of (6.85) is replaced by R. Finally, the term -1 is due to the ghost sector; as we have discussed, we need to subtract one chiral complex ghost. If one is dealing with real chiral spin-$\frac{3}{2}$ fields, one needs to divide the result by 2.

6.7 Lorentz anomalies for self-dual antisymmetric tensor fields coupled to gravity in $4k + 2$ dimensions

In addition to chiral fermions, also self-dual (or antiself-dual) antisymmetric tensor gauge fields in $4k+2$ dimensions can produce gravitational (Lorentz) anomalies. From string theory one already knows an example: a chiral boson in two dimensions is a self-dual antisymmetric tensor ($\partial_\mu\varphi = \epsilon_{\mu\nu}\partial^\nu\varphi$ implies $(\partial_0 - \partial_1)\varphi = 0$), and in string theory such a field has a gravitational anomaly. To discuss the higher-dimensional case, we first need some formalism for the antisymmetric tensor (AT) gauge fields.

The field strength and the Lagrangian for an arbitrary real antisymmetric tensor gauge field with p indices (a p-form) in Minkowski space are defined by

$$\begin{aligned} F_{\mu_1\cdots\mu_{p+1}} &= \partial_{\mu_1}A_{\mu_2\cdots\mu_{p+1}} \pm p \text{ cyclic permutations} \\ \mathcal{L} &= -\frac{e}{2\cdot(p+1)!}F_{\mu_1\cdots\mu_{p+1}}F_{\nu_1\cdots\nu_{p+1}}\, g^{\mu_1\nu_1}\ldots g^{\mu_{p+1}\nu_{p+1}} \end{aligned} \tag{6.174}$$

with $e = \det e_\mu^m$. For a scalar and a vector field these definitions yield the Klein–Gordon and Maxwell actions, respectively. The stress tensor follows from the coupling to gravity

$$\mathcal{L}_{int} = \frac{1}{2}h^{\mu\nu}T_{\mu\nu}(F) + O(h^2) \tag{6.175}$$

and reads in flat space as

$$T_{\mu\nu}(F) = \frac{1}{p!}F_{\mu\mu_1\cdots\mu_p}F_\nu{}^{\mu_1\cdots\mu_p} - \frac{1}{2\cdot(p+1)!}\eta_{\mu\nu}(F_{\mu_1\cdots\mu_{p+1}})^2 \tag{6.176}$$

where $g_{\mu\nu} = \eta_{\mu\nu} + h_{\mu\nu}$. With these normalizations the kinetic term has the standard form $\mathcal{L} = \frac{1}{2\cdot p!}(\partial_t A_{\mu_1\ldots\mu_p})^2 + \cdots$ and the stress tensor for a field strength with $2k+1$ indices in $4k+2$ dimensions is traceless. The action has the abelian gauge symmetry

$$\delta A_{\mu_1\cdots\mu_p} = \partial_{\mu_1}\lambda_{\mu_2\cdots\mu_p} \pm (p-1) \text{ cyclic permutations.} \tag{6.177}$$

Generalizing the Lorentz gauge for a vector field, we add a gravitationally covariant gauge-fixing term for this abelian gauge symmetry,

$$\mathcal{L}_{fix} = -\frac{e}{2\cdot(p-1)!}(g^{\mu\mu_1}D_\mu A_{\mu_1\cdots\mu_p})^2 \tag{6.178}$$

and then find a diagonal kinetic term

$$\mathcal{L} + \mathcal{L}_{fix} = -\frac{e}{2\cdot p!}(D_\mu A_{\mu_1\cdots\mu_p})^2 + \text{terms with curvatures.} \tag{6.179}$$

We shall use tensors with flat indices, and in that case the covariant derivatives will contain spin connections instead of Christoffel connections. Faddeev–Popov ghosts will also be needed in general, but they are not self-dual so they will not contribute to the chiral anomalies (they do contribute to the trace anomalies).

Consider a one-loop graph with an antisymmetric tensor field in the loop coupled to external gravity. Let the field strength be self-dual; we shall denote such fields by self-dual AT, or sometimes by SAT. In Minkowski spacetime this is only possible in $4k+2$ dimensions

$$F_{\mu_1\cdots\mu_{2k+1}} = \frac{e}{(2k+1)!}\epsilon_{\mu_1\cdots\mu_{2k+1}\nu_1\cdots\nu_{2k+1}}F^{\nu_{2k+1}\cdots\nu_1}. \tag{6.180}$$

(On the other hand, in Euclidean space this is only possible in $4k$ dimensions, instantons being an example with $k=1$.[19]) We shall consider real self-dual AT in Minkowski space; if $A_{\mu_1\cdots\mu_{2k}}$ is complex, the anomaly is twice as large. In general, no covariant action is known for a self-dual antisymmetric tensor field which on one hand can be easily quantized and on the other hand allows for easy calculations.[20] However, there exist actions which are not (manifestly) covariant [102–104] but for which the field equations are the covariant duality conditions (which, together with the Bianchi identity $\partial_{[\mu}F_{\mu_1\ldots\mu_{2k+1}]} = 0$ imply the field equation $\nabla^{\mu_1}F_{\mu_1\ldots\mu_{2k+1}} = 0$). We shall discuss these actions shortly.

The first question that one would like to be answered is: are there really anomalies in the conservation of the stress tensor for self-dual antisymmetric tensor fields? We have already mentioned an example: chiral bosons in two dimensions. In 1+1 dimensions a self-dual scalar satisfies the equation $\partial_\mu\varphi = e\epsilon_{\mu\nu}\partial^\nu\varphi$, which in flat-space light-cone coordinates becomes $\partial_-\varphi = (\partial_t - \partial_x)\varphi = 0$. This defines what is known as a chiral boson. In conformal field theory a real chiral boson can be fermionized

[19] In Euclidean space $\frac{1}{2}\epsilon_{\mu\nu\alpha\beta}\epsilon^{\alpha\beta\sigma\tau} = +(\delta^\sigma_\mu\delta^\tau_\nu + \cdots)$, but also in Minkowski space $\frac{1}{3!}\epsilon_{\mu\nu\rho\alpha\beta\gamma}\epsilon^{\alpha\beta\gamma\sigma\tau\kappa} = +(\delta^\sigma_\mu\delta^\tau_\nu\delta^\kappa_\rho + \cdots)$ because interchanging $\alpha\beta\gamma$ and $\sigma\tau\kappa$ yields a minus sign that compensates the minus sign in $\epsilon_{012345} = -\epsilon^{012345}$.

[20] A classical covariant action with scalar fields in the denominator does exist [105], but it is not clear how to covariantly gauge-fix it.

to a complex chiral fermion, and since chiral fermions do have gravitational (Lorentz) anomalies in $4k+2$ dimensions, as we saw in Section 6.3, we have obtained an example of a self-dual antisymmetric tensor with a gravitational anomaly. One can also directly compute the diagram which contains the anomaly, similar to the calculation of the anomaly of a spin-$\frac{1}{2}$ field in two dimensions which we performed in Section 5.2. The corresponding diagram for a scalar in the loop reads as

$V(p) =$ 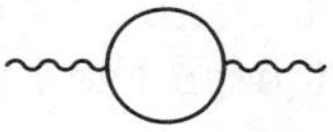 $= \int d^2x \, \langle T_{++}(x) T_{++}(y) \rangle e^{ip(x-y)}.$ (6.181)

If $\partial_- \varphi = 0$, on-shell the coupling to gravity reduces to $\frac{1}{2} h^{++} T_{++}$ where $T_{++} = \partial_+ \varphi \partial_+ \varphi$, and we can compute $V(p)$ using either x-space methods of conformal field theory or momentum space methods [1].

A not-manifestly covariant action describing a chiral boson in two dimensions has been introduced by Floreanini and Jackiw in [102] and coupled to gravity in [103] and to supergravity in [154]. The coupling of chiral bosons to gravity in higher dimensions appeared in [103]. The calculation of gravitational anomalies in higher dimensions was given in [104]. Here we consider the coupling of chiral bosons to gravity in $1+1$ dimensions. It can be used to prove explicitly the existence of a gravitational anomaly for this bosonic system, as we shall now briefly review. The action describing a chiral boson coupled to gravity in Minkowski space is

$$\mathcal{L} = \dot{\varphi}\varphi' - F\varphi'\varphi' \tag{6.182}$$

where the dot and the prime indicate derivatives with respect to time $x^0 = \tau$ and space $x^1 = \sigma$, respectively, and $F = e_0{}^+/e_1{}^+ = -E_-{}^1/E_-{}^0$, where $+$ and $-$ are flat indices. We have denoted by $E_a{}^\mu$ the inverse of the vielbein $e_\mu{}^a$. The equation of motion reads as

$$\frac{\partial}{\partial\sigma}\left(\dot{\varphi} - F\varphi'\right) = 0 \tag{6.183}$$

and with suitable spacelike boundary conditions it gives the correct chiral equation in curved space

$$\dot{\varphi} - F\varphi' = 0 \quad \rightarrow \quad E_-{}^\mu \partial_\mu \varphi = 0. \tag{6.184}$$

The Lagrangian is not manifestly covariant. Nevertheless it is invariant under the following general coordinate transformations:

$$\begin{aligned} \delta\varphi &= (\xi^1 + F\xi^0)\varphi' \\ \delta e_\mu{}^a &= \xi^\nu \partial_\nu e_\mu{}^a + (\partial_\mu \xi^\nu) e_\nu{}^a \end{aligned} \tag{6.185}$$

from which follows

$$\delta F = \xi^\nu \partial_\nu F + \partial_0 \xi^1 + F(\partial_0 \xi^0 - \partial_1 \xi^1) - F^2 \partial_1 \xi^0. \tag{6.186}$$

Note that on-shell the transformation rule of the chiral boson φ coincides with the usual transformation rule of a scalar field.

To compute the gravitational anomaly it is convenient to express F in terms of the variable

$$h_{--} = \frac{e_1{}^+ - e_0{}^+}{e_1{}^+ + e_0{}^+} = -\frac{e_{=}{}^+}{e_{\ddagger}{}^+},$$

already used in Section 5.1, see (5.56). One then finds $F = 1 - h_{--}/1 + h_{--}$. It is now easy to extract from the Lagrangian the linearized coupling to h_{--},

$$\mathcal{L} = \mathcal{L}_0 + \mathcal{L}_{int} = (\dot{\varphi}\varphi' - \varphi'\varphi') + 2h_{--}\varphi'\varphi' + \cdots . \tag{6.187}$$

The free propagator is given by

$$\begin{aligned}\langle\varphi(x)\varphi(y)\rangle &= -\frac{i}{2}\big[\partial_1(\partial_1 - \partial_0)\big]^{-1}\delta^2(x-y)\\ &= \int \frac{d^2p}{(2\pi)^2}\,\mathrm{e}^{ip\cdot(x-y)}\,\frac{i}{\sqrt{2}}\frac{p_+}{p_1}\frac{1}{p^2 - i\epsilon}\end{aligned} \tag{6.188}$$

where $p \cdot x = p_+x^+ + p_-x^-$, $p^2 \equiv p_\mu p^\mu = -2p_+p_-$, $p_\pm = \frac{1}{\sqrt{2}}(p_0 \pm p_1)$ and $-i\epsilon$ is the Feynman prescription giving the correct causal boundary conditions. The leading term of the effective action is then (for the normalization see (5.11))

$$\begin{aligned}W^{(2)}[h] &= \frac{i}{2}\langle S_{int}^2\rangle\\ &= \frac{i}{2}\iint d^2x\, d^2y\, 2h_{--}(x)\big\langle\varphi'(x)\varphi'(x)\,\varphi'(y)\varphi'(y)\big\rangle 2h_{--}(y)\\ &= \frac{i}{2}\int\frac{d^2p}{(2\pi)^2}\,h_{--}(p)U(p)h_{--}(-p)\end{aligned} \tag{6.189}$$

where

$$U(p) \equiv 4\int d^2x\,\mathrm{e}^{-ip\cdot x}\,\varphi'(x)\varphi'(x)\,\varphi'(0)\varphi'(0) =$$

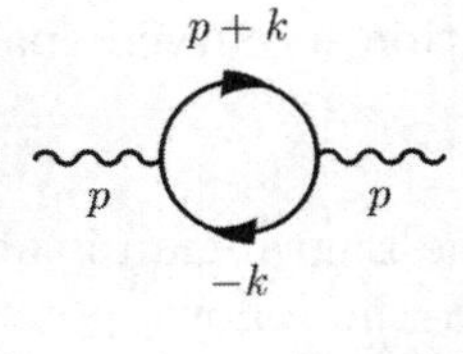

$$= -4\int\frac{d^2k}{(2\pi)^2}\,\frac{(p_1+k_1)(p_+ + k_+)}{(p+k)^2 - i\epsilon}\,\frac{k_1k_+}{k^2 - i\epsilon}. \tag{6.190}$$

Analytic regularization can be employed as in Section 5.2 to obtain

$$U(p) = \frac{i}{24\pi}\left(\frac{p_+^3}{p_-} - 3p_+^2\right). \tag{6.191}$$

Up to the local term $-3p_+^2$, which can be canceled by a counterterm, the resulting effective action

$$W_{\mathit{eff}}^{(2)} = -\frac{1}{48\pi}\int \frac{d^2p}{(2\pi)^2} h_{--}(p)\left(\frac{p_+^3}{p_-} - 3p_+^2\right) h_{--}(-p) \qquad (6.192)$$

produces the expected gravitational anomaly, as in (5.62) for the chiral fermion (there we looked at the opposite chirality). These calculations confirm that there is a genuine gravitational anomaly for this self-dual antisymmetric tensor field. The calculation of the gravitational anomaly for the chiral boson in two dimensions and using the Fujikawa approach can be found in [155]. The previous construction can be extended to $4k+2$ dimensions to calculate the correct gravitational anomalies for self-dual AT fields using the Feynman rules obtained from an action [104].

Let us now sketch how we are going to compute the gravitational anomaly of the real self-dual AT in $n = 4k+2$ dimensions, using quantum mechanics and following [1]. First we add a whole array of other real AT which are not self-dual and which therefore have no anomalies: $F = 0, F_\mu = \partial_\mu A, F_{\mu\nu} = \partial_\mu A_\nu - \partial_\nu A_\mu, \ldots, F_{\mu_1\cdots\mu_n} = 0$. The reason we add these AT is based on a simple but useful fact: one can use bispinors[21] $\psi_{\alpha\beta}$ to describe their field strengths all at once

$$\psi_{\alpha\beta} = \frac{1}{2^{n/4}}\sum_{l=0}^{n}\frac{1}{l!}(\gamma^{m_l\cdots m_1})_{\alpha\beta}\, F_{m_1\ldots m_l}, \quad n = 4k+2$$

$$F_{m_1\ldots m_l} = \frac{1}{2^{n/4}}\,\psi_{\alpha\beta}(\gamma_{m_1\ldots m_l})^{\beta\alpha}. \qquad (6.193)$$

For example, in two dimensions we have $\psi_{\alpha\beta} = \frac{1}{\sqrt{2}}\gamma^\mu_{\alpha\beta}\partial_\mu\varphi$. Chirality of $\psi_{\alpha\beta}$, defined by $(\gamma_5)_\alpha{}^{\alpha'}\psi_{\alpha'\beta} = \psi_{\alpha\beta}$, implies self-duality of the AT. Since the AT only couple to gravity by means of their field strength $F_{m_1\cdots m_l}$ we can build Feynman graphs if we know the vertices for the interaction of $F_{m_1\cdots m_l}$ with gravity and the propagators of $F_{m_1\cdots m_l}$. Knowing the vertices and propagators of $F_{m_1\cdots m_l}$, we can construct those for $\psi_{\alpha\beta}$. If one is interested in the diagram with $n/2+1$ external lines, one can evaluate the propagators in flat space, and take only the vertices with one external field and two quantum fields. So one may drop the curvature terms in (6.179), and only retain the linearized interaction in (6.175). Gauge artifacts (such

[21] Bispinors are also called Dirac–Kähler fermions. They were introduced by Kähler in 1960 [156] and discussed by Banks *et al.* [157]. Actually, already in 1929 C. Lanczos, nowadays best known for his work in classical mechanics, had studied a modification of the Dirac equation of 1928 with quaternions [158]. He found that this modified theory described an antisymmetric tensor. Quaternions can be represented by the four Pauli matrices $(I, \vec{\sigma})$ as the bispinor $\sigma^\mu{}_{\alpha\dot{\alpha}}$.

as vertices due to the gauge-fixing term or the ghost action) should not contribute to the result for the anomaly, and are then dropped. The calculation of Feynman graphs along these lines was performed in [1], and discussed in a textbook [3]. Here we are interested in the QM approach to these problems. We shall write down a covariant transformation law for $\tilde{\psi}_{\alpha\beta} \equiv g^{1/4}\psi_{\alpha\beta}$, compute the corresponding Jacobian, and use a regulator $\exp(\beta\mathcal{R})$ with $\mathcal{R} = \tilde{\not{D}}\tilde{\not{D}}$ in the trace, just as for the chiral spin-$\frac{1}{2}$ and spin-$\frac{3}{2}$ fields. Here $\tilde{\not{D}}$ is a Dirac operator for the bispinor which will be described below (see (6.210)). For the Jacobian the two indices α and β of $\psi_{\alpha\beta}$ are treated differently, just as for the spinor and vector indices of the gravitino. We call the corresponding spaces the α space and the β space. We again find a covariant translation in both the α space and the β space, while in the β space we find an additional Lorentz transformation which acts on the spinor index. As in the spin-$\frac{3}{2}$ case, one needs this extra term in the transformation law in order so that the action of the nonlinear sigma model only contains the combination $\tilde{R}_{ij} = R_{ij} - 2(D_i\xi_j - D_j\xi_i)$. Hence, the only difference from the spin-$\frac{3}{2}$ case is that the extra Lorentz transformation acts on a spinor index instead of a vector index. We use ghosts c^* and c in β space, and fermions ψ_1^a in α space, again as in the case of spin $\frac{3}{2}$.[22] Finally, we reduce the trace to Feynman graphs in quantum mechanics and again find the anomaly as a product of a factor for the α trace and a factor for the β trace.

We now give the details. First, we discuss the properties of the gamma matrices γ_m in $4k+2$ dimensions. We consider $n = 4k+2$ dimensional Euclidean spaces because the regulator regulates in all directions in momentum space only if we use Euclidean space. In the space with 2, 10, 18, ... dimensions a symmetric Majorana representation exists [148]: all Dirac matrices γ_m can be chosen as real and symmetric $2^{n/2}\times 2^{n/2}$ matrices satisfying $\{\gamma_m, \gamma_n\} = \delta_{mn}$ with $\delta_{mn} = (+1, \ldots, +1)$.[23] In all

[22] In [1] the extra spinor index β is treated on an equal footing with the spinor index α, by introducing a second set of Grassmann variables ψ_2^a in addition to the Grassmann variables ψ_1^a in α space, rather then treating β space as an internal space and using ghosts c_α^* and c^β. In the susy model underlying the approach of [1], ψ_1^a and ψ_2^a appear symmetrically, and this yields an $N = 2$ model. It can be obtained by dimensional reduction from the $N = (1,1)$ model in $1+1$ dimensions, see Appendix D.

[23] From supergravity or string theory one knows that there exists a Majorana representation in ten-dimensional Minkowski space with a real matrix γ_5 [159]. Interchanging the matrix γ_0 and γ_5, one obtains a Majorana representation in Euclidean space. The chirality matrix in Euclidean space is equal to $i\gamma_0$, and hence is purely imaginary and antisymmetric. The same results hold in two dimensions with σ_1 and σ_3. In six Euclidean dimensions, an example of a purely imaginary antisymmetric representation is given by the set of matrices $\gamma_k \otimes \sigma_2$, $i\gamma_1\gamma_2\gamma_3 \otimes \sigma_1$, $i\gamma_1\gamma_2\gamma_3 \otimes \sigma_3$ and $\gamma_5 \otimes I$, where the γ_μ form a Majorana representation in four Minkowski dimensions.

even dimensions there is a charge conjugation matrix C_+ satisfying $C_+\gamma_m C_+^{-1} = +\gamma_m^T$, which is related to the usual charge conjugation matrix C_- satisfying $C_-\gamma_m C_-^{-1} = -\gamma_m^T$ by $C_+ = C_-\gamma_5$, as one can easily check. This matrix C_+ is the unit matrix in our case, and we will use it to raise and lower spinor indices. Hence, we need not be careful whether the spinor indices are up or down. Furthermore, it is now clear that $\psi_{\alpha\beta}$ in (6.193) is real for real $F_{m_1\cdots m_l}$ (actually, as explained below, in Euclidean space we use complex field strengths). The chirality matrix denoted by γ_5 is given by $\gamma_5 = (-i)^{n/2}\gamma_1\cdots\gamma_n$ (where $\gamma_n = -i\gamma_0$ such that $\gamma_n^2 = 1$) so that γ_5 is purely imaginary and antisymmetric with a square of unity. In the opposite case, Euclidean spaces with $6, 14, 22, \ldots$ dimensions, one can choose an antisymmetric purely imaginary representation of the Dirac matrices. Then C_- equals the unit matrix. We shall continue below with the case of $2, 10, 18, \ldots$ dimensions, but there is a parallel treatment for the cases where $n = 6, 14, 22, \ldots$.

We define

$$\gamma_{m_1\cdots m_p} = \frac{1}{p!}(\gamma_{m_1}\cdots\gamma_{m_p} \pm (p!-1) \text{ permutations}). \qquad (6.194)$$

So $\gamma_{m_1\cdots m_p}$ is the totally antisymmetric part of the product of p Dirac matrices with strength one. Two formulas one needs are

$$\begin{aligned}(\gamma_{m_1\cdots m_p})^T &= \gamma_{m_p\cdots m_1} &(6.195)\\ \operatorname{Tr}\gamma_{m_1\cdots m_p}\gamma_{n_p\cdots n_1} &= 2^{n/2}[\delta_{m_1n_1}\cdots\delta_{m_pn_p} \pm (p!-1)\,\text{perms of}\,(n_1\cdots n_p)].\end{aligned}$$

For example, for $n = 2$ and $p = 2$ one has $(\gamma_{12})^T = \gamma_{21}$ and

$$\operatorname{Tr}\gamma_{mn}\gamma_{rs} = 2(\delta_{ms}\delta_{nr} - \delta_{mr}\delta_{ns}). \qquad (6.196)$$

The propagator of an arbitrary AT in Euclidean space follows from (6.179) and is given by

$$\begin{aligned}\langle A_{\mu_1\cdots\mu_p}(x)A_{\nu_1\cdots\nu_p}(y)\rangle &= \int\frac{d^nk}{(2\pi)^n}\mathrm{e}^{ik(x-y)}\frac{1}{k^2}\\ &\times(\delta_{\mu_1\nu_1}\ldots\delta_{\mu_p\nu_p} \pm\ (p!-1)\ \text{perms of}\ (\nu_1\ldots\nu_p)). \qquad (6.197)\end{aligned}$$

(and $\langle\varphi(x)\varphi(y)\rangle = \int\frac{d^nk}{(2\pi)^n}\mathrm{e}^{ik(x-y)}\frac{1}{k^2}$ for a scalar). Our metric is such that $k^2 = k_1^2 + \cdots + k_n^2$ in n-dimensional Euclidean space. Then it follows that the propagator of two field strengths in flat space is given in momentum space by

$$\begin{aligned}\langle F_{\mu_1\cdots\mu_{p+1}}(k)F_{\nu_1\cdots\nu_{p+1}}(-k)\rangle &= \frac{1}{k^2}(k_{\mu_1}k_{\nu_1}\delta_{\mu_2\nu_2}\cdots\delta_{\mu_{p+1}\nu_{p+1}} \qquad (6.198)\\ &\pm \text{all permutations of}\ \mu_i\ \text{and all cyclic permutations of}\ \nu_i).\end{aligned}$$

This propagator has $(p+1)(p+1)!$ terms.

The propagator of the tensors $F_{\mu_1\cdots\mu_{p+1}}$ determines the propagator of the bispinor. We claim that the latter is given by

$$\langle\psi_{\alpha\beta}(k)\psi_{\gamma\delta}(-k)\rangle = \frac{1}{2k^2}\left[(\gamma_5\not{k})_{\alpha\gamma}(\gamma_5\not{k})_{\beta\delta} + k^2\delta_{\alpha\gamma}\delta_{\beta\delta}\right]. \tag{6.199}$$

To prove this formula we insert the definition of the bispinors in (6.193) into the left-hand side

$$\begin{aligned}
&\frac{1}{2^{n/2}}\sum_{l=0}^{n}\langle F^{\mu_1\cdots\mu_l}(k)F_{\nu_1\cdots\nu_l}(-k)\rangle(\gamma_{\mu_l\cdots\mu_1})_{\alpha\beta}(\gamma^{\nu_l\cdots\nu_1})_{\gamma\delta}\frac{1}{l!}\frac{1}{l!}\\
&\quad = \frac{1}{2^{n/2}}\sum_{l=0}^{n}\frac{1}{k^2}\left(k^{\mu_1}k_{\nu_1}\delta^{\mu_2}_{\nu_2}\cdots\delta^{\mu_l}_{\nu_l}\right)(\gamma_{\mu_l\cdots\mu_1})_{\alpha\beta}(\gamma^{\nu_l\cdots\nu_1})_{\gamma\delta}\frac{1}{(l-1)!}\\
&\quad = \frac{1}{2^{n/2}}\frac{1}{k^2}\sum_{l=0}^{n}k^{\mu}k^{\nu}(\gamma_{\mu_l\cdots\mu_2\mu})_{\alpha\beta}(\gamma^{\mu_l\cdots\mu_2}{}_{\nu})_{\gamma\delta}\frac{1}{(l-1)!}.
\end{aligned} \tag{6.200}$$

On the other hand, a Fierz rearrangement of the first term on the right-hand side of (6.199) yields[24]

$$\begin{aligned}
&\frac{1}{2k^2}(\gamma_5\not{k})_{\alpha\gamma}(\gamma_5\not{k})_{\beta\delta} = \frac{-1}{2k^2}(\gamma_5\not{k})_{\alpha\gamma}(\not{k}\gamma_5)_{\delta\beta}\\
&\quad = \frac{1}{2^{n/2}}\frac{-1}{2k^2}\sum_{l=0}^{n}\frac{1}{l!}(\gamma_{\mu_1\cdots\mu_l})_{\alpha\beta}(\not{k}\gamma_5\gamma^{\mu_l\cdots\mu_1}\gamma_5\not{k})_{\delta\gamma}\\
&\quad = \frac{1}{2^{n/2}}\frac{-1}{2k^2}\sum_{l=0}^{n}\frac{1}{l!}(\gamma_{\mu_1\cdots\mu_l})_{\alpha\beta}k^2(\gamma^{\mu_l\cdots\mu_1})_{\delta\gamma}\\
&\qquad + \frac{1}{2^{n/2}}\frac{-1}{2k^2}\sum_{l=0}^{n}\frac{(-1)^l}{(l-1)!}2k^{\mu_1}(\gamma_{\mu_1\cdots\mu_l})_{\alpha\beta}(\not{k}\gamma^{\mu_l\cdots\mu_2})_{\delta\gamma}.
\end{aligned} \tag{6.201}$$

The first term is a Fierz rearrangement of $-\frac{1}{2k^2}k^2\delta_{\alpha\gamma}\delta_{\delta\beta}$ and the second term becomes equal to (6.200) after using

$$(\not{k}\gamma^{\mu_l\cdots\mu_2})_{\delta\gamma} = (-)^{l-1}(\not{k}\gamma^{\mu_2\cdots\mu_l})_{\gamma\delta} = (-)^{l-1}k_{\nu}(\gamma^{\nu\mu_2\cdots\mu_l})_{\gamma\delta}. \tag{6.202}$$

This proves the expression for the $\psi_{\alpha\beta}$ propagator.

The sum of the stress tensors of the AT fields in terms of a bispinor is given by

$$T_{\mu\nu}(\psi) = \frac{1}{4}\psi_{\alpha\beta}\psi_{\gamma\delta}(\gamma_\mu\gamma_5)^{\alpha\gamma}(\gamma_\nu\gamma_5)^{\beta\delta} + (\mu\leftrightarrow\nu). \tag{6.203}$$

---

[24] The general formula reads as $M_{\alpha\gamma}N_{\delta\beta} = \frac{1}{2^{n/2}}\sum_{l=0}^{n}\frac{1}{l!}(\gamma_{\mu_1\cdots\mu_l})_{\alpha\beta}(N\gamma^{\mu_l\cdots\mu_1}M)_{\delta\gamma}$ and can be proven by taking the trace with $(\gamma^{\nu_l\cdots\nu_1})_{\delta\alpha}$ using $\frac{1}{2^{n/2}}\mathrm{Tr}\,\gamma_{\mu_1\cdots\mu_l}\gamma^{\nu_l\cdots\nu_1} = \delta^{\nu_1}_{\mu_1}\cdots\delta^{\nu_l}_{\mu_l} \pm (l!-1)$ permutations.

Note that the two types of indices α and β are propagated independently in (6.199) and do not get mixed by the interactions in (6.203).

We shall again calculate the anomaly in Euclidean space, but here we run into the problem that tensors which are self-dual in Minkowski space, are no longer self-dual in Euclidean space because the square of the duality operator $F \to {}^*F$ equals -1 in Euclidean space. To still be able to use Euclidean space we therefore **complexify** the AT in Minkowski space, and divide the final answer for the anomaly by a factor of 2 to undo the complexification.

The duality operation from one field strength to the dual of another becomes $F \to i\,{}^*F$ in Euclidean space. It corresponds to multiplication of one of the indices of $\psi_{\alpha\beta}$ by γ_5, for example

$$\psi_{\alpha\beta} \to (\gamma_5)_\alpha{}^{\alpha'}\psi_{\alpha'\beta}. \tag{6.204}$$

(Recall that γ_5 is purely imaginary in Euclidean space.) In particular, the field strength with $2k+1$ indices is mapped into i times its own dual. Hence, the matrix $\frac{1}{2}(1+\gamma_5)_\alpha{}^{\alpha'}$ projects this field strength onto to its self-dual part. Consider a Feynman graph with gravitational couplings to $T_{\mu\nu}(F)$ at all vertices, except at one vertex where one couples to a self-dual AT as $\frac{1}{2}h^{\mu\nu}T_{\mu\nu}[\frac{1}{2}(F+i\,{}^*F)]$. It corresponds in the bispinor approach to a loop with couplings to $T_{\mu\nu}(\psi)$ at all vertices, except at one vertex where one projects onto a chiral bispinor (which we denote by ψ_L). The stress tensor at this vertex reads

$$\begin{aligned} T_{\mu\nu}(\psi_L) = \frac{1}{4}\left(\frac{1+\gamma_5}{2}\right)_\alpha{}^{\alpha'}\psi_{\alpha'\beta}\left(\frac{1+\gamma_5}{2}\right)_\gamma{}^{\gamma'}\psi_{\gamma'\delta}(\gamma_\mu\gamma_5)^{\alpha\gamma}(\gamma_\nu\gamma_5)^{\beta\delta} \\ +(\mu \leftrightarrow \nu). \end{aligned} \tag{6.205}$$

The transformation rule δ_{sym} of the AT is, as in the spin-$\frac{3}{2}$ case, a sum of a covariant translation and a Lorentz transformation; the latter acts on the flat vector indices of the AT,

$$\delta_{sym}(\xi)\tilde{F}_{m_1\cdots m_n} = \left[\frac{1}{2}(D_\mu\xi^\mu + \xi^\mu D_\mu) + \delta_{lL}(D_{[m}\xi_{n]})\right]\tilde{F}_{m_1\cdots m_n}. \tag{6.206}$$

In the bispinor approach this corresponds to

$$\begin{aligned} \delta_{sym}(\xi)\tilde{\psi}_{\alpha\beta} &= \frac{1}{2}(D_\mu\xi^\mu + \xi^\mu D_\mu)\tilde{\psi}_{\alpha\beta} \\ &+ \frac{1}{4}D_{[m}\xi_{n]}(\gamma^{mn})_\alpha{}^{\alpha'}\tilde{\psi}_{\alpha'\beta} + \frac{1}{4}D_{[m}\xi_{n]}(\gamma^{mn})_\beta{}^{\beta'}\tilde{\psi}_{\alpha\beta'} \end{aligned} \tag{6.207}$$

where the Lorentz transformation now acts both on the α and β indices. The last term can also be written as $-\frac{1}{4}D_{[m}\xi_{n]}\psi_{\alpha\beta'}(\gamma^{mn})^{\beta'}{}_\beta$ because spinor indices are raised and lowered by the charge conjugation matrix which is the unit matrix. Now we would like to compute the gravitational anomaly due to $2\delta_{sym}(\xi)$, which is precisely the transformation used both in the spin-$\frac{1}{2}$ and spin-$\frac{3}{2}$ cases. To rewrite this transformation in a useful form which will make the calculation easy, we again derive a lemma, this time applied to the bispinor instead of the spin-$\frac{3}{2}$ field. The lemma uses the regulator $\mathcal{R} = \tilde{\not{D}}\tilde{\not{D}}$, where the Dirac matrices γ^μ which contract D_μ act in α space (i.e. as matrix multiplication from the left). The lemma states that for the anomaly calculation $\delta_{cov}(\xi)$ equals $-2\delta_{lL}^{(\alpha\,\text{space})}(D_{[m}\xi_{n]})$, where the Lorentz transformation $\delta_{lL}^{(\alpha\,\text{space})}$ acts in α space and is analogous to the $\delta_{lL}^{(1/2)}$ acting in the spinor space of the gravitino. Thus with this regulator one finds the relation

$$2\delta_{sym}(\xi) = \delta_{cov}(\xi) + 2\delta_{lL}^{(\beta\,\text{space})}(D_{[m}\xi_{n]}). \tag{6.208}$$

Then the transformation law of the bispinor density which produces the Jacobian can be written as

$$\begin{aligned}\delta_{AGW}\tilde{\psi}_{\alpha\beta} &= [\delta_{cov}(\xi) + 2\delta_{lL}^{(\beta-\text{space})}(D_{[m}\xi_{n]})]\tilde{\psi}_{\alpha\beta} \\ &= \frac{1}{2}(D_\mu\xi^\mu + \xi^\mu D_\mu)_{\alpha\beta}{}^{\alpha'\beta'}\tilde{\psi}_{\alpha'\beta'} + 2\delta_\alpha{}^{\alpha'}\frac{1}{4}D_{[m}\xi_{n]}(\gamma^{mn})_\beta{}^{\beta'}\tilde{\psi}_{\alpha'\beta'}.\end{aligned} \tag{6.209}$$

The covariant derivative $(D_\mu)_{\alpha\beta}{}^{\alpha'\beta'}$ contains spin connection terms which act on both spinor indices of $\psi_{\alpha\beta}$. (The Lorentz transformations can be transferred from $F_{m_1\ldots m_n}$ to $\psi_{\alpha\beta}$ because the Dirac matrices are Lorentz invariant tensors.) The regulator is proportional to $\tilde{\not{D}}\tilde{\not{D}}$ where

$$\begin{aligned}\tilde{\not{D}} = g^{1/4}(\gamma^\mu)^\alpha{}_{\alpha''}\bigg[&\delta^{\alpha''}_{\alpha'}\partial_\mu\delta^\beta_{\beta'} + \frac{1}{4}\omega_{\mu mn}(\gamma^{mn})^{\alpha''}{}_{\alpha'}\delta^\beta_{\beta'} \\ &+ \frac{1}{4}\omega_{\mu mn}(\gamma^{mn})^\beta{}_{\beta'}\delta^{\alpha''}_{\alpha'}\bigg]g^{-1/4}.\end{aligned} \tag{6.210}$$

The connection in the β sector can be treated as a Yang–Mills field, so adding anticommuting ghosts we write

$$\gamma^{mn} = c^*_\beta(\gamma^{mn})^\beta{}_{\beta'}c^{\beta'}, \quad \{c^\beta, c^*_\gamma\} = \delta^\beta_\gamma \quad (\beta \text{ sector}). \tag{6.211}$$

This is analogous to the replacement of the internal symmetry generators $(T_\alpha)^M{}_N$ which we discussed previously.

The term in the α sector is treated as for the spin-$\frac{1}{2}$ case, hence in the α sector we replace

$$\gamma^{mn} \to 2\psi_1^a\psi_1^b, \quad \{\psi_1^a, \psi_1^b\} = \delta^{ab} \quad (\alpha \text{ sector}). \tag{6.212}$$

The regulator $\tilde{\not{D}}\tilde{\not{D}}$ leads to a term with $D_\mu D^\mu$ and a term with $\gamma^{\mu\nu}[D_\mu, D_\nu]$. The latter contains curvatures in the α sector and curvatures in the β sector,

$$\tilde{\not{D}}\tilde{\not{D}} = g^{-1/4} D_\mu \sqrt{g} g^{\mu\nu} D_\nu g^{-1/4} + \psi_1^a \psi_1^b \left(\frac{1}{2} R_{abcd} \psi_1^c \psi_1^d + \frac{1}{4} R_{abmn} c^* \gamma^{mn} c \right). \tag{6.213}$$

The operator D_μ is given by the expression inside the square brackets in (6.210).

The covariant translation in (6.209) yields a term $D_{[i}\xi_{j]} q^i \dot{q}^j$ in the α sector, while the extra Lorentz transformation of the β indices (with the parameter $D_{[m}\xi_{n]}$) produces a term $D_{[m}\xi_{n]} c^* \gamma^{mn} c$ in the β sector. In both the α sector and the β sector we again encounter the combination

$$\frac{1}{4} R_{mnab} \psi_1^a \psi_1^b - D_{[m}\xi_{n]}. \tag{6.214}$$

The trace in the α sector leads to $q\dot{q}$ loops which produce a factor

$$\exp \frac{1}{2} \mathrm{tr} \ln \left(\frac{\tilde{R}/4}{\sinh \tilde{R}/4} \right). \tag{6.215}$$

where the trace tr is over the vector indices of $\tilde{R}_{ij} = \tilde{R}_{ijab} \psi_1^a \psi_1^b$. This was discussed in Section 6.3. The trace in the β sector produces a trace

$$\mathrm{Tr}\, e^{F/2} = \mathrm{Tr} \exp \frac{1}{8} \psi_1^a \psi_1^b R_{abmn} \gamma^{mn} \tag{6.216}$$

where the trace Tr is over the spinor indices of γ^{mn}. This was discussed in Section 6.2 (the antihermitian matrices $\frac{1}{2}\gamma^{mn}$ $(m < n)$ correspond to the antihermitian Yang–Mills generators T_α).

Putting these factors together, we find for **the gravitational anomaly due to a self-dual real antisymmetric tensor** field in n dimensions

$$An(\text{grav, AT}) = \frac{(-i)^{n/2}}{(2\pi)^{n/2}} \int \left(d^n x_0^i \sqrt{g(x_0)} \right) \left(\prod_{a=1}^{n} d\psi_{1,bg}^a \right) \left(-\frac{1}{4} \right) \mathrm{Tr} \exp \left(\frac{1}{8} \psi_1^a \psi_1^b \tilde{R}_{abmn} \gamma^{mn} \right) \exp \frac{1}{2} \mathrm{tr} \ln \left[\frac{\tilde{R}/4}{\sinh(\tilde{R}/4)} \right]. \tag{6.217}$$

In the conclusions of Section 6.8 we make some comments on the overall normalization of this result. The minus sign in the factor of $-\frac{1}{4}$ is due to the fact that we are now computing a loop with bosonic bispinors instead of fermionic fields. The factor of $\frac{1}{4}$ is due to the factor of $\frac{1}{2}$ from the chiral projection operator $\frac{1}{2}(1+\gamma_5)$ which appears in the Jacobian, and the factor of $\frac{1}{2}$ one needs to undo the complexification of the AT which was needed to be able to go to Euclidean space. Since the symmetrized trace of an odd number of Lorentz generators vanishes,[25] the first factor only yields products of an even number of $\tilde{R}$ terms, and so does the second factor because $x/\sinh x$ is even in x. It follows that there is **only a gravitational anomaly in $\mathbf{n = 4k+2}$ dimensions**, as in the case of spin-$\frac{1}{2}$ and spin-$\frac{3}{2}$ fields.

The trace over the spinor indices in $\mathrm{Tr}\exp\left(\frac{1}{8}\tilde{R}_{mn}\gamma^{mn}\right)$ with $\tilde{R}_{mn} = R_{mnab}\psi_1^a\psi_1^b - 4D_{[m}\xi_{n]}$ can be rewritten as a trace over the vector indices of $\tilde{R}_{mn}$ as follows. We can skew diagonalize the real antisymmetric matrix $\tilde{R}_{mn}$ so that it attains the 2×2 block form,

$$\tilde{R}_{mn} = \begin{pmatrix} & x_1 & & & \\ -x_1 & & & & \\ & & & x_2 & \\ & & -x_2 & & \\ & & & & \ddots \end{pmatrix} \tag{6.218}$$

with real x_j. We then decompose the 2^{2k+1}-dimensional spinor space as a direct product of $2k+1$ two-dimensional spinor spaces, and we can choose γ^{mn} such that γ^{12} acts nontrivially only in the first two-dimensional subspace, γ^{34} in the second two-dimensional subspace, etc. Then the exponent of the direct sum becomes the direct product of the exponents

$$\exp\frac{1}{8}\tilde{R}_{mn}\gamma^{mn} = \bigotimes_{l=1}^{n/2}\exp\frac{1}{4}\tilde{R}_{2l-1,2l}\gamma^{2l-1,2l}. \tag{6.219}$$

In each subspace the trace yields $2\cosh\tilde{R}_{2l-1,2l}$ since the square of $\gamma^{2l-1,2l}$ equals minus unity and $\tilde{R}^2_{mn}$ has $-x_j^2$ along the diagonal. Except for the factor of 2^{2k+1}, which is the dimension of the spinor space, this is the same result as one obtains from $\exp\mathrm{tr}\ln\cosh\tilde{R}$. Hence, we can make the following replacement in the expression for the anomaly:

$$\mathrm{Tr}\,e^{\left(\frac{1}{8}\tilde{R}_{mn}\gamma^{mn}\right)} = 2^{2k+1}\exp\mathrm{tr}\ln\cosh\tilde{R}/4. \tag{6.220}$$

[25] From group theory we know that $SO(2n)$ has Casimir operators of rank $2, 4, \ldots, 2n-2, n$. However the trace over n generators in the spinor representation vanishes. In 10 dimensions the first factor yields a contributions with six $\tilde{R}$ factors (a hexagon graph), but not a contribution with five $\tilde{R}$ factors; for example, the product of five Lorentz generators with all indices different from each other is proportional to the trace of the chirality matrix γ_5, which vanishes.

In the end, we need $2k+2$ factors $\tilde{R}$ because we need one factor with $D_{[m}\xi_{n]}$ and $2k+1$ factors $R_{mnab}\psi_1^a\psi_1^b$ to saturate the integral over the fermionic zero modes. We can then absorb a factor of 2^{2k+2} into $\tilde{R}_{mn}$, which leads to another overall factor of $\frac{1}{2}$. The overall factor is then $-\frac{1}{8}$.

Our final answer for the gravitational anomaly of a real self-dual AT is given by

$$An(\text{grav, AT}) = \frac{(-i)^{n/2}}{(2\pi)^{n/2}} \int \left(d^n x_0 \sqrt{g(x_0)} \right) \left(\prod_{a=1}^{n} d\psi_{1,bg}^a \right) \times \left(-\frac{1}{8} \right) \exp \frac{1}{2} \text{tr} \ln \left[\frac{\tilde{R}/2}{\tanh(\tilde{R}/2)} \right] \quad (6.221)$$

where $\tilde{R} = R_{mnab}\psi_1^a\psi_1^b - 2(D_m\xi_n - D_n\xi_m)$. We need $\frac{1}{2}n+1$ factors of $\tilde{R}$ in n dimensions to saturate the Grassmann integral (one of the $\tilde{R}$ should yield the $D\xi$). The α index has yielded the usual result for spin-$\frac{1}{2}$ with the sinh, while the β index has yielded a similar result but with cosh. Together they yield tanh. The prefactor $-\frac{1}{8}$ comes from the $\frac{1}{2}$ in $\frac{1}{2}(1+\gamma_5)$, from the fact that we consider real AT fields, and from the conversion of the trace over β spinor space into a trace over vector indices.

The reason β space gives a different result from α space can be traced to the fact that we acted with the operator $\frac{1}{2}(1+\gamma_5)$ only in the α space to project out the self-dual part of the AT fields. In the approach of [1] this means that the fermions ψ_1^a have periodic boundary conditions, whereas the fermions ψ_2^a are antiperiodic (see (2.185) and (2.187) and footnote 21).

6.8 Cancellation of gravitational anomalies in IIB supergravity

The gravitational anomaly for a complex chiral spin-$\frac{1}{2}$ field, a complex chiral spin-$\frac{3}{2}$ field, and a real self-dual antisymmetric tensor field are given by

$$An\left(\text{grav, spin } \frac{1}{2}\right) = \int \exp \frac{1}{2} \text{tr} \ln \left[\frac{\tilde{R}/4}{\sinh(\tilde{R}/4)} \right]$$

$$An\left(\text{grav, spin } \frac{3}{2}\right) = \int \left[\left(\text{tr}\, e^{\tilde{R}/2} \right) - 1 \right] \exp \frac{1}{2} \text{tr} \ln \left[\frac{\tilde{R}/4}{\sinh(\tilde{R}/4)} \right]$$

$$An\,(\text{grav, AT}) = \int \left(-\frac{1}{8} \right) \exp \frac{1}{2} \text{tr} \ln \left[\frac{\tilde{R}/2}{\tanh(\tilde{R}/2)} \right] \quad (6.222)$$

where we recall that $\tilde{R} = R_{ijab}\psi^a_{1,bg}\psi^b_{1,bg} - 4D_{[i}\xi_{j]}$. The symbol $\int$ denotes the measure $\frac{(-i)^{n/2}}{(2\pi)^{n/2}} \int \prod_{i=1}^n dx_0^i \sqrt{g(x_0)} \prod_{i=a}^n d\psi^a_{1,bg}$.

As a first application we check that in 1+1 dimensions the gravitational anomaly for a complex chiral spin-$\frac{1}{2}$ field is equal to the gravitational anomaly of a real self-dual antisymmetric tensor (chiral boson[26]). This result is well known in string theory where it is used in the calculation of the central charge [159]. For this purpose we need the term quadratic in $\tilde{R}$ (one of these $\tilde{R}$ yields the contribution proportional to $D_{[i}\xi_{j]}$). We find for a complex chiral spin-$\frac{1}{2}$ field

$$\exp \frac{1}{2}\mathrm{tr}\ln\left[\frac{\tilde{R}/4}{\sinh(\tilde{R}/4)}\right] = \exp\left\{-\frac{1}{2}\mathrm{tr}\ln\left[1+\frac{1}{3!}(\tilde{R}/4)^2+\cdots\right]\right\}$$
$$= \cdots - \frac{1}{12}\mathrm{tr}(\tilde{R}/4)^2 + \cdots \tag{6.223}$$

while for the chiral boson we obtain, using $\tanh x = x - \frac{1}{3}x^3 + \cdots$,

$$-\frac{1}{8}\exp\frac{1}{2}\mathrm{tr}\ln\left[\frac{\tilde{R}/2}{\tanh(\tilde{R}/2)}\right] = -\frac{1}{8}\exp\left\{-\frac{1}{2}\mathrm{tr}\ln\left[1-\frac{1}{3}(\tilde{R}/2)^2+\cdots\right]\right\}$$
$$= \cdots - \frac{1}{48}\mathrm{tr}(\tilde{R}/2)^2 + \cdots. \tag{6.224}$$

Clearly the anomalies are equal.

A less obvious case is type IIB supergravity. This theory contains: a complex chiral spin-$\frac{3}{2}$ field, a complex antichiral spin-$\frac{1}{2}$ field, and a real five-index self-dual antisymmetric field strength. To check that the sum of these anomalies cancels too, we must expand the formulas for the anomalies to sixth order in $\tilde{R}$. Let us simplify the notation and denote $\tilde{R}/4$ by y and $\mathrm{tr}\, y^n$ by t_n.

The spin-$\frac{1}{2}$ field yields

$$An\left(\frac{1}{2}\right) = \exp\left[-\frac{1}{2}\mathrm{tr}\ln\left(1+\frac{1}{3!}y^2+\frac{1}{5!}y^4+\frac{1}{7!}y^6+\cdots\right)\right]$$
$$= \exp\left(-\frac{1}{12}t_2+\frac{1}{360}t_4-\frac{1}{5670}t_6+\cdots\right)$$
$$= 1+\left(-\frac{1}{12}t_2\right)+\left(\frac{1}{360}t_4+\frac{1}{288}t_2^2\right)$$
$$+\left(-\frac{1}{5670}t_6-\frac{1}{4320}t_2t_4-\frac{1}{10368}t_2^3\right)+\cdots. \tag{6.225}$$

[26] A self-dual antisymmetric tensor in two dimensions satisfies $\partial_\mu\varphi = \epsilon_{\mu\nu}\partial^\nu\varphi$ or $(\partial_0 + \partial_1)\varphi = 0$, and thus describes a left-moving (chiral) boson.

The spin-$\frac{3}{2}$ field yields

$$\begin{aligned}
An\left(\frac{3}{2}\right) &= \left[\operatorname{tr}\left(1 + 2y^2 + \frac{2}{3}y^4 + \frac{4}{45}y^6 + \cdots\right) - 1\right] An\left(\frac{1}{2}\right) \\
&= \left[(n-1) + 2t_2 + \frac{2}{3}t_4 + \frac{4}{45}t_6 + \cdots\right] An\left(\frac{1}{2}\right) \\
&= (n-1) + \left[\left(2 - \frac{n-1}{12}\right)t_2\right] \\
&\quad + \left[\left(\frac{2}{3} + \frac{n-1}{360}\right)t_4 + \left(-\frac{1}{6} + \frac{n-1}{288}\right)t_2^2\right] \\
&\quad + \left[\left(\frac{4}{45} - \frac{n-1}{5670}\right)t_6 + \left(-\frac{1}{20} - \frac{n-1}{4320}\right)t_2 t_4\right. \\
&\quad \left. + \left(\frac{1}{144} - \frac{n-1}{10\,368}\right)t_2^3\right] + \cdots .
\end{aligned} \tag{6.226}$$

The self-dual AT field yields

$$\begin{aligned}
An(\mathrm{AT}) &= -\frac{1}{8}\exp\left[-\frac{1}{2}\operatorname{tr}\ln\left(1 - \frac{1}{3}4y^2 + \frac{2}{15}16y^4 - \frac{17}{315}64y^6 + \cdots\right)\right] \\
&= -\frac{1}{8}\exp\left(\frac{2}{3}t_2 - \frac{28}{45}t_4 + \frac{1984}{2835}t_6 + \cdots\right) \\
&= -\frac{1}{8} + \left(-\frac{1}{12}t_2\right) + \left(\frac{7}{90}t_4 - \frac{1}{36}t_2^2\right) \\
&\quad + \left(-\frac{248}{2835}t_6 + \frac{7}{135}t_2 t_4 - \frac{1}{162}t_2^3\right) + \cdots .
\end{aligned} \tag{6.227}$$

One may check that for $n = 10$ all terms of sixth order in $\tilde{R}$ (the terms with t_6, $t_2 t_4$ and t_2^3) cancel in the following combination:

$$An\left(\frac{3}{2}\right) - An\left(\frac{1}{2}\right) + An(\mathrm{AT}) = 0. \tag{6.228}$$

Indeed,

$$\begin{aligned}
&\left(\frac{4}{45} - \frac{9}{5670}\right) + \frac{1}{5670} - \frac{248}{2835} = 0 \\
&\left(-\frac{1}{20} - \frac{9}{4320}\right) + \frac{1}{4320} + \frac{7}{135} = 0 \\
&\left(\frac{1}{144} - \frac{9}{10\,368}\right) + \frac{1}{10\,368} - \frac{1}{162} = 0.
\end{aligned} \tag{6.229}$$

This corresponds to the cancellation of gravitational anomalies in type IIB supergravity [1].

In $4k+2$ dimensions, the conditions for vanishing of the overall coefficients of $(\mathrm{tr}\tilde{R}^2)^{k+1}$, $(\mathrm{tr}\tilde{R}^2)^{k-1}\mathrm{tr}\tilde{R}^4$ and $(\mathrm{tr}\tilde{R}^2)^{k-2}\mathrm{tr}\tilde{R}^6$ due to loops with $n_{1/2}$ complex chiral spin-$\frac{1}{2}$ fields, loops with $n_{3/2}$ complex chiral spin-$\frac{3}{2}$ fields, and loops with n_{AT} real self-dual antisymmetric tensor fields, are given by, respectively

$$\begin{aligned} n_{1/2} - (20k+23)n_{3/2} + (-8)^k n_{AT} &= 0 \\ n_{1/2} - (20k-265)n_{3/2} - \frac{7}{2}(-8)^k n_{AT} &= 0 \\ n_{1/2} - (20k+455)n_{3/2} + \frac{31}{4}(-8)^k n_{AT} &= 0. \end{aligned} \tag{6.230}$$

These equations are linearly dependent, and give a unique solution

$$\begin{aligned} n_{1/2} &= (20k-41)(-8)^{k-2} n_{AT} \\ n_{3/2} &= (-8)^{k-2} n_{AT}. \end{aligned} \tag{6.231}$$

For $k = 0, 1, 2$ these equations have solutions for theories in $2, 6$ and 10 dimensions. However, there are no further solutions for $k = 3, 4, 5, \ldots$ because there are then further constraints which cannot be met [160].

6.9 Cancellation of anomalies in $N = 1$ supergravity

As we mentioned in the introduction, Alvarez-Gaumé and Witten derived compact expressions for chiral and gravitational anomalies in any dimensions in 1983 [1]. Then they applied these formulas to IIB supergravity in $9+1$ dimensions where gravitational anomalies are present, and found that they cancel. We discussed this in the preceding section. They also applied these formulas to $N = 1$ supergravity in $9+1$ dimensions coupled to Yang–Mills theory, but in this case the sum of all anomalies did not cancel, and they concluded that the $N = 1$ theory is anomalous. Green and Schwarz [4] noted that even if anomalies do not seem to cancel, it is sometimes still possible to construct a local counterterm in the action for which the variation cancels the anomalies. In such cases one has candidate anomalies which are not genuine anomalies. They indeed were able to construct such counterterms, but only for certain choices of the gauge group of the Yang–Mills theory, namely $SO(32)$ and $E_8 \times E_8$. Thus, this constituted a double success: $N = 1$ supergravity was also non anomalous, and, in addition, the gauge groups were determined. In this section we show that anomalies in $N = 1$ supergravity can indeed be canceled; this is straightforward and only the expressions of the gravitational anomalies which we obtained before are needed. It should be noted that Green and Schwarz also showed that in string theory the anomalies in the Yang–Mills sector (the open string sector) cancel. They considered the Neven–Schwarz–Ramond (NSR) string. In this case there are three sets of string loop diagrams to be computed: a planar graph, nonorientable graphs (graphs with an odd number of twists) and nonplanar graphs (graphs with an even number of twists). The latter graphs do not produce anomalies, while the first two graphs contain anomalies which sum up to a complicated expression multiplied by a factor of $(1 + 32\eta/n)$, where $\eta = -1$ for $SO(n)$, $\eta = +1$ for $Usp(n)$ and $\eta = 0$ for $U(n)$ [161]. Thus, also in string theory the anomalies of the open string cancel, but only for $SO(32)$. The cancellation of the Yang–Mills anomalies in $N = 1$ supergravity for the group $E_8 \times E_8$ corresponds in string theory to cancellation of anomalies of the heterotic string. The analysis of the closed string, which should lead to cancellations of anomalies involving external gravitons has never been worked out to the best of our knowledge.

The anomaly cancellation in the dual version of $N = 1$ supergravity (with a 6-form instead of a 2-form field B) was given in [162].

After this work on $(9 + 1)$-dimensional supergravity, similar work was done in other models. For example, in six dimensions the authors of [163] studied cancellation of gravitational anomalies for supergravity coupled to several matter multiplets, and found several solutions. There are again $B \wedge R \wedge R$ counterterms [162]. There exist auxiliary fields for $N = 2$ supergravity in six dimensions [164], so one could in principle construct supergravity actions with Chern–Simons terms (using the tensor calculus of supergravity). More recently, anomaly cancellation on $K_3 \times S_1/Z_2$ has been discussed [165].

The field content of $N = 1$ supergravity coupled to $N = 1$ supersymmetric Yang–Mills theory is given by

$$(e_\mu{}^m, \psi_{\mu L}, \chi_R, B_{\mu\nu}, \varphi) \text{ and } (A_\mu^a, \lambda_L^a) \tag{6.232}$$

where $\psi_{\mu L}$ is a Majorana–Weyl (real chiral) gravitino, χ_R is a real antichiral "dilatino" and $e_\mu{}^m$ is the real vielbein; $B_{\mu\nu}$ is a real antisymmetric tensor, while φ is the real dilaton. Furthermore A_μ^a is the Yang–Mills field with gauge group G and λ_L^a are real chiral "gauginos" (partners of the gauge fields). Note that the gauginos are in the same representation of G as the gauge fields, namely the adjoint representation. This is due to supersymmetry which requires that the fermionic partner of the gauge field be in the same representation of the gauge group G as the gauge field. The chirality of the gaugino is the same as that of the gravitino but opposite to that of the dilatino. Furthermore, all fermionic fields are real fields in Minkowski space, so we should add an extra factor of $\frac{1}{2}$ to our formulas for anomalies because they were given for complex chiral fermions. However, since we shall require that anomalies cancel, we shall not keep these overall factors of $\frac{1}{2}$. Ahead of time we mention that the antisymmetric tensor field $B_{\mu\nu}$, though neither self-dual nor antiself-dual, will play a crucial role in the construction of counterterms, and precisely because the representation of the gauginos is the adjoint representation, it is possible to cancel the Yang–Mills anomalies.

For readers not familiar with supersymmetry and supergravity, we mention that the $N = 1$ in "$N = 1$ supergravity" refers to the fact that there is only one real chiral gravitino and one real chiral supersymmetry parameter. The IIB theory has two real chiral gravitinos which one often combines into one complex gravitino (as we did in the previous section). There is also a IIA supergravity theory with one real chiral and one real antichiral gravitino; this is a "vector theory" with one real nonchiral gravitino, which is free from anomalies. (One might call this theory $N = (1,1)$ supergravity, and the previous one $N = (2,0)$ supergravity, but this terminology is not common.) The number of bosonic states matches the number of fermionic states, both for the $N = 1$ Yang–Mills theory and for the $N = 1$ supergravity theory: $8 = 8$ for the Yang–Mills theory and $(\frac{1}{2}8 \times 9 - 1) + \frac{1}{2}8 \times 7 + 1 = \frac{1}{2}(8 - 1) \times 16 + \frac{1}{2}16 = 64$ for the supergravity theory.

There are three sets of anomalies to be dealt with, which we now first briefly introduce.

(I) Purely Yang–Mills anomalies

These are due to a hexagon loop with the gaugino λ in the loop coupled to external Yang–Mills fields.

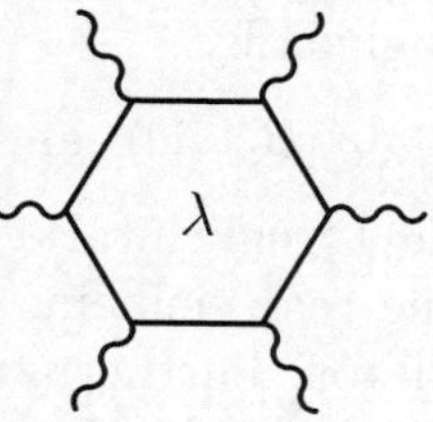

Wiggly lines denote external Yang–Mills gauge fields. There are no loops with a dilatino or gravitino because these fields have no minimal couplings to the Yang–Mills fields. (There are nonminimal couplings of the form $\bar{\psi}^{\mu}\gamma^{\nu}\lambda F_{\mu\nu}$ but these do not lead to anomalies.) We shall see that the counterterm which cancels these anomalies in the case of $G = SO(32)$ has the form

$$\Delta\mathcal{L}^{SO(32)} \sim B\,\mathrm{tr}F^4 + \omega_{3Y}^{(0)}\,\omega_{7Y}^{(0)}. \tag{6.233}$$

For $E_8 \times E_8$ the counterterm is different, and we construct it in Appendix F. The counterterm is a 10-form which is integrated over ten-dimensional space. The symbols $\omega_k^{(0)}$ denote Chern–Simons k-forms, and the subscript Y denotes Yang–Mills. The reasons that Chern–Simons actions appear has to do with the fact that the anomalies we derived before are **covariant anomalies**, whereas the counterterm which we shall construct cancels **consistent anomalies**. The latter are obtained by using the descent equations as we shall discuss, and the descent equations produce Chern–Simons terms. If the covariant anomalies cancel, then the consistent anomalies also cancel and vice versa. This is discussed in general articles on anomalies [2] and we refer to these articles for proofs. However, the fundamental anomalies are the consistent anomalies because they yield the variation of the effective action. The consistent anomalies can be constructed in two steps: first the general form is given by the descent equations and then the coefficients are fixed by matching the leading term of the consistent and covariant anomalies (up to an overall constant [2]). Thus, we shall cancel consistent anomalies by counterterms, while the covariant anomalies are merely a technical tool.

The hexagon graph is the first graph which can be anomalous (just like the triangle graph is the first graph which is anomalous in four spacetime dimensions). There are also polygon graphs with seven vertices, eight vertices, etc., but these graphs merely complete the leading expression from the hexagon graph. The complete consistent anomaly must satisfy the so-called consistency conditions which are so strong that if one knows the leading term of the consistent anomaly (corresponding to hexagon graphs), the other terms are completely fixed. (Because the transformation law $\delta A_\mu = \partial_\mu \Lambda + [A_\mu, \Lambda]$ is nonlinear in A_μ, one obtains relations between terms with different numbers of A fields.) The leading term in the consistent anomaly is equal to the leading term in the covariant anomaly up to an overall factor of $(d/2+1)^{-1}$, but since we are concerned with the question of when anomalies cancel, we shall not keep track of this overall constant. Thus, if the leading terms in the one-loop anomaly cancel, all nonleading terms also cancel. We shall actually obtain the complete formulas directly for the consistent anomalies and the complete counterterms, so we shall not restrict ourselves to only the leading terms.

(II) Purely gravitational anomalies

The counterpart of the purely Yang–Mills anomalies are the one-loop graphs with only external gravitons. Since all fields couple minimally to gravity, but only chiral fermions yield anomalies (there are no self-dual antisymmetric tensor fields in $N=1$ supergravity), the graphs to be studied are the following:

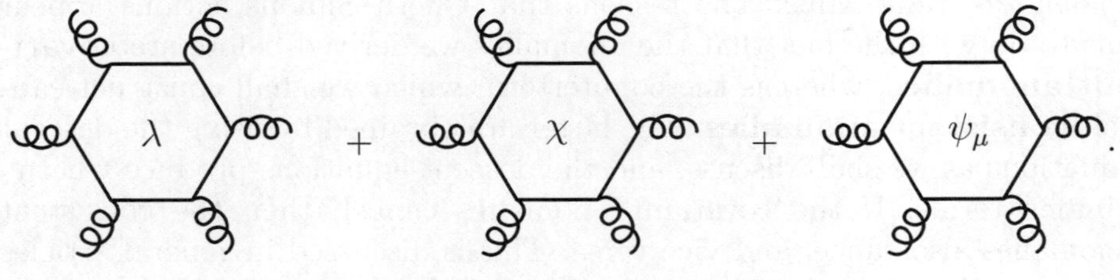

Curly lines denote gravitons. The counterterm which cancels these anomalies will be derived below and has the generic form

$$\Delta\mathcal{L}^{grav} \sim B[\mathrm{tr}R^4 + (\mathrm{tr}R^2)^2] + \omega_{3L}^{(0)}\,\omega_{7L}^{(0)}. \tag{6.234}$$

The subscript L denotes Lorentz. This counterterm is of course independent of the gauge group G, so it is the same for the $SO(32)$ theory and the $E_8 \times E_8$ theory. Its structure is very similar to the counterterm in the pure Yang–Mills case, but note that we now need Lorentz Chern–Simons terms, instead of Yang–Mills Chern–Simons terms. In the $N=1$ supergravity theory one encounters Yang–Mills Chern–Simons terms in the action and in the transformation rules, but no Lorentz Chern–Simons terms [166]. Thus in order to cancel anomalies one has to go beyond the minimal $N = 1$ supergravity theory. It is not known whether one can construct an extended supergravity theory with a finite number of fields that contain Lorentz Chern–Simons terms in the action. Most experts believe that this is not possible, and that by adding a Lorentz Chern–Simons term to minimal $N = 1$ supergravity, and adding further terms to obtain local supersymmetry, the answer is the full string effective action (whatever that means). Formally, Lorentz Chern–Simons terms are similar to Yang–Mills Chern–Simons terms. The only difference is that the Lorentz group $SO(9,1)$ is noncompact, whereas we shall only consider compact Lie groups for the gauge fields. The anomaly cancellation is a local phenomenon (local in spacetime) at the perturbative (one-loop) level, so issues of compactness or noncompactness do not matter as far as anomaly cancellation is concerned. (We shall, however, sometimes perform partial integrations, so to be precise we should state that we restrict ourselves to manifolds without boundaries, such as compactified R^{10} space.)

(III) Mixed anomalies

The third and last class of anomalies are the mixed anomalies: hexagon graphs with at least one graviton and at least one gauge field:

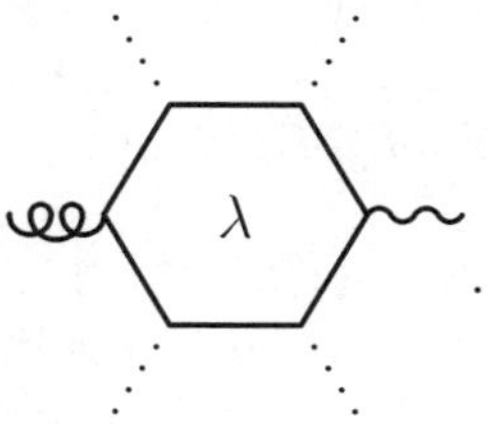

We may distinguish the cases with r gauge fields where $r = 1, 2, 3, 4, 5$. (The cases of $r = 0$ and $r = 6$ correspond to purely gravitational and purely Yang–Mills anomalies, respectively, and for given r the gravitons and the gauge fields may appear in any order.) The structure of the counterterm which cancels these anomalies is of a form which we might expect in view of the counterterms previously given. For $SO(32)$ the counterterm has the following form:

$$\Delta\mathcal{L}^{mixed,SO(32)} \sim B(\mathrm{tr}F^2\mathrm{tr}R^2) + \omega_{3L}^{(0)}\,\omega_{3Y}^{(0)}(\mathrm{tr}F^2 + \mathrm{tr}R^2) \\ + \omega_{3L}^{(0)}\,\omega_{7Y}^{(0)} + \omega_{7L}^{(0)}\,\omega_{3Y}^{(0)}. \tag{6.235}$$

The counterterm for the $E_8 \times E_8$ case is again different. There are no counterterms of the form $\omega_{3L}^{(0)}\,\omega_{3L}^{(0)}(\mathrm{tr}F^2 + \mathrm{tr}R^2)$ or $\omega_{3Y}^{(0)}\,\omega_{3Y}^{(0)}(\mathrm{tr}F^2 + \mathrm{tr}R^2)$ because they vanish. (The 6-forms $\omega_{3L}^{(0)}\,\omega_{3L}^{(0)}$ and $\omega_{3Y}^{(0)}\,\omega_{3Y}^{(0)}$ vanish since interchanging two 3-forms yields one overall minus sign.) Actually, we shall not separately construct the counterterm for the mixed anomalies but rather construct the whole counterterm in one fell swoop. The reason is that the field $B_{\mu\nu}$ transforms simultaneously into gauge fields and into gravitational fields; as we shall discuss

$$\delta_{gauge}B = \omega_{2Y}^{(1)} - \omega_{2L}^{(1)}. \tag{6.236}$$

Here $\omega_{2Y}^{(1)}$ and $\omega_{2L}^{(1)}$ are 2-forms which are constructed from the variation of $\omega_{3Y}^{(0)}$ and $\omega_{3L}^{(0)}$, as we shall discuss. It is clear that substituting $\delta_{gauge}B$ in the counterterms $\Delta\mathcal{L}^{grav}$ and $\Delta\mathcal{L}^{YM}$ yields variations which contain simultaneously gravitational fields and gauge fields ("mixed variations"). Rather than first constructing $\Delta\mathcal{L}^{YM}$ and $\Delta\mathcal{L}^{grav}$, and then using their mixed variations in the construction of $\Delta\mathcal{L}^{mixed}$, it is easier to construct the complete $\Delta\mathcal{L}^{total}$ at once.

However, to isolate the salient points where one finds the restrictions on the gauge group, we shall first construct $\Delta\mathcal{L}^{grav}$ and $\Delta\mathcal{L}^{YM}$ separately. Then, as we have already said, we shall construct $\Delta\mathcal{L}^{total}$.

The covariant gauge anomalies for a complex spin-$\frac{1}{2}$ field were derived earlier,

$$An_{YM} = \int \mathrm{Tr}\, \mathrm{e}^{\frac{1}{2}\tilde{F}}$$
$$\frac{1}{2}\tilde{F} \equiv \frac{1}{2} F^{\alpha}_{ab} T_{\alpha} \psi^{a}_{1,bg} \psi^{b}_{1,bg} + \eta^{\alpha} T_{\alpha}. \tag{6.237}$$

The symbol $\int$ was defined below (6.222). In 10 dimensions we need the terms with six $\tilde{F}$ (recall that we need terms linear in η^{α}, so in one curvature we must take the term $\eta^{\alpha} T_{\alpha}$; we are then left with five curvatures, i.e. a 10-form). Thus, the anomaly is proportional to

$$An_{YM} \sim \mathrm{Tr}\, \tilde{F}^{6}. \tag{6.238}$$

(The precise coefficient in front does not concern us here; later when we construct $\Delta\mathcal{L}^{total}$ we shall be careful with coefficients.) There are now two issues we must deal with:

(i) the relation between traces of expressions in the adjoint representation (in particular, $\mathrm{Tr}\, \tilde{F}^{6}$) and traces in the vector representation (which we shall denote by the symbol tr);

(ii) the construction of consistent anomalies from the descent equations.

We now briefly discuss these issues, and then return to the construction of counterterms.

Traces in group theory

Consider the adjoint representation of $SO(n)$. (We begin with $SO(n)$ because this is the simplest example, but we shall also discuss the other groups.) The carrier space (the space on which the group acts) is given by a "vector" $v_{kl} = -v_{lk}$, where $k, l = 1, \ldots, n$. The group $SO(n)$ acts on v_{kl} as follows:

$$v_{kl} \to v'_{kl} = (\Omega v)_{kl}; \quad (\Omega v)_{kl} = \sum_{m<n} \Omega_{kl}{}^{mn} v_{mn}. \tag{6.239}$$

Thus, the pair of indices $I = (k, l)$ with $k < l$ runs over $N = \frac{1}{2}n(n-1)$ values, and we can also write

$$v_I \to v'_I = \Omega_I{}^J v_J, \quad I, J = 1, \ldots, N. \tag{6.240}$$

On the other hand, the adjoint transformation can also be written in terms of the defining representation of $SO(n)$ (the $n \times n$ real orthogonal matrices denoted by $O_n{}^{n'}$). Namely,

$$(\Omega v)_I = (\Omega v)_{kl} = \sum_{k',l'} O_k{}^{k'} O_l{}^{l'} v_{k'l'} = 2 \sum_{k'<l'} O_k{}^{k'} O_l{}^{l'} v_{k'l'}. \quad (6.241)$$

Note that we discuss here group elements for finite transformations, thus $O^T = O^{-1}$ and O itself is not antisymmetric; however, v_{kl} is antisymmetric. We can then write the following relation between the adjoint and vector representation of $SO(n)$:

$$\Omega_{kl}{}^{mn} = O_k{}^m O_l{}^n - O_k{}^n O_l{}^m \quad (6.242)$$

where $k < l$ and $m < n$. Readers who are not sure whether one should add a factor of $\frac{1}{2}$ or not may check this relation for the case of $SO(3)$.

We need relations between traces of products of generators in the adjoint representation, and similar traces in the defining representation. They can all be derived by taking the trace of the group elements in (6.242). Namely, set $m = k$ and $n = l$, and sum over k and l from 1 to n. This yields

$$\begin{aligned}\operatorname{Tr}\Omega &= \sum_I \Omega_I{}^I = \sum_{k<l} \Omega_{kl}{}^{kl} = \frac{1}{2}\sum_{k,l} \Omega_{kl}{}^{kl} \\ &= \frac{1}{2}\sum_{k=1}^{N}\sum_{l=1}^{N}\left(O_k{}^k O_l{}^l - O_k{}^l O_l{}^k\right) \\ &= \frac{1}{2}\Big[(\operatorname{tr} O)^2 - \operatorname{tr}(O^2)\Big]. \end{aligned} \quad (6.243)$$

To find expressions for traces over elements in the enveloping Lie algebra (products of elements in the Lie algebra such as $\operatorname{Tr}\tilde{F}^6$), we write $\Omega = \mathrm{e}^{\mathcal{A}}$ and $O = \mathrm{e}^{A}$, where $\mathcal{A}$ lies in the adjoint representation and A lies in the vector representation of the Lie algebra of $SO(n)$. From (6.243) one finds (after multiplying by 2 to simplify the notation)

$$\begin{aligned} 2\operatorname{Tr}\mathrm{e}^{\mathcal{A}} &= 2\operatorname{Tr}\left[1 + \mathcal{A} + \frac{1}{2!}(\mathcal{A})^2 + \frac{1}{3!}(\mathcal{A})^3 + \cdots\right] \\ &= \left\{\operatorname{tr}\left[1 + A + \frac{1}{2!}(A)^2 + \frac{1}{3!}(A)^3 + \cdots\right]\right\}^2 \\ &\quad -\left\{\operatorname{tr}\left[1 + 2A + \frac{1}{2!}(2A)^2 + \cdots\right]\right\}. \end{aligned} \quad (6.244)$$

Comparing terms with the same number of factors yields a hierarchy of relations for $SO(n)$,

$$2\operatorname{Tr}1 = (\operatorname{tr}1)^2 - \operatorname{tr}1 = n(n-1)$$

$$\begin{aligned}
\mathrm{Tr}\,\mathcal{A} &= 0 \qquad (\text{because } \mathrm{tr}\,A = 0)\\
\mathrm{Tr}\,\mathcal{A}^2 &= (n-2)\,\mathrm{tr}\,A^2\\
\mathrm{Tr}\,\mathcal{A}^4 &= (n-8)\,\mathrm{tr}\,A^4 + 3(\mathrm{tr}\,A^2)^2\\
\mathrm{Tr}\,\mathcal{A}^6 &= (n-32)\,\mathrm{tr}\,A^6 + 15(\mathrm{tr}\,A^2)(\mathrm{tr}\,A^4).
\end{aligned} \tag{6.245}$$

The first line gives the dimension of the adjoint representation.

For $Sp(n)$ one finds the same formulas, but with $+$ signs instead of $-$ signs because the adjoint representation for $Sp(n)$ is given by a tensor v_{kl} which is symmetric[27] ($v_{kl} = v_{lk}$). As we shall see, anomalies which are factorized (such as $\mathrm{tr}F^2\,\mathrm{tr}F^4$) can be canceled by counterterms, but non factorized expressions (such as $\mathrm{tr}F^6$) can never be canceled. This immediately rules out the $Sp(n)$ groups, and of the $SO(n)$ groups only $SO(32)$ needs to be kept.

What about $SU(n)$ or the exceptional groups? For $SU(n)$ the carrier space for the adjoint representation in terms of the vector (i.e. defining) representation is given by vectors ${v_i}^j$ which transform as follows:

$$(Uv)_i{}^j = U_i{}^{j;k}{}_l\, v_k{}^l = u_i{}^k (u^*)^j{}_l\, v_k{}^l \tag{6.246}$$

where $(u^*)^j{}_l = (u_j{}^l)^*$. This adjoint representation is obtained by taking the direct product of the $\boldsymbol{n}$ and $\boldsymbol{n}^*$ of $SU(n)$ and removing the trace,[28] hence v is traceless, $v_k{}^k = 0$. Thus the dimension of the adjoint representation of $SU(n)$ is $n^2 - 1$. We then have

$$U_i{}^{j;k}{}_l = u_i{}^k (u^*)^j{}_l - \frac{1}{n}\delta_i^j \delta_l^k \tag{6.247}$$

and taking the trace we obtain

$$\mathrm{Tr}\,U = (\mathrm{tr}\,u)(\mathrm{tr}\,u^*) - 1. \tag{6.248}$$

Setting $U = \mathrm{e}^{\mathcal{A}}$ and $u = \mathrm{e}^{A}$, with $\mathcal{A}$ in the adjoint representation of the

---

[27] The group $Sp(n)$ leaves the bilinear form $x^i \Omega_{ij} y^j$ $(i, j = 1, \ldots, n)$ invariant, where Ω_{ij} is nondegenerate and antisymmetric $\Omega_{ij} = -\,\Omega_{ji}$. Then for infinitesimal transformations $x \longrightarrow Mx$ one obtains $M^T\Omega + \Omega M = 0$. The matrices ΩM are symmetric, and define $Sp(n)$ (they are the generators of $Sp(n)$ in the defining representation). The matrices $M_i{}^j$ act on v_{kl} just as in (6.241), but now (6.242) obtains a $+$ sign instead of a $-$ sign. Actually in supergravity and string theory one uses the group $Usp(n)$ which is the intersection of $U(n)$ and $Sp(n, C)$. It has the same symmetry properties as $Sp(n)$.

[28] One may identify each $v_i{}^j$ with an $n \times n$ matrix with only an entry in the ith row and jth column; if the sum of these matrices $v_i{}^j$ is a matrix v which is antihermitian then the transformation rule for v correspond to a commutator of v with the generators in the fundamental representation. Clearly the carrier space defined by v is $n^2 - 1$ dimensional.

Lie algebra of $SU(n)$, and A in the vector representation, leads to

$$\begin{aligned} &\mathrm{Tr}\left[1+\mathcal{A}+\frac{1}{2!}(\mathcal{A})^2+\frac{1}{3!}(\mathcal{A})^3+\cdots\right] \\ &\quad = \mathrm{tr}\left[1+A+\frac{1}{2!}(A)^2+\cdots\right]\mathrm{tr}\left[1+A+\frac{1}{2!}(A)^2+\cdots\right]^* - 1. \quad (6.249) \end{aligned}$$

Equating terms with the same number of generators yields

$$\begin{aligned} \mathrm{Tr}\,1 &= n^2-1 \\ \mathrm{Tr}\,\mathcal{A} &= 0 \qquad\qquad (\text{because } \mathrm{tr}\,A = 0) \\ \mathrm{Tr}\,\mathcal{A}^2 &= n(\mathrm{tr}\,A^2 + \mathrm{tr}\,A^{*2}) \\ \mathrm{Tr}\,\mathcal{A}^3 &= n(\mathrm{tr}\,A^3 + \mathrm{tr}\,A^{*3}) \\ \mathrm{Tr}\,\mathcal{A}^4 &= n(\mathrm{tr}\,A^4 + \mathrm{tr}\,A^{*4}) + 6(\mathrm{tr}\,A^2)(\mathrm{tr}\,A^{*2}) \\ \mathrm{Tr}\,\mathcal{A}^6 &= n(\mathrm{tr}\,A^6 + \mathrm{tr}\,A^{*6}) + 15(\mathrm{tr}\,A^2)(\mathrm{tr}\,A^{*4}) \\ &\quad +15(\mathrm{tr}\,A^4)(\mathrm{tr}\,A^{*2}) + 20(\mathrm{tr}\,A^3)(\mathrm{tr}\,A^{*3}). \qquad (6.250) \end{aligned}$$

The first line again gives the dimension of the adjoint representation. Because the generators A of $SU(n)$ in the fundamental representation are antihermitian $n\times n$ matrices and the trace is invariant under transposition, we may replace $\mathrm{tr}(A^*)^k$ by $(-)^k\,\mathrm{tr}A^k$. Hence, all $SU(n)$ groups must be rejected because the coefficient of the leading term in $\mathrm{Tr}\mathcal{A}^6$ $(2n)$ never vanishes. Also $U(n)$ must be rejected because the only difference with respect to $SU(n)$ is the relation $\mathrm{Tr}\,I = n^2 - 1$ which becomes $\mathrm{Tr}\,I = n^2$ for $U(n)$.

Finally, we consider the exceptional groups. Here the coefficient of the leading term sometimes vanishes identically, due to the properties of the Casimir invariants C_k. Let us first list these Casimir invariants for the simple Lie groups

$$\begin{aligned} SU(n)&:\ C_2, C_3, \ldots, C_n \\ SO(2n+1)&:\ C_2, C_4, \ldots, C_{2n} \\ Sp(2n)&:\ C_2, C_4, \ldots, C_{2n} \\ SO(2n)&:\ C_2, C_4, \ldots, C_{2n-2}, C_n \\ G_2&:\ C_2, C_6 \\ F_4&:\ C_2, C_6, C_8, C_{12} \\ E_6&:\ C_2, C_5, C_6, C_8, C_9, C_{12} \\ E_7&:\ C_2, C_6, C_8, C_{10}, C_{12}, C_{14}, C_{18} \\ E_8&:\ C_2, C_8, C_{12}, C_{14}, C_{18}, C_{20}, C_{24}, C_{30}. \qquad (6.251) \end{aligned}$$

The Casimir operators C_k for a representation R are obtained by contracting a totally symmetric irreducible tensor in the adjoint representation

$d^{a_1 \ldots a_k}$ with the generators in the representation R

$$C_k(R) = d^{a_1 \ldots a_k} T_{a_1}^{(R)} \cdots T_{a_k}^{(R)}. \tag{6.252}$$

By irreducible we mean that "traces" (contractions with lower-order invariant tensors) have been removed. The usual Casimir operator corresponds to the quadratic Casimir operator, with d^{ab} equal to the inverse of the Killing metric $g_{ab} = f_{pa}{}^q f_{qb}{}^p$. (According to the definition of semisimple groups, these groups have an invertible Killing metric.) For example, for $SU(3)$ one has $C_2(R) = g^{ab} T_a^{(R)} T_b^{(R)}$ and $C_3(R) = d^{abc} T_a^{(R)} T_b^{(R)} T_c^{(R)}$, where d^{abc} are the "d-symbols" which yield the chiral triangle anomalies in four dimensions.[29] For $SO(6)$ one has $C_2 = g^{ab} T_a T_b$, $C_3 = \epsilon^{ijklmn} T_{ij} T_{kl} T_{mn}$ (where T_{ij} with $i < j$ corresponds to T_a) and $C_4 = T_{ij} T^{jk} T_{kl} T^{li}$.

One can construct invariant tensors by taking traces over products of generators, $\mathrm{tr}(T_{a_1}^{(R)} \cdots T_{a_k}^{(R)}) \equiv T_{a_1 \ldots a_k}^{(R)}$. These $T_{a_1 \ldots a_k}^{(R)}$ are invariant tensors. Again we may restrict our attention to $T_{a_1 \ldots a_k}^{(R)}$ which are totally symmetric and irreducible. Since for given k there is at most one such invariant tensor,[30] we have $T_{a_1 \ldots a_k}^{(R)} = T_k(R) d_{a_1 \ldots a_k}$. For the quadratic Casimir operators one has $T_{ab}^{(R)} = T_2(R) g_{ab}$, where $T_2(R)$ are called the Dynkin labels. There is a simple relation between $C_2(R)$ and $T_2(R)$, obtained by tracing $C_2(R)$

$$\dim G \, T_2(R) = \dim R \, C_2(R) \tag{6.253}$$

where $\dim G$ denotes the number of generators of the gauge group and $\dim R$ denotes the dimension of the representation R.

For certain representations of certain groups it may happen that for certain k the trace $\mathrm{tr} F^k$ does not contain only a term with the irreducible $d_{a_1 \ldots a_k}$ but also products of terms with lower-dimensional $d_{a_1 \ldots a_l}$. Suppose this happens for two representations R_1 and R_2. Then one finds a relation of the form $\mathrm{tr}_1 F^k = a \, \mathrm{tr}_2 F^k + b \, (\mathrm{tr}_2 F^{k-m})(\mathrm{tr}_2 F^m) + \cdots$. In particular, for our purposes it will be crucial to find cases (to find groups) for which the trace of F^6 in the adjoint representation, denoted by $\mathrm{Tr} F^6$, does not contain a term with the maximal Casimir invariant $d_{a_1 \ldots a_6}$. As we now discuss in more detail, this means that $\mathrm{Tr} F^6$ factorizes, and factorization will permit the construction of counterterms which cancel anomalies. The trace $\mathrm{Tr} F^6$ can be written in terms Casimir invariants which are

[29] For $SU(n)$ one may define the d symbols by $\{T_a^{(F)}, T_b^{(F)}\} = \frac{1}{n} g_{ab} + i d_{abc} g^{cd} T_d^{(F)}$, where (F) denotes the fundamental representation. One usually normalizes the generators such that $g_{ab} = -\delta_{ab}$. In that case tr $T_a^{(F)} T_b^{(F)} = -\frac{1}{2} \delta_{ab}$.

[30] Except for $SO(4n)$ in which case there are two invariants tensors with $2n$ indices. For example, for $SO(8)$ one has $\mathrm{tr}\, T_{ij} T^{jk} T_{kl} T^{li}$ and $\epsilon^{i_1 \cdots i_8} \mathrm{tr}\, T_{i_1 i_2} \cdots T_{i_7 i_8}$. Parity can be used to distinguish between them.

irreducible and totally symmetric invariant tensors in the adjoint representation. For example, for $SO(n)$ one should symmetrize $\mathrm{Tr}\, T_{a_1} \cdots T_{a_6}$ and remove traces to obtain C_6. (The Kronecker symbol δ^{ab} is an invariant tensor of $SO(n)$.) For other groups one should subtract contractions with all invariant tensors if they exist. Then $\mathrm{Tr}F^6$ becomes a polynomial in the invariant tensors $d_{a_1 \cdots a_k}$ contracted with curvatures $F^{a_1} \cdots F^{a_k}$. Next, note that $\mathrm{Tr}F^k$ contains a term with $d_{a_1 \cdots a_k}$ if the latter exists, but the coefficient of this term may vanish.

Consider now E_8. Since it has no Casimir invariants of rank less than eight except the quadratic Casimir invariant, it follows that $\mathrm{Tr}F^6$ must factorize into a constant times $(\mathrm{Tr}F^2)^3$. For E_7 one will still be left with a term involving C_6 and for E_6, F_4 and G_2 the same situation holds. We analyze these groups in Appendix F, but we mention here that their anomalies cannot be canceled: these groups must be rejected. So, only E_8 has a chance to be anomaly-free. At this point we mention ahead of time that from an analysis of anomalies in the purely gravitational sector it will follow that the number of generators should be 496. Miraculously, this is the number of generators of $E_8 \times E_8$ and of $SO(32)$. So, our analysis of traces in the fundamental representation of the gauge groups has narrowed the choice of the gauge groups down to $E_8 \times E_8$ and $SO(32)$.

Actually there are two further solutions. One possibility is $[U(1)]^{496}$; here anomalies can be canceled but since it not known how to construct a string theory which has this gauge group and also produces the Standard Model group $SU(3) \times SU(2) \times U(1)$ by some other mechanism, this case has received little attention. Another possibility is $E_8 \times [U(1)]^{248}$, but again no string theory is known which can accommodate this group, and we shall not pursue this possibility further.

Descent equations

Here we summarize the construction of the consistent Yang–Mills and gravitational anomalies in n dimensions from invariant polynomials in $n + 2$ dimensions by means of the descent equations. For proofs see [2] or [137]. The construction proceeds in five steps. Afterwards we shall give examples.

(1) One starts from an invariant $(n + 2)$-form I_{n+2} with n even, for example, $\mathrm{tr}F^{n/2+1}$, $\mathrm{tr}R^{n/2+1}$, $\mathrm{tr}R^{n/2-1}\mathrm{tr}R^2$ or $\mathrm{tr}F^{n/2-1}\mathrm{tr}R^2$, etc. The curvatures are defined by $F = dA + AA$ and $R = d\omega + \omega\omega$. Any representation of the Yang–Mills generators can be used.

(2) Since $dI_{n+2} = 0$ (which is easy to prove, using $dF = [F, A]$ and $dR = [R, \omega]$ and cyclicity of the trace), I_{n+2} is closed and therefore

exact, $I_{n+2} = d\omega^{(0)}_{n+1}$ (closure implies exactness, at least locally in $n+2$ dimensions). The $(n+1)$-form $\omega^{(0)}_{n+1}$ is the Chern–Simons term.

(3) The gauge variation of the Chern–Simons term is an exact form

$$\delta_{gauge}\, \omega^{(0)}_{n+1} = d\omega^{(1)}_n \quad \text{where} \quad \omega^{(1)}_n = \operatorname{tr} \Lambda d(\cdots). \tag{6.254}$$

The proof of this relation can be found in [2]. This shows that the Chern–Simons term is invariant under rigid (constant-Λ) gauge transformations, and the Chern–Simons action (the integrated Chern–Simons term) is invariant under infinitesimal gauge transformations (if there are no boundaries).

(4) The consistent anomaly G is given by

$$G(\Lambda) = (n+2) \int \omega^{(1)}_n. \tag{6.255}$$

Note that G is linear in the local gauge parameter $\Lambda(x)$. To indicate this we write $G(\Lambda)$. Because we can choose $\Lambda(x)$ to be nonvanishing only in a small region, the integral is always well defined. If one begins with a compact $(n+1)$-dimensional manifold B for which the boundary is an n-dimensional compactified spacetime Σ, then the gauge variation of the Chern–Simons term in B is the consistent anomaly in Σ,

$$\delta_{gauge} \int_B \omega^{(0)}_{n+1} = \int_\Sigma \omega^{(1)}_n. \tag{6.256}$$

(5) The consistent anomaly $G(\Lambda)$ must satisfy the consistency conditions[31]

$$\delta_{gauge}(\Lambda_1) G(\Lambda_2) - \delta_{gauge}(\Lambda_2) G(\Lambda_1) = G([\Lambda_1, \Lambda_2]). \tag{6.257}$$

The reason for this is that the consistent anomaly is the response of the effective action Γ under a gauge variation:

$$G(\Lambda) = \delta_{gauge}(\Lambda)\Gamma = \int dx \left[\delta_{gauge}(\Lambda) A^a_\mu(x)\right] \frac{\delta}{\delta A^a_\mu(x)} \Gamma. \tag{6.258}$$

---

[31] In the BRST formalism one replaces Λ by a ghost c. Then the BRST variation removes the terms $G([\Lambda_1, \Lambda_2])$ on the right-hand side and the consistency conditions reduce to the statement that the consistent anomaly must be BRST invariant: $QG(c) = 0$.

The consistency conditions state that two ordinary derivatives $\delta/\delta A^a_\mu$ of the effective action Γ commute. Let us prove that G in (6.256) indeed satisfies the consistency conditions in (6.257). Imagine that the n-dimensional space Σ over which $G \equiv \int_\Sigma g$ is integrated is the boundary of an $(n+1)$-dimensional ball B. Then, using the fact that $dg = \delta_{gauge}\, \omega^{(0)}_{n+1}$, we obtain

$$G(\Lambda) = \int_B dg(\Lambda) = \delta_{gauge}(\Lambda) \int_B \omega^{(0)}_{n+1}. \tag{6.259}$$

Since $[\delta_{gauge}(\Lambda_1), \delta_{gauge}(\Lambda_2)] = \delta_{gauge}([\Lambda_1, \Lambda_2])$, the consistency conditions are satisfied.

Let us now give two examples; these examples will be used in the construction of the counterterms.

Example 1: d = 2

In this case one begins with $I_4 = \mathrm{tr} F^2$ (or $\mathrm{Tr} F^2$; it does not matter which representation one uses for the descent equations). Then the Chern–Simons form is

$$\omega^{(0)}_3 = \mathrm{tr}\Big(FA - \frac{1}{3}A^3\Big). \tag{6.260}$$

It is easy to check that $d\omega^{(0)}_3 = \mathrm{tr}F^2$. Since $\delta_{gauge}\, \omega^{(0)}_3 = d(\mathrm{tr}\,\Lambda\, dA)$, as one readily verifies by using $\delta_{gauge} A = D\Lambda = d\Lambda + [A, \Lambda]$ and $\delta_{gauge} F = [F, \Lambda]$, we find

$$\omega^{(1)}_2 = \mathrm{tr}\,\Lambda\, dA. \tag{6.261}$$

So the consistent anomaly in two dimensions is

$$G(\Lambda) = 4 \int d^2x\, \mathrm{tr}\Lambda\, dA. \tag{6.262}$$

The consistency conditions reduce to

$$\int d^2x \Big[\mathrm{tr}(\Lambda_2\, dD\Lambda_1) - \mathrm{tr}(\Lambda_1\, dD\Lambda_2)\Big] = \int d^2x\, \mathrm{tr}([\Lambda_1, \Lambda_2]\, dA) \tag{6.263}$$

which is clearly true since $dd = 0$, and

$$\begin{aligned} &\mathrm{tr}\Lambda_2 d[A, \Lambda_1] - \mathrm{tr}\Lambda_1 d[A, \Lambda_2] \\ &\quad = 2\,\mathrm{tr}[\Lambda_1, \Lambda_2]dA - \mathrm{tr}A d[\Lambda_1, \Lambda_2] \\ &\quad = \mathrm{tr}[\Lambda_1, \Lambda_2]dA + d(\mathrm{tr}[\Lambda_1, \Lambda_2]A). \end{aligned} \tag{6.264}$$

Example 2: d = 6

We start from $I_8 = \mathrm{tr} F^4$. Then the Chern–Simons term is

$$\begin{aligned} \omega_7^{(0)} &= \mathrm{tr}\left[(dA)^3 A + \frac{8}{5}(dA)^2 A^3 + \frac{4}{5} dA A^2 dA A + 2 dA A^5 + \frac{4}{7} A^7 \right] \\ &= \mathrm{tr}\left(F^3 A - \frac{2}{5} F^2 A^3 - \frac{1}{5} F A^2 F A + \frac{1}{5} F A^5 - \frac{1}{35} A^7 \right). \end{aligned} \tag{6.265}$$

One may check that $d\omega_7^{(0)} = \mathrm{tr} F^4$. To compute the gauge variation of this expression, all terms due to $\delta_{gauge} A = [A, \Lambda]$ and $\delta_{gauge} F = [F, \Lambda]$ cancel in the trace due to cyclicity, and one only needs to use $\delta_{gauge} A = d\Lambda$. (We could have already used this observation in the previous example to simplify the derivation of (6.261)). One finds that

$$\delta_{gauge}\, \omega_7^{(0)} = \mathrm{tr}[F^3\, d\Lambda + \cdots] = d\mathrm{tr}[\Lambda\, dA\, dA\, dA + \cdots]. \tag{6.266}$$

Since $\delta_{gauge}\, \omega_7^{(0)} = d\omega_6^{(1)}$ one obtains

$$\omega_6^{(1)} = \mathrm{tr}\Lambda d(dA\, dA\, A + \cdots) \tag{6.267}$$

and the consistent anomaly is now

$$G(\Lambda) = 8 \int d^6 x\, \mathrm{tr}\Lambda d(dA\, dA\, A + \cdots). \tag{6.268}$$

The reader may complete the terms denoted by ellipses, but we shall not need the explicit form of these terms.

Cancellation of pure gravitational anomalies

The covariant purely gravitational anomalies due to spin-$\frac{3}{2}$ gravitinos spin-$\frac{1}{2}$ gauginos and a spin-$\frac{1}{2}$ dilatino were given in (6.225)–(6.227). The gaugino consists of $\dim G$ spin-$\frac{1}{2}$ fields, because gauginos are in the adjoint representation. The hexagon graphs correspond to terms with six curvatures, and extracting an overall minus sign we obtain

$$\begin{aligned} An_{grav} &= (\dim G - 1) An\left(\frac{1}{2}\right) + An\left(\frac{3}{2}\right) \\ &= (\dim G - 1)\left(\frac{1}{5670} t_6 + \frac{1}{4320} t_2 t_4 + \frac{1}{10\,368} t_2^3 \right) \\ &\quad + \left[\left(-\frac{4}{45} + \frac{9}{5670} \right) t_6 + \left(\frac{1}{20} + \frac{9}{4320} \right) t_2 t_4 \right. \\ &\quad \left. + \left(-\frac{1}{144} + \frac{9}{10368} \right) t_2^3 \right] \\ &= \frac{\dim G - 496}{5670} t_6 + \frac{\dim G + 224}{4320} t_2 t_4 + \frac{\dim G - 64}{10368} t_2^3 \end{aligned} \tag{6.269}$$

where $t_n = \mathrm{tr}\, y^n$ and $y = \tilde{R}/4\pi$. (All fields are real, so we should add an overall factor of $\frac{1}{2}$, but at this point we are not interested in overall factors). As we explained before, products of traces have a chance of being canceled. Hence, since $\mathrm{tr}\tilde{R}^6$ is nonvanishing (it is easy to write down a 10×10 antisymmetric matrix A for which $\mathrm{tr}A^6$ is nonvanishing), we must restrict the number of generators $\dim G$ of the gauge group G by

$$\dim G = 496. \tag{6.270}$$

The remaining terms then simplify considerably,

$$An_{grav} = c\,\mathrm{tr}\tilde{R}^2 \left[\mathrm{tr}\tilde{R}^4 + \frac{1}{4}(\mathrm{tr}\tilde{R}^2)^2\right] \tag{6.271}$$

where the constant c $(= \frac{1}{6})$ does not interest us at this point. We now want to apply the descent equations to find the consistent anomaly and the counterterm, but for this we must first find the invariant 12-form I_{12} from which to start. One obtains I_{12} by replacing $\tilde{R}$ by R in An_{grav}; indeed, it will reproduce the leading terms in An_{grav} as we shall see. Thus,

$$I_{12} = c\,\mathrm{tr}R^2 \left[\mathrm{tr}R^4 + \frac{1}{4}(\mathrm{tr}R^2)^2\right]. \tag{6.272}$$

Since $dI_{12} = 0$, I_{12} itself is $d\omega^{(0)}_{11L}$ where

$$\begin{aligned}\omega^{(0)}_{11L} &= \alpha\, c\, \omega^{(0)}_{3L}\left[\mathrm{tr}R^4 + \frac{1}{4}(\mathrm{tr}R^2)^2\right] \\ &\quad + (1-\alpha)c\,\mathrm{tr}R^2\left[\omega^{(0)}_{7L} + \frac{1}{4}\omega^{(0)}_{3L}\mathrm{tr}R^2\right].\end{aligned} \tag{6.273}$$

As a check note that $d\omega^{(0)}_{3L} = \mathrm{tr}R^4$ and $d(\omega^{(0)}_{7L} + \frac{1}{4}\omega^{(0)}_{3L}\mathrm{tr}R^2)$ equals $\mathrm{tr}R^4 + \frac{1}{4}(\mathrm{tr}R^2)^2$, so we have a free parameter α in $\omega^{(0)}_{11L}$. We now show that as far as cancellation of anomalies by counterterms is concerned, any value of α can be taken.

Since the terms proportional to α are annihilated by d, they are d-exact. (They are given by $\alpha d(-\omega^{(0)}_{3L}\omega^{(0)}_{7L}\cdot)$) Since any term dX in $\omega^{(0)}_{11L} = \cdots + dX$ will lead to a term $\delta_{gauge} X$ in the anomaly which can be removed by a counterterm $\Delta\mathcal{L} = -X$, any choice of α will be allowed. We shall impose Bose symmetry $(\alpha = \frac{4}{12})$ because this will yield an expression for the consistent anomaly for which the leading term agrees with the leading term of the covariant anomaly

$$\omega^{(0)}_{11L} = \frac{c}{12}\left\{4\omega^{(0)}_{3L}\left[\mathrm{tr}\,R^4 + \frac{1}{4}(\mathrm{tr}\,R^2)^2\right] + 8\,\mathrm{tr}\,R^2\left(\omega^{(0)}_{7L} + \frac{1}{4}\omega^{(0)}_{3L}\mathrm{tr}\,R^2\right)\right\}. \tag{6.274}$$

The gauge variation of this Chern–Simons term is then the following exact form:

$$\begin{aligned}\delta_{gauge}\,\omega^{(0)}_{11L} &= d\omega^{(1)}_{10L}\\ &= \frac{4c}{12}d\omega^{(1)}_{2L}\left[\operatorname{tr}R^4+\frac{1}{4}(\operatorname{tr}R^2)^2\right]\\ &\quad+\frac{8c}{12}d\left[\omega^{(1)}_{6L}\operatorname{tr}R^2+\frac{1}{4}\omega^{(1)}_{2L}(\operatorname{tr}R^2)^2\right].\end{aligned} \tag{6.275}$$

Hence the consistent anomaly (with all subleading terms included) reads as

$$G_{cons}=c'\int dx\left\{\omega^{(1)}_{2L}\left[\operatorname{tr}R^4+\frac{1}{4}(\operatorname{tr}R^2)^2\right]+2\left[\omega^{(1)}_{6L}+\frac{1}{4}\omega^{(1)}_{2L}\operatorname{tr}R^2\right]\operatorname{tr}R^2\right\}. \tag{6.276}$$

The 2-form $\omega^{(1)}_{2L}$ and the 6-form $\omega^{(1)}_{6L}$ were given in (6.261) and (6.267).

Let us now compare this result with the covariant anomaly in (6.271) by expanding $\tilde{R}_{ij}=R_{ij}-2(D_i\xi_j-D_j\xi_i)$ and taking the terms linear in $(D_i\xi_j-D_j\xi_i)$

$$G_{cov}=c''\int dx\left\{(\operatorname{tr}D\xi R)\left[\operatorname{tr}R^4+\frac{3}{4}(\operatorname{tr}R^2)^2\right]+2(\operatorname{tr}D\xi R^3)\operatorname{tr}R^2\right\}. \tag{6.277}$$

Since the leading terms in $\omega^{(1)}_{2L}$ and $(\operatorname{tr}D\xi R)$ are $\Lambda^{ij}dA^{ij}$ and $D^i\xi^j d\omega^{ij}$, respectively, the expressions for G_{cons} and G_{cov} agree if we note that $A=\omega$ for the Lorentz group, and identify Λ with $D\xi$. Similarly, $\omega^{(1)}_{6L}$ and $\operatorname{tr}D\xi R^3$ agree as far as the leading terms are concerned. Hence, the consistent and covariant anomalies agree, and this was done by fixing α according to Bose symmetry.

The counterterm, the variation of which is equal to minus the consistent anomaly, is given by

$$\int\Delta\mathcal{L}_{grav}=c'\int\left\{\alpha B[\operatorname{tr}R^4+\beta(\operatorname{tr}R^2)^2]+\gamma\omega^{(0)}_{3L}\omega^{(0)}_{7L}\right\} \tag{6.278}$$

where the constants α, β and γ are still to be determined. To construct the gauge variation of this counterterm, one must first discuss how the 2-form B transforms.

In the $N=1$ supergravity theory coupled to Yang–Mills theory, the action contains the modified field strength $H=dB+\omega^{(0)}_{3Y}$ which should be gauge invariant. Supersymmetry requires this combination [161, 166], but we can rescue Yang–Mills gauge invariance by defining that $\delta_{gauge} B = -\omega_{2Y}^{(1)}$. We now extend the definition of $\delta_{gauge} B$ to include a term $\omega_{2L}^{(1)}$ because then we shall be able to cancel anomalies,

$$\delta_{gauge} B = \omega_{2L}^{(1)} - \omega_{2Y}^{(1)}. \tag{6.279}$$

At this point it is not clear that one should take the difference $\omega_{2L}^{(1)} - \omega_{2Y}^{(1)}$; however, in order to cancel the mixed anomalies one needs this difference, see the discussion below (6.299). Although we shall not need it, let us mention that one can also introduce a modified field strength

$$H = dB + \omega_{3Y}^{(0)} - \omega_{3L}^{(0)}. \tag{6.280}$$

It is invariant both under gauge transformations and under local Lorentz transformations. One then finds from $\int dH = 0$ over a compact space that $\int (\mathrm{tr}\, R^2 - \mathrm{tr}\, F^2) = 0$. This is used for Kaluza–Klein compactifications to Calabi–Yau manifolds.

With the gauge variation of B fixed, we can now construct the variation of the counterterm. It varies giving the following purely gravitational expression:

$$\delta \Delta \mathcal{L}_{grav} = c' \left\{ \alpha \omega_{2L}^{(1)} [\mathrm{tr}\, R^4 + \beta (\mathrm{tr}\, R^2)^2] + \gamma d\omega_{2L}^{(1)} \omega_{7L}^{(0)} + \gamma \omega_{3L}^{(0)} d\omega_{6L}^{(1)} \right\} \tag{6.281}$$

where we have used the fact that $\delta_{gauge} B = \omega_{2L}^{(1)} + \cdots$. Partially integrating to make curvatures out of the Chern–Simons terms yields

$$\delta \Delta \mathcal{L}_{grav} = c' \left\{ \alpha \omega_{2L}^{(1)} [\mathrm{tr}\, R^4 + \beta (\mathrm{tr}\, R^2)^2] - \gamma \left[\omega_{2L}^{(1)} \mathrm{tr}\, R^4 - \omega_{6L}^{(1)} \mathrm{tr}\, R^2 \right] \right\}. \tag{6.282}$$

Let us now compare this expression with the consistent anomaly in (6.276). Choosing $-\alpha + \gamma = 1$, $\gamma = -2$ and $\alpha\beta = -\frac{3}{4}$, the variation of the counterterm cancels the gravitational consistent anomaly. Hence, the purely gravitational anomalies can always be canceled by a suitable counterterm, as long as the gauge group G has 496 generators.

Cancellation of pure Yang–Mills anomalies

We now consider the opposite case, namely the pure Yang–Mills anomalies (anomalies due to Yang–Mills gauge transformations of terms in the action that only depend on Yang–Mills fields). The covariant Yang–Mills anomalies due to hexagon graphs with a gaugino in the loop are proportional to $\mathrm{Tr}\tilde{F}^6$. We have already discussed the fact that for $SO(32)$ the anomaly factorizes into $(\mathrm{tr}\, \tilde{F}^2)(\mathrm{tr}\, \tilde{F}^4)$ where the trace tr is now over the defining (vector) representation. Hence, we start from $I_{12} = \mathrm{tr}F^2\mathrm{tr}F^4$ and obtain the Chern–Simons term in 11 dimensions by extracting d,

$$\omega^{(0)}_{11Y} = \alpha\omega^{(0)}_{3Y}\mathrm{tr}F^4 + (1-\alpha)\omega^{(0)}_{7Y}\mathrm{tr}F^2. \tag{6.283}$$

Bose symmetry sets $\alpha = \frac{4}{12}$ and $(1-\alpha) = \frac{8}{12}$, but this time we keep α to see where it ends up. The consistent anomaly is then

$$G_{cons} = \int dx \Big[\alpha\omega^{(1)}_{2Y}\mathrm{tr}F^4 + (1-\alpha)\omega^{(1)}_{6Y}\mathrm{tr}F^2\Big]. \tag{6.284}$$

The counterterm is now of the form

$$\Delta\mathcal{L}_{YM} = aB[\mathrm{tr}F^4 + b(\mathrm{tr}F^2)^2] + c\omega^{(0)}_{3Y}\omega^{(0)}_{7Y} \tag{6.285}$$

where the constants a, b and c are to be determined. Variation yields

$$\begin{aligned}\delta_{gauge}\,\Delta\mathcal{L}_{YM} &= \Big\{-a\omega^{(1)}_{2Y}[(\mathrm{tr}F^4 + b(\mathrm{tr}F^2)^2] + c\left[d\omega^{(1)}_{2Y}\,\omega^{(0)}_{7Y} + \omega^{(0)}_{3Y}\,d\omega^{(1)}_{6Y}\right]\Big\}\\ &= \Big\{-a\omega^{(1)}_{2Y}[\mathrm{tr}F^4 + b(\mathrm{tr}F^2)^2] - c\omega^{(1)}_{2Y}\mathrm{tr}F^4 + c\omega^{(1)}_{6Y}\mathrm{tr}F^2\Big\}\end{aligned} \tag{6.286}$$

where we have used the fact $\delta_{gauge}B = -\omega^{(1)}_{2Y}$. Hence, for $a + c = \alpha$, $c = \alpha - 1$ and $b = 0$, the Yang–Mills anomalies are also canceled. These equations have a solution for any α. Hence, the purely gauge (Yang–Mills) anomalies can be canceled for $SO(32)$. In Appendix F we obtain the same result for $E_8 \times E_8$.

Cancellation of mixed anomalies: the complete counterterm

Finally, we consider mixed anomalies. In fact, as explained before, we consider all anomalies together. The covariant anomalies come from $An_{3/2}(\psi_\mu)$, $An_{1/2}(\chi)$ and $An_{1/2}(\lambda)$, but only $An_{1/2}(\lambda)$ depends on $\tilde{F}$. The purely gravitational anomalies are recorded in (6.225)–(6.227). We need to multiply the result for $An_{1/2}$ in (6.225) by $\mathrm{Tr}\,e^{\tilde{F}} - 1$, where the -1 refers to the dilatino. We absorb a factor of $-i/2\pi$ into each $\tilde{R}$ and $\tilde{F}$. The total result for the terms which contribute to the anomaly in 10 dimensions reads as

$$\begin{aligned}&An_{3/2}(\psi_\mu) + An_{1/2}(\chi) + An_{1/2}(\lambda)\\ &\quad = \int dx\left[\left(-\frac{4}{45} + \frac{9}{5670}\right)\mathrm{tr}(\tilde{R}/8\pi)^6\right.\\ &\qquad + \left(\frac{1}{20} + \frac{9}{4320}\right)\mathrm{tr}(\tilde{R}/8\pi)^2\mathrm{tr}(\tilde{R}/8\pi)^4\end{aligned}$$

$$
\begin{aligned}
&+\left(-\frac{1}{144}+\frac{9}{10\,368}\right)[\mathrm{tr}(\tilde{R}/8\pi)^2]^3 \\
&+\left\{1+\frac{1}{12}\mathrm{tr}(\tilde{R}/8\pi)^2+\frac{1}{360}\mathrm{tr}(\tilde{R}/8\pi)^4+\frac{1}{288}[\mathrm{tr}(\tilde{R}/8\pi)^2]^2\right. \\
&+\frac{1}{5670}\mathrm{tr}(\tilde{R}/8\pi)^6+\frac{1}{4320}\mathrm{tr}(\tilde{R}/8\pi)^4\mathrm{tr}(\tilde{R}/8\pi)^2 \\
&\left.+\frac{1}{10\,368}[\mathrm{tr}(\tilde{R}/8\pi)^2]^3\right\} \\
&\times\left[(\dim G-1)-\frac{1}{2}\mathrm{Tr}(\tilde{F}/4\pi)^2+\frac{1}{24}\mathrm{Tr}(\tilde{F}/4\pi)^4\right. \\
&\left.-\frac{1}{720}\mathrm{Tr}(\tilde{F}/4\pi)^6\right]\Bigg]. \qquad (6.287)
\end{aligned}
$$

The -1 in $(\dim G-1)$ accounts for $An_{1/2}(\chi)$. The mixed anomalies involve $\mathrm{Tr}\tilde{F}^2$ and $\mathrm{Tr}\tilde{F}^4$. Substituting $\dim G = 496$, but not yet using any other properties of the gauge group, yields for the total covariant anomaly in ten dimensions

$$
\begin{aligned}
An(\mathrm{total}) = \frac{1}{(4\pi)^6}\int dx\,\frac{1}{48}&\left\{\frac{1}{8}\mathrm{tr}\tilde{R}^2\mathrm{tr}\tilde{R}^4+\frac{1}{32}(\mathrm{tr}\tilde{R}^2)^3\right. \\
&-\left[\frac{1}{240}\mathrm{tr}\tilde{R}^4+\frac{1}{192}(\mathrm{tr}\tilde{R}^2)^2\right]\mathrm{Tr}\tilde{F}^2 \\
&\left.+\frac{1}{24}\mathrm{tr}\tilde{R}^2\mathrm{Tr}\tilde{F}^4-\frac{1}{15}\mathrm{Tr}\tilde{F}^6\right\}. \qquad (6.288)
\end{aligned}
$$

In order for the total set of anomalies to be canceled by a counterterm involving the field B, the anomaly must factorize as follows:

$$
An(\mathrm{total}) = \frac{1}{(4\pi)^6}\int dx\,\frac{1}{48}(\mathrm{tr}\tilde{R}^2+a\mathrm{tr}\tilde{F}^2)X \qquad (6.289)
$$

where a is a constant and X is a polynomial in $\tilde{R}$ and $\tilde{F}$. This is only possible if $\mathrm{Tr}\tilde{F}^6$ factorizes: it should be possible to write it as a linear combination of $\mathrm{Tr}\tilde{F}^4\,\mathrm{Tr}\tilde{F}^2$ and $(\mathrm{Tr}\tilde{F}^2)^3$,

$$
\mathrm{Tr}\tilde{F}^6 = b\,\mathrm{Tr}\tilde{F}^4\,\mathrm{Tr}\tilde{F}^2 + c\,(\mathrm{Tr}\tilde{F}^2)^3. \qquad (6.290)
$$

Note that so far we have only been dealing with traces Tr over the adjoint representation. Later we shall express the traces $\mathrm{Tr}F^4$ and $\mathrm{Tr}F^2$ for $SO(32)$ in terms of the traces $\mathrm{tr}F^4$ and $\mathrm{tr}F^2$ over the defining representation. For $E_8 \times E_8$ all results will be given in the adjoint representation, because in that case the adjoint representation is equal to the defining representation.

The purely gravitational and purely Yang–Mills terms can always be factorized when (6.290) holds, and then the anomaly must be of the form

$$An \sim (\mathrm{tr}\tilde{R}^2 + a\mathrm{Tr}\tilde{F}^2)\left[\frac{1}{8}\mathrm{tr}\tilde{R}^4 + \frac{1}{32}(\mathrm{tr}\tilde{R}^2)^2 + d\,\mathrm{tr}\tilde{R}^2\mathrm{Tr}\tilde{F}^2 \right.$$
$$\left. - \frac{b}{15a}\mathrm{Tr}\tilde{F}^4 - \frac{c}{15a}(\mathrm{Tr}\tilde{F}^2)^2\right]. \tag{6.291}$$

Note that a new constant appears in X, namely d.

In order for this formula to also correctly reproduces the cross terms, the following conditions should be satisfied:

$$\begin{aligned}
&R^4F^2 \text{ terms:} && -\frac{1}{240} = \frac{a}{8} && \Rightarrow\ a = -\frac{1}{30}\\
&R^2F^4 \text{ terms:} && -\frac{1}{24} = -\frac{b}{15a} && \Rightarrow\ b = \frac{1}{48}\\
&(R^2)^2F^2 \text{ terms:} && -\frac{1}{192} = \frac{a}{32} + d && \Rightarrow\ d = -\frac{1}{240}\\
&R^2(F^2)^2 \text{ terms:} && 0 = -\frac{c}{15a} + ad && \Rightarrow\ c = -\frac{1}{(120)^2}.
\end{aligned} \tag{6.292}$$

Hence, anomaly cancellation is only possible if

$$\mathrm{Tr}\tilde{F}^6 = \frac{1}{48}\mathrm{Tr}\tilde{F}^4\,\mathrm{Tr}\tilde{F}^2 - \frac{1}{120^2}(\mathrm{Tr}\tilde{F}^2)^3. \tag{6.293}$$

If this relation holds, factorization of the anomaly is possible, and the anomaly is given by

$$An = \frac{1}{(4\pi)^6}\int dx\left(\mathrm{tr}\tilde{R}^2 - \frac{1}{30}\mathrm{Tr}\tilde{F}^2\right)\left[\frac{1}{8}\mathrm{tr}\tilde{R}^4 + \frac{1}{32}(\mathrm{tr}\tilde{R}^2)^2\right.$$
$$\left. - \frac{1}{240}\mathrm{tr}\tilde{R}^2\mathrm{Tr}\tilde{F}^2 + \frac{1}{24}\mathrm{Tr}\tilde{F}^4 - \frac{1}{7200}(\mathrm{Tr}\tilde{F}^2)^2\right]. \tag{6.294}$$

We now specialize to the case $SO(32)$ for which

$$\begin{aligned}
\mathrm{Tr}F^2 &= 30\mathrm{tr}F^2\\
\mathrm{Tr}F^4 &= 24\mathrm{tr}F^4 + 3(\mathrm{tr}F^2)^2\\
\mathrm{Tr}F^6 &= 15\mathrm{tr}F^4\mathrm{tr}F^2.
\end{aligned} \tag{6.295}$$

Expressing $\mathrm{Tr}F^6$ in terms of $\mathrm{Tr}F^4$ and $\mathrm{Tr}F^2$, one finds that (6.290) is satisfied. In terms of the traces over the fundamental representation (6.294) reduces to

$$I_{12} = \frac{1}{(4\pi)^6} \int dx \left\{ \frac{1}{48} \left(\mathrm{tr}\tilde{R}^2 - \mathrm{tr}\tilde{F}^2 \right) \right.$$
$$\left. \times \left[\mathrm{tr}\tilde{F}^4 - \frac{1}{8}\mathrm{tr}\tilde{F}^2\mathrm{tr}\tilde{R}^2 + \frac{1}{8}\mathrm{tr}\tilde{R}^4 + \frac{1}{32}(\mathrm{tr}\tilde{R}^2)^2 \right] \right\}. \qquad (6.296)$$

Note that no term with $(\mathrm{tr}\tilde{F}^2)^3$ are present. The consistent anomaly is thus

$$G_{cons} \sim \int dx \left\{ \beta(\omega^{(1)}_{2Y} - \omega^{(1)}_{2L}) \left[\mathrm{tr}F^4 - \frac{1}{8}\mathrm{tr}F^2\mathrm{tr}R^2 + \frac{1}{8}\mathrm{tr}R^4 + \frac{1}{32}(\mathrm{tr}R^2)^2 \right] \right.$$
$$+ (1-\beta)(\mathrm{tr}F^2 - \mathrm{tr}R^2) \left[\omega^{(1)}_{6Y} - \frac{\alpha}{8}\omega^{(1)}_{2Y}\mathrm{tr}R^2 - \frac{(1-\alpha)}{8}\mathrm{tr}F^2\omega^{(1)}_{2L} \right.$$
$$\left. \left. + \frac{1}{8}\omega^{(1)}_{6L} + \frac{1}{32}\omega^{(1)}_{2L}\mathrm{tr}R^2 \right] \right\} \qquad (6.297)$$

where α and β are free parameters.

We can now construct the counterterm. We have a two-parameter solution, depending on α and β, but using Bose symmetry we set $\alpha = \frac{1}{2}$ and $\beta = \frac{1}{3}$. Then the counterterm is given by

$$\Delta\mathcal{L}_{total} \sim \frac{1}{3}B \left[\mathrm{tr}F^4 + -\frac{1}{8}\mathrm{tr}F^2\mathrm{tr}R^2 + \frac{1}{8}\mathrm{tr}R^4 + \frac{1}{32}(\mathrm{tr}R^2)^2 \right]$$
$$+ \frac{2}{3}(\omega_{3Y} - \omega_{3L})X_7 \qquad (6.298)$$

where $dX_7 = X_8$ with X_8 being the coefficient of $\frac{1}{3}B$,

$$X_7 = \omega_{7Y} - \frac{1}{16}\omega_{3Y}\mathrm{tr}R^2 - \frac{1}{16}\mathrm{tr}F^2\omega_{3L} + \frac{1}{8}\omega_{7L} + \frac{1}{32}\omega_{3L}\mathrm{tr}R^2. \quad (6.299)$$

Variation of this counterterm indeed cancels the consistent anomaly. Note that the transformation rule of B is fixed by requiring that anomalies cancel; it reads as $\delta B = \omega^{(1)}_{2L} - \omega^{(1)}_{2Y}$, and, for example, $\delta B = \omega^{(1)}_{2L} + \omega^{(1)}_{2Y}$, would have made anomaly cancellation impossible.

6.10 The $SO(16) \times SO(16)$ string

As a last example of anomaly cancellation in a field theory we consider the $SO(16) \times SO(16)$ heterotic string theory. The massless sector leads to a field theory in 10 dimensions with gauge group $SO(16)\times SO(16)$ and chiral spin-$\frac{1}{2}$ fermions in the $\mathbf{16} \times \mathbf{16}$ vector representation of $SO(16) \times SO(16)$, and further antichiral spin-$\frac{1}{2}$ fermions in the $\mathbf{128}\times\mathbf{1}$ and the $\mathbf{1}\times\mathbf{128}$ spinor representations of $SO(16)\times SO(16)$. Hence, there are again Yang–Mills anomalies and gravitational anomalies, and we shall apply the general formulas to check whether these anomalies cancel. The anomalies are contained in the expressions

$$\left(\mathrm{tr}_{16\times16}\,\mathrm{e}^{-i\tilde{F}/4\pi}-\mathrm{tr}_{128\times1}\,\mathrm{e}^{-i\tilde{F}/4\pi}-\mathrm{tr}_{1\times128}\,\mathrm{e}^{-i\tilde{F}/4\pi}\right)\\ \times\exp\left\{\frac{1}{2}\mathrm{tr}\log\left[\frac{-i\tilde{R}/8\pi}{\sinh(-i\tilde{R}/8\pi)}\right]\right\}. \tag{6.300}$$

Denoting $\mathrm{tr}_{16\times16}-\mathrm{tr}_{128\times1}-\mathrm{tr}_{1\times128}$ by Tr, the relevant terms are

$$An_{total}=(\mathrm{Tr}I)An_{1/2}-\left(\frac{1}{2}\mathrm{Tr}\tilde{F}^2\right)\left(\frac{1}{16}\right)\left[\frac{1}{360}\mathrm{Tr}\tilde{R}^4+\frac{1}{288}(\mathrm{Tr}\tilde{R}^2)^2\right]\\ +\left(\frac{1}{24}\mathrm{Tr}\tilde{F}^4\right)\left(\frac{1}{48}\mathrm{Tr}\tilde{R}^2\right)-\frac{1}{720}\mathrm{Tr}\tilde{F}^6. \tag{6.301}$$

First of all we note that $\mathrm{Tr}I=256-128-128=0$, hence the purely gravitational anomalies cancel: $-32=16(-2)$. Also $\mathrm{Tr}F^2=0$, but to prove this relation, and deduce other relations for $\mathrm{Tr}F^4$ and $\mathrm{Tr}F^6$, we must first express $\mathrm{Tr}F^p$ in terms of $\mathrm{tr}_{16}F^p$, where by $\mathrm{tr}_{16}F^p$ we mean the trace over the defining vector representation of $SO(16)$. We shall now first derive expressions for $\mathrm{Tr}F^2$, $\mathrm{Tr}F^4$ and $\mathrm{Tr}F^6$, and then return to the issue of whether one can find a counterterm to cancel anomalies.

To begin with, consider the spinor representation of $SO(16)$, denoted by its dimension **128**. We claim that

$$\mathrm{tr}_{128}F^2=16\mathrm{tr}_{16}F^2. \tag{6.302}$$

To show this, we consider the generator A for a rotation in the x–y plane, corresponding to $A=\frac{1}{2}\gamma_1\gamma_2$ in the spinor representation,

$$A_{16}=\begin{pmatrix}0&1&0&\cdot\\-1&0&0&\cdot\\0&0&0&\cdot\\\cdot&\cdot&\cdot&\cdot\end{pmatrix}_{16\times16}\qquad A_{128}=\frac{1}{2}\gamma^{12}. \tag{6.303}$$

Since chiral spinors of $SO(16)$ have a $\frac{1}{2}\times256=128$ component, while $\mathrm{tr}A_{16}^2=-2$ and $\mathrm{tr}A_{128}^2=-\frac{1}{4}128$, we see that (6.302) holds: $-32=16(-2)$. Consider next $\mathrm{Tr}F^2$. It contains $\mathrm{tr}_{16\times16}F^2=16(\mathrm{tr}_{16}F_1^2+\mathrm{tr}_{16}F_2^2)$ and $\mathrm{tr}_{128\times1}F^2=16\mathrm{tr}_{16}F_1^2$ and $\mathrm{tr}_{1\times128}F^2=16\mathrm{tr}_{16}F_2^2$. Thus, indeed $\mathrm{Tr}F^2=0$.

We turn to the expression $\mathrm{Tr}F^4$. It contains the following contributions:

$$\mathrm{tr}_{16\times16}F^4=16\mathrm{tr}_{16}F_1^4+16\mathrm{tr}_{16}F_2^4+6\mathrm{tr}_{16}F_1^2\mathrm{tr}_{16}F_2^2. \tag{6.304}$$

Note that $F = F_1 \otimes I_2 + I_1 \otimes F_2$, hence F^4 contains cross terms

$$\mathrm{tr}_{128\times 1} F^4 = 16\mathrm{tr}_{128} F_1^4, \quad \mathrm{tr}_{1\times 128} F^4 = 16\mathrm{tr}_{128} F_2^4. \qquad (6.305)$$

To proceed we must express $\mathrm{tr}_{128} F^4$ in terms of $\mathrm{tr}_{16} F^4$. We do this as follows: we assume

$$\mathrm{tr}_{128} F^4 = a\mathrm{tr}_{16} F^4 + b(\mathrm{tr}_{16} F^2)^2 \qquad (6.306)$$

and evaluate these expression for two suitable generators. The first choice is obviously given by (6.303). The second choice of a suitable generator is the simultaneous rotation in the x–y and x–z planes,

$$A'_{16} = \begin{pmatrix} 0 & 1 & 1 \\ -1 & 0 & 0 \\ -1 & 0 & 0 \end{pmatrix}_{16\times 16} \qquad A'_{128} = \frac{1}{2}\gamma_1\gamma_2 + \frac{1}{2}\gamma^1\gamma^3. \qquad (6.307)$$

The first generator in (6.303) satisfies

$$(A_{128})^4 = \frac{1}{16}, \quad (A_{16})^2 = -I, \quad (A_{16})^4 = I \qquad (6.308)$$

and (6.306) yields $\frac{1}{16} \times 128 = 2a + 4b$. The second generator satisfies $(A'_{128})^2 = -\frac{1}{2}$, hence

$$(A'_{128})^4 = \frac{1}{4}, \quad (A'_{16})^2 = \begin{pmatrix} -2 & 0 & 0 \\ 0 & -1 & -1 \\ 0 & -1 & -1 \end{pmatrix}_{16\times 16}$$

$$(A'_{16})^4 = \begin{pmatrix} 4 & 0 & 0 \\ 0 & 2 & 0 \\ 0 & 2 & 0 \end{pmatrix}_{16\times 16} \quad . \qquad (6.309)$$

Then (6.306) yields $\frac{1}{4} \times 128 = 8a + 16b$. From these relations one finds a and b

$$\mathrm{tr}_{128} F^4 = -8\mathrm{tr}_{16} F^4 + 6(\mathrm{tr}_{16} F^2)^2. \qquad (6.310)$$

From here it is easy to obtain $\mathrm{Tr}F^4$ in terms of $\mathrm{tr}_{16} F^4$ and $\mathrm{tr}_{16} F^2$. Namely,

$$\begin{aligned} \mathrm{Tr}F^4 &= \mathrm{tr}_{16\times 16} F^4 - \mathrm{tr}_{128} F_1^4 - \mathrm{tr}_{128} F_2^4 \\ &= 24\mathrm{tr}_{16} F_1^4 + 24\mathrm{tr}_{16} F_2^4 - 6(\mathrm{tr}_{16} F_1^2)^2 - 6(\mathrm{tr}_{16} F_2^2)^2 \\ &\quad + 6\,\mathrm{tr}_{16} F_1^2 \mathrm{tr}_{16} F_2^2. \end{aligned} \qquad (6.311)$$

The last expression we need is the relation between $\mathrm{Tr}F^6$ and $\mathrm{tr}F^6$, $\mathrm{tr}F^4$ and $\mathrm{tr}F^2$. We claim that

$$\mathrm{Tr}F^6 = a\mathrm{tr}_{16} F^6 + b\mathrm{tr}_{16} F^4 \mathrm{tr}_{16} F^2 + c(\mathrm{tr}_{16} F^2)^3 \qquad (6.312)$$

with $a = 16$, $b = -15$ and $c = \frac{15}{4}$. This relation again follows by postulating this expression and then evaluating it for three suitable generators. As such we take the rotation in the x–y plane in (6.303), the simultaneous rotation in the x–y and x–z planes given in (6.307), and finally the simultaneous rotation in the x–y, x–z and y–z planes,

$$A''_{16} = \begin{pmatrix} 0 & 1 & -1 \\ -1 & 0 & 1 \\ 1 & -1 & 0 \end{pmatrix}_{16\times 16} \qquad (A''_{16})^2 = \begin{pmatrix} -2 & 1 & -1 \\ 1 & -2 & 1 \\ 1 & 1 & -2 \end{pmatrix}_{16\times 16}$$
$$(A''_{16})^4 = \begin{pmatrix} 6 & -3 & -3 \\ -3 & 6 & -3 \\ -3 & -3 & 6 \end{pmatrix}_{16\times 16}$$
$$A''_{128} = \frac{1}{2}\gamma_1\gamma_2 + \frac{1}{2}\gamma^1\gamma^3 + \frac{1}{2}\gamma^2\gamma^3 \qquad (A''_{128})^4 = -\frac{3}{4}. \tag{6.313}$$

One then finds the following three equations for a, b, c:

$$\left(-\frac{1}{4}\right)^3 128 = a(-2) + b(-4) + c(-8)$$
$$\left(-\frac{1}{2}\right)^3 128 = a(-16) + b(-32) + c(-64)$$
$$\left(-\frac{3}{4}\right)^3 128 = a(-54) + b(-108) + c(-216). \tag{6.314}$$

This yields

$$\mathrm{tr}_{128}F^6 = 16\mathrm{tr}_{16}F^6 - 15\mathrm{tr}_{16}F^4\mathrm{tr}_{16}F^2 + \frac{15}{4}(\mathrm{tr}_{16}F^2)^3. \tag{6.315}$$

From here it is straightforward to evaluate $\mathrm{Tr}F^6$,

$$\begin{aligned}\mathrm{Tr}F^6 &= \mathrm{tr}_{16\times 16}F^6 - \mathrm{tr}_{128}F_1^6 - \mathrm{tr}_{128}F_2^6 \\ &= \left(16\mathrm{tr}_{16}F_1^6 + 15\mathrm{tr}_{16}F_1^4\mathrm{tr}_{16}F_2^2 + 15\mathrm{tr}_{16}F_1^2\mathrm{tr}_{16}F_2^4 + 16\mathrm{tr}_{16}F_2^6\right) \\ &\quad - \Bigg[16\mathrm{tr}_{16}F_1^6 - 15\mathrm{tr}_{16}F_1^4\mathrm{tr}_{16}F_1^2 + \frac{15}{4}(\mathrm{tr}_{16}F_1^2)^3 \\ &\quad + 16\mathrm{tr}_{16}F_2^6 - 15\mathrm{tr}_{16}F_2^4\mathrm{tr}_{16}F_2^2 + \frac{15}{4}(\mathrm{tr}_{16}F_2^2)^3\Bigg] \\ &= 15(\mathrm{tr}_{16}F_1^4 + \mathrm{tr}_{16}F_2^4)(\mathrm{tr}_{16}F_1^2 + \mathrm{tr}_{16}F_2^2) \\ &\quad - \frac{15}{4}\left[(\mathrm{tr}_{16}F_1^2)^3 + (\mathrm{tr}_{16}F_2^2)^3\right]. \end{aligned} \tag{6.316}$$

Note that the dangerous nonfactorized terms $\mathrm{tr}_{16}F_1^6$ and $\mathrm{tr}_{16}F_2^6$ have canceled.

Finally, we add up all contributions to the anomalies

$$An_{total} = \left(\frac{1}{24}\mathrm{Tr}\tilde{F}^4\right)\left(\frac{1}{48}\mathrm{Tr}\tilde{R}^2\right) - \frac{1}{720}\mathrm{Tr}\tilde{F}^6. \tag{6.317}$$

We expect this expression to factorize. Indeed, it does,

$$An_{total} = \frac{1}{24 \times 48}\left(\mathrm{Tr}\tilde{R}^2 - \mathrm{tr}_{16}\tilde{F}_1^2 - \mathrm{tr}_{16}\tilde{F}_2^2\right)\mathrm{tr}_{16}\tilde{F}^4. \tag{6.318}$$

From here on, we follow the same path as before: we omit the twiddles in An_{total} to obtain I_{12}, extract an exterior derivative d to obtain the 11-dimensional Chern–Simons term, and vary it to obtain the integrand for the consistent anomaly. One expression for the latter is

$$\begin{aligned} G = &\frac{1}{3}c\int dx\,\left[\omega_{2L}^{(1)} - \omega_{2Y_1}^{(1)} - \omega_{2Y_2}^{(1)}\right]\mathrm{Tr}F^4 \\ &+\frac{2}{3}c\int dx\,\left(\mathrm{tr}R^2 - \mathrm{tr}_{16}F_1^2 - \mathrm{tr}_{16}F_2^2\right)\omega_{6Y}^{(1)} \end{aligned} \tag{6.319}$$

where $\mathrm{Tr}F^4 = dX_7$ and $\delta X_7 = d\omega_{6Y}^{(1)}$. The last term is rewritten as $\frac{2}{3}c\int dx\,[\omega_{3L}^{(0)} - \omega_{3Y_1}^{(0)} - \omega_{3Y_2}^{(0)}]\,\delta X_7$. The counterterm whose variation is $-G$, and thus cancels the anomaly, is given by

$$\Delta\mathcal{L}_{total} = B\mathrm{Tr}F^4 - \frac{2}{3}\left[\omega_{3L}^{(0)} - \omega_{3Y_1}^{(0)} - \omega_{3Y_2}^{(0)}\right]X_7. \tag{6.320}$$

Thus the gravitational, Yang–Mills and mixed anomalies of the Majorana–Weyl fermions in the $SO(16) \times SO(16)$ string can be canceled by a suitable counterterm. This is a nontrivial result because this string is not finite (there are infrared divergences due to dilaton tadpoles). However, it is modular invariant (large diffeomorphism anomalies on the worldsheet cancel).

6.11 Index theorems and path integrals

The rigid $U(1)$ chiral anomaly for a Dirac fermion has a beautiful mathematical interpretation as the index of the Dirac operator. In this section we use the technology developed in Part I of this book to give a path integral proof of the original Atiyah–Singer index theorem [167], following the works of Alvarez-Gaumé [35] and Friedan and Windey [38].

The Atiyah–Singer index theorem equates the index of the Dirac operator defined on a compact manifold with Euclidean signature to a topological invariant. The Dirac operator contains the minimal coupling of a Dirac fermion to external gravitational and gauge fields. The topological invariant is expressed as the integral of a polynomial in terms of Riemann curvatures and gauge field strengths. This polynomial is exactly the expression for the $U(1)$ chiral anomaly of the Dirac field.

The index of an operator O is an integer number, defined as the difference between the number of zero modes (eigenvectors with zero eigenvalue, so solutions of the equation $O\psi = 0$) of the operator O and the number of zero modes of its adjoint $O^\dagger$. In supersymmetric theories Witten has also introduced the concept of an index, defined as the number of bosonic states minus the number of fermionic states [168]. Owing to a special property of supersymmetry, which pairs together bosonic and fermionic states of nonvanishing energy, this index also counts the number of bosonic zero-energy states minus the number of fermionic zero-energy states (and if this index is nonvanishing, the existence of at least one zero-energy state is guaranteed, so that supersymmetry is not spontaneously broken). As we will soon describe, the index of the Dirac operator can be identified with the Witten index of a suitable supersymmetric quantum mechanical system. The latter can be computed using the path integral description of this system. This provides a nice derivation of the Atiyah–Singer index theorem.

We start by recalling the definition of the Witten index. Let us consider a quantum mechanical system with hermitian conserved charges H, Q and $(-1)^F$, where H is the Hamiltonian, Q is the supersymmetry charge and $(-1)^F$ is the fermion parity operator (F is the fermion number operator, but we only need the parity operator $(-1)^F$ which assigns the values $+1$ to bosonic states and -1 to fermionic states). These operators satisfy the $N = 1$ algebra

$$Q^2 = H \tag{6.321}$$

and

$$\{(-1)^F, Q\} = 0, \quad [(-1)^F, H] = 0. \tag{6.322}$$

(Other (anti)commutators defining the superalgebra vanish.)

One immediate consequence of this algebra is that the energy cannot be negative. In fact, using $H = Q^2$ and positive definiteness of the physical Hilbert space (we use energy eigenstates $|E\rangle$ normalized to $||E\rangle|^2 = 1$), one can compute

$$E = \langle E|H|E\rangle = |Q|E\rangle|^2 \geq 0. \tag{6.323}$$

From this result one also deduces that the zero-energy states $|0, i\rangle$ labeled by the index i are also vacuum states (states with the lowest energy) and supersymmetric, i.e. annihilated by the supersymmetry charge, $Q|0, i\rangle = 0$. In addition, one can deduce that states with positive energy $E > 0$ are degenerate. Namely, for every bosonic eigenstate $|E, b\rangle$ (which by definition satisfies $H|E, b\rangle = E|E, b\rangle$ and $(-1)^F|E, b\rangle = |E, b\rangle$) there exists a fermionic state

$$|E, f\rangle \equiv \frac{1}{\sqrt{E}} Q|E, b\rangle \tag{6.324}$$

which satisfies $H|E, f\rangle = E|E, f\rangle$ and $(-1)^F|E, f\rangle = -|E, f\rangle$, and vice versa.

The Witten index is defined as $\mathrm{Tr}(-1)^F$. Since positive energy states cancel pairwise in the trace, this index counts the number of bosonic zero-energy states ($n_b^{E=0}$) minus the number of fermionic zero-energy states ($n_f^{E=0}$). It can be regulated by smoothly cutting off states with sufficiently high energy

$$\mathrm{Tr}\,(-1)^F = \mathrm{Tr}\,(-1)^F \mathrm{e}^{-\beta H} = n_b^{E=0} - n_f^{E=0} \tag{6.325}$$

but since the result does not depend on the regulating parameter β, one may use the latter to find simple ways of computing the index. Under smooth deformations of the parameters of the theory, states can only reach or leave zero energy in pairs. Thus the Witten index is expected to be a topological invariant which characterizes the supersymmetric quantum theory.

Let us now consider the hermitian Dirac operator coupled to a non-abelian gauge field and to gravity in n even dimensions,

$$i\not{D} \equiv i\gamma^m e_m{}^\mu \left(\partial_\mu + \frac{1}{4}\omega_{\mu pq}\gamma^p\gamma^q + A_\mu^\alpha T_\alpha\right) \tag{6.326}$$

where T_α are the antihermitian generators of the gauge group. The chirality matrix γ_5 squares to one and can always be chosen to be diagonal, so that the Dirac operator must be off-diagonal since it anticommutes with γ_5,

$$\gamma_5 = \begin{pmatrix} 1 & 0 \\ 0 & -1 \end{pmatrix}, \quad i\not{D} = \begin{pmatrix} 0 & iD_R \\ iD_L & 0 \end{pmatrix}. \tag{6.327}$$

Note that iD_L maps left-handed spinors λ_L ($\gamma_5\lambda_L = \lambda_L$) to right-handed spinors λ_R ($\gamma_5\lambda_R = -\lambda_R$), while $iD_R = (iD_L)^\dagger$ does the opposite. The index of the Dirac operator is now defined as the number of zero modes of iD_L minus the number of zero modes of its hermitian conjugate iD_R,

$$I = \dim\,\ker\, iD_L - \dim\,\ker\, iD_R \tag{6.328}$$

(the mathematical notation "dim ker iD_L" denotes the dimension of the kernel of the operator iD_L, and thus gives the number of its zero modes).[32]

[32] Mathematically one prefers to define the index by $I = \dim\,\ker\, D_R D_L - \dim\,\ker\, D_L D_R$, because these operators do not map between different spaces but act in a given space. However, $D_L = 0$ is equivalent to $D_R D_L = 0$ and $D_R = 0$ is equivalent to $D_L D_R = 0$, because $(iD_L)^\dagger = iD_R$. Hence the two definitions are equivalent.

Since eigenvectors with nonzero eigenvalues always appear pairwise and have opposite chirality (in fact, γ_5 anticommutes with $i\not{D}$, and $i\not{D}$ is used to map eigenvectors of one chirality to eigenvectors of the other chirality), the index can be written as the trace of the chirality operator γ_5 on the full spinor space. In addition, it can be regulated using an arbitrary parameter β as

$$I = \mathrm{Tr}\,\gamma_5 = \mathrm{Tr}\,\gamma_5\,\mathrm{e}^{-\frac{1}{2}\beta(i\not{D})^2}. \tag{6.329}$$

It is now clear that the susy algebra described earlier is realized by the Dirac system as follows:

$$Q = \frac{i}{\sqrt{2}}\not{D}, \quad H = \frac{1}{2}(i\not{D})^2, \quad (-1)^F = \gamma_5 \tag{6.330}$$

and the index of the Dirac operator can be identified with the Witten index of the corresponding supersymmetric quantum mechanical model

$$I = \mathrm{Tr}\,\gamma_5\,\mathrm{e}^{-\frac{1}{2}\beta(i\not{D})^2} = \mathrm{Tr}\,(-1)^F\,\mathrm{e}^{-\beta H}. \tag{6.331}$$

One can now use the path integral representation of the supersymmetric model and calculate the index in the perturbatively accessible limit $\beta \to 0$

$$I = \mathrm{Tr}\,(-1)^F\,\mathrm{e}^{-\beta H} = \int_{PBC} DxD\psi \int_{ABC,\,1p.s.} D\bar{c}Dc\,\mathrm{e}^{-S} \tag{6.332}$$

where as described earlier (see Sections 5.2, 6.1, 6.2, and also Section 6.4 and Appendix E) $(-1)^F$ turns the natural antiperiodic boundary conditions of the worldline fermions ψ into periodic ones. The trace over the ghost fields $\bar{c}, c$ needed to create the gauge group quantum numbers is restricted to the one-particle states ($1\,p.s.$), and the action of the model with H as Hamiltonian and Q as susy charge is given by

$$\begin{aligned} S = \int_{-\beta}^{0} dt \Big\{ &\frac{1}{2}[g_{ij}(x)\dot{x}^i\dot{x}^j + i\psi_a\dot{\psi}^a + i\dot{x}^i\omega_{iab}\psi^a\psi^b] \\ &+ i\bar{c}_A\dot{c}^A + i\dot{x}^i A_i^\alpha(x)\bar{c}_A(T_\alpha)^A{}_B c^B \Big\}. \end{aligned} \tag{6.333}$$

The final result can be read off by combining the results from Sections 6.1 and 6.2 for constant $U(1)$ parameter α,

$$I = \frac{(-i)^{n/2}}{(2\pi)^{n/2}} \int \mathrm{Tr}\,\mathrm{e}^F \det{}^{1/2}\Big(\frac{R/2}{\sinh R/2}\Big) \tag{6.334}$$

where the differential forms F and R are defined by

$$F \equiv \frac{1}{2}F_{\mu\nu}\,dx^\mu\,dx^\nu, \quad R \equiv R^\alpha{}_\beta = \frac{1}{2}R^\alpha{}_{\beta\mu\nu}\,dx^\mu\,dx^\nu. \tag{6.335}$$

To obtain the index of the Dirac operator in n dimensions one should expand the integrand and extract the differential n-form that can be integrated over the n-dimensional manifold (the "top-form").

In the mathematical literature one often introduces the Chern character

$$\mathrm{Ch}(F) \equiv \mathrm{Tr}\, \mathrm{e}^{F} \tag{6.336}$$

and the Dirac genus (also called the A-roof genus)

$$\hat{A}(R) \equiv \det{}^{1/2}\left(\frac{R/2}{\sinh R/2}\right). \tag{6.337}$$

The Dirac index can then be written in the following compact form:

$$I = \frac{1}{(2\pi i)^{n/2}} \int \mathrm{Ch}(F)\hat{A}(R). \tag{6.338}$$

Index theorems related to other differential operators can be derived along similar lines using suitable supersymmetric systems. For example, using an $N=2$ system one can obtain index theorems for the Euler number and the Hirzebruch signature [35]. In fact, these other index theorems can also be deduced from the index theorem for the Dirac field, specifying suitably the nature of the gauge field and reinterpreting the internal gauge quantum numbers as additional Lorentz quantum numbers [38].

Any regularization scheme may be used to compute the index densities. The complete result is saturated at one-loop, and at this level calculations are identical in any regularization scheme developed so far, since different counterterms start contributing only at two-loops. The susy algebra guarantees that no corrections of order β should arise. This is most easily checked in dimensional regularization, since then the counterterm is covariant [89]. However, time slicing has also been used for the same purpose [169]. These checks can be interpreted as tests on the correctness of the regularized path integrals for supersymmetric quantum mechanical systems.

We conclude with a brief sketch of the derivation of the index theorem related to the Gauss–Bonnet formula (the latter expresses the Euler number $\chi(M)$ of a compact manifold M as the integral of a local expression constructed from Riemann curvatures). We shall use the $N=2$ model described in Appendix D to obtain this relation. The $N = 2$ Euclidean action reads as

$$S = \int_{-\beta}^{0} dt\, \left[\frac{1}{2} g_{ij}\dot{x}^i\dot{x}^j + \bar{\psi}_a(\dot{\psi}^a + \dot{x}^i\omega_i{}^{ab}\psi_b) + \frac{1}{4}R_{abcd}\bar{\psi}^a\bar{\psi}^b\psi^c\psi^d\right]. \tag{6.339}$$

The Hilbert space of this model is easily constructed by canonical methods and it is given by the sum of all differential forms (or by a bispinor if one uses the Majorana basis ψ_1^a and ψ_2^a). Let us quickly explain this statement. Consider first the flat case, so that indices $i, j, \ldots$ and $a, b, \ldots$ can be identified. The basic (anti)commutation relations of the canonical variables are (we use $\hbar = 1$)

$$[\hat{x}^i, \hat{p}_j] = i\delta^i_j \quad , \quad \{\hat{\psi}^i, \hat{\psi}^\dagger_j\} = \delta^i_j. \tag{6.340}$$

Any state $|\phi\rangle$ of the Hilbert space can then be described by a wave function

$$\phi(x, \psi) \equiv (\langle x| \otimes \langle \psi|)|\phi\rangle \tag{6.341}$$

where $|x\rangle$ are the usual position eigenstates and $\langle\psi|$ are fermionic coherent bra states for the annihilation operator (i.e. left eigenstates of the annihilation operator)

$$\langle\psi| = \langle\Omega|\mathrm{e}^{\psi^i \hat{\psi}^\dagger_i}, \quad \langle\Omega|\hat{\psi}^i = \hat{\psi}^\dagger_i|\Omega\rangle = 0 \tag{6.342}$$

where ψ_i are complex Grassmann variables, $\hat{\psi}^i$ and $\hat{\psi}^\dagger_i$ are operators, and $|\Omega\rangle$ is the completely filled Fock vacuum. These coherent states satisfy

$$\langle\psi|\hat{\psi}^i = \langle\psi|\psi^i. \tag{6.343}$$

Since ψ are Grassmann variables, any wave function has the following finite expansion:

$$\phi(x, \psi) = \phi(x) + \phi_i(x)\psi^i + \frac{1}{2}\phi_{i_1 i_2}(x)\psi^{i_1}\psi^{i_2} + \cdots + \frac{1}{n!}\phi_{i_1 \ldots i_n}(x)\psi^{i_1}\cdots\psi^{i_n}. \tag{6.344}$$

Thus, we see that the Hilbert space is described by the collection of all possible differential forms of the n-dimensional space (ψ^i play the role of dx^i). On such wave functions the basic charges of the susy algebra ($\{\hat{Q}, \hat{Q}^\dagger\} = 2\hat{H}$) are realized as differential operators,

$$\begin{aligned} H = \frac{1}{2}p_i p^i \quad &\to \quad \hat{H} = -\frac{1}{2}\partial_i\partial^i \\ Q = p_i\psi^i \quad &\to \quad \hat{Q} = -i\psi^i\partial_i \\ \bar{Q} = p_i\bar{\psi}^i \quad &\to \quad \hat{Q}^\dagger = -i\partial_i\frac{\partial}{\partial\psi_i}. \end{aligned} \tag{6.345}$$

The supersymmetry charges $\hat{Q}$ and $\hat{Q}^\dagger$ can be identified with the exterior derivative d and its adjoint $d^\dagger$ (the divergence operator) acting on forms, while the Hamiltonian $\hat{H}$ is proportional to the Laplacian $dd^\dagger + d^\dagger d$. The curved space case brings in the expected covariantization of these formulas. Note that the fermion number $(-1)^F$ separates the forms into even and odd. In fact, one can choose the fermion number operator as $F=\hat{\psi}_i^\dagger\hat{\psi}^i$ and assign a zero fermion number to the state $|\Omega\rangle$. Then by expanding the exponent in the definition of the coherent bra $\langle\psi|$ one can compute

$$\langle x|\otimes\langle\psi|(-1)^F|\phi\rangle = \phi(x) - \phi_i(x)\psi^i + \frac{1}{2}\phi_{i_1 i_2}(x)\psi^{i_1}\psi^{i_2} + \cdots \\ \cdots + (-1)^n\frac{1}{n!}\phi_{i_1\ldots i_n}(x)\psi^{i_1}\cdots\psi^{i_n}. \quad (6.346)$$

The Euler character $\chi(M)$ is given by the alternating sum of the Betti numbers and this corresponds to the difference between the numbers of even and odd **harmonic** forms (forms α that satisfy $(dd^\dagger + d^\dagger d)\alpha = 0$). Thus, it is equal to the Witten index $\mathrm{Tr}\,(-1)^F$ of the $N=2$ model since only zezo-energy states contribute to the trace. Each harmonic form has vanishing energy (is annihilated by H), and contributes $+1$ or -1 to the trace. We may now calculate the regularized trace with a path integral

$$\chi(M) = \mathrm{Tr}\,(-1)^F\,\mathrm{e}^{-\beta H} = \int_{PBC} Dx\, D\psi\, D\bar{\psi}\,\mathrm{e}^{-S}. \quad (6.347)$$

The calculation is carried out at small β by separating the zero modes x_0^i, ψ_0^a, $\bar{\psi}_0^a$ which are strongly coupled from the weakly coupled (small-β) quantum fluctuations, and calculating the one-loop determinants of the quantum fluctuations. The latter, in fact, do not contribute as the determinant of the bosons is exactly compensated by the determinant of the fermions.

Thus, one is left with an integration over the zero modes of the exponential of the only interaction left, the curvature term in (6.339) evaluated on the zero modes,

$$\int \frac{d^n x_0\,\sqrt{g(x_0)}}{(2\pi\beta)^{n/2}}\left(\frac{\beta}{i}\right)^{2n/2} d^n\psi_0\, d^n\bar{\psi}_0\,\mathrm{e}^{-\frac{1}{4\beta}R_{abcd}(x_0)\bar{\psi}_0^a\bar{\psi}_0^b\psi_0^c\psi_0^d}. \quad (6.348)$$

Here we have written the integration of the bosonic zero modes with the standard Feynman measure plus a similar expression for the fermionic zero modes (each fermion zero mode has $(\beta/i)^{1/2}$ as the corresponding Feynman measure, as can be seen by comparing with the time slicing formulas[33]). Expanding the exponent to saturate the Grassmann integrations one obtains a nonvanishing result only in n even dimensions

$$\chi(M) = \frac{(-1)^{n/2}}{(8\pi)^{n/2}}\int d^n x_0\,\sqrt{g(x_0)}\;\epsilon R\ldots R\epsilon \quad (6.349)$$

[33]In the action $\int_{-\beta}^{0} L\,dt = \beta\int_{-1}^{0} L\,d\tau$ the $R\psi^4$ term has a factor of β, but after rescaling the fermions by a factor of $\beta^{-1/2}$ the $R\psi^4$ gets a factor of $1/\beta$ and the total path integral measure of the $N=2$ model then has a factor of $\beta^{n/2}$.

where we have used the notation

$$\epsilon R \ldots R \epsilon \equiv \epsilon^{a_1 a_2 \cdots a_{n-1} a_n} \underbrace{R_{a_1 a_2}{}^{b_1 b_2 \cdots} R_{a_{n-1} a_n}{}^{b_{n-1} b_n}}_{n/2 \text{ terms}} \epsilon_{b_1 b_2 \cdots b_{n-1} b_n}. \quad (6.350)$$

This is indeed the standard expression for the Gauss–Bonnet formula.

Similarly one could use the same $N=2$ model to compute the Hirzebruch signature, which counts the difference between the number of self-dual and the number of antiself-dual harmonic forms. The corresponding index theorem relates this difference to the rigid abelian axial anomaly for self-dual forms, and it is also closely related to the calculation presented in Section 6.7, so it will not be repeated here.

7

Trace anomalies from ordinary and susy quantum mechanics

We now turn to a second class of anomalies, namely the trace anomalies. These are anomalies in the local scale invariance of actions for scalar fields, spin-$\frac{1}{2}$ fields and certain vector and antisymmetric tensor fields (vectors in $n=4$, antisymmetric tensors with two indices in $n=6$, etc.). One needs gauge-fixing terms and ghosts, but the trace anomalies of these vector and antisymmetric tensor fields are independent of the gauge chosen [170]. From a technical point of view, these anomalies are very interesting, because one needs higher loop graphs on the worldline to compute them. In fact, due to the β dependence of the measure of the quantum mechanical path integrals, $A=(2\pi\hbar\beta)^{-n/2}$, one needs $(\frac{1}{2}n+1)$-loop calculations in quantum mechanics for the one-loop trace anomalies of n-dimensional quantum field theories. Already in two dimensions one needs two-loop graphs and in four dimensions three-loop graphs. Another interesting technical point regards the fermions. In the path integral they now have antiperiodic boundary conditions. Originally we devised a path integral approach in which fermions were still treated by an operator formalism and in which actions were operator-valued [40]. Here we shall instead present a complete path integral approach, with ordinary actions, in which the fermions are described in the path integral by Grassmann fields. Contrary to the case of chiral anomalies, we shall not use a background field formalism for the fermions because the background fermions are constant and thus cannot accomodate anti-periodic boundary conditions. Instead we shall directly use fermionic quantum fields with anti-periodic boundary conditions. The results we find agree with the results in the literature for trace anomalies obtained by different methods (see [139], for example).

We shall separately discuss the anomalies for spin-0, spin-$\frac{1}{2}$ and spin-1 fields. In these calculations we use the decomposition $x^i(\tau)=x_0^i+q^i(\tau)$

with $q^i(\tau)$ vanishing at the endpoints $\tau = 0$ and $\tau = -1$. We end this chapter by discussing another approach, sometimes called the string-inspired approach, in which one again sets $x^i(\tau) = x_0^i + q^i(\tau)$, but now with $\int_{-1}^0 d\tau\, q^i(\tau)=0$. In fact, we generalize to $\int_{-1}^0 d\tau\, \rho(\tau)q^i(\tau)=0$, where $\int_{-1}^0 d\tau\, \rho(\tau)=1$. This leads to a shift symmetry which is treated with BRST methods and yields new ghost vertices.

7.1 Trace anomalies for scalar fields in two and four dimensions

The classical action of a massless real scalar field φ in n dimensions, which we take with a Euclidean signature for definiteness, is Weyl invariant after one adds a so-called improvement term to the action,

$$S = \int d^n x\, \sqrt{g}\frac{1}{2}\left(g^{\mu\nu}\partial_\mu\varphi\partial_\nu\varphi - \xi R\varphi^2\right), \qquad \xi = \frac{(n-2)}{4(n-1)}$$
$$\delta_W\varphi(x) = \frac{1}{4}(2-n)\sigma(x)\varphi(x), \qquad \delta_W g_{\mu\nu}(x) = \sigma(x)g_{\mu\nu}(x). \tag{7.1}$$

The transformation laws are easily derived by first considering the case where σ is constant. For local $\sigma(x)$ the proof of the Weyl invariance is straightforward if one uses $\delta_W R = -\sigma R + (n-1)D^\mu\partial_\mu\sigma$, see Appendix A. In the path integral for the quantum fields one integrates over $\tilde{\varphi} = g^{1/4}\varphi$ and then obtains a functional Z of the metric

$$Z[g_{\mu\nu}] = \int \mathcal{D}\tilde{\varphi}\ \mathrm{e}^{-\frac{1}{\hbar}S[\tilde{\varphi}, g_{\mu\nu}]}. \tag{7.2}$$

It is easy to check that the field $\tilde{\varphi}$ transforms under Weyl rescaling as $\delta_W\tilde{\varphi} = \frac{1}{2}\sigma\tilde{\varphi}$ in any dimension. Under an infinitesimal local scale transformation of the metric $g_{\mu\nu} \to g_{\mu\nu} + \delta_W g_{\mu\nu}$, combined with a compensating change of the integration variable $\tilde{\varphi} \to \tilde{\varphi} + \delta_W\tilde{\varphi}$ such that the action remains invariant, one obtains the following Weyl anomaly from the Jacobian:

$$An_W(\text{spin } 0) \equiv \int d^n x\, (\delta_W g^{\mu\nu})\frac{\delta}{\delta g^{\mu\nu}}\ln Z[g] = \frac{1}{2\hbar}\int d^n x\, \sqrt{g}\,\sigma g^{\mu\nu}\langle T_{\mu\nu}\rangle$$
$$= \lim_{\beta\to 0}\mathrm{Tr}\frac{\partial\delta_W\tilde{\varphi}}{\partial\tilde{\varphi}}\mathrm{e}^{-\beta\mathcal{R}} = \lim_{\beta\to 0}\mathrm{Tr}\,\frac{1}{2}\sigma\mathrm{e}^{-\beta\mathcal{R}}. \tag{7.3}$$

We defined the stress tensor by $T_{\mu\nu} = \frac{2}{\sqrt{g}}\frac{\delta}{\delta g^{\mu\nu}}S$. This normalization yields the usual expression $T_{\mu\nu} = \partial_\mu\varphi\partial_\nu\varphi + \cdots$ for a scalar field. Classically the action is Weyl invariant, $S[g_{\mu\nu} + \delta_W g_{\mu\nu}, \tilde{\varphi} + \delta_W\tilde{\varphi}] = S[g_{\mu\nu}, \tilde{\varphi}]$, hence on-shell (when the field equation of $\tilde{\varphi}$ holds) the trace of the classical stress tensor vanishes, $2\delta_W g_{\mu\nu}\frac{\delta}{\delta g_{\mu\nu}}S = -\sqrt{g}\sigma T_\mu{}^\mu = 0$. At the quantum level, there is an anomaly: the Jacobian

$$\mathrm{Det}\,\frac{\partial[\tilde{\varphi}(x)+\delta_W\tilde{\varphi}(x)]}{\partial\tilde{\varphi}(y)} = 1 + \mathrm{Tr}\,\frac{\partial\delta_W\tilde{\varphi}(x)}{\partial\tilde{\varphi}(y)} \tag{7.4}$$

differs from unity by the trace of the infinite-dimensional operator

$$\frac{\partial\delta_W\tilde{\varphi}(x)}{\partial\tilde{\varphi}(y)} = \frac{1}{2}\sigma(x)\delta^n(x-y). \tag{7.5}$$

This trace must be regulated. The regulator $\mathcal{R}$ is the field operator of $\tilde{\varphi}$ and reads as

$$\mathcal{R} = -g^{-1/4}\partial_\mu\sqrt{g}g^{\mu\nu}\partial_\nu g^{-1/4} - \xi R. \tag{7.6}$$

We discussed previously that this regulator preserves Einstein invariance. It can also be derived from the algorithm of [13]: one adds a mass term $\sqrt{g}m^2\varphi^2$ to the Weyl invariant action which preserves Einstein symmetry but breaks Weyl invariance, and one constructs $\mathcal{R}$ from the terms quadratic in quantum fields. The fact that one cannot write down a mass term which is simultaneously Einstein and Weyl invariant implies that there may be an anomaly in these symmetries. In fact, we shall see that an anomaly appears in even dimensions. We choose again to preserve the Einstein symmetry, thus locating the anomaly in the Weyl symmetry. However, unlike the case of chiral anomalies where it was convenient to consider transformations which were a suitable linear combinations of local Lorentz and Einstein symmetries, now we consider no linear combination of Weyl and Einstein transformations, but only Weyl transformations.

To evaluate the anomaly we shall use the same quantum mechanical approach as for the chiral anomalies, and consider for the scalar field

$$An_W(\text{spin } 0) = \lim_{\beta\to 0}\mathrm{Tr}\,\sigma_S(\hat{x})e^{-\frac{\beta}{\hbar}\hat{H}}$$
$$\hat{H} = \frac{1}{2}g^{-1/4}p_i\sqrt{g}g^{ij}p_j g^{-1/4} - \frac{1}{2}\hbar^2\xi R \tag{7.7}$$

where $\sigma_S(x) = \frac{1}{2}\sigma(x)$ is the product of the Weyl weight of $\tilde{\varphi}$ and the Weyl rescaling parameter normalized to $\delta_W g_{ij}(x) = \sigma(x)g_{ij}(x)$. We again use the notation that curved indices in quantum field theory are written as μ,ν, while in the quantum mechanical case we use i,j.

Trace anomalies for scalars in two dimensions

To evaluate the trace (local scale) anomaly for a real scalar field in n dimensions, we use the transition element discussed in Part I of this book.

In terms of $\sigma_S = \frac{1}{2}\sigma$ we obtain, using the trace formula

$$\mathrm{Tr}A = \int d^n x_0 \sqrt{g(x_0)} \langle x_0 | A | x_0 \rangle \tag{7.8}$$

which follows from the completeness relation as discussed in (2.184),

$$\begin{aligned} An_W(\text{spin } 0) &= \mathrm{Tr}\, \sigma_S(x) \mathrm{e}^{-\frac{\beta}{\hbar}H} \\ &= \int d^n x_0 \sqrt{g(x_0)} \sigma_S(x_0) \langle x_0 | e^{-\frac{\beta}{\hbar}H} | x_0 \rangle \\ &= \int d^n x_0 \sqrt{g(x_0)} \sigma_S(x_0) \frac{1}{(2\pi\beta\hbar)^{n/2}} \langle \mathrm{e}^{-\frac{1}{\hbar}S^{(int)}} \rangle, \end{aligned} \tag{7.9}$$

where the limit $\beta \to 0$ is understood and with

$$\begin{aligned} -\frac{1}{\hbar}S^{int} &= -\frac{1}{\beta\hbar}\int_{-1}^{0} d\tau\, \frac{1}{2}[g_{ij}(x_0+q) - g_{ij}(x_0)](\dot{q}^i\dot{q}^j + b^i c^j + a^i a^j) \\ & -\beta\hbar \int_{-1}^{0} d\tau \left(\frac{1}{8}R + \frac{1}{8}g^{ij}\Gamma^l_{ik}\Gamma^k_{jl} - \frac{1}{2}\xi R \right). \end{aligned} \tag{7.10}$$

One can use any regularization discussed in the first part of this book and for definiteness we have chosen time slicing, so that we have included the corresponding counterterm in the action. In $n = 2$ dimensions the action for a scalar is Weyl invariant by itself, $\xi = 0$, which has enormous implications for string theory. We must extract the term proportional to $\beta\hbar$ from $\langle \exp(-\frac{1}{\hbar}S^{int}) \rangle$ to cancel the factor of $(\beta\hbar)^{-1}$ in the measure. Since propagators are proportional to $\beta\hbar$ while vertices are proportional to $(\beta\hbar)^{-1}$ or $\beta\hbar$, we need tree graphs with one vertex proportional to $\beta\hbar$, or loop graphs with one more propagator than vertices. There are also singular terms proportional to $(2\pi\beta\hbar)^{-1}$. These should be removed by renormalization of the quantum field theory: a theory which is renormalizable in flat space remains renormalizable in an external gravitational field.

To facilitate the computation, we introduce normal coordinates in which the symmetrized derivatives $\partial_{(i}\partial_{j \ldots}\partial_l \Gamma_{mn)}{}^p$ at $x = x_0$ vanish [171]. Then,

$$\begin{aligned} g_{ij}(x_0+q) = g_{ij}(x_0) &- \frac{1}{3}R_{iklj}(x_0)q^k q^l - \frac{1}{6}D_m R_{iklj}(x_0)q^m q^k q^l \\ &- R_{mniklj}(x_0)q^m q^n q^k q^l + \cdots \end{aligned} \tag{7.11}$$

where

$$R_{mniklj} = \frac{1}{20}D_m D_n R_{iklj} + \frac{2}{45}R_{ikmp}R_{ljn}{}^p. \tag{7.12}$$

All Riemann curvatures in this chapter are curvatures in terms of Christoffel symbols and not spin connections. We refer to Appendix A for definitions.

Only the two-loop graph with the topology of the number 8, and the tree graph with the R and $\Gamma\Gamma$ vertex (but with $\Gamma^i_{jk}(x_0) = 0$ in normal coordinates) contribute. We find

$$
\begin{aligned}
An_W(\text{spin } 0, n=2) &= \int d^2x_0 \sqrt{g(x_0)}\sigma_S(x_0)\frac{1}{(2\pi\beta\hbar)} \\
&\times \int_{-1}^{0} d\tau \left[\left(-\frac{1}{2}\frac{1}{\beta\hbar}\right)\left\langle -\frac{1}{3}R_{iklj}q^k q^l(\dot{q}^i\dot{q}^j + b^i c^j + a^i a^j)\right\rangle + (-\beta\hbar)\frac{1}{8}R\right] \\
&= \int \frac{d^2x}{2\pi}\sqrt{g(x)}\sigma_S(x) \\
&\times \left[\frac{1}{6}\frac{1}{(\beta\hbar)^2}R_{iklj}\int_{-1}^{0} d\tau\, \langle q^k q^l(\dot{q}^i\dot{q}^j + b^i c^j + a^i a^j)\rangle - \frac{1}{8}R\right]. \qquad (7.13)
\end{aligned}
$$

The propagators are given by

$$
\begin{aligned}
&\langle q^i(\sigma)q^j(\tau)\rangle = -\beta\hbar g^{ij}(z)\Delta(\sigma,\tau) \\
&\Delta(\sigma,\tau) = \sigma(\tau+1)\theta(\sigma-\tau) + \tau(\sigma+1)\theta(\tau-\sigma) \qquad (7.14)
\end{aligned}
$$

and

$$
\begin{aligned}
&\langle a^i(\sigma)a^j(\tau)\rangle + \langle b^i(\sigma)c^j(\tau)\rangle = -\beta\hbar g^{ij}(z)\,{}^{\bullet\bullet}\!\Delta(\sigma,\tau) \\
&{}^{\bullet\bullet}\!\Delta(\sigma,\tau) = \partial_\sigma^2\Delta(\sigma,\tau). \qquad (7.15)
\end{aligned}
$$

They cancel the factor of $(\beta\hbar)^{-2}$.

Using $R_{iklj}g^{kl} = -R_{ij}$, the two-loop graph yields the following integral over equal-time propagators:

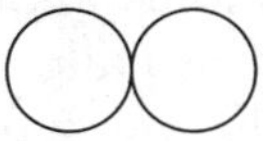

$$
\begin{aligned}
&= \frac{1}{6}R\int_{-1}^{0} d\tau\,[-\Delta({}^{\bullet}\!\Delta^{\bullet} + {}^{\bullet\bullet}\!\Delta) + {}^{\bullet}\!\Delta\,{}^{\bullet}\!\Delta] \\
&= \frac{1}{6}R\int_{-1}^{0} d\tau\left(-\tau(\tau+1) + \left(\tau+\frac{1}{2}\right)^2\right) = \frac{1}{24}R. \qquad (7.16)
\end{aligned}
$$

Here we used time slicing, according to which ${}^{\bullet}\!\Delta^{\bullet}(\sigma,\tau) = 1 - \delta(\sigma,\tau)$ and ${}^{\bullet\bullet}\!\Delta(\sigma,\tau) = \delta(\sigma,\tau)$, where $\delta(\sigma,\tau)$ is a Kronecker delta at equal times $\sigma = \tau$. Furthermore, the $\theta(\sigma,\tau)$ in ${}^{\bullet}\!\Delta(\sigma,\tau) = \tau + \theta(\sigma,\tau)$ equals $\frac{1}{2}$ at equal-time contractions. One then obtains the nonsingular integrals in (7.16).

Other schemes give the same result. For example, in mode regularization the $\Gamma\Gamma$ term, though different, again does not contribute since $\Gamma_{ij}^k(x_0) = 0$ in normal coordinates, and using the properties of $\Delta(\sigma,\tau)$ in mode regularization to partially integrate, see (3.76), one finds the same result.

Altogether one finds in terms of $\sigma_S = \frac{1}{2}\sigma$,

$$
An_W(\text{spin } 0, n=2) = \int \frac{d^2x_0}{2\pi}\sqrt{g(x_0)}\sigma_S(x_0)\left(-\frac{1}{12}\right)R. \qquad (7.17)
$$

Trace anomalies for scalars in four dimensions

For a real scalar field we now must evaluate

$$
\begin{aligned}
An_W(\text{spin } 0, n=4) = &\int d^4x \sqrt{g(x_0)}\sigma_S(x_0)\frac{1}{(2\pi\beta\hbar)^2} \\
&\times \Bigg\langle \exp\Bigg[-\frac{1}{\beta\hbar}\frac{1}{2}\int_{-1}^{0} d\tau \Bigg[-\frac{1}{3}R_{iklj}(x_0)q^k q^l - \frac{1}{6}D_m R_{iklj}(x_0)q^m q^k q^l \\
&- R_{mniklj}(x_0)q^m q^n q^k q^l + \cdots\Bigg](\dot{q}^i\dot{q}^j + b^i c^j + a^i a^j) \\
&- \beta\hbar\int_{-1}^{0} d\tau\left(\frac{1}{8}R + \frac{1}{8}g^{ij}\Gamma^l_{ik}\Gamma^k_{jl} - \frac{1}{2}\xi R\right)\Bigg]\Bigg\rangle \qquad (7.18)
\end{aligned}
$$

where $\xi = \frac{1}{6}$. Clearly we now need the terms proportional to $(\beta\hbar)^2$ from $\langle \exp(-\frac{1}{\hbar}S^{int})\rangle$. In particular, one finds a contribution from the $\Gamma\Gamma$ vertices, due to expanding both $\Gamma(x_0+q)$ into $q^m\partial_m\Gamma(x_0)$ and contracting with an equal-time loop. In earlier articles this contribution has been overlooked; we stress that we cannot drop the term quadratic in $\Gamma^k_{ij}(x_0+q)$ even when one uses normal coordinates, although $\Gamma^k_{ij}(x_0)$ of course vanishes in such coordinates.

One must also expand the counterterm with the scalar curvature terms to order q^2, and contract the two q fields to a loop. Since there are no three-point vertices in normal coordinates (because in normal coordinates $\Gamma^k_{ij}(x_0) = 0$), we do not need the five-point vertices. However, one needs the six-point vertices which yield "clover-leaf graphs", and four-point vertices which yield "three-bubble graphs" and "eye-graphs". Finally, there are also disconnected diagrams: one-half of the square of the $n = 2$ result; however with $\xi = \frac{1}{6}$ in $n = 4$ their contribution cancels, as one may easily check from (7.10) and (7.16),

$$\frac{1}{2!}\Big($$  $$+\ \bullet\Big)^2 = 0. \qquad (7.19)$$

The other contributions, together with the graphs from which they are obtained, follow. First there are the contributions from two $Rqq(\dot{q}\dot{q}+bc+aa)$ vertices. These yield "three-bubble graphs"

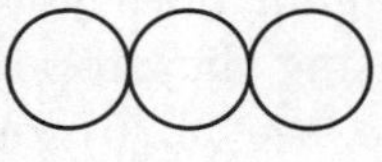 $$= \frac{1}{72}(-\beta\hbar)^2\left(-\frac{1}{6}R_{ij}^2\right) \qquad (7.20)$$

and "eye-graphs"

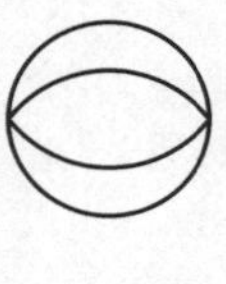 $$= \frac{1}{72}(-\beta\hbar)^2\left(-\frac{1}{4}R_{ijkl}^2\right). \qquad (7.21)$$

Then there are the contributions from one $Rq^4(\dot{q}\dot{q}+bc+aa)$ vertex. They yield various "clover-leaf graphs"

$$= \frac{1}{72}(-\beta\hbar)^2\left(\frac{3}{20}D^2R+\frac{1}{10}R_{ijkl}^2+\frac{1}{15}R_{ij}^2\right). \tag{7.22}$$

The propagators in these graphs denote q or a, b, c ghost propagators.

Finally, there are the one-loop graph contributions from the R and $\Gamma\Gamma$ vertices; using $\xi=\frac{1}{6}$ they yield

$$\begin{aligned} &= \left[-\frac{\beta\hbar}{24}\left(\frac{1}{2}D^2R\right)-\frac{\beta\hbar}{8}g^{mn}g^{ij}\partial_m\Gamma_{ik}^l\partial_n\Gamma_{jl}^k\right] \\ &\quad\times(-\beta\hbar)\int_{-1}^{0}d\tau\,\Delta(\tau,\tau). \end{aligned} \tag{7.23}$$

Since in Riemann normal coordinates

$$\begin{aligned} \partial_m\Gamma_{ik}^l &= \frac{1}{3}(\partial_m\Gamma_{ik}^l+\partial_i\Gamma_{km}^l+\partial_k\Gamma_{mi}^l) \\ &\quad+\frac{1}{3}(\partial_m\Gamma_{ik}^l-\partial_i\Gamma_{mk}^l)+\frac{1}{3}(\partial_m\Gamma_{ik}^l-\partial_k\Gamma_{im}^l) \\ &= \frac{1}{3}R_{mik}{}^l+\frac{1}{3}R_{mki}{}^l \end{aligned} \tag{7.24}$$

we obtain for (7.23), using $\int_{-1}^{0}d\tau\,\Delta(\tau,\tau)=-\frac{1}{6}$,

$$\left(-\frac{1}{6}\right)(\beta\hbar)^2\left[\frac{1}{48}D^2R+\frac{1}{72}(R_{mikl}+R_{mkil})(R^{milk}+R^{mlik})\right]. \tag{7.25}$$

Using the cyclic identity for the Riemann curvatures to obtain $R_{milk}R^{mlik}=\frac{1}{2}(R_{milk})^2$, the R_{ijkl}^2 terms acquire a factor of $(-1-\frac{1}{2}-\frac{1}{2}+\frac{1}{2})=-\frac{3}{2}$ and the one-loop graphs yield

$$= \frac{1}{72}(\beta\hbar)^2\left(-\frac{1}{4}D^2R+\frac{1}{4}R_{ijkl}^2\right). \tag{7.26}$$

Adding all terms, the contributions from the eye-graphs cancel those from the $\Gamma\Gamma$ term, and the R_{ijkl}^2 terms only come from the clover-leaf graphs. We obtain

$$\begin{aligned} An_W(\text{spin }0, n=4) &= \int\frac{d^4x}{(2\pi)^2}\sqrt{g(x)}\sigma_S(x) \\ &\quad\times\left(aR_{ijkl}^2+bR_{ij}^2+cR^2+dD^2R\right) \end{aligned} \tag{7.27}$$

with $\sigma_S = \frac{1}{2}\sigma$ and

$$a = \frac{1}{720}, \qquad b = -\frac{1}{720}, \qquad c = 0, \qquad d = -\frac{1}{720}. \tag{7.28}$$

This is the correct result.

Let us now compare this calculation based on time slicing with the equivalent one in dimensional regularization (DR). In DR the counterterm is covariant, $V_{DR} = \frac{1}{8}R$, so that we should drop the $\Gamma\Gamma$ term in the last line of (7.18), but of course the Feynman graphs should be evaluated using the rules of DR. The counterterm graph needs no regularization and dropping the $\Gamma\Gamma$ term we obtain

$$= \frac{1}{72}(\beta\hbar)^2\left(-\frac{1}{4}D^2R\right). \tag{7.29}$$

In DR all the other graphs give the same contribution as in TS, except for the "eye-graph" which yields

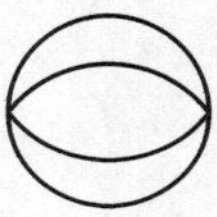

$$= \frac{1}{72}(\beta\hbar)^2\, 3R_{ijkl}^2 K \tag{7.30}$$

where

$$K = \int_{-1}^{0} d\tau \int_{-1}^{0} d\sigma\, \{\Delta^2[({}^{\bullet}\!\Delta^{\bullet})^2 - (\Delta^{\bullet\bullet})^2] - 2\,\Delta\, {}^{\bullet}\!\Delta\, \Delta^{\bullet}\, {}^{\bullet}\!\Delta^{\bullet} + ({}^{\bullet}\!\Delta)^2(\Delta^{\bullet})^2\}. \tag{7.31}$$

This graph should vanish because the $\Gamma\Gamma$ term is absent, and the "eye-graph" canceled the contribution from $\Gamma\Gamma$ in time slicing. Clearly the first three terms in K need regularization (also recall that in the sum of the first two terms divergences cancel). Using DR we obtain for the first three terms

$$\begin{aligned}
&\int d^{D+1}t \int d^{D+1}s\, \{\Delta^2[({}_{\mu}\Delta_{\nu})^2 - (\Delta_{\nu\nu})(\Delta_{\mu\mu})] - 2\,\Delta({}_{\mu}\Delta)(\Delta_{\nu})({}_{\mu}\Delta_{\nu})\} \\
&\quad = \int\!\!\int\, [2\,\Delta(\Delta_{\nu})^2({}_{\mu\mu}\Delta) - 4\,\Delta({}_{\mu}\Delta)(\Delta_{\nu})({}_{\mu}\Delta_{\nu})] \\
&\quad = \int\!\!\int\, [4\,\Delta\,(\Delta_{\nu})^2({}_{\mu\mu}\Delta) + 2\,({}_{\mu}\Delta)^2(\Delta_{\nu})^2] \\
&\quad \to 4\int_{-1}^{0} d\tau\, \Delta(\Delta^{\bullet})^2|_{\tau} + 2\int_{-1}^{0} d\tau \int_{-1}^{0} d\sigma\, ({}^{\bullet}\!\Delta)^2(\Delta^{\bullet})^2.
\end{aligned} \tag{7.32}$$

We have twice integrated by parts in the first term of the first line so that the second term due to the ghosts is canceled. In the second line we have integrated by parts the derivative μ in $(\Delta_{\nu})({}_{\mu}\Delta_{\nu}) = \frac{1}{2}\,{}_{\mu}(\Delta_{\nu}^2)$. Finally, in the third line we have used the fact that ${}_{\mu\mu}\Delta(t,s) = \delta^{D+1}(t,s)$, used the Dirac delta function in $D+1$ dimensions and then removed the regulating parameter $D \to 0$. The final expression in the fourth line is then evaluated at $D = 0$. In fact, the limiting values of the various functions (like $({}^{\bullet}\!\Delta^{\bullet})|_\tau$) given as Fourier series have no ambiguities when multiplied together and can be safely used inside integrals. Adding to this result the last term in (7.31) and using (3.89)–(3.93) we obtain

$$K = 4\int_{-1}^{0} d\tau\, \Delta({}^{\bullet}\!\Delta^{\bullet})^2|_\tau + 3\int_{-1}^{0} d\tau \int_{-1}^{0} d\sigma\, ({}^{\bullet}\!\Delta)^2({}^{\bullet}\!\Delta^{\bullet})^2 = -\frac{1}{30} + \frac{1}{30} = 0. \tag{7.33}$$

Thus the "eye diagrams" indeed vanish and the total result for the anomaly is the same as obtained previously in time slicing. We see that DR is computationally simpler, since one does not have to expand the non-covariant $\Gamma\Gamma$ counterterm in normal coordinates (which is rather laborious in higher dimensions). This simplification turns out to be quite useful for the calculation at fourth-loop order, needed to calculate the trace anomaly for a scalar in six dimensions. The QM path integral method [87, 88], again produces the expected result [172].

7.2 Trace anomalies for spin-$\frac{1}{2}$ fields in two and four dimensions

For complex spin-$\frac{1}{2}$ fields (Dirac fields) in n dimensions, the action

$$S = \int d^n x\, \sqrt{g}\, \bar{\psi}\gamma^m e_m{}^\mu D_\mu \psi\,; \qquad D_\mu\psi = \partial_\mu\psi + \frac{1}{4}\omega_\mu{}^{mn}\gamma_m\gamma_n\psi \tag{7.34}$$

is locally scale invariant in any dimension under[1]

$$\delta_W \psi(x) = \frac{1}{4}(1-n)\sigma(x)\psi(x), \qquad \delta_W e_m{}^\mu(x) = -\frac{1}{2}\sigma(x)e_m{}^\mu(x). \tag{7.35}$$

We again use a Euclidean signature so that ψ and $\bar{\psi}$ are independent complex Grassmann variables. So for fermions there is no improvement term.

We use $\tilde{\psi} = g^{1/4}\psi$ and $\tilde{\bar{\psi}} = g^{1/4}\bar{\psi}$ as integration variables in the path integral. In any dimension $\delta_W \tilde{\psi} = \frac{1}{4}\sigma(x)\tilde{\psi}$. We introduce the parameter

$$\sigma_F = -2\frac{1}{4}\sigma = -\frac{1}{2}\sigma \tag{7.36}$$

which contains the Weyl weight of ψ and $\bar{\psi}$ and the minus sign for the fermionic Jacobian. One then finds

$$An_W(\text{Dirac}) = \lim_{\beta\to 0} \text{Tr}\, \sigma_F(x) e^{-\beta\mathcal{R}}. \tag{7.37}$$

[1] The proof is easy: for constant σ the weights clearly cancel, while for local $\sigma(x)$ the terms with $\partial_\mu\sigma$ produced by $D_\mu\psi$ cancel if one uses (A.24).

The regulator $\mathcal{R}$ is the square of the Dirac operator as in (6.7). Hence,

$$\hat{H} = \frac{1}{2} g^{-1/4} \pi_i \sqrt{g} g^{ij} \pi_j g^{-1/4} - \frac{1}{8}\hbar^2 R$$
$$\pi_i = p_i - \frac{i\hbar}{2}\omega_{iab}\psi_1^a\psi_1^b. \tag{7.38}$$

To take the trace, we could proceed step-by-step in the same way as for chiral anomalies, namely use the trace formula and insert a complete set of bosonic and fermionic states between σ and $e^{-\beta\mathcal{R}}$. This would lead to four sets of Grassmann integrals, over χ, $\bar{\chi}$, η and $\bar{\eta}$, of which three would again be easy. However, since $\sigma(\hat{x})$ does not depend on fermions, we need not insert unity, and we can simplify the calculation by using only the trace formula

$$An_W(\text{Dirac}) = \int d^n x_0 \sqrt{g(x_0)}\sigma_F(x_0)$$
$$\int \prod_{a=1}^{n} d\chi^a \, d\bar{\chi}_a \, e^{\bar{\chi}_a\chi^a} \langle \bar{\chi}, x_0 | e^{-\frac{\beta}{\hbar}\hat{H}} | \chi, x_0\rangle. \tag{7.39}$$

As shown in Section 2.4, the transition element yields in the continuum limit the following result:

$$\langle \bar{\chi}, x_0 | e^{-\frac{\beta}{\hbar}\hat{H}} | \chi, x_0\rangle = e^{\bar{\chi}_a\chi^a(0)} \langle e^{-\frac{1}{\hbar}S^{int}}\rangle. \tag{7.40}$$

At the discretized level, $\chi^a(0)$ appeared as χ^a_{N-1} and integration over χ and $\bar{\chi}$ led to $\chi^a_{N-1} = -\chi$, hence the **anti**-periodic boundary conditions. We do not use a background field formalism, and directly work with propagators for Majorana fermions which are anti-periodic

$$\langle \psi_1^a(\sigma)\psi_1^b(\tau)\rangle = \frac{1}{2}\beta\hbar\epsilon(\sigma - \tau). \tag{7.41}$$

One may notice that this propagator coincides with the propagator for Majorana fermions with antiperiodic boundary conditions, see (4.41) and (4.42). The integral over χ and $\bar{\chi}$ yields a factor of $2^{n/2}$, which is, of course, the dimension of the spinor space[2] (we consider here even n spacetime dimensions), and in this way the calculation of the Weyl anomaly is reduced to the following formula:

$$An_W(\text{Dirac}) = \int d^n x_0 \sqrt{g(x_0)}\sigma_F(x_0) \frac{2^{n/2}}{(2\pi\beta\hbar)^{n/2}} \langle e^{-\frac{1}{\hbar}S^{int}}\rangle \tag{7.42}$$

with

[2] For $\hat{H} = 0$ this factor is easy to check: the integration over χ and $\bar{\chi}$ yields a factor 2^n, but one must divide by the dimension $2^{n/2}$ of the Hilbert space of the extra spinors ψ_2^a, see Section 2.4.

$$
\begin{aligned}
-\frac{1}{\hbar}S^{int} = & -\frac{1}{\beta\hbar}\int_{-1}^{0} d\tau\, \frac{1}{2}[g_{ij}(x_0+q) - g_{ij}(x_0)](\dot{q}^i\dot{q}^j + b^i c^j + a^i a^j) \\
& -\frac{1}{\beta\hbar}\int_{-1}^{0} d\tau\, \frac{1}{2}\dot{q}^i \omega_{iab}\psi_1^a\psi_1^b \\
& -\beta\hbar\int_{-1}^{0} d\tau\, \frac{1}{8}\Big(g^{ij}\Gamma^l_{ik}\Gamma^k_{jl} + \frac{1}{2}g^{ij}\omega_{iab}\omega_j{}^{ab}\Big)
\end{aligned} \tag{7.43}
$$

where ψ_1^a are Majorana spinors with the propagator $\langle\psi_1^a(\sigma)\psi_1^b(\tau)\rangle = \frac{1}{2}\beta\hbar\delta^{ab}\epsilon(\sigma-\tau)$.

There is no term with the scalar curvature in the last line in (7.43) because the Riemann term from the Weyl reordering of the bosonic part of the Hamiltonian cancels with the R term from expanding $\not{D}\not{D}$, while for fermions there is no improvement term. The term with $\Gamma\Gamma$ comes from the Weyl ordering of bosonic part of the regulator while the term $\frac{1}{2}\omega\omega$ is due to Weyl ordering of ψ_1^a. (For complex ψ, $\bar{\psi}$ one obtains a factor of unity in front of the $\omega\omega$ term, but for Majorana fermions one finds a factor of $\frac{1}{2}$, see Appendix C.)

Trace anomalies for fermions in two dimensions

The trace anomaly for complex spin-$\frac{1}{2}$ fields is thus given by

$$
\begin{aligned}
An_W(\text{Dirac}, n=2) = & \int d^2x_0\, \sqrt{g(x_0)}\sigma_F(x_0)\frac{2}{2\pi\beta\hbar} \\
& \Big\langle \exp\Big\{-\frac{1}{\beta\hbar}\int_{-1}^{0} d\tau\, \frac{1}{2}\left[g_{ij}(x_0+q) - g_{ij}(x_0)\right]\left(\dot{q}^i\dot{q}^j + b^i c^j + a^i a^j\right) \\
& -\frac{1}{\beta\hbar}\int_{-1}^{0} d\tau\, \frac{1}{2}\dot{q}^i\omega_{iab}\psi_1^a\psi_1^b - \beta\hbar\int_{-1}^{0} d\tau\, \frac{1}{8}\left(\Gamma\Gamma + \frac{1}{2}\omega\omega\right)\Big\}\Big\rangle.
\end{aligned} \tag{7.44}
$$

The terms with ψ_1 do not contribute if one uses a gauge in which $\omega_{iab}(x_0) = 0$ (the Fock–Schwinger gauge), because the graph with the topology of number 8 with one q loop and one ψ_1 loop vanishes (ω_{iab} is traceless). So, the spin-$\frac{1}{2}$ anomaly in $n = 2$ dimensions also comes only from the purely bosonic sector, namely from the two-loop graph in (7.16) which yielded $\frac{1}{24}R$,

$$
An_W(\text{Dirac } n=2) = 2\int \frac{d^2x}{2\pi}\sqrt{g(x)}\sigma_F(x)\frac{1}{24}R. \tag{7.45}
$$

Hence, as is well known from string theory, the trace anomaly for real spin-0 fields and complex spin-$\frac{1}{2}$ fields in two dimensions are equal[3] (recall that $\sigma_F = -\sigma_S$).

[3]In Euclidean space independent complex ψ and $\bar{\psi}$ correspond to a Dirac spinor in Minkowski space. A Majorana spinor in Minkowski space thus has a trace anomaly which is a factor of $\frac{1}{2}$ smaller then a real scalar field (but with the same sign).

Trace anomalies for fermions in four dimensions

Next, we consider the trace anomaly for Dirac fermions in $n = 4$ dimensions. The expression for the anomaly to be evaluated is given by

$$An_W(\text{Dirac } n = 4) = \int d^4x_0 \sqrt{g(x_0)}\sigma_F(x_0)\frac{2^2}{(2\pi\beta\hbar)^2}\langle e^{-\frac{1}{\hbar}S^{int}}\rangle \quad (7.46)$$

where as before the expectation value is unity if S^{int} vanishes, while the factor of 2^2 yields the dimensions of the fermionic part of the Hilbert space. Using the Fock–Schwinger gauge, in which $\partial_{(j}\omega_{i)ab}(x_0) = 0$, S^{int} reduces to

$$\begin{aligned}
-\frac{1}{\hbar}S^{int} = &-\frac{1}{\beta\hbar}\int_{-1}^{0} d\tau \Big(\frac{1}{3}R_{iklj}q^k q^l - \frac{1}{6}D_m R_{iklj}q^m q^k q^l \\
&- R_{mniklj}q^m q^n q^k q^l + \cdots\Big)(\dot{q}^i\dot{q}^j + b^i c^j + a^i a^j) \\
&- \frac{1}{4}\frac{1}{\beta\hbar}\int_{-1}^{0} d\tau \left[\dot{q}^i q^j R_{jiab}\left(\omega(x_0)\right)\psi_1{}^a\psi_1{}^b + \cdots\right] \\
&- \frac{\beta\hbar}{8}\int_{-1}^{0} d\tau \left(g^{ij}\Gamma^l_{ik}\Gamma^k_{jl} + \frac{1}{2}g^{ij}\omega_{iab}\omega_j{}^{ab}\right). \qquad (7.47)
\end{aligned}$$

Curvatures depending on Γ^l_{ik} are denoted with indices i, j, k, l, while curvatures depending on ω_{iab} are denoted by R_{ijab}. Since only squares of each appear below and they only differ by a sign, one need not be careful about the sign difference, but using a different notation helps to identify the corresponding Feynman graphs.

The total contribution to $\langle e^{-\frac{1}{\hbar}S^{int}}\rangle$ is a sum of the following terms:

(i) the contributions from the first two lines in (7.47); they are the sum of (7.20), (7.21) and (7.22),

 + 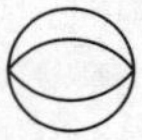 + 

$$= \frac{(\beta\hbar)^2}{720}\left(-\frac{3}{2}R^2_{ijkl} - R^2_{ij} + \frac{3}{2}D^2 R\right) \quad (7.48)$$

(ii) the contributions from the $\Gamma\Gamma$ and $\omega\omega$ terms; they yield

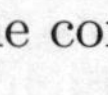

$$\begin{aligned}
&= -\frac{\beta\hbar}{8}\left[\frac{1}{9}\left(-\frac{3}{2}\right)R^2_{ijkl} + \frac{1}{8}R^2_{ijab}\right](-\beta\hbar)\left[\int_{-1}^{0} d\tau\Delta(\tau,\tau)\right] \\
&= \frac{-(\beta\hbar)^2}{192}R^2_{ijkl}\left(-\frac{1}{6}\right) \qquad (7.49)
\end{aligned}$$

(iii) disconnected graphs: one-half of the square of the two-loop graphs in (7.19),

$$\frac{1}{2!}\Big(\ \Big)^2 = \frac{(\beta\hbar)^2}{2!}\Big(-\frac{1}{24}R\Big)^2 \tag{7.50}$$

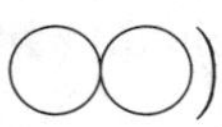

(iv) the contribution from the graphs with fermions

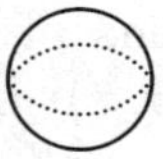

$$= \frac{1}{32}\frac{1}{(\beta\hbar)^2} \tag{7.51}$$

$$\times \left\langle \int_{-1}^{0} d\sigma (\dot{q}^i q^j R_{jiab}\psi_1^a\psi_1^b)(\sigma) \int_{-1}^{0} d\tau\, (\dot{q}^k q^l R_{lkcd}\psi_1^c\psi_1^d)(\tau) \right\rangle$$

$$= -\frac{(\beta\hbar)^2}{16} R_{ijab}^2 \int_{-1}^{0} d\sigma \int_{-1}^{0} d\tau\, ({}^{\bullet}\Delta^{\bullet}\Delta - {}^{\bullet}\Delta\Delta^{\bullet})\frac{1}{4}\epsilon^2(\sigma-\tau).$$

Solid lines lines indicate scalars and dotted lines denote fermions. Other contractions vanish since R_{ijab} is traceless in ij and in ab. However, this graph vanishes, since ${}^{\bullet}\Delta\Delta^{\bullet} = \Delta$ and with time slicing ${}^{\bullet}\Delta^{\bullet} = 1 - \delta(\sigma-\tau)$. Hence,

$${}^{\bullet}\Delta^{\bullet}\Delta - {}^{\bullet}\Delta\Delta^{\bullet} = {}^{\bullet}\Delta^{\bullet}\Delta - \Delta = -\delta(\sigma-\tau)\Delta(\sigma,\tau) \tag{7.52}$$

and $\delta(\sigma-\tau)\epsilon^2(\sigma-\tau)$ vanishes according to our rule that $\delta(\sigma-\tau)$ is a Kronecker delta. Hence, again no fermion loops contribute in time slicing.

The final result is

$$An_W(\text{Dirac}, n=4) = \int \frac{d^4x}{(2\pi)^2}\sqrt{g(x)}\sigma_F(x) 4\Big[\Big(\frac{-3/2}{720}+\frac{1}{288}\Big)R_{ijkl}^2$$

$$-\frac{1}{720}R_{ij}^2 + \frac{1}{2!}\Big(-\frac{1}{24}R\Big)^2 + \frac{1}{480}D^2R\Big]$$

$$= \int \frac{d^4x}{(2\pi)^2}\sqrt{g(x)}\sigma_F(x)\Big(-\frac{7}{1440}R_{ijkl}^2 - \frac{1}{180}R_{ij}^2 + \frac{1}{288}R^2 + \frac{1}{120}D^2R\Big). \tag{7.53}$$

This is the correct result.

Let us check once more this final result by employing dimensional regularization instead of time slicing. In DR the fermions do not modify the counterterm form the bosonic sector, and thus the last line of the action in (7.47) is now absent. As we have seen from the previous computation in TS, apart form the coefficient 2^2 (which is due to the normalization of the fermionic path integral) only the counterterm $\frac{1}{2}\omega\omega$ gave an additional contribution with respect to the anomaly of a scalar field with the coupling

$\xi = \frac{1}{4}$. In DR the $\frac{1}{2}\omega\omega$ counterterm is absent and thus the extra contribution can only come from a nonvanishing fermionic loop. This fermionic loop is the one in (7.52) that was vanishing in TS,

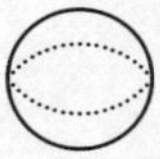

$$= -\frac{(\beta\hbar)^2}{16} R^2_{ijab} \int_{-1}^{0} d\tau \int_{-1}^{0} d\sigma \, [{}^{\bullet}\!\Delta^{\bullet}\, \Delta - {}^{\bullet}\!\Delta \Delta^{\bullet}]\, \Delta^2_{AF} \qquad (7.54)$$

where we recall that all functions Δ and Δ_{AF} are functions of τ and σ (recall that Δ_{AF} is antisymmetric, $\Delta_{AF}(\tau,\sigma) = -\Delta_{AF}(\sigma,\tau)$; in fact, $\Delta_{AF}(\tau,\sigma) = \frac{1}{2}\epsilon(\tau-\sigma)$, where $\epsilon(x)$ is the sign function $\epsilon(x) = x/|x|$).

We regulate the first term in (7.54) with DR. The second contribution in (7.54) does not need regularization and could be computed directly by integrating the sums of the Fourier mode expansions defining the propagators, but we carry it along anyway. In order to apply DR we must generalize propagators and interactions as discussed in Chapter 4. We obtain

$$\begin{aligned}
&\int_{-1}^{0} d\tau \int_{-1}^{0} d\sigma \, [{}^{\bullet}\!\Delta^{\bullet}\, \Delta - {}^{\bullet}\!\Delta \Delta^{\bullet}]\Delta^2_{AF} \\
&\to \int d^{D+1}t \int d^{D+1}s \, ({}_{\mu}\Delta_{\nu}\, \Delta - {}_{\mu}\Delta\, \Delta_{\nu}) \,\mathrm{tr}[-\gamma^{\mu}\Delta_{AF}(t,s)\gamma^{\nu}\Delta_{AF}(s,t)] \\
&= \int d^{D+1}t \int d^{D+1}s \, \Delta_{\nu}\, \Delta \,\mathrm{tr}\Big[2 \underbrace{\Big(\gamma^{\mu}\frac{\partial}{\partial t^{\mu}}\Delta_{AF}(t,s)\Big)}_{\delta^{D+1}(t,s)} \gamma^{\nu}\Delta_{AF}(s,t)\Big] \\
&\quad -2\int d^{D+1}t \int d^{D+1}s \; {}_{\mu}\Delta\, \Delta_{\nu} \,\mathrm{tr}[-\gamma^{\mu}\Delta_{AF}(t,s)\gamma^{\nu}\Delta_{AF}(s,t)] \\
&= 0 - 2\int d^{D+1}t \int d^{D+1}s \; {}_{\mu}\Delta\, \Delta_{\nu} \,\mathrm{tr}[-\gamma^{\mu}\Delta_{AF}(t,s)\gamma^{\nu}\Delta_{AF}(s,t)] \\
&\to -2\int_{-1}^{0} d\tau \int_{-1}^{0} d\sigma \; {}^{\bullet}\!\Delta\Delta^{\bullet}\Delta^2_{AF} = \frac{1}{24}
\end{aligned} \qquad (7.55)$$

where in the second line we integrated by parts the μ derivative in ${}_{\mu}\Delta_{\nu}$, which when acting on fermions produces a delta function ("equation of motion terms"). The delta function is integrated in $D+1$ dimensions and gives a vanishing contribution as $\Delta_{AF}(0) = 0$. The remaining terms are then computed at $D \to 0$, where ${}^{\bullet}\!\Delta\Delta^{\bullet} = \Delta$ and $\Delta^2_{AF} = \frac{1}{4}$.

This gives the same contribution as the $\omega\omega$ term in (7.49), and thus one also obtains the correct answer for the trace anomaly in dimensional regularization.

7.3 Trace anomalies for a vector field in four dimensions

The Maxwell action is Weyl invariant in four dimensions if one does not transform A_{μ} but only the metric. However, to quantize one must add a gauge-fixing term and a ghost action, which themselves are not Weyl invariant. We compute the trace anomaly from the general formula $An = \mathrm{Tr}\,(\sigma_V\, e^{-\frac{\beta}{\hbar}\mathcal{R}_V}) + \mathrm{Tr}\,(\sigma_{gh}\, e^{-\frac{\beta}{\hbar}\mathcal{R}_{gh}})$ where the second term yields the contribution from the ghosts. The sum of the contributions from the Maxwell field and its ghosts will satisfy the Wess–Zumino consistency conditions, justifying to some extent this procedure of adding the contributions from Weyl noninvariant actions to compute the Weyl anomaly of a classically Weyl invariant system. The sum of the gauge-fixing term and the ghost action is BRST exact, and as the vacuum is BRST invariant, one could rigorously justify our procedure.[4] This approach has been used before [139, 174], and we follow it here. We take as path integral variables $\tilde{A}_m = g^{1/4} e_m{}^\mu A_\mu$ and again represent the vector indices by ghosts, just as in the case of the Yang–Mills anomalies of Section 7.2.

The classical Maxwell action

$$\mathcal{L} = -\frac{1}{4}\sqrt{g} g^{\mu\rho} g^{\nu\sigma} F_{\mu\nu} F_{\rho\sigma} \tag{7.56}$$

is Weyl invariant under $\delta_W g_{\mu\nu} = \sigma(x) g_{\mu\nu}$ and $\delta_W A_\mu = 0$. For definiteness we again use a Euclidean signature. The field $\tilde{A}_m = g^{1/4} e_m{}^\mu A_\mu$ transforms as

$$\delta_W \tilde{A}_m = \frac{1}{2}\sigma \tilde{A}_m \qquad \text{in} \quad n = 4. \tag{7.57}$$

As a gauge-fixing term we use the Fermi–Feynman term in curved space $\mathcal{L} = -\frac{1}{2}\sqrt{g}(D^\mu A_\mu)^2$ with $D^\mu A_\mu = g^{\mu\nu}(\partial_\mu A_\nu - \Gamma^\rho_{\mu\nu} A_\rho)$. The gauge-fixed action then becomes

$$\mathcal{L} = -\frac{1}{2}\sqrt{g}(D_\mu A_\nu)(D_\rho A_\sigma) g^{\mu\rho} g^{\nu\sigma} + \frac{1}{2}\sqrt{g} A_\mu R^{\mu\nu} A_\nu \tag{7.58}$$

where $R_{\mu\nu}$ is the Ricci tensor defined in Appendix A. As a regulator in the space with $\tilde{A}_m$ we obtain

$$(\mathcal{R}_V)^m{}_n = (g^{-1/4} D_\mu \sqrt{g} g^{\mu\nu} D_\nu g^{-1/4})^m{}_n + R^m{}_n \tag{7.59}$$

where both D_μ and D_ν only contain a spin connection for the vector index of $\tilde{A}_m$

$$(D_\mu)^m{}_n = \partial_\mu \delta^m_n + \omega_\mu{}^m{}_n. \tag{7.60}$$

The Hamiltonian for the corresponding quantum mechanical model is given by

$$\begin{aligned} \hat{H}_V &= \frac{1}{2} g^{-1/4} \pi_i \sqrt{g} g^{ij} \pi_j g^{-1/4} - \frac{\hbar^2}{2} c^*_m R^m{}_n c^n \\ \pi_i &= p_i - i\hbar (c^*_m \omega_i{}^m{}_n c^n) \end{aligned} \tag{7.61}$$

where the vector ghosts satisfy the equal-time canonical commutation relations $\{c^n, c^*_m\} = \delta^n_m$. We continue to denote the indices for the internal vector space by $m, n, p, q, \ldots$ and the indices for the coordinates and

[4] This situation reminds one of the quantization of susy Yang–Mills theories, where in x-space gauge-fixing and ghost terms break susy. One can use an extension of the Batalin–Vilkovisky method to derive Ward identities which treat susy and gauge symmetry on a par [173].

momenta in the QM model by $i, j, k, l, \ldots$. This is useful for keeping track of the various contributions, but there is of course no intrinsic difference between both kinds of indices.

In the path integral we need the Hamiltonian in Weyl-ordered form. However, as explained in Section 7.2 where we evaluated the traces over the ghosts, **one should not Weyl order the ghosts**, rather products of $c^*\omega c$ correspond to products of matrices. One obtains

$$H_V = \left(\frac{1}{2}g^{ij}\hat{\pi}_i\hat{\pi}_j\right)_W + \frac{\hbar^2}{8}(R + g^{ij}\Gamma^k_{il}\Gamma^l_{jk}) - \frac{\hbar^2}{2}(c^*_m R^m{}_n c^n). \quad (7.62)$$

The first three terms are the same as for a scalar, except that π contains $c^*\omega c$ terms. We repeat that one should not Weyl order the terms proportional to $(c^*\omega c)$ and $(c^*\omega c)(c^*\omega c)$.

The anomaly comes from a trace over the space of $\tilde{A}_m$ and a trace over the space of the ghosts. We first discuss the former. The contribution to the anomaly from $\tilde{A}_m$ reads as

$$An(\tilde{A}_m) = \mathrm{Tr}\,\sigma_V\, \mathrm{e}^{-\frac{\beta}{\hbar}\hat{H}_V} = \int d^4x_0\,\sqrt{g(x_0)}\sigma_V(x_0)\frac{1}{(2\pi\beta\hbar)^2}\,\langle \mathrm{e}^{-\frac{1}{\hbar}S^{int}_V}\rangle \quad (7.63)$$

where $\sigma_V = \frac{1}{2}\sigma$ and

$$\begin{aligned} -\frac{1}{\hbar}S^{int}_V = &-\frac{1}{\beta\hbar}\int_{-1}^0 \frac{1}{2}[g_{ij}(x_0+q) - g_{ij}(x_0)](\dot{q}^i\dot{q}^j + b^ic^j + a^ia^j)\,d\tau \\ &-\int_{-1}^0 \dot{q}^i(c^*_m\,\omega_i{}^m{}_n c^n)\,d\tau + \frac{\beta\hbar}{2}\int_{-1}^0 c^*_m R^m{}_n c^n\,d\tau \\ &-\frac{\beta\hbar}{8}\int_{-1}^0\left(R + g^{ij}\Gamma^l_{ik}\Gamma^k_{jl}\right)d\tau. \end{aligned} \quad (7.64)$$

We need all graphs of order $(\beta\hbar)^2$ to cancel the factor of $(\beta\hbar)^{-2}$ in the Feynman measure. The q and a, b, c propagators are of order $\beta\hbar$, but the $\langle c^m c^*_n\rangle$ ghost propagator is $\beta\hbar$-independent,

$$\begin{aligned} \langle q^i(\sigma)q^j(\tau)\rangle &= -\beta\hbar g^{ij}(x_0)\Delta(\sigma,\tau) \\ \langle c^m(\sigma)c^*_n(\tau)\rangle &= \delta^m_n\theta(\sigma-\tau). \end{aligned} \quad (7.65)$$

As already explained in Section 7.2 we only get trees for the ghosts, and integration over the Grassmann variables at the front and at the back of the tree leads to a trace over the indices at the ends of the tree.

Using Riemann normal coordinates, the expansion of $g_{ij}(x_0+q) - g_{ij}(x_0)$ contains terms with two, three, four, ... q-fields, leading to four, five, six, ... point functions for the q fields and a, b, c ghosts. The contributions from this term alone were already determined when we calculated the trace anomaly of a scalar in four dimensions. There are no contributions from the product of this vertex and the second vertex to order $(\beta\hbar)^2$ because the Riemann tensor is traceless, but products of the first and last vertex lead to disconnected graphs, the contribution of which is proportional to the product of the two-loop graph for a scalar in $n = 2$ with the scalar curvature.

The next vertex is the $\dot{q}(c^*\omega c)$ vertex. In a frame were $\omega_i{}^m{}_n(x_0) = 0$ and $\partial_{(j}\omega_{i)}{}^m{}_n(x_0) = 0$, it becomes

$$-\int_{-1}^{0} d\tau\, q^i \dot{q}^j \left(c^*_m \frac{1}{2} R_{ij}{}^m{}_n c^n \right). \tag{7.66}$$

The square of this vertex yields an R^2_{ijmn} term. To order $(\beta\hbar)^2$ there are no contributions from the product of this vertex with the vertices in the last line of (7.64).

Finally, in the last line of S^{int} the (c^*Rc) vertex can either be squared to yield an R^2_{mn} term or the R_{mn} inside (c^*Rc) can be expanded to second order in q to yield a D^2R term or it can be multiplied with the $-\frac{1}{8}(R+\Gamma\Gamma)$ terms to yield a R^2 term. The $-\frac{1}{8}(R+\Gamma\Gamma)$ terms can be squared to yield an R^2 term or the R and $\Gamma\Gamma$ terms can be expanded to second order to yield a D^2R and an R^2_{ijkl} term.

The contributions proportional to $(\hbar\beta)^2$ are as follows:[5]

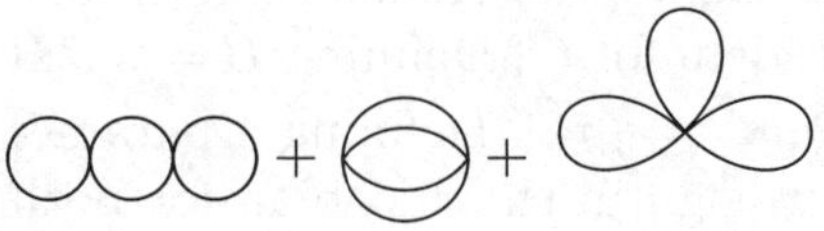

$$= 4\left(-\frac{1}{480} R^2_{ijkl} - \frac{1}{720} R^2_{ij} + \frac{1}{480} D^2 R \right)$$

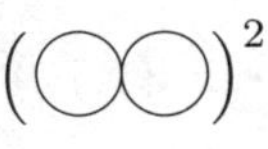

$$= \frac{1}{2!} 4 \left(\frac{1}{24} R \right)^2$$

$$= \left(\frac{1}{24} R \right) \left(\frac{1}{2} R - \frac{4}{8} R \right) = 0$$

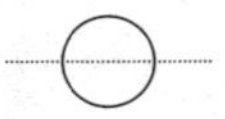

$$= -\frac{1}{48} R^2_{ijmn}$$

$$= \frac{1}{8} R^2_{mn}$$

$$= \frac{1}{24} D^2 R$$

[5] Recall that one obtains an integral $\int d\chi_{gh}\, d\bar{\eta}_{gh}\, P^{gh}_{\bar{\eta},\chi}$ while the transition element contains a factor of $e^{\bar{\eta}_{gh}\chi^{gh}}$. For the interactions which are independent of the ghost fields the factor $\bar{\eta}_{gh}\chi^{gh}$, obtained by expanding $e^{\bar{\eta}_{gh}\chi^{gh}}$, saturates the integral over $d\chi_{gh}$ and $d\bar{\eta}_{gh}$, and one obtains a factor $\mathrm{Tr}\, I = 4$.

$($$)\ (\bullet) = -\frac{1}{16}R^2$

$$(\bullet)^2 = 4\frac{1}{2!}\left(\frac{1}{8}R\right)^2 = \frac{1}{32}R^2$$

 $= 4\left(-\frac{1}{96}D^2R + \frac{1}{288}R^2_{ijkl}\right).$ (7.67)

Dotted lines indicate ghosts, and external ghosts are traced over. To evaluate the fourth diagram we used

$$\begin{aligned}\int\!\!\int d\sigma\, d\tau\, ({}^{\bullet}\Delta^{\bullet}\Delta - {}^{\bullet}\Delta\Delta^{\bullet})\,\theta &= -\int\!\!\int d\sigma\, d\tau\ \delta(\sigma-\tau)\Delta(\sigma-\tau)\theta(\sigma-\tau)\\ &= -\frac{1}{2}\int d\tau\, \tau(\tau+1) = \frac{1}{12}.\end{aligned} \tag{7.68}$$

The QFT Faddeev–Popov ghosts, denoted by B and C, contribute, too. Their action reads

$$\mathcal{L}(\text{ghosts}) = \sqrt{g}\, Bg^{\mu\nu}D_\mu\partial_\nu C. \tag{7.69}$$

The regulator which follows from this action is the same as for scalar fields but without an improvement term. Under rigid scale transformation one has in four dimensions $\delta_W B = -\frac{1}{2}\sigma B$ and idem for C. Defining $\tilde{B} = g^{1/4}B$ and $\tilde{C} = g^{1/4}C$ one has $\delta_W\tilde{B} = \frac{1}{2}\sigma\tilde{B}$ and $\delta_W\tilde{C} = \frac{1}{2}\sigma\tilde{C}$. Defining a parameter σ_{gh} which takes into account the minus sign in the Jacobian for both ghosts

$$\sigma_{gh} = (-2)\frac{1}{2}\sigma \tag{7.70}$$

we obtain

$$\begin{aligned}An(\text{ghosts}) &= \mathrm{Tr}\,\sigma_{gh}\, \mathrm{e}^{-\frac{\beta}{\hbar}\mathcal{R}_{gh}}\\ &= \int d^4x_0\,\sqrt{g(x_0)}\sigma_{gh}(x_0)\frac{1}{(2\pi\beta\hbar)^2}\,\langle \mathrm{e}^{-\frac{1}{\hbar}S^{int}_{gh}}\rangle\end{aligned} \tag{7.71}$$

where

$$\begin{aligned}-\frac{1}{\hbar}S^{int}_{gh} =& -\frac{1}{\beta\hbar}\int_{-1}^{0} d\tau\,\frac{1}{2}[g_{ij}(x_0+q) - g_{ij}(x_0)](\dot{q}^i\dot{q}^j + b^ic^j + a^ia^j)\\ &- \frac{\beta\hbar}{8}\int_{-1}^{0} d\tau\left(R + g^{ij}\Gamma^l_{ik}\Gamma^k_{jl}\right).\end{aligned} \tag{7.72}$$

The contributions of the ghosts are not multiplied by a factor of 4 because there is no integration over internal symmetry ghosts (but since $\sigma_{gh} = (-2)\frac{1}{2}\sigma$, the ghosts still subtract two degrees of freedom from $\tilde{A}_m$). One finds to order $(\beta\hbar)^2$ the following contributions from the Faddeev–Popov ghosts to the trace anomaly:

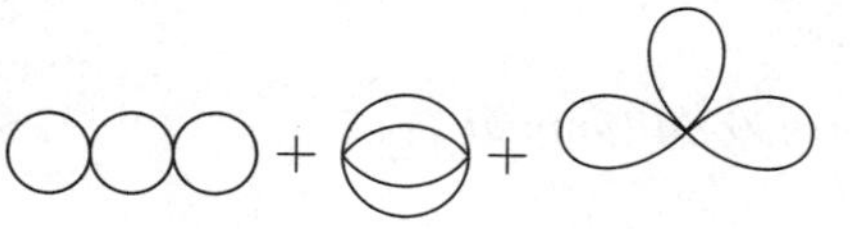 $= \left(-\frac{1}{480} R_{ijkl}^2 - \frac{1}{720} R_{ij}^2 + \frac{1}{480} D^2 R \right)$

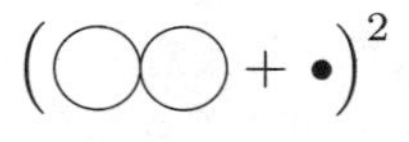 $= \frac{1}{2!} \left(\frac{1}{24} R - \frac{1}{8} R \right)^2 = \frac{1}{288} R^2$

 $= -\frac{1}{96} D^2 R + \frac{1}{288} R_{ijkl}^2. \qquad (7.73)$

Adding the contributions of the ghosts to the contributions of the vector, not forgetting the factor of -2 from $\sigma_{gh}(x_0) = -2\sigma_V(x_0)$, one obtains the total result

$$An_W(\text{Maxwell}, n = 4)$$
$$= \int \frac{d^4x}{(2\pi)^2} \sqrt{g(x)} \sigma_V(x) \left(-\frac{13}{720} R_{ijkl}^2 + \frac{11}{90} R_{ij}^2 - \frac{5}{144} R^2 + \frac{1}{40} D^2 R \right). \tag{7.74}$$

This is the correct result. The coefficient of D^2R is scheme-dependent (a counterterm R^2 can change it). We find the same result as DeWitt (second reference in [18]), while Duff [139] finds a coefficient of $-\frac{1}{60}$. For an alternative calculation of the trace anomaly of a spin-1 field based on the $N = 2$ nonlinear sigma model, see [31]

7.4 String-inspired approach to trace anomalies

In the previous calculation of trace anomalies we have used the path integral expression for the transition amplitude $\langle x_0 | e^{-\frac{\beta}{\hbar} H} | x_0 \rangle$ to obtain the trace

$$\operatorname{tr} \sigma \, e^{-\frac{\beta}{\hbar} H} = \int d^n x_0 \sqrt{g(x_0)} \, \sigma(x_0) \, \langle x_0 | e^{-\frac{\beta}{\hbar} H} | x_0 \rangle. \tag{7.75}$$

This transition amplitude was calculated with a quantum-background split in the path integral, $x^i(\tau) = x_0^i + q^i(\tau)$, where x_0^i is constant and $q^i(\tau)$ obeys the Dirichlet boundary conditions $q^i(-1) = q^i(0) = 0$. At the point x_0^i, the function $x^i(\tau)$ is continuous, but its derivative need not be continuous ($x^i(\tau)$ may have a kink at $\tau = 0$). However, in any discretized approach this subtlety constitutes no problem. Thus $x^i(\tau)$ satisfies periodic boundary conditions (PBC). This corresponds to a loop starting and ending at a fixed point x_0^i in target space (see Fig. 7.1). The integration over x_0^i then places the loop everywhere in target space (with weight factor $\sigma(x_0)$).

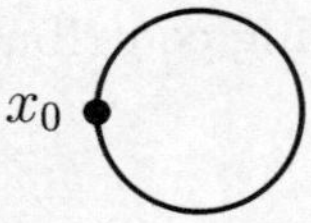

Fig. 7.1 *Dirichlet boundary conditions at x_0.*

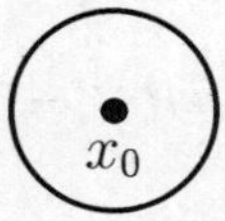

Fig. 7.2 *Periodic boundary conditions with fixed average position x_0.*

The path integral with PBC is geometrically a path integral on the circle (or one-dimensional torus), and can also be computed in other ways, for example using the so-called "string-inspired" approach [27, 175]. In this approach the periodic functions $x^i(\tau)$ are again split into $x^i(\tau) = x_0^i + y^i(\tau)$, but now the coordinate x_0^i represents the average position of the loop,

$$x_0^i = \int_{-1}^{0} d\tau \, x^i(\tau). \tag{7.76}$$

Hence, $y^i(\tau)$ satisfies the condition

$$\int_{-1}^{0} d\tau \, y^i(\tau) = 0. \tag{7.77}$$

This situation is depicted in Fig. 7.2.

The need to split $x^i(\tau)$ into two parts is due to the fact that in perturbative calculations the quadratic part of the action has a kinetic operator $\delta_{ij}\partial_\tau^2$ which is not invertible on the torus because of the presence of zero modes. Zero modes are by definition normalizable eigenfunctions of the linearized field operator for the fluctuations with zero eigenvalue, and the operator ∂_τ^2 indeed has periodic solutions which are normalizable, namely $x^i(\tau) = x_0^i$. The separation and subsequent integration over the zero modes can be achieved in various ways. We first describe an extension of an approach of Strassler [175] to curved space in which $x^i(\tau)$ is decomposed into $x_0^i + y^i(\tau)$, where $y^i(\tau)$ satisfies (7.77). As we shall discuss, in this approach one has different propagators from those used in Part I of this book. Using general coordinates, the global (integrated) trace anomaly (with constant parameter σ) comes out correctly, but the local trace anomaly is off by noncovariant total derivatives (which are not only proportional to $\Box R$ in $d = 4$). Furthermore, the naive use of Riemann normal coordinates yields an incorrect result for the integrated trace anomaly. With hindsight, it is not so astonishing that the decomposition of $x^i(\tau)$ into $x_0^i + y^i(\tau)$ with $\int_{-1}^{0} d\tau \, y^i(\tau) = 0$ yields noncovariant total derivatives in curved target space, as a consequence of which one cannot extract the local trace anomaly by this method. The numbers x_0^i in (7.76) and (7.77) are defined independently of the metric in target space, and in curved space one can think of other, perhaps better, ways of defining x_0^i (for example, through invariant line elements).

To remedy these shortcomings, we present a different approach to quantum mechanics on the circle [89]. We introduce an **extra** constant mode into $x^i(\tau)$; this leads to a shift symmetry which one fixes with BRST methods. This was first worked out in the context of general nonlinear sigma models by Friedan [176], see also [91, 177–179]. One chooses Riemann normal coordinates, and this leads to new ghosts for the shift symmetry which yield further vertices. In this approach the local trace anomaly comes out correctly (except for the coefficient of the term $\Box R$ in $d = 4$, which can be modified by the addition of the local counterterm R^2 in the action).

Let us now first discuss the straightforward approach to quantum mechanics on a circle with curved target space and then return to the approach with shift symmetry. We discretize $x^i(\tau)$ on the circle into $n \times N$ integrations variables x_k^i for periodic functions on a circle,

$$x_k^i = \sum_{p=0}^{N/2} \frac{2}{\sqrt{N}} \cos\left(\frac{2kp\pi}{N}\right) r_p^{i,c} + \sum_{p=1}^{N/2-1} \frac{2}{\sqrt{N}} \sin\left(\frac{2kp\pi}{N}\right) r_p^{i,s} \quad (7.78)$$

where $k = 0, \ldots, N$ and $i = 1, \ldots, n$. (We take N even.) We decompose x_k^i into a "center-of-mass" part and fluctuations around it

$$x_k^i = \frac{2}{\sqrt{N}} r_0^{i,c} + q_k^i. \quad (7.79)$$

Coupling q_k^i to external sources, one finds the propagators for the string-inspired method in closed form [64],

$$\Delta^{SI}(\sigma, \tau) = \frac{1}{2}|\sigma - \tau| - \frac{1}{2}(\sigma - \tau)^2 - \frac{1}{12}. \quad (7.80)$$

As expected, this propagator is translation invariant, and it is easy to check that it satisfies $\partial_\sigma^2 \Delta^{SI}(\sigma, \tau) = \delta(\sigma - \tau) - 1$, which yields the Dirac delta function in the space of periodic functions $q^i(\tau)$ orthogonal to the constant function. The factor of $-\frac{1}{12}$ achieves that $\int_{-1}^0 d\sigma\, \Delta^{SI}(\sigma, \tau) = 0$, which should hold as $\langle \int_{-1}^0 d\sigma\, q^i(\sigma)\, q^i(\tau) \rangle$ should vanish. One can now use these propagators, and compute the trace anomaly in $d = 2$ and $d = 4$. The results of this approach were mentioned at the beginning of this section, and we refer to the literature for further details [65].

Let us now turn to the approach with an extra shift symmetry. We first make some general comments about shift symmetries and their BRST quantization, and then specialize to Riemann normal coordinates. Suppose one begins by setting

$$x^i(\tau) = x_0^i + y^i(\tau) \tag{7.81}$$

but one does not restrict $y^i(\tau)$. Then a shift symmetry is introduced

$$\delta x_0^i = \epsilon^i, \qquad \delta y^i(\tau) = -\epsilon^i. \tag{7.82}$$

This symmetry resembles the shift symmetry in background Yang–Mills theory [180], where $A^a_{\mu,bg}(x) + a^a_{\mu,qu}(x)$ is invariant under the local shift gauge symmetry

$$\delta A^a_{\mu,bg}(x) = M^a_\mu(x), \qquad \delta a^a_{\mu,qu}(x) = -M^a_\mu(x). \tag{7.83}$$

(This symmetry then becomes part of the BRST symmetry with anticommuting external sources $M^a_\mu(x)$ [180].) However, there is a difference: in the Yang–Mills case one does not integrate over the background Yang–Mills fields $A^a_{\mu,bg}(x)$, whereas in our case we shall integrate both over $y^i(\tau)$ and over x_0^i. The use of BRST methods for a point particle may seem like overkill, but it solves the problem of obtaining the right answer for local trace anomalies on a circle. We shall see that in one particular gauge this approach is reduced to the formulation which we discussed in the first part of the book. In another gauge we find a formulation on the circle (with extra vertices with ghosts).

Consider the partition function for the one-dimensional bosonic nonlinear sigma model (for notational simplicity here we set $\hbar = 1$),

$$\begin{aligned} Z(\beta) &= \int_{PBC} \mathcal{D}x \; \mathrm{e}^{-S[x]} \\ S[x] &= \frac{1}{\beta}\int_{-1}^{0} d\tau \left[\frac{1}{2} g_{ij}(x)\dot{x}^i\dot{x}^j + \beta^2 V(x)\right] \end{aligned} \tag{7.84}$$

where PBC denotes the periodic boundary conditions $x^i(-1) = x^i(0)$. As anticipated, we introduce a redundant variable x_0^i by setting $x^i(\tau) = x_0^i + y^i(\tau)$ in the action, $S[x(\tau)] = S[x_0 + y(\tau)]$. This automatically introduces the shift symmetry in (7.82), as in the background field method. We need to consider only constant x_0^i so that the shift symmetry requires a constant parameter ϵ^i. However, we take both x_0^i and $y^i(\tau)$ as dynamical variables (i.e. to be integrated over in the path integral). We use BRST methods and introduce a constant ghost field η^i together with the following BRST transformation rules:

$$\begin{aligned} \delta x_0^i &= \eta^i \Lambda \\ \delta y^i(\tau) &= -\eta^i \Lambda \\ \delta \eta^i &= 0 \end{aligned} \tag{7.85}$$

where Λ is the usual constant anticommuting purely imaginary BRST parameter. The gauge algebra is abelian and, consequently, the ghost is inert under BRST. This makes the BRST symmetry nilpotent, $[\delta(\Lambda_1), \delta(\Lambda_2)] = 0$. To fix a gauge one must also introduce constant non-minimal fields, namely an anticommuting hermitian $\bar{\eta}_i$ and a commuting real π_i, with the BRST rules

$$\delta\bar{\eta}_i = i\pi_i\Lambda, \qquad \delta\pi_i = 0. \tag{7.86}$$

The gauge-fixing fermion $\Psi = \bar{\eta}_i x_0^i$ leads back to (7.84), and then one can use the approach we have followed in most of the book. Instead we choose the gauge-fixing fermion

$$\Psi = \bar{\eta}_i \int_{-1}^{0} d\tau\, \rho(\tau) y^i(\tau) \tag{7.87}$$

which depends on an arbitrary function $\rho(\tau)$, sometimes called the background charge, normalized to one, $\int_{-1}^{0} d\tau \rho(\tau) = 1$. It produces the following quantum action:

$$\begin{aligned} S_{qu}[x_0, y, \eta, \bar{\eta}, \pi] &= S[x_0 + y] + \frac{\delta}{\delta\Lambda}\Psi \\ &= S[x_0 + y] + i\pi_i \int_{-1}^{0} d\tau\, \rho(\tau) y^i(\tau) - \bar{\eta}_i \eta^i \end{aligned} \tag{7.88}$$

where $\delta/\delta\Lambda$ denotes a BRST variation with the anticommuting parameter Λ removed from the left. This quantum action is BRST invariant. In this gauge the ghosts can be trivially integrated out, while the integration over the auxiliary variable π_i produces a delta function which constrains the fields y^i to satisfy

$$\int_{-1}^{0} d\tau\, \rho(\tau) y^i(\tau) = 0. \tag{7.89}$$

With this constraint the perturbative kinetic term for the periodic fields $y^i(\tau)$ can be inverted to obtain the propagator. One expands $y^i(\tau)$ into sines and cosines, including a constant mode y_0^i. The propagator for $y^i(\tau)$ is obtained by first expressing y_0^i in terms of the other modes y_n^i using the constraint, and then using the Gaussian path integral to obtain the correlator of the remaining y_n^i. The propagator for general ρ is then given by

$$\langle y^i(\tau) y^j(\sigma) \rangle = -\beta g^{ij}(x_0) \mathcal{B}_{(\rho)}(\tau, \sigma) \tag{7.90}$$

where

$$\mathcal{B}_{(\rho)}(\tau, \sigma) = \Delta^{SI}(\tau - \sigma) - F_\rho(\tau) - F_\rho(\sigma) + C_\rho \tag{7.91}$$

with

$$\Delta^{SI}(\tau-\sigma)=\frac{1}{2}|\tau-\sigma|-\frac{1}{2}(\tau-\sigma)^2-\frac{1}{12}$$

$$F_\rho(\tau)=\int_{-1}^{0}dx\,\Delta^{SI}(\tau-x)\rho(x)$$

$$C_\rho=\int_{-1}^{0}dx\,F_\rho(x)\rho(x). \tag{7.92}$$

It satisfies

$$\frac{d^2}{d\tau^2}\mathcal{B}_{(\rho)}(\tau,\sigma)=\delta(\tau-\sigma)-\rho(\tau) \tag{7.93}$$

and

$$\int_{-1}^{0}d\tau\,\rho(\tau)\,\mathcal{B}_{(\rho)}(\tau,\sigma)=0. \tag{7.94}$$

The propagator with arbitrary background charge was discussed in [181].

Using this propagator, one finds for the partition function

$$Z(\beta)=\int d^n x_0\,\frac{\sqrt{g(x_0)}}{(2\pi\beta)^{n/2}}\,z^{(\rho)}(x_0,\beta) \tag{7.95}$$

where

$$z^{(\rho)}(x_0,\beta)=1-\beta\left(V+\frac{1}{12}R\right)+\frac{\beta}{2\sqrt{g}}\left(C_\rho+\frac{1}{12}\right)\partial_k(\sqrt{g}g^{ij}\Gamma_{ij}^k)+O(\beta^2). \tag{7.96}$$

For $\rho(\tau)=\delta(\tau)$ one has $C_\rho=-\frac{1}{12}$ and then recovers the results obtained in Part I of the book, while for $\rho(\tau)=1$ one has $C_\rho=0$ and recovers the results of the string-inspired approach [65]. That the noncovariant total derivative term has the simple form in (7.96) was shown in [28]. Of course, noncovariant total derivative terms are also expected at higher order in β. So far we have worked with arbitrary coordinates. Let us now turn to the question of how to introduce Riemann normal coordinates.

We write $y^i(\tau)$ in Riemann normal coordinates $\xi^i(\tau)$ centered at x_0^i. This means that $y^i(\tau)$ depends on x_0^i and $\xi^i(\tau)$, so $y^i(\tau)=y^i(x_0,\xi(\tau))$. The fact that ξ^i in $y^i(x_0,\xi)$ are Riemann normal coordinates centered at the point x_0^i leads to the following expression for y^i:

$$y^i=\xi^i-\frac{1}{2}\Gamma_{ij}^i(x_0)\xi^j\xi^k-\frac{1}{3!}\bar{\nabla}_j\Gamma_{kl}^i(x_0)\xi^j\xi^k\xi^l+\cdots \tag{7.97}$$

where $\bar{\nabla}_j$ only acts on lower indices. In order for $\delta y^i = -\epsilon^i$ we now need a nonlinear transformation rule for $\xi^i(\tau)$

$$\begin{aligned} \delta x_0^i &= \epsilon^i \\ \delta \xi^i(\tau) &= -Q^i{}_j(x_0, \xi(\tau))\epsilon^j \end{aligned} \tag{7.98}$$

which is a reformulation of (7.82) in these new coordinates. By varying x_0^i and ξ^i in the expression for y^i by using (7.98), one should reproduce $\delta y^i = -\epsilon^i$. In this way the expression of $Q^i{}_j(x_0, \xi)$ can be calculated explicitly [176]. It reads as

$$\begin{aligned} Q^i{}_j(x_0, \xi) &= \delta^i_j - \frac{1}{3} R^i{}_{mnj}\, \xi^m \xi^n - \frac{1}{12} \nabla_m R^i{}_{npj}\, \xi^m \xi^n \xi^p \\ &- \left(\frac{1}{60} \nabla_m \nabla_n R^i{}_{pqj} + \frac{1}{45} R^i{}_{mns} R^s{}_{pqj} \right) \xi^m \xi^n \xi^p \xi^q + O(\xi^5). \end{aligned} \tag{7.99}$$

Now we can introduce ghosts and auxiliary fields as usual. The BRST symmetry for the nonlinear shift symmetry is

$$\begin{aligned} \delta x_0^i &= \eta^i \Lambda \\ \delta \xi^i(\tau) &= -Q^i{}_j(x_0, \xi(\tau))\, \eta^j \Lambda \\ \delta \eta^i &= 0 \\ \delta \bar{\eta}_i &= i \pi_i \Lambda \\ \delta \pi_i &= 0. \end{aligned} \tag{7.100}$$

It is nilpotent since $Q^i{}_j(x_0, \xi)$ satisfies certain relations arising from the abelian nature of the shift symmetry. The gauge fermion

$$\Psi = \bar{\eta}_i \int_{-1}^{0} d\tau\, \rho(\tau) \xi^i(\tau) \tag{7.101}$$

produces the gauge-fixed quantum action

$$\begin{aligned} S_{qu}[x_0, \xi, \eta, \bar{\eta}, \pi] &= S[x_0, \xi] + \frac{\delta}{\delta \Lambda} \Psi \\ &= S[x_0, \xi] + i \pi_i \int_{-1}^{0} d\tau\, \rho(\tau) \xi^i(\tau) \\ &\quad - \bar{\eta}_i \int_{-1}^{0} d\tau\, \rho(\tau) Q^i{}_j(x_0, \xi(\tau))\, \eta^j. \end{aligned} \tag{7.102}$$

The integration over the auxiliary variable π_i again gives a delta function which constrains the fields ξ^i to satisfy

$$\int_{-1}^{0} d\tau\, \rho(\tau) \xi^i(\tau) = 0 \tag{7.103}$$

so that their propagator can be obtained. This constraint has a simple tensorial transformation under change of coordinates of x_0^i, in fact Riemann normal coordinates coordinates transform as vectors under a reparametrization of the origin x_0^i. Note, however, that the ghosts now give a nontrivial contribution, i.e. a nontrivial Faddeev–Popov determinant. This is given by the last line in (7.102).

The above gauge-fixed action can be used in the path integral (one may check that the measure is both reparametrization and BRST invariant). The auxiliary field π^i can be integrated out to obtain

$$Z(\beta) = \int dx_0\, d\eta\, d\bar{\eta} \oint D\xi\, Da\, Db\, Dc\ \delta\left[\int_0^1 d\tau\, \rho(\tau)\xi^i(\tau)\right] \mathrm{e}^{-\bar{S}_{qu}} \quad (7.104)$$

with

$$\begin{aligned}\bar{S}_{qu} = \frac{1}{\beta}\int_{-1}^{0} d\tau \Big\{ &\frac{1}{2} g_{ij}(x_0,\xi)(\dot{\xi}^i\dot{\xi}^j + a^i a^j + b^i c^j) \\ &+ \beta^2[V(x_0,\xi) + V_{CT}(x_0,\xi)]\Big\} \\ &- \bar{\eta}_i \int_{-1}^{0} d\tau\, \rho Q^i{}_j(x_0,\xi)\, \eta^j \end{aligned} \quad (7.105)$$

where $g_{ij}(x_0,\xi) = g_{ij}(x_0 + y(x_0,\xi))$, etc. We have added the counterterm V_{CT} appropriate to the regularization scheme one chooses as well as the a, b, c ghosts. The corresponding perturbative expansion can then be cast as follows:

$$\begin{aligned} Z(\beta) &= \int d^n x_0 \frac{\sqrt{g(x_0)}}{(2\pi\beta)^{n/2}}\, z^{(\rho)}(x_0,\beta) \\ z^{(\rho)}(x_0,\beta) &= \left\langle \exp(-\bar{S}_{qu}^{(int)}) \right\rangle \end{aligned} \quad (7.106)$$

where $\bar{S}_{qu}^{(int)}$ is obtained from (7.105) with the substitution $g_{ij}(x_0,\xi) \to g_{ij}(x_0,\xi) - g_{ij}(x_0)$ and $Q^i{}_j(x_0,\xi) \to Q^i{}_j(x_0,\xi) - \delta^i_j$. The expectation value of the interactions are computed with the propagators

$$\begin{aligned} \langle \xi^i(\tau)\xi^j(\sigma)\rangle &= -\beta g^{ij}(x_0)\mathcal{B}_{(\rho)}(\tau,\sigma) \\ \langle a^i(\tau)a^j(\sigma)\rangle &= \beta g^{ij}(x_0)\Delta_{gh}(\tau-\sigma) \\ \langle b^i(\tau)c^j(\sigma)\rangle &= -2\beta g^{ij}(x_0)\Delta_{gh}(\tau-\sigma) \\ \langle \eta^i \bar{\eta}_j\rangle &= -\delta^i_j. \end{aligned} \quad (7.107)$$

The terms in (7.106) have the following origin: the factor $\sqrt{g(x_0)}$ is due to the a, b, c ghosts which contain the constant "zero modes", and the factor of $(2\pi\beta)^{-n/2}$ is the usual free particle measure which corresponds to the determinant of $-\frac{1}{2\beta}\partial_\tau^2$ on the circle with zero modes excluded.

The Green functions appearing in the propagators are as follows. The function $\mathcal{B}_{(\rho)}(\tau,\sigma)$ is the Green function of the operator ∂_τ^2 acting on fields constrained by the equation $\int_{-1}^0 d\tau\rho(\tau)\xi^i(\tau) = 0$ and was given before. We just recall that a simple way to obtain this propagator is to use the mode expansions $\xi^i(\tau) = \sum_n \xi_n^i e^{2\pi i n\tau}$ and reconstruct the function $\mathcal{B}_{(\rho)}(\tau,\sigma)$ from the correlators of the mode coefficients $\langle \xi_n^i \xi_m^j\rangle$. We leave this as an exercise for the interested reader. The Green function for the ghosts is given by the delta function on the torus

$$\Delta_{gh}(\tau-\sigma) = \delta(\tau-\sigma) \tag{7.108}$$

but we keep writing Δ_{gh} to indicate that in perturbative calculations it appears in a regulated form.

As an application one may compute the trace anomalies for a scalar field in two and four dimensions using the string-inspired propagator ($\rho(\tau)=1$) and dimensional regularization. One obtains

$$\begin{aligned} z^{(SI)}(x_0,\beta) = 1 + \beta\Big(\frac{1}{24}R - V_q\Big) + \beta^2\bigg[\frac{1}{2}\Big(\frac{1}{24}R - V_q\Big)^2 \\ + \frac{1}{720}R_{ijkl}^2 - \frac{1}{720}R_{ij}^2 + \frac{1}{1728}\Box R - \frac{1}{24}\Box V_q\bigg] + \cdots \end{aligned} \tag{7.109}$$

with $V_q \equiv V + V_{DR} = V + \frac{1}{8}R$. The trace anomalies in $n = 2$ arise from the β term in (7.109). In this case we have to set $V = 0$ and the expected result of $-\frac{1}{12}R$ is readily obtained (compare with the local term multiplying σ_S in (7.17)). For $n = 4$ we have to include the improvement term $V = -\frac{1}{2}\xi R = -\frac{1}{12}R$. Thus $V_q = \frac{1}{24}R$. The trace anomalies are now given by the β^2 term in (7.109) and read as

$$\frac{1}{720}(R_{ijkl}^2 - R_{ij}^2) - \frac{1}{864}\Box R. \tag{7.110}$$

This result reproduces the answer in (7.27) and (7.28) up to the total derivative term $\Box R$. As we have discussed, its coefficient is in general ρ-dependent, but it corresponds to a trivial anomaly which is also scheme-dependent. The answer for arbitrary background charge ρ can be found in the literature [89]. The conclusion is that it is legal to use the string-inspired approach to compute trace anomalies.

8
Conclusions and summary

In this book we have discussed in detail the construction of path integrals for quantum mechanical nonlinear sigma models and applied the results to compute various one-loop anomalies of quantum field theories in n dimensions. There are, of course, many more applications of quantum mechanical path integrals to physical problems [20–27]. The anomalies we computed were chiral anomalies due to single loops with complex chiral spin $\frac{1}{2}$, complex chiral spin $\frac{3}{2}$ and real self-dual antisymmetric tensor fields in the loop, coupled to external gravity and/or external gauge fields. For the computation of these anomalies we only needed one-loop graphs on the worldline. We also computed trace anomalies due to one-loop graphs in quantum field theory, but here we needed higher-loop graphs on the worldline.

We can now answer the questions about the time slicing method raised at the end of Section 1.2.

(i) Given a quantum mechanical Hamiltonian $H(\hat{x},\hat{p})$ depending on a set of operators $\hat{x}^i, \hat{p}_i$ (with $i = 1,\ldots,n$) with definite ordering of the $\hat{x}^i$ and $\hat{p}_i$ operators, the Hamiltonian $H_W(x,p)$ which enters the phase-space path integral is obtained by rewriting $H(\hat{x},\hat{p})$ in Weyl-ordered form, and then replacing the operators $\hat{x}^i, \hat{p}_i$ by functions $x^i(t), p_i(t)$. So the operator $H(\hat{x},\hat{p})$ is the same as the Weyl-ordered operator $(H(\hat{x},\hat{p}))_W$; it is only written in a different form. Having rewritten $H(\hat{x},\hat{p})$ in the form $(H(\hat{x},\hat{p}))_W$, we replace the operator $\hat{H}$ by the function $H_W(x,p)$. For example, $\hat{x}\hat{p} = (\hat{x}\hat{p})_S + \frac{1}{2}i\hbar \equiv (\hat{x}\hat{p})_W$ where $(\hat{x}\hat{p})_S$ is the symmetrized expression $\frac{1}{2}(\hat{x}\hat{p} + \hat{p}\hat{x})$, and thus $(xp)_W = xp + \frac{1}{2}i\hbar$ with $xp = px$. In particular, the operator

$$
\begin{aligned}
\hat{H}_1(\hat{x},\hat{p}) &= \frac{1}{2} g^{-1/4}(\hat{x})\hat{p}_i g^{1/2}(\hat{x}) g^{ij}(\hat{x})\hat{p}_j g^{-1/4}(\hat{x}) \\
&= \frac{1}{2}(\hat{g}^{ij}\hat{p}_i\hat{p}_j)_S + \frac{\hbar^2}{8}(\hat{g}^{il}\hat{\Gamma}^k_{ij}\hat{\Gamma}^j_{lk} + \hat{R}) \\
&\equiv (\hat{H}_1(\hat{x},\hat{p}))_W
\end{aligned}
\tag{8.1}
$$

where $(\hat{g}^{ij}\hat{p}_i\hat{p}_j)_S \equiv \frac{1}{4}\hat{p}_i\hat{p}_j\hat{g}^{ij} + \frac{1}{2}\hat{p}_i\hat{g}^{ij}\hat{p}_j + \frac{1}{4}\hat{g}^{ij}\hat{p}_i\hat{p}_j$ leads to the function

$$H_{1,W}(x,p) = \frac{1}{2}g^{ij}p_ip_j + \frac{\hbar^2}{8}(g^{il}\Gamma^k_{ij}\Gamma^j_{lk} + R). \tag{8.2}$$

A particularly simple Hamiltonian is

$$\hat{H}_2(\hat{x},\hat{p}) = \frac{1}{2}(\hat{g}^{ij}\hat{p}_i\hat{p}_j)_S = \frac{1}{8}\hat{p}_i\hat{p}_j\hat{g}^{ij} + \frac{1}{4}\hat{p}_i\hat{g}^{ij}\hat{p}_j + \frac{1}{8}\hat{g}^{ij}\hat{p}_i\hat{p}_j \tag{8.3}$$

which is already Weyl-ordered. The corresponding $H_{2,W}(x,p)$ is $\frac{1}{2}g^{ij}p_ip_j$. However, $\hat{H}_2$ is not general coordinate invariant. In our applications we needed $\hat{H}_1$.

(ii) The measure in phase space is by definition given in the time-sliced approach by the discretized phase-space expression

$$d\mu(x,p) = \frac{\prod_{j=1}^n (dp_{1,j}\cdots dp_{N,j})(dx_1^j\cdots dx_{N-1}^j)}{(2\pi\hbar)^{nN}g^{1/4}(z)g^{1/4}(y)} \tag{8.4}$$

where z and y are the endpoints of $\langle z|\mathrm{e}^{-(\beta/\hbar)\hat{H}}|y\rangle$. The normalizations were defined in (2.4) and (2.5). With these normalizations the transition element becomes a biscalar. Integrating out the momenta, we found the discretized configuration-space measure

$$d\mu(x) = \frac{\prod_{j=1}^n (dx_1^j\cdots dx_{N-1}^j)\prod_{k=1}^N g^{1/2}\left(\frac{1}{2}(x_k + x_{k-1})\right)}{(2\pi\epsilon\hbar)^{nN/2}g^{1/4}(z)g^{1/4}(y)}\,. \tag{8.5}$$

We then exponentiated the factors $g^{1/2}(\frac{1}{2}(x_k + x_{k-1}))$ by introducing ghosts $a^i_{k-1/2}$, $b^i_{k-1/2}$, $c^i_{k-1/2}$, decomposing x^i_k into a background part $x^i_{bg,k}$ and a quantum part q^i_k, where in the continuum limit $x^i_{bg}(t) = z^i + \frac{t}{\beta}(z^i - y^i)$. Next, we expanded the q^i_k into $\sin(km\pi/N)$ with $k,m = 1,\ldots,N-1$ to diagonalize the kinetic term, and then performed the remaining integrations. The result reads as

$$\langle z|\exp\Big(-\frac{\beta}{\hbar}\hat{H}\Big)|y\rangle = \left[\frac{g(z)}{g(y)}\right]^{1/4}\frac{1}{(2\pi\beta\hbar)^{n/2}}\langle \mathrm{e}^{-\frac{1}{\hbar}S^{(int)}}\rangle\,. \tag{8.6}$$

Because there are N integrations over $a_{k-1/2}$, $b_{k-1/2}$, $c_{k-1/2}$, but only $N-1$ integrations over q_k, one obtains an extra factor of $g^{1/2}(z)$ which converts the prefactor $[g(z)g(y)]^{-1/4}$ into $[g(z)/g(y)]^{1/4}$. This is a consequence of the fact that there is one more p integration than x integration. Because we decomposed the action into a kinetic part $S^{(0)}$ and an interaction part $S^{(int)}$ and used the metric at z in $S^{(0)}$, the coordinates y and z do not appear symmetrically in the discretized expressions.

Our conclusion concerning the measure is the following: the precise form of the intermediate measures with $d\mu(x,p)$ and $d\mu(x)$ was crucial for obtaining the correct transition element. Of relevance is the final prefactor $[g(z)/g(y)]^{1/4}(2\pi\beta\hbar)^{-n/2}$ in (8.6). All of the physics is contained in $\langle \mathrm{e}^{-\frac{1}{\hbar}S^{(int)}}\rangle$. For perturbation theory, this object is to be expanded into powers of $S^{(int)}$, and the Wick contraction rules are applied, with propagators derived in Part I of the book. Since all expressions were deductions from the operatorial starting point, these results are unambiguous. Equal-time contractions and products of factors $\delta(\sigma-\tau)$ and $\theta(\sigma-\tau)$ are well defined. The final result for the transition element for the bosonic case up to and including two worldline loops can be cast in the following form (see (2.228)):

$$\begin{aligned}\langle z|\mathrm{e}^{-\frac{\beta}{\hbar}\hat{H}}|y\rangle &= \frac{1}{(2\pi\hbar\beta)^{n/2}}\mathrm{e}^{-\frac{1}{\hbar}S_{cl}[z,y;\beta]}\\ &\quad\times\left\{1-\frac{1}{24}\beta\hbar\left[R(z)+R(y)\right]+\mathcal{O}(\beta^{3/2})\right\}\\ &\quad\times\left[\beta^{n/2}g^{-1/4}(z)\left(\det-\frac{\partial}{\partial z^i}\frac{\partial}{\partial y^j}S_{cl}[z,y;\beta]\right)^{1/2}g^{-1/4}(y)\right].\end{aligned} \tag{8.7}$$

Here $S_{cl}[z,y;\beta]$ is the classical action evaluated on a geodesic from y to z in time β. The second line yields the trace anomaly in $n=2$. The last line gives the Morette–Van Vleck determinant; expanding to order β it adds terms of the form $\beta R_{ij}(z-y)^i(z-y)^j$ in the second line, and is needed in order for the transition element to satisfy factorization,

$$\langle z|\mathrm{e}^{-\frac{\beta}{\hbar}\hat{H}}|y\rangle = \int d^n w\,\langle z|\mathrm{e}^{-\frac{\beta}{2\hbar}\hat{H}}|w\rangle\sqrt{g(w)}\,\langle w|\mathrm{e}^{-\frac{\beta}{2\hbar}\hat{H}}|y\rangle. \tag{8.8}$$

The Einstein invariance of the transition element in (8.7) is manifest: each of the three factors is a biscalar. In this final result the coordinates z and y clearly appear symmetrically.

(iii) In configuration space the background trajectories $x^i_{bg}(t)$ satisfy the boundary conditions $x^i_{bg}(-\beta)=y^i$ and $x^i_{bg}(0)=z^i$. In phase space we decomposed $x^i=x^i_{bg}+q^i$, and we imposed the same boundary conditions on x^i as in configuration space, but we did not decompose p_i. No problems with integrations over p_i were encountered because the integrations over p_i were Gaussian at all N points (there is no zero mode for p_i). So it is possible to have boundary conditions in phase space at both ends of the time interval, but only on half of the variables. Although in our case we imposed boundary conditions at both ends only on x, we could also impose boundary conditions on x at one end and on p at the other end. Adding a mass term for x^i, one can even impose boundary conditions on only p at both ends.

(iv) One calculates the path integrals (the transition element) by using propagators and vertices as constructed from the discretized approach. If $z \neq y$, the factor of $[g(z)/g(y)]^{1/4}$ from the measure does contribute to the transition element $\langle z|\exp(-\frac{\beta}{\hbar}\hat{H})|y\rangle$. Of course, the a, b, c ghosts also contribute in configuration space, and they, too, are to be considered as part of the measure.

(v) In most calculations we have used the decomposition $x^i(t) = x^i_{bg}(t) + q^i(t)$, where the background field $x^i_{bg}(t)$ satisfies the boundary conditions and $q^i(t)$ vanishes at the boundary. If one is dealing with periodic boundary conditions (as in $\text{Tr}J\text{e}^{-\beta\mathcal{R}}$), one obtains $x^i(t) = x^i_0 + q^i(t)$ and then one should integrate over x^i_0. In this case, one obtains the same results for integrated anomalies if one uses the string-inspired approach in which one uses the decomposition $x^i(t) = x^i_0 + q^i(t)$ where now $\int_{-1}^{0} q^i(t)\, dt = 0$. This latter approach corresponds to quantization on a circle, and is used in string theory. In order to be able to use Riemann normal coordinates in the string-inspired approach, we introduced a shift symmetry $x^i_0 \to x^i_0 + \epsilon^i$, $q^i(t) \to q^i(t) - \epsilon^i$ and quantized with BRST methods.

(vi) For Majorana spinors $\psi^a(t)$ with $a = 1, \ldots, n$ we developed two methods for constructing path integrals: "fermion doubling" (by adding a second set of free Majorana spinors $\psi^a_2(t)$) and "fermion halving" (by combining two real ψ spinors into one complex spinor ψ and its complex conjugate $\bar{\psi}$). This allowed us to define creation and annihilation operators, which were used to construct coherent states. We imposed one boundary condition on the complex spinors ψ at $t = -\beta$ and another boundary condition on the spinors $\bar{\psi}$ at $t = 0$. So, the number of boundary conditions is correct for a first-order differential equation: one for each fermion. It did not matter whether $\bar{\psi}$ was the complex conjugate of ψ or not because Berezin (Grassmann) integration does not see the difference. Similarly, it did not matter whether the Grassmann parameters η and $\bar{\eta}$ in the coherent states were independent or not.

(vii) The final expression for chiral anomalies involves an integration over the background zero modes of the fermions, $\int \prod_a d\psi^a_{1,bg}$. We began with four Grassmann integrations, two for the trace formula and two for inserting a complete set of coherent states for fermions. Three of the four integrations were trivial, leaving us with the integration over the n anticommuting $\psi^a_{1,bg}$. For trace anomalies there is no integration over fermionic zero modes. This is to be expected: since there is no matrix γ_5 in the Jacobian for local scale transformations, the fermions have antiperiodic boundary conditions, hence in a mode expansion there are no constant modes ψ_0^a. Technically the absence of an integration over the background fermionic zero modes came about because the Jacobian did not depend on fermions (for trace anomalies the Jacobian is proportional to the Weyl weight of the fields, so it does not contain Dirac matrices), hence we only needed to integrate over the two Grassmann variables for the trace formula. The integral over these Grassmann variables was trivial, and merely yielded the factor of $2^{n/2}$ which accounts for the dimension of the space of spinors (for even n spacetime dimensions).

These answers hopefully clarify the construction and the evaluation of the path integrals in time slicing regularization. We have also discussed alternative constructions of the path integral which are based on mode regularization and dimensional regularization. These schemes have been divised directly for configuration-space path integrals. The action that appears in the exponent contains three terms: the classical action, an action for ghost fields that exponentiate the nontrivial path integral measure and a counterterm that is fixed by the request of satisfying the renormalization conditions (which are in this case the requirement that the transition amplitude should satisfy a Schrödinger equation with a Hamiltonian containing the covariant Laplacian without any scalar curvature term R if the classical Hamiltonian does not contain such a term). These renormalization conditions ensure that any regularization scheme will produce the same final answer.

In particular, mode regularization starts by expanding all fields, including the ghosts, into Fourier sums. The regularization is achieved by truncating these sums at a fixed mode M, so that all distributions that appear in Feynman graphs become well-behaved functions. Then one performs all computations at finite M, as now they are completely unambiguous. Eventually one takes the limit $M \to \infty$, thus obtaining a unique finite result. In practice one can proceed faster: one may perform all manipulations that are valid at the regulated level (for example, partial integration) to cast the integrands in alternative forms that can be computed directly in the $M \to \infty$ limit. For this purpose one must combine the ghost graphs with corresponding graphs without ghosts. This scheme requires the addition of a local counterterm V_{MR} to the action to satisfy the renormalization conditions mentioned earlier. This local counterterm is given by

$$V_{MR} = \frac{1}{8}R - \frac{1}{24}g^{ij}g^{mn}g_{kl}\Gamma^k_{im}\Gamma^l_{jn}. \tag{8.9}$$

The covariant term with R is the same as in (8.2), but the noncovariant term is different. This noncovariant piece is necessary to restore covariance (which is broken at the regulated level), so that the complete final result is covariant.

Dimensional regularization is a perturbative regularization which uses an adaptation of standard dimensional regularization to regulate distributions that are defined on the compact space $I = [-1, 0]$. One adds D extra infinite dimensions $I \to I \times R^D$ and one could now perform all computations of ambiguous. Feynman graphs in $D + 1$ dimensions. We assume that extra dimensions act as a regulator when D is extended analytically to the complex plane, as usual in dimensional regularization. After evaluation of the integrals one should take the $D \to 0$ limts. Actually, this way of proceeding would be quite difficult since the compact space I produces sums over discrete momenta which are hard to evaluate. However, there is no need to compute at arbitrary complex D. One may use manipulations valid at the regulated level, like differential equations satisfied by the Green functions and partial integration, to cast the integrand in equivalent forms that can be unambiguously computed in the $D \to 0$ limit. This method requires a covariant counterterm

$$V_{DR} = -\frac{1}{8}R. \tag{8.10}$$

This approach can be extended to include fermions, as we have discussed in Chapter 4.

Having thus clarified the construction of quantum mechanical path integrals for bosonic point particles $x^i(t)$ and Dirac or Majorana fermionic point particles $\psi^a(t)$, we turned to the calculation of anomalies. The anomaly was written as $An = \text{Tr}Je^{-\beta H}$, and suitable decompositions of unity for bosons and fermions as well as trace formulas were used to rewrite the anomaly as a path integral (see (6.12) and (6.13)). For spin-$\frac{1}{2}$ and spin-$\frac{3}{2}$ fields we took Einstein densities as basic variables in the quantum field theory, namely $\tilde{\lambda} = g^{1/4}\lambda$ for spin $\frac{1}{2}$, and $\tilde{\psi}_m = g^{1/4}\psi_m$ for spin $\frac{3}{2}$, where m is a flat vector index. The reason for this choice was that in this basis Einstein anomalies cancel manifestly (the Jacobian became the integral of a total space derivative). As a regulator we took the square of the field operator for nonchiral $\tilde{\lambda}$ and $\tilde{\psi}_m$. To account for the chirality of the spin-$\frac{1}{2}$ and spin-$\frac{3}{2}$ fields, we inserted a factor of $(1 + \gamma_5)/2$ into the Jacobian, but only the term $\gamma_5/2$ contributed to the anomaly. For anomalies due to loops with external gauge fields (Yang–Mills fields) the generators $(T_\alpha)^A{}_B$ of internal symmetries were replaced by $c^*_A(T_\alpha)^A{}_B c^B$, where c^*_A and c^B are internal antighosts and ghosts. At first we were constrained to work only in the one-particle sector of the Fock space of the ghosts and antighosts, but then we used an explicit expression for a projection operator which achieved this, and then one could evaluate unconstrained traces and corresponding path integrals. We showed that a simple change of basis reduced the complicated spin-$\frac{3}{2}$ operator of supergravity into the Dirac operator. We considered the gravitational anomalies in

$2\delta_{sym}$, where δ_{sym} is a particular sum of Einstein and local Lorentz transformations (see (6.162)). We discussed three transformation laws which frequently appear in the literature: covariant translations, symmetrized covariant translations and local Lorentz transformations. They are linearly dependent; one combination is sufficient to cover all anomalies. We choose that linear combination in which only covariant derivatives with spin connections but no Christoffel connection appeared. The corresponding Jacobian had the virtue that it could be cast into a simple quantum mechanical operator. Using an identity which only holds for the regulators mentioned above (see (6.166) and (6.161)), the transformations law $2\delta_{sym}$ could be rewritten as δ_{cov} for spin $\frac{1}{2}$, and $\delta_{cov}+\delta_{lL}^{(1)}$ for spin $\frac{3}{2}$, where δ_{cov} is a covariant translation (see (6.164)). Here $\delta_{lL}^{(1)}$ denotes a local Lorentz transformation with parameter $(D_\mu\xi_\nu - D_\nu\xi_\mu)$ which only acts on the vector index of the gravitino. The result is equation (1) of the preface. We treated the vector index of the gravitino as a kind of Yang–Mills index, by replacing the spin-1 Lorentz generators M_{mn} by $c_m^* M_{mn} c^n$, where c_m^* and c^n are treated in the same way as internal symmetry ghosts c_A^* and c^B. In this way the problem for spin $\frac{3}{2}$ became factorized into a problem for spin $\frac{1}{2}$ and a problem for spin 1, with the latter treated as being due to an internal symmetry.

For the self-dual antisymmetric tensor gauge field (SAT) there was (and is) the complication that no covariant action exists. Following AGW [1], we replaced the SAT by a commuting bispinor $\psi_{\alpha\beta}$ and then performed the same calculations as for ψ_m, only replacing the vector index m of the gravitino by the spinor index β of the bispinor. This readily yielded a general formula for the gravitational (local Lorentz) anomalies of SAT in $4k+2$ dimensions, but although one can give plausible arguments to justify this approach, a rigorous deductive proof is still lacking. In particular, there is a problem with the overall normalization of the SAT anomaly as we now discuss.

If one uses Feynman graphs with the stress tensor for arbitrary antisymmetric tensor fields at all vertices, and only at one vertex the stress tensor of a SAT, one obtains the correct covariant anomaly [1, 3]. One can check that it is correct by showing that it gives the same anomaly for a chiral scalar as for a chiral fermion in $1+1$ dimensions (a basic result of string theory). However, in the quantum mechanical calculations of the anomaly for SAT as extracted from the bispinors, one is off by a factor of $\frac{1}{2}$. We inserted this factor $\frac{1}{2}$ by hand, and then all derivations and results became as clear and simple as for spin $\frac{1}{2}$ and spin $\frac{3}{2}$.

However, inserting by hand a factor $\frac{1}{2}$ cannot be the final word. We quote Witten [183] "...For selfdual tensor fields matters are trickier, unfortunately. First of all, there is not a convenient Lagrangian for the field. For that reason, in [1] the normalization of the anomaly for this field was found by examining Feynman diagrams; this gave an unexpected numerical factor (see footnote 6 of that paper) . . . ". This footnote 6 of [1] (which, however, dealt with bispinors in Feynman diagrams, not bispinors in quantum mechanics) reads: "The interested reader should note the following points. One has, of course, a minus sign for the regulator loop. In matrix elements (or Feynman vertices) of the energy–momentum tensor one must include a factor of two from Bose statistics. The γ_5's in the propagator and vertices cancel harmlessly, leaving only γ_5's from the projectors $\frac{1}{2}(1+\gamma^5)$. The γ_5 in the α trace goes into making the kinematical factor R, but the F in the β trace can be dropped (it gives terms that vanish as $M \to \infty$). So the $\frac{1}{2}(1+\gamma^5)$ in the β trace gives a factor $\frac{1}{2}$. Another factor $\frac{1}{2}$ occurs because the fields are real, but there is a factor of 2 from choosing which Dirac trace is the α trace and which is the β trace. Because the two factors of $\frac{1}{2}$ and one factor 2, we get eventually $\frac{1}{2}$ the amplitude for a charged Dirac particle in an external field".

So one has the following situation: if one follows the same procedure as for spin $\frac{3}{2}$, adding only one matrix γ_5 to the trace to take into account chirality, one is repeating the same steps as for spin $\frac{3}{2}$ and one is off by a factor of $\frac{1}{2}$. If, on the other hand, one argues that one needs chiral projections in both spinor sectors, and first decides which index of the bispinor is like the vector index of the gravitino, the γ_5 in this sector does not contribute according to Witten, and then one indeed obtains the correct result. Endo (second reference in [151]) has argued that one can begin with two antisymmetric tensor fields with an action of the form $A\partial B$, where A is self-dual. In the transition from (A, B) to $\psi_{\alpha\beta}$ he finds in the relation $\xi^\mu D^\nu T_{\mu\nu} = An$ a factor of unity for SAT but $\frac{1}{2}$ for $\psi_{\alpha\beta}$. This does not seem to help us because in our quantum mechanical approach we calculate the Jacobian directly, and any scale of the action cancels since one takes the limit $\beta \to 0$ in $\exp(-\beta H)$.

Our point of view is that the most straightforward way to demonstrate the presence of the extra factor of $\frac{1}{2}$ in the approach with bispinors based on quantum mechanics is to take the known noncovariant (but local and polynomial) actions for abelian SAT in $4k+2$ dimensions coupled to gravity [103, 104], and then to apply deductively step-by-step the methods developed in this book. Calculations with noncovariant action are tedious, but this approach yields unambiguous results.

Appendix A
Riemann curvatures

To define the Riemann curvatures in terms of a connection Γ^k_{ij} or spin connection $\omega_i{}^a{}_b$, we begin with the "vielbein postulate"

$$D_i e_j{}^a \equiv \partial_i e_j{}^a - \Gamma^k_{ij} e_k{}^a + \omega_i{}^a{}_b e^b_j = 0. \tag{A.1}$$

The vielbein field $e_i{}^a(x)$ is related to the metric $g_{ij}(x)$ by

$$g_{ij}(x) = e_i{}^a(x) e_j{}^b(x) \delta_{ab} \tag{A.2}$$

in Euclidean space (or $g_{ij}(x) = e_i{}^a(x) e_j{}^b(x) \eta_{ab}$ in Minkowski space). It gives the orientation of the coordinates $X^a_P(x)$ of an inertial frame at the point P with respect to the curvilinear coordinates x^i, namely $e_i{}^a(x) = (\partial X^a_P / \partial x^i)_{x=x_P}$ [184]. From this definition it is clear that $e_i{}^a(x)$ transforms as a covariant vector in general relativity with index i. Moreover, the strong equivalence principle states that the physics should look the same in any local inertial frame. Hence, the theory should be invariant under local Lorentz rotations of the inertial frames. Thus, there is a second symmetry in the theory under which vielbeins transform as $\tilde{e}_i{}^a(x) = \Lambda^a{}_b(x) e_i{}^b(x)$, where $\Lambda^a{}_b(x)$ is a Lorentz matrix.

The equation in (A.1) has a geometrical meaning. Consider a vector field $v^i(x)$ with curved indices, and use the vielbein field $e_i{}^a(x)$ to construct a corresponding vector field $v^a(x) \equiv v^i(x) e_i{}^a(x)$ with flat indices. Parallel transport of $v^i(x)$ along a distance Δx^j with an arbitrary connection field $\Gamma^k_{ij}(x)$ leads to a vector field $\tilde{v}^i(x+\Delta x)$ at $x+\Delta x$ defined by $\tilde{v}^i(x+\Delta x) = v^i(x) - \Delta x^j \Gamma^i_{jk}(x) v^k(x)$. Similarly parallel transport of $v^a(x)$ along the distance Δx^j with an arbitrary connection field $\omega_i{}^a{}_b(x)$ yields a vector field $\tilde{v}^a(x+\Delta x)$ at $x+\Delta x$ defined by $\tilde{v}^a(x+\Delta x) = v^a(x) - \Delta x^j \omega_j{}^a{}_b(x) v^b(x)$. The vielbein postulate in (A.1) implies that $\tilde{v}^i(x)$ and $\tilde{v}^a(x)$ are related to each other in the same way as $v^i(x)$ and $v^a(x)$ are related, $\tilde{v}^a(x+\Delta x) \equiv \tilde{v}^i(x+\Delta x) e_i{}^a(x+\Delta x)$. Indeed, expansion to first order in Δx reproduces (A.1). Thus, there is only one vector field which one can either write as $v^i(x)$ or as $v^a(x)$, and only one connection, which is given by $\Gamma^k_{ij}(x)$ if one writes the vector field as $v^i(x)$, or given by $\omega_i{}^a{}_b(x)$ if one uses $v^a(x)$ to represent the vector field. In other words, the operations of parallel transport and conversion from curved to flat indices (or vice versa) commute.

Next, we require that length is preserved by parallel transport. The square of the length of a vector v^a with flat index is by definition $v^a\delta_{ab}v^b$ in Euclidean space (or $v^a\eta_{ab}v^b$ in Minkowski space). It follows that length is preserved if and only if $\omega_{iab} \equiv \omega_i{}^c{}_b\delta_{ca}$ is antisymmetric in ab. This we shall always assume to be the case. For a vector v^i with curved index we define the square of the length by $v^ig_{ij}v^j$ where the metric is related to the vielbein by $g_{ij} = e_i{}^a e_j{}^b\delta_{ab}$ in Euclidean space (and $g_{ij} = e_i{}^a e_j{}^b\eta_{ab}$ in Minkowski space). The connection Γ^k_{ij} preserves length if it satisfies $\Gamma_{ij;k} + \Gamma_{ik;j} - \partial_i g_{jk} = 0$ (with $\Gamma_{ij;k} \equiv \Gamma^l_{ij}g_{lk}$), which is equivalent to the antisymmetry of ω_{iab} if the vielbein postulate holds.

The covariant derivatives of vectors are proportional to the difference of the original vector field and the parallel transported vector field

$$\begin{aligned} v^i(x+\Delta x) - \tilde{v}^i(x+\Delta x) &= \Delta x^j D_j v^i(x) \\ \Rightarrow \quad D_j v^i &= \partial_j v^i + \Gamma^i_{jk}v^k \\ v^a(x+\Delta x) - \tilde{v}^a(x+\Delta x) &= \Delta x^j D_j v^a(x) \\ \Rightarrow \quad D_j v^a &= \partial_j v^a + \omega_j{}^a{}_b v^b. \end{aligned} \tag{A.3}$$

By requiring that $D_j(w_iv^i) = \partial_j(w_iv^i) = D_j(w_av^a)$ one also derives that $D_jw_i = \partial_jw_i - \Gamma^k_{ji}w_k$ and $D_jw_a = \partial_jw_a + \omega_{ja}{}^b w_b$.

From (A.1) we can express $\omega_i{}^a{}_b$ in terms of Γ^k_{ij} and $e_i{}^a$. If Γ^k_{ij} has an antisymmetric part $\frac{1}{2}(\Gamma^k_{ij} - \Gamma^k_{ji}) = T_{ij}{}^k$, this part is called the torsion tensor. If there is no torsion, Γ^k_{ij} is the usual Christoffel symbol

$$\left\{ {k \atop i\,j} \right\} = \frac{1}{2}g^{kl}(\partial_i g_{jl} + \partial_j g_{il} - \partial_l g_{ij}) \tag{A.4}$$

as one can easily show by solving $D_ig_{jk} = 0$. The corresponding torsionless spin connection follows from (A.1),

$$\begin{aligned} \omega_i{}^{ab} = {} & \frac{1}{2}e^{aj}(\partial_i e_j{}^b - \partial_j e_i{}^b) - \frac{1}{2}e^{bj}(\partial_i e_j{}^a - \partial_j e_i{}^a) \\ & - \frac{1}{2}e^{aj}e^{bk}(\partial_j e_{kc} - \partial_k e_{jc})e_i{}^c. \end{aligned} \tag{A.5}$$

Torsion preserves length if $T_{ijl} \equiv T_{ij}{}^k g_{kl}$ is totally antisymmetric. Until (A.12) we do not make the assumption that torsion is absent.

From $[D_i, D_j]e^a_k = 0$ one finds

$$[D_i, D_j]e^a_k = -R_{ijk}{}^l(\Gamma)e^a_l + R_{ij}{}^a{}_b(\omega)e_k{}^b = 0 \tag{A.6}$$

where evidently

$$R_{ijk}{}^l(\Gamma) = \partial_i\Gamma^l_{jk} + \Gamma^l_{im}\Gamma^m_{jk} - (i \leftrightarrow j) \tag{A.7}$$

$$R_{ij}{}^a{}_b(\omega) = \partial_i\omega_j{}^a{}_b + \omega_i{}^a{}_c\omega_j{}^c{}_b - (i \leftrightarrow j). \tag{A.8}$$

Hence,

$$R_{ijk}{}^a(\Gamma) = R_{ij}{}^a{}_k(\omega). \tag{A.9}$$

The variation of a curvature is the covariant derivative of the variation of the corresponding connection

$$\delta R_{ijk}{}^l(\Gamma) = D_i\delta\Gamma^l_{jk} - D_j\delta\Gamma^l_{ik}$$
$$\delta R_{ij}{}^a{}_b(\omega) = D_i\delta\omega_j{}^a{}_b - D_j\delta\omega_i{}^a{}_b \tag{A.10}$$

where

$$D_i\delta\Gamma^l_{jk} = \partial_i\delta\Gamma^l_{jk} - \Gamma^m_{ij}\delta\Gamma^l_{mk} - \Gamma^m_{ik}\delta\Gamma^l_{jm} + \Gamma^l_{im}\delta\Gamma^m_{jk}$$
$$D_i\delta\omega_j{}^a{}_b = \partial_i\delta\omega_j{}^a{}_b - \Gamma_{ij}{}^m\delta\omega_m{}^a{}_b + \omega_i{}^a{}_c\delta\omega_j{}^c{}_b + \omega_{ib}{}^c\delta\omega_j{}^a{}_c. \tag{A.11}$$

From (A.8) it follows that $R_{ijab}(\omega)$ is antisymmetric in its last two indices. Then (A.9) can also be written as

$$R_{ijka}(\Gamma) = -R_{ijka}(\omega). \tag{A.12}$$

From now on in this appendix we assume that Γ^k_{ij} is the Christoffel symbol. The symmetries of the Riemann tensor are then

$$R_{ijkl}(\Gamma) = -R_{jikl}(\Gamma)$$
$$R_{ijkl}(\Gamma) = -R_{ijlk}(\Gamma)$$
$$R_{ijkl}(\Gamma) + R_{jkil}(\Gamma) + R_{kijl}(\Gamma) = 0$$
$$R_{ijkl}(\Gamma) = R_{klij}(\Gamma). \tag{A.13}$$

The first property is obvious from the definition of the Riemann curvature in (A.6). The second one is obvious from (A.8), but also follows from considering that the metric is covariantly constant

$$0 = [D_i, D_j]g_{kl} = R_{ijkl}(\Gamma) + R_{ijlk}(\Gamma). \tag{A.14}$$

The third one, known as the cyclic identity, can be obtained by considering the vanishing of the second derivative of a covariant vector v_i (a 1-form) totally antisymmetrized ($d^2v = 0$)

$$0 = \partial_{[i}\partial_j v_{k]} = D_{[i}D_j v_{k]} = [D_i, D_j]v_k + \text{cyclic terms}$$
$$= \Big(R_{ijk}{}^l(\Gamma) + R_{jki}{}^l(\Gamma) + R_{kij}{}^l(\Gamma)\Big)v_l \tag{A.15}$$

where $[\cdots]$ denotes antisymmetrization and the Christoffel connections can be put in for free as they drop out. Finally, the last property can be proven by transforming to an inertial frame in which Γ^k_{ij} vanishes.

The Ricci tensor is defined by

$$R_{ij} = R_{ikj}{}^k(\Gamma) = R_{ik}{}^k{}_j(\omega) \tag{A.16}$$

and the scalar curvature is given by

$$R = g^{ij} R_{ij}. \tag{A.17}$$

With these conventions the scalar curvature for the sphere is negative.

At the linearized level $g_{ij} \simeq \eta_{ij} + h_{ij}$, and

$$R^{\rm lin}_{ijkl}(\Gamma) = \frac{1}{2}(\partial_i\partial_k h_{jl} + \partial_j\partial_l h_{ik} - \partial_j\partial_k h_{il} - \partial_i\partial_l h_{jk}) \tag{A.18}$$

$$R^{\rm lin}_{ij} = \frac{1}{2}(-\partial_i h_j - \partial_j h_i + \partial_i\partial_j h + \Box h_{ij}) \tag{A.19}$$

where $h_i = \partial^j h_{ij}$, $h = \eta^{ij} h_{ij}$, while $\Box = \eta^{ij}\partial_j\partial_j$. Clearly,

$$R^{\rm lin} = -\partial^i h_i + \Box h = (-\partial^i\partial^j + \partial^k\partial_k\eta^{ij})h_{ij}. \tag{A.20}$$

In Section 2.6 we need the full nonlinear expression of R in terms of $\Box g \equiv g^{ij}g^{kl}\partial_k\partial_l g_{ij}$, $\partial^j g_j \equiv g^{ik}g^{jl}\partial_k\partial_l g_{ij}$, $\partial_k g \equiv g^{ij}\partial_k g_{ij}$, $g^i \equiv g^{ij}g^{kl}\partial_k g_{lj}$ and $g_k \equiv g^{ij}\partial_i g_{jk}$. Straightforward evaluation yields

$$\begin{aligned} R &= \frac{1}{2}g^{ik}g^{jl}(\partial_i\partial_k g_{jl} - \partial_i\partial_l g_{jk} - \partial_j\partial_k g_{il} + \partial_j\partial_l g_{ik}) + (\partial g)(\partial g) \text{ terms} \\ &= \Box g - \partial^j g_j - \frac{3}{4}(\partial_k g_{ij})^2 + \frac{1}{2}(\partial_i g_{jk})\,\partial_j g_{ik} + \frac{1}{4}(\partial_j g)^2 - (\partial_j g)\, g^j + g_j^2. \end{aligned} \tag{A.21}$$

In Section 2.6 we also need the full nonlinear expression for the Riemann curvature in the form $R_{ikjl}(\Gamma) = \frac{1}{2}(\partial_i\partial_j g_{kl} + \text{three terms}) + (\partial g)^2$ terms. Straightforward evaluation yields

$$\begin{aligned} R_{ikjl}(\Gamma) = &\frac{1}{2}(\partial_i\partial_j g_{kl} - \partial_i\partial_l g_{kj} - \partial_k\partial_j g_{il} + \partial_k\partial_l g_{ij}) \\ &+ (\Gamma^m_{ij}\Gamma^n_{kl} - \Gamma^m_{kj}\Gamma^n_{il})g_{mn}. \end{aligned} \tag{A.22}$$

As a check we note that the expression on the right-hand side has all the symmetries of the left-hand side: antisymmetry in each pair, symmetry under pair exchange and cyclic identity.

For the calculations of trace anomalies we need to know how the scalar curvature transforms under local Weyl rescalings

$$\begin{aligned} \delta_W\, g_{ij} &= \sigma(x) g_{ij} \\ \delta_W\, e_i{}^a &= \frac{1}{2}\sigma(x) e_i{}^a. \end{aligned} \tag{A.23}$$

Using the vielbein postulate and (A.4), one finds

$$\begin{aligned}\delta_W\,\omega_i{}^a{}_b &= \delta_W(\Gamma^k_{ij}\,e_k{}^a e_b{}^j) - \frac{1}{2}\delta^a_b\partial_i\sigma \\ &= \frac{1}{2}(e_i{}^a e_b{}^j - e_{ib}e^{ja})\partial_j\sigma. \end{aligned} \tag{A.24}$$

Substitution into (A.10) then yields in n dimensions

$$\delta_W R = -\sigma R + (n-1)D^i D_i\sigma. \tag{A.25}$$

Appendix B

Weyl ordering of bosonic operators

In this appendix we discuss the concept of Weyl ordering. For a detailed account, see for example [9].

To evaluate matrix elements of the form $M = \langle z|\hat{O}|y\rangle$ with $\langle z|$ and $|y\rangle$ eigenstates of the position operator $\hat{x}^i$ and $\hat{O}(\hat{x}, \hat{p})$ an arbitrary operator, we first insert a complete set of momentum eigenstates $I = \int |p\rangle\langle p| d^n p$,

$$M = \int \langle z|\hat{O}|p\rangle\langle p|y\rangle \, d^n p. \tag{B.1}$$

It is then very convenient to **rewrite** $\hat{O}$ as a Weyl-ordered operator $\hat{O}_W$, because, as we shall prove, one can then replace in $\hat{O}_W(\hat{x}, \hat{p})$ the operators $\hat{x}^i$ and $\hat{p}_j$ by the c-number values $\frac{1}{2}(z^i + y^i)$ and p_j, respectively. If we denote the corresponding function of $\frac{1}{2}(z^i + y^i)$ and p_i by $O_W(\frac{1}{2}(z+y), p)$, we can prove the following:

Theorem: $$M = \int \langle z|p\rangle O_W\left(\frac{1}{2}(z+y), p\right) \langle p|y\rangle \, d^n p. \tag{B.2}$$

We must clearly first define what the Weyl ordering is, and in particular how to construct the operator $\hat{O}_W$ from a given operator $\hat{O}$. In this construction we shall also need the notion of a symmetrized operator $\hat{O}_S$. We shall show that, in general, an operator $\hat{O}$ can be rewritten as a sum of the corresponding symmetrized operator $\hat{O}_S$ and further terms,

$$\hat{O} = \hat{O}_S + \text{further terms} = \hat{O}_W. \tag{B.3}$$

As the notation indicates, the operator $\hat{O}_W$ is equal to the original operator $\hat{O}$, but it is written in such a way that $\hat{x}$ and $\hat{p}$ appear symmetrically.

Let us first give a few examples. The operator $\hat{x}\hat{p}$ can clearly be written as the sum of $\frac{1}{2}(\hat{x}\hat{p} + \hat{p}\hat{x})$ and $\frac{1}{2}(\hat{x}\hat{p} - \hat{p}\hat{x})$. The latter term is equal to $\frac{1}{2}i\hbar$.

The former term is the symmetrized form of $\hat{x}\hat{p}$, so $(\hat{x}\hat{p})_S \equiv \frac{1}{2}(\hat{x}\hat{p}+\hat{p}\hat{x})$. We then have

$$\hat{x}\hat{p} = (\hat{x}\hat{p})_S + \frac{1}{2}i\hbar \equiv (\hat{x}\hat{p})_W, \qquad \hat{p}\hat{x} = (\hat{p}\hat{x})_S - \frac{1}{2}i\hbar \equiv (\hat{p}\hat{x})_W. \tag{B.4}$$

Clearly, $(\hat{x}\hat{p})_S$ is equal to $(\hat{p}\hat{x})_S$, but $(\hat{x}\hat{p})_W$ is not equal to $(\hat{p}\hat{x})_W$. As a second example consider the operator $\hat{x}\hat{x}\hat{p}\hat{p}$. Its symmetrized form, as we shall discuss below, can be written as

$$(\hat{x}\hat{x}\hat{p}\hat{p})_S = \frac{1}{4}\hat{x}\hat{x}\hat{p}\hat{p} + \frac{1}{2}\hat{p}\hat{x}\hat{x}\hat{p} + \frac{1}{4}\hat{p}\hat{p}\hat{x}\hat{x}. \tag{B.5}$$

One may then check that

$$\hat{x}\hat{x}\hat{p}\hat{p} - (\hat{x}\hat{x}\hat{p}\hat{p})_S = 2i\hbar(\hat{x}\hat{p})_S - \frac{1}{2}\hbar^2. \tag{B.6}$$

So in this example we have

$$\hat{x}\hat{x}\hat{p}\hat{p} = (\hat{x}\hat{x}\hat{p}\hat{p})_S + 2i\hbar(\hat{x}\hat{p})_S - \frac{1}{2}\hbar^2 \equiv (\hat{x}\hat{x}\hat{p}\hat{p})_W. \tag{B.7}$$

The term $2i\hbar(\hat{x}\hat{p})_S - \frac{1}{2}\hbar^2$ corresponds to the term denoted by "further terms" in (B.3). So, as an operator $(\hat{x}\hat{x}\hat{p}\hat{p})_W$ is equal to $\hat{x}\hat{x}\hat{p}\hat{p}$, but for our purposes it is useful to use the fundamental commutation relations to write $\hat{x}\hat{x}\hat{p}\hat{p}$ in such a way that all $\hat{x}$ and $\hat{p}$ appear symmetrically, and if this is the case we call this expression the Weyl-ordered form.

In general, we call an operator $\hat{A}(\hat{x},\hat{p})$ symmetrized, if all operators $\hat{x}^i$ and $\hat{p}_j$ appear in all possible orderings with equal weights. The symmetrized form of monomials is produced by the formula

$$(\alpha^i\hat{p}_i + \beta_j\hat{x}^j)^N = \sum_{m_i,n_j} N! \prod_{i,j} \frac{1}{\Pi\, m_i!\, \Pi\, n_j!}(\alpha^i)^{m_i}(\beta_j)^{n_j}\left[(\hat{p}_i)^{m_i}(\hat{x}^j)^{n_j}\right]_S$$
$$\sum m_i + \sum n_j = N \tag{B.8}$$

or, equivalently,

$$N!\prod_{i,j}(\hat{p}_i^{m_i}(\hat{x}^j)^{n_j})_S = \prod_{i,j}\Big(\frac{\partial}{\partial\alpha^i}\Big)^{m_i}\Big(\frac{\partial}{\partial\beta_j}\Big)^{n_j}(\alpha^i\hat{p}_i + \beta_j\hat{x}^j)^N$$
$$\sum m_i + \sum n_j = N. \tag{B.9}$$

We shall discuss this result by first considering one pair of variables $\hat{x}$ and $\hat{p}$.

For one set of operators $\hat{p}$ and $\hat{x}$ one has (omitting hats for notational simplicity)

$$\begin{aligned}
(xp)_S &= \frac{1}{2}(xp + px) = (px)_S \\
(x^2p)_S &= \frac{1}{3}(x^2p + xpx + px^2) \\
&= \frac{1}{4}(x^2p + 2xpx + px^2) \\
&= \frac{1}{2}(x^2p + px^2).
\end{aligned} \tag{B.10}$$

To derive the last two relations from the first one, one may repeatedly use the basic identity $2xpx = x^2p + px^2$. More generally,

$$\begin{aligned}
(x^np)_S &= \frac{1}{n+1}\sum_{l=0}^{n} x^{n-l}px^l \\
&= \frac{1}{2^n}\sum_{l=0}^{n}\binom{n}{l} x^{n-l}px^l \\
&= \frac{1}{2}(x^np + px^n).
\end{aligned} \tag{B.11}$$

To prove these relations one may combine the terms with $x^{n-l}px^l$ and x^lpx^{n-l} and move the p past the x^l (or past the x^{n-l}). The commutators then cancel again. (As is clear from (B.11), when n is even, there is a term for which $n - l = l$ which one should first split into two terms. For example, for $n = 2$ one first rewrites $2xpx$ as $xpx + xpx$ and then the commutators $[x,p]x + x[p,x]$ cancel again.) Of course, by the same argument one also has

$$(xp^2)_S = \frac{1}{3}(xp^2 + pxp + p^2x) = \frac{1}{4}(xp^2 + 2pxp + p^2x) = \frac{1}{2}(xp^2 + p^2x). \tag{B.12}$$

The next term with two p in this series is

$$\begin{aligned}
(x^2p^2)_S &= \frac{1}{6}\left(\begin{aligned} x^2p^2 + xpxp + xp^2x + p^2x^2 \\ + pxpx + px^2p \end{aligned}\right) \\
&= \frac{1}{4}(x^2p^2 + 2xp^2x + p^2x^2) \\
&= \frac{1}{4}(x^2p^2 + 2px^2p + p^2x^2).
\end{aligned} \tag{B.13}$$

Note that in the second line, the p^2 are kept together, while in the third line the x^2 are kept together. We shall achieve this for all cases which follow below. To obtain this result, we wrote the second term in the first line as $\frac{1}{12}(x\underline{px}p + xp\underline{xp})$ and used the $[x,p]$ commutation relations for the underlined operators; similarly for the term $\frac{1}{6}pxpx$. Then, we used the fact that in the remainder $-xp^2x + px^2p = -x[p^2,x] + [p,x^2]p = 0$.

In a similar manner one shows that

$$\begin{aligned}(x^3p^2)_S &= \binom{5}{2}^{-1}[x^3p^2 + \cdots] \quad \text{(10 terms)}\\ &= \frac{1}{8}(x^3p^2 + 3x^2p^2x + 3xp^2x^2 + p^2x^3) \quad (\text{keeping } p^2 \text{ together})\\ &= \frac{1}{4}(x^3p^2 + 2px^3p + p^2x^3) \quad (\text{keeping } x^3 \text{ together}). \end{aligned} \tag{B.14}$$

(We obtained the last line by writing the term $3x^2p^2x$ as $\frac{2}{3}$ times $3(p^2x^3 + [x^2,p^2]x)$ plus $\frac{1}{3}$ times $3(x^3p^2 + x^2[p^2,x])$, and similarly for $3xp^2x^2$, since in this way the commutators cancel.)

In general,

$$\begin{aligned}(x^np^2)_S &= \frac{2}{(n+1)(n+2)}\sum_{l,m=0}^{n} x^{n-l-m}px^lpx^m\\ &= \frac{1}{2^n}\sum_{l=0}^{n}\binom{n}{l}x^{n-l}p^2x^l\\ &= \frac{1}{4}(x^np^2 + 2px^np + p^2x^n). \end{aligned} \tag{B.15}$$

The most general formula for one pair of canonical variables is then

$$\begin{aligned}(x^mp^r)_S &= \frac{1}{2^m}\sum_{l=0}^{m}\binom{m}{l}x^{m-l}p^rx^l \quad (\text{keeping } p^r \text{ together})\\ &= \frac{1}{2^r}\sum_{k=0}^{r}\binom{r}{k}p^{r-k}x^mp^k \quad (\text{keeping } x^m \text{ together}). \end{aligned} \tag{B.16}$$

Consider now the matrix element $M = \langle z|(\hat{x}^m\hat{p}^r)_S|y\rangle$ where we have reinstated the hats. Inserting a complete set of p-states and using the symmetrized expression with p^r kept together one finds,

$$\begin{aligned}M &= \int \langle z|\frac{1}{2^m}\sum_{l=0}^{m}\binom{m}{l}\hat{x}^{m-l}\hat{p}^r|p\rangle\langle p|\hat{x}^l|y\rangle\, d^np\\ &= \int \langle z|p\rangle\frac{1}{2^m}\sum_{l=0}^{m}\binom{m}{l}z^{m-l}y^lp^r\langle p|y\rangle\, d^np\\ &= \int \langle z|p\rangle\left(\frac{z+y}{2}\right)^m p^r\langle p|y\rangle\, d^np. \end{aligned} \tag{B.17}$$

This is enough to prove the theorem in (B.2) because any Weyl-ordered operator is a sum of symmetrized terms (see, for example, (B.7)). In particular, for any Weyl-ordered Hamiltonian operator $\hat{H}_W(\hat{x},\hat{p})$ we find the midpoint rule

$$\langle z|\hat{H}_W(\hat{x},\hat{p})|y\rangle = \int \langle z|p\rangle H_W\left(\frac{1}{2}(z+y),p\right)\langle p|y\rangle\, d^n p. \tag{B.18}$$

Consider now as a particular case the operator

$$\hat{H} = \frac{1}{2} g^{-1/4} p_i g^{1/2} g^{ij} p_j g^{-1/4}. \tag{B.19}$$

To write it in Weyl-ordered form we first simplify this expression by moving p_i to the left and p_j to the right. The result is

$$\begin{aligned}\hat{H} &= \frac{1}{2}\left(p_i - \frac{1}{4} i\hbar\partial_i \ln g\right) g^{-1/4} g^{1/2} g^{ij} g^{-1/4}\left(p_j + \frac{1}{4} i\hbar\partial_j \ln g\right)\\ &= \frac{1}{2} p_i g^{ij} p_j + \frac{\hbar^2}{8}\partial_i\left(g^{ij}\partial_j \ln g\right) + \frac{\hbar^2}{32} g^{ij}(\partial_i \ln g)(\partial_j \ln g).\end{aligned} \tag{B.20}$$

The first term is not yet Weyl-ordered, hence we rewrite it using its symmetrized form, keeping the x operators in $g^{ij}(x)$ together. The symmetrized form of $\frac{1}{2} p_i g^{ij} p_j$ is

$$\frac{1}{2}(p_i g^{ij} p_j)_S = \frac{1}{8}(p_i p_j g^{ij} + 2 p_i g^{ij} p_j + g^{ij} p_i p_j). \tag{B.21}$$

The difference between $\frac{1}{2} p_i g^{ij} p_j$ and its symmetrized form is given by

$$\begin{aligned}\frac{1}{2}(p_i g^{ij} p_j) - \frac{1}{2}(p_i g^{ij} p_j)_S &= \frac{1}{8} p_i [g^{ij}, p_j] + \frac{1}{8}[p_i, g^{ij}] p_j\\ &= \frac{1}{8}[p_i, [g^{ij}, p_j]] = \frac{\hbar^2}{8}\partial_i\partial_j g^{ij}.\end{aligned} \tag{B.22}$$

Hence, $\hat{H}$ reads in Weyl-ordered form as

$$\begin{aligned}\hat{H} &= \frac{1}{2}(p_i g^{ij} p_j)_S + \frac{\hbar^2}{8}\left[\partial_i\partial_j g^{ij} + \partial_i(g^{ij}\partial_j \ln g) + \frac{1}{4} g^{ij}(\partial_i \ln g)(\partial_j \ln g)\right]\\ &= \frac{1}{2}(p_i g^{ij} p_j)_S + \frac{\hbar^2}{8}[(\partial_i\partial_j g^{ij} + g^{-1/4}\partial_i(g^{1/4} g^{ij}\partial_j \ln g)].\end{aligned} \tag{B.23}$$

The last two terms can be written in terms of Christoffel symbols and the scalar curvature. To find the coefficient of R we evaluate the leading terms of the form $\partial_i\partial_j g_{kl}$ and find with (A.20)

$$-\frac{\hbar^2}{8} g^{ik}(\partial_i\partial_j g_{kl}) g^{jl} + \frac{\hbar^2}{8} g^{ij} g^{kl}\partial_i\partial_j g_{kl} + \cdots = \frac{\hbar^2}{8} R + \cdots. \tag{B.24}$$

The final result reads as

$$\hat{H} = \frac{1}{2}(p_i g^{ij} p_j)_S + \frac{\hbar^2}{8}(R + g^{ij}\Gamma^l_{ik}\Gamma^k_{jl}) = \hat{H}_W \tag{B.25}$$

with $(p_i g^{ij} p_j)_S$ defined in (B.21). An easy way to check the coefficient of the Christoffel term is to consider the one-dimensional case where R vanishes. On the other hand, an easy way to check the coefficient of the term with R is to go to a frame where $\partial_i g_{jk}$ vanishes at a given point. However, the fact that $\hat{H}$ is of the form given in (B.25) only follows from an explicit computation.

Appendix C
Weyl ordering of fermionic operators

In this appendix we extend the Weyl ordering of bosonic canonical variables discussed in the previous appendix to the case of fermionic canonical variables [58, 185].

Consider operators $O(\hat{\psi}^a, \hat{\psi}_b^\dagger)$ depending on fermionic canonical variables $\hat{\psi}^a$ and $\hat{\psi}_b^\dagger$ $(a, b = 1, n)$ which satisfy

$$\{\hat{\psi}^a, \hat{\psi}^b\} = 0, \quad \{\hat{\psi}^a, \hat{\psi}_b^\dagger\} = \hbar\delta_b^a, \quad \{\hat{\psi}_a^\dagger, \hat{\psi}_b^\dagger\} = 0. \tag{C.1}$$

We define the antisymmetric ordering, which we still denote by a subscript S, as the ordering which results if one expands $(\alpha_a\hat{\psi}^a + \beta^b\hat{\psi}_b^\dagger)^N$ where α_a and β^b are Grassmann variables. Hence

$$\prod_{a,b} N! \left((\hat{\psi}^a)^{m_a}(\hat{\psi}_b^\dagger)^{n_b}\right)_S = \prod_{a,b}\left(\frac{\partial}{\partial\alpha_a}\right)^{m_a}\left(\frac{\partial}{\partial\beta^b}\right)^{n_b}(\alpha_a\hat{\psi}^a + \beta^b\hat{\psi}_b^\dagger)^N \tag{C.2}$$

where $N = \sum_{\hat{a}} m_a + \sum_{\hat{b}} n_b$ and the order in which the $\hat{\psi}^a$ and $\hat{\psi}_b^\dagger$ appear on the left-hand side is the same as the order in which the $\partial/\partial\alpha_a$ and $\partial/\partial\beta^b$ appear on the right-hand side. For example,

$$(\hat{\psi}^a\hat{\psi}_b^\dagger)_S = \frac{1}{2}(\hat{\psi}^a\hat{\psi}_b^\dagger - \hat{\psi}_b^\dagger\hat{\psi}^a) \tag{C.3}$$

and

$$\begin{aligned}(\hat{\psi}^a\hat{\psi}^b\hat{\psi}_c^\dagger)_S &= \frac{1}{6}(\hat{\psi}^a\hat{\psi}^b\hat{\psi}_c^\dagger - \hat{\psi}^b\hat{\psi}^a\hat{\psi}_c^\dagger - \hat{\psi}^a\hat{\psi}_c^\dagger\hat{\psi}^b + \hat{\psi}^b\hat{\psi}_c^\dagger\hat{\psi}^a \\ &\quad + \hat{\psi}_c^\dagger\hat{\psi}^a\hat{\psi}^b - \hat{\psi}_c^\dagger\hat{\psi}^b\hat{\psi}^a) \\ &= \frac{1}{3}\left(\hat{\psi}^a\hat{\psi}^b\hat{\psi}_c^\dagger - \frac{1}{2}\hat{\psi}^a\hat{\psi}_c^\dagger\hat{\psi}^b + \frac{1}{2}\hat{\psi}^b\hat{\psi}_c^\dagger\hat{\psi}^a + \hat{\psi}_c^\dagger\hat{\psi}^a\hat{\psi}^b\right)\end{aligned}$$

$$= \frac{1}{2}(\hat{\psi}^a\hat{\psi}^b\hat{\psi}_c^\dagger + \hat{\psi}_c^\dagger\hat{\psi}^a\hat{\psi}^b)$$
$$= \frac{1}{4}(\hat{\psi}^a\hat{\psi}^b\hat{\psi}_c^\dagger - \hat{\psi}^a\hat{\psi}_c^\dagger\hat{\psi}^b + \hat{\psi}^b\hat{\psi}_c^\dagger\hat{\psi}^a + \hat{\psi}_c^\dagger\hat{\psi}^a\hat{\psi}^b). \qquad \text{(C.4)}$$

As in the bosonic case one has for any function $f(\hat{\psi})$

$$(\psi_a^\dagger\psi_b^\dagger f(\psi))_S = \frac{1}{4}\Big[\psi_a^\dagger\psi_b^\dagger f(\psi) \pm \psi_a^\dagger f(\psi)\psi_b^\dagger \mp \psi_b^\dagger f(\psi)\psi_a^\dagger + f(\psi)\psi_a^\dagger\psi_b^\dagger\Big] \qquad \text{(C.5)}$$

where the upper (lower) signs apply when $f(\psi)$ is commuting (anticommuting). For any function $g(\hat{\psi}^\dagger)$ one has the equivalent results

$$(\hat{\psi}^a\hat{\psi}^b g(\psi^\dagger))_S = \frac{1}{4}\Big[\hat{\psi}^a\hat{\psi}^b g(\hat{\psi}^\dagger) \pm \hat{\psi}^a g(\hat{\psi}^\dagger)\hat{\psi}^b \mp \hat{\psi}^b g(\psi^\dagger)\hat{\psi}^a + g(\psi^\dagger)\hat{\psi}^a\psi^b\Big]. \qquad \text{(C.6)}$$

Let us now write the $N = 2$ supersymmetric Hamiltonian in Weyl-ordered form. The action in Minkowski time reads as (see Appendix D)

$$L = \frac{1}{2}g_{ij}\dot{x}^i\dot{x}^j + \frac{i}{2}\psi_\alpha^a(\dot{\psi}_\alpha^a + \dot{x}^i\omega_i{}^a{}_b\psi_\alpha^b) + \frac{1}{8}R_{abcd}(\omega)\psi_\alpha^a\psi_\alpha^b\psi_\beta^c\psi_\beta^d \qquad \text{(C.7)}$$

where summation over $\alpha = 1, 2$ and $\beta = 1, 2$ is understood, $a = 1, \ldots, n$ and $i = 1, \ldots, n$, and $\omega_i{}^a{}_b(x)$ is the spin connection. The curvature $R_{abcd}(\omega)$ is defined in Appendix A.

The momentum p_i conjugate to x^i is given by

$$p_i = g_{ij}\dot{x}^j + \frac{i}{2}\omega_{iab}\psi_\alpha^a\psi_\alpha^b. \qquad \text{(C.8)}$$

Furthermore, one has for a fixed value of a the relation $\frac{1}{2}\psi_\alpha^a\dot{\psi}_\alpha^a = \bar{\psi}_a\dot{\psi}^a +$ *total derivative*, with $\psi^a = (\psi_1 + i\psi_2)/\sqrt{2}$ and $\bar{\psi}_a = (\psi_1 - i\psi_2)/\sqrt{2}$. Then, using left derivatives to define anticommuting canonically conjugate momenta

$$p(\psi)_a = \frac{\partial}{\partial\dot{\psi}^a}S = -i\bar{\psi}_a \quad \Rightarrow \quad \{\bar{\psi}_a, \psi^b\} = \hbar\delta_a^b. \qquad \text{(C.9)}$$

The classical Hamiltonian then reads as

$$\begin{aligned} H &= \dot{x}^i p_i + \dot{\psi}^a p(\psi)_a - L \\ &= g^{ij}p_i\left(p_j - \frac{i}{2}\omega_{jab}\psi_\alpha^a\psi_\alpha^b\right) + i\bar{\psi}^a\dot{\psi}^a \\ &\quad - \frac{1}{2}g^{ij}\left(p_i - \frac{i}{2}\omega_{iab}\psi_\alpha^a\psi_\alpha^b\right)\left(p_j - \frac{i}{2}\omega_{jcd}\psi_\beta^c\psi_\beta^d\right) - i\bar{\psi}^a\dot{\psi}^a \end{aligned}$$

$$-\frac{i}{2}g^{ij}\left(p_j-\frac{i}{2}\omega_{jab}\psi^a_\alpha\psi^b_\alpha\right)\omega_{icd}\psi^c_\beta\psi^d_\beta-\frac{1}{8}R_{abcd}\psi^a_\alpha\psi^b_\alpha\psi^c_\beta\psi^d_\beta$$
$$=\frac{1}{2}g^{ij}\left(p_i-\frac{i}{2}\omega_{iab}\psi^a_\alpha\psi^b_\alpha\right)\left(p_j-\frac{i}{2}\omega_{jcd}\psi^c_\beta\psi^d_\beta\right)$$
$$-\frac{1}{8}R_{abcd}\psi^a_\alpha\psi^b_\alpha\psi^c_\beta\psi^d_\beta. \tag{C.10}$$

In terms of ψ^a and $\bar{\psi}^a$ this becomes

$$H=\frac{1}{2}g^{ij}(p_i-i\omega_{iab}\bar{\psi}^a\psi^b)(p_j-i\omega_{jcd}\bar{\psi}^c\psi^d)$$
$$-\frac{1}{2}R_{abcd}\bar{\psi}^a\psi^b\bar{\psi}^c\psi^d. \tag{C.11}$$

We now define the corresponding Hamiltonian operator. We fix the operator ordering by requiring that $\hat{H}$ be general coordinate (Einstein) invariant and locally Lorentz invariant, namely it should commute with the generators of general coordinate and local Lorentz transformations. To achieve this, the same factors $g^{\pm 1/4}$ as in the bosonic case are needed. The $N=2$ Hamiltonian operator thus becomes

$$\hat{H}=\frac{1}{2}g^{-1/4}\left(p_i-i\omega_{iab}\bar{\psi}^a\psi^b\right)g^{1/2}g^{ij}\left(p_j-i\omega_{jcd}\bar{\psi}^c\psi^d\right)g^{-1/4}$$
$$-\frac{1}{2}R_{abcd}\bar{\psi}^a\psi^b\bar{\psi}^c\psi^d \tag{C.12}$$

where $\{\psi^a,\bar{\psi}_b\}=\hbar\delta^a_b$.

To write this operator in Weyl-ordered form, i.e. to **rewrite** it such that all canonical variables appear symmetrized or antisymmetrized, we note that the two-fermion terms are already antisymmetrized, since $\{\bar{\psi}_a,\psi^b\}$ is proportional to δ^b_a and ω_{iab} is traceless. For the same reason we can write the four-fermion term in $\hat{H}$ as follows:

$$\hat{H}(\psi^4)=-\frac{1}{2}[g^{ij}\omega_{iab}\omega_{jcd}+R_{abcd}(\omega)]\bar{\psi}^a\psi^b\bar{\psi}^c\psi^d$$
$$=-\frac{1}{16}[g^{ij}\omega_{iab}\omega_{jcd}+R_{abcd}(\omega)]\{[\bar{\psi}^a,\psi^b],[\bar{\psi}^c,\psi^d]\}. \tag{C.13}$$

We now first prove the following lemma.

Lemma: $$\frac{1}{8}\{[\bar{\psi}^a,\psi^b],[\bar{\psi}^c,\psi^d]\}-(\bar{\psi}^a\psi^b\bar{\psi}^c\psi^d)_S=\frac{\hbar^2}{4}\delta^{ad}\delta^{bc}. \tag{C.14}$$

Proof: $(\bar{\psi}^a\psi^b\bar{\psi}^c\psi^d)_S=-(\bar{\psi}^a\bar{\psi}^c\psi^b\psi^d)_S$. Next, rewrite this term, once keeping $\psi^b\psi^d$ together and once keeping $\bar{\psi}^a\bar{\psi}^c$ together

$$(\bar{\psi}^a\bar{\psi}^c\psi^b\psi^d)_S=\frac{1}{8}\left[\begin{array}{l}\bar{\psi}^a\bar{\psi}^c\psi^b\psi^d+\bar{\psi}^a\psi^b\psi^d\bar{\psi}^c-\bar{\psi}^c\psi^b\psi^d\bar{\psi}^a+\psi^b\psi^d\bar{\psi}^a\bar{\psi}^c\\ \quad+\bar{\psi}^a\bar{\psi}^c\psi^b\psi^d+\psi^b\bar{\psi}^a\bar{\psi}^c\psi^d-\psi^d\bar{\psi}^a\bar{\psi}^c\psi^b+\psi^b\psi^d\bar{\psi}^a\bar{\psi}^c\end{array}\right].$$

Adding this to

$$
\begin{aligned}
&\frac{1}{8}\{[\bar{\psi}^a, \psi^b], [\bar{\psi}^c, \psi^d]\} \\
&= \frac{1}{8}\left[\begin{array}{l} +\bar{\psi}^a\psi^b\bar{\psi}^c\psi^d - \psi^b\bar{\psi}^a\bar{\psi}^c\psi^d - \bar{\psi}^a\psi^b\psi^d\bar{\psi}^c + \psi^b\bar{\psi}^a\psi^d\bar{\psi}^c \\ +\bar{\psi}^c\psi^d\bar{\psi}^a\psi^b - \psi^d\bar{\psi}^c\bar{\psi}^a\psi^b - \bar{\psi}^c\psi^d\psi^b\bar{\psi}^a + \psi^d\bar{\psi}^c\psi^b\bar{\psi}^a \end{array}\right]
\end{aligned}
$$

one finds by combining corresponding pairs of terms

$$
\frac{\hbar}{8}(\bar{\psi}^a\delta^{bc}\psi^d + 0 + 0 + \psi^b\delta^{ad}\bar{\psi}^c + \bar{\psi}^c\delta^{ad}\psi^b + 0 + 0 + \psi^d\delta^{bc}\bar{\psi}^a) = \frac{\hbar^2}{4}\delta^{ad}\delta^{bc}.
$$

It follows that

$$
\begin{aligned}
\hat{H}(\psi^4) &= \hat{H}_S(\psi^4) - \frac{\hbar^2}{8}\delta^{ad}\delta^{bc}[g^{ij}\omega_{iab}\omega_{jcd} + R_{abcd}(\omega)] \\
&= \hat{H}_S(\psi^4) + \frac{\hbar^2}{8}(g^{ij}\omega_{iab}\omega_j{}^{ab} - R).
\end{aligned} \tag{C.15}
$$

The terms in $\hat{H}$ without fermions yield back the Hamiltonian of the bosonic model obtained in (B.25)

$$
\hat{H}(\text{no }\psi) = \frac{1}{2}(g^{ij}p_ip_j)_S + \frac{\hbar^2}{8}(R + g^{ij}\Gamma^k_{il}\Gamma^l_{jk}). \tag{C.16}
$$

The two-fermion terms are also Weyl-ordered in the sector with x and p

$$
\begin{aligned}
&-\frac{i}{4}g^{-1/4}p_ig^{1/4}g^{ij}\omega_{jcd}\psi^c_\beta\psi^d_\beta - \frac{i}{4}g^{1/4}\omega_{iab}\psi^a_\alpha\psi^b_\alpha g^{ij}p_jg^{-1/4} \\
&= -\frac{i}{4}\{p_i, g^{ij}\omega_{jcd}\psi^c_\beta\psi^d_\beta\}.
\end{aligned} \tag{C.17}
$$

We therefore conclude that rewriting (C.12) in Weyl-ordered form yields

$$
\begin{aligned}
\hat{H}_W &= \left(\frac{1}{2}g^{ij}\pi_i\pi_j - \frac{1}{2}R_{abcd}(\omega)\bar{\psi}^a\psi^b\bar{\psi}^c\psi^d\right)_S \\
&\quad + \frac{\hbar^2}{8}g^{ij}(\Gamma^l_{ik}\Gamma^k_{jl} + \omega_{iab}\omega_j{}^{ab}) \qquad \text{(C.18)} \\
\pi_i &= p_i - i\omega_{iab}\bar{\psi}^a\psi^b.
\end{aligned}
$$

We note that the terms with R from the bosonic sector and the $(\psi)^4$ sector cancel.

One can even achieve a formulation of the $N = 2$ model where the $\Gamma\Gamma$ and $\omega\omega$ terms cancel. Although the choice of ψ^a as a basic variable is very suitable (since it yields as a kinetic term simply $i\psi^a_\alpha\dot{\psi}^a_\alpha$) one can make a different choice, namely ψ^i with curved index i. Since $\psi^a = e^a_i\psi^i$, the action becomes $\frac{1}{2} i g_{ij} \psi^i_\alpha (\dot{\psi}^j_\alpha + \dot{x}^k \Gamma^j_{kl} \psi^l_\alpha)$. We used the "vielbein postulate" $\partial_i e^a_j - \Gamma_{ij}{}^k e^a_k + \omega_i{}^a{}_b e^b_j = 0$ of (A.1). The momentum conjugate to $\psi^i \equiv (\psi^i_1 + i\psi^i_2)/\sqrt{2}$ is then $p(\psi)_j = -i\psi^\dagger_j \equiv -ig_{ij}(\psi^j_1 - i\psi^j_2)/\sqrt{2}$ and the canonical commutation relations become

$$\{\psi^i, \psi^\dagger_j\} = \hbar \delta^i_j. \tag{C.19}$$

(Note that the Jacobi identity for $(p_j(\psi), \psi^i, \bar{\psi}_k)$ is still satisfied because the bracket of ψ^i and $\bar{\psi}_k$ is a constant.) The conjugate momentum of x^i becomes $p_i = g_{ij}\dot{x}^j + \frac{1}{2} i \Gamma_{il;k} \psi^k_\alpha \psi^l_\alpha$. The Hamiltonian now reads as

$$\hat{H} = \frac{1}{2} g^{-1/4} \left(p_i - \frac{i}{2} \Gamma_{il;k} \psi^k_\alpha \psi^l_\alpha \right) g^{ij} g^{1/2} \left(p_j - \frac{i}{2} \Gamma_{jn;m} \psi^m_\beta \psi^n_\beta \right) g^{-1/4}$$
$$- \frac{1}{2} R_{abcd}(\omega) \bar{\psi}^a \psi^b \bar{\psi}^c \psi^d \tag{C.20}$$

and the four Fermi terms now read as

$$-\frac{1}{2} \left[g^{ij} \Gamma_{il;k} \Gamma_{jn;m} + R_{klmn}(\omega) \right] \bar{\psi}^k \psi^l \bar{\psi}^m \psi^n. \tag{C.21}$$

The antisymmetrization of $\hbar^2 \bar{\psi}^k \psi^l \bar{\psi}^m \psi^n$ yields a factor of $\frac{1}{4} g^{kn} g^{lm}$ and this produces a term

$$-\frac{\hbar^2}{8} (g^{ij} \Gamma^k_{il} \Gamma^l_{jk} + R). \tag{C.22}$$

Adding (C.16) and (C.22), all $\hbar^2$ terms now cancel.

For practical calculations, the choice of ψ^a as a basis variable is preferred, even at the expense of the extra $\hbar^2$ terms.

Appendix D

Nonlinear susy sigma models and $d = 1$ superspace

As explained in the introduction, for the computation of anomalies in certain d-dimensional quantum field theories, which themselves need not be supersymmetric field theories, supersymmetric $d = 1$ nonlinear σ-models (a particular class of supersymmetric quantum mechanical models) are needed. One can write down these models in "t space", beginning with kinetic terms $g_{ij}(\varphi)\dot{\varphi}^i\dot{\varphi}^j$ for the bosonic fields $\varphi^i(t)$ and finding further terms with fermionic $\psi^i(t)$, by using the so-called Noether method. This is instructive if one is interested in the structure of the theory, in particular the leading terms of the action and transformation rules. To obtain the complete answer, the Noether method is somewhat cumbersome (although it always yields the complete answer if such a complete answer exists). We shall begin by using this Noether method for the models we are interested in; the procedure becomes clear along the way and has pedagogical value. Then, however, we shall follow the superspace approach, and obtain the complete answer at once. The superspace method is less intuitive, but once one has understood the overall structure of a theory, the superspace approach gives complete answers while avoiding the tedious labor of the Noether approach. The Noether method and the superspace method are complementary. We end by deriving the $N = 1$ and $N = 2$ models in Minkowski time.

The Noether method

In the Noether method we start with a bosonic and fermionic kinetic term

$$L(\text{kin}) = \frac{1}{2} g_{ij}(\varphi)\Big(\dot{\varphi}^i\dot{\varphi}^j + i\psi^i\dot{\psi}^j\Big). \tag{D.1}$$

By a dot we indicate a d/dt derivative (we are in Minkowskian time) and $\varphi^i(t)$ is a real function. The anticommuting $\psi^i(t)$ are by definition real under hermitian conjugation, and we added the factor i in order that for the action to be hermitian,

$$(i\psi^i\dot{\psi}^j)^\dagger = -i\dot{\psi}^j\psi^i = +i\psi^i\dot{\psi}^j. \tag{D.2}$$

For superspace applications it is natural that φ^i and ψ^i have the same index i, because then one can construct a superfield $\varphi^i + i\theta\psi^i$. However, one can use either ψ^i or $\psi^a = e_i^a\psi^i$, where $e_i^a(\varphi)$ are the vielbein fields which are square roots of the metric, $g_{ij} = e_i^a e_j^b \eta_{ab}$ (in target space). At the end of this appendix we shall mention the changes which occur if one uses ψ^a, but for now we use ψ^i.

Under supersymmetry transformations, φ^i should transform into ψ^i and vice versa. From the action we see that the dimension of φ^i, denoted by $[\varphi^i]$, differs from that of ψ^i by half the dimension of t (which has by definition a dimension of -1, $[t] = -1$); namely, $2[\varphi] - 2[t] = 2[\psi] - [t]$. Since the action is dimensionless (if $\hbar = 1$), we find $[\varphi] = -\frac{1}{2}$ and $[\psi] = 0$. It follows that if $\delta\varphi^i \sim \epsilon\psi^i$, then $[\epsilon] = -\frac{1}{2}$. Consequently, in $\delta\psi^i \sim \epsilon\varphi^i$ we need a derivative to make the dimensions come out right: $\delta\psi^i \sim \epsilon\dot{\varphi}^i$. We do not consider terms with $\dot{\epsilon}$ because we consider at this point rigid susy models which have by definition a constant ϵ. Since ψ^i and φ^i are real, ϵ must also be real or purely imaginary. Choosing ϵ to be real we need a factor of i in $\delta\varphi^i \sim i\epsilon\psi^i$ in order for $\delta\varphi^i$ to also be real. To obtain $\delta\varphi^i = -i\epsilon\psi^i$, one can scale ϵ appropriately. Then $\delta\psi^i = \beta\dot{\varphi}^i\epsilon$, and the value $\beta = 1$ will be shown to follow by requiring invariance of the action or closure of the supersymmetry algebra. Hence, we assume the following transformation rules:

$$\delta\varphi^i = -i\epsilon\psi^i, \quad \delta\psi^i = \beta\dot{\varphi}^i\epsilon \quad (\beta = 1). \tag{D.3}$$

The parameter ϵ is constant. We could study locally supersymmetric theories (supergravity theories[1]) with a local parameter $\epsilon(t)$ and a gauge field for supersymmetry, but we shall only need rigidly supersymmetric theories with constant ϵ.

We now begin the Noether procedure. We vary $L(\text{kin})$ in (D.1) using $\delta\varphi^i$ and $\delta\psi^i$ given above, and find

$$\begin{aligned}\delta L(\text{kin}) &= \frac{1}{2}(\partial_k g_{ij})(-i\epsilon\psi^k)\left(\dot{\varphi}^i\dot{\varphi}^j + i\psi^i\dot{\psi}^j\right) \\ &\quad + \frac{1}{2}g_{ij}\left[2\dot{\varphi}^i\frac{d}{dt}(-i\epsilon\psi^j) + i\beta(\dot{\varphi}^i\epsilon)\dot{\psi}^j + i\beta\psi^i\frac{d}{dt}(\dot{\varphi}^j\epsilon)\right].\end{aligned} \tag{D.4}$$

[1] A discussion of supergravity in quantum mechanics is given in [186]. One can couple the matter fields φ and ψ to the supergravity gauge fields (the vielbein and the gravitino), but no gauge action for supergravity itself exists in $d = 1$ dimensions.

In the action, this expression is integrated over t, and if we partially integrate the last term to remove double derivatives, all $\dot{\varphi}\dot{\psi}$ terms cancel if $\beta = 1$. We are left with the variation

$$\delta L(\text{kin}) = \left(-\frac{i}{2}\partial_k g_{ij}\right)\epsilon(\psi^k \dot{\varphi}^i \dot{\varphi}^j + i\psi^k \psi^i \dot{\psi}^j - \dot{\varphi}^k \psi^i \dot{\varphi}^j) \qquad \text{(D.5)}$$

where we have used $\beta = 1$ and ϵ and ψ^i anticommute: $\epsilon\psi^i = -\psi^i\epsilon$. There are no terms with $\dot{\epsilon}$ because ϵ is constant.

We now observe that the combination $-i\epsilon\psi^k$ can be written as $\delta\varphi^k$ and one might be tempted to write the first term in $\delta L(\text{kin})$ as $(\frac{1}{2}\partial_k g_{ij})\delta\varphi^k \dot{\varphi}^i \dot{\varphi}^j$. However, this is just how we found this term, so we would be going backwards. We can also consider another combination, for example $\dot{\varphi}^i\epsilon$, and replace it by $\delta\psi^i$. We shall choose the latter alternative for reasons to become clear. Our aim is to find a new term in the action, $L(\text{extra})$, such that $\delta L(\text{extra}) = -\delta L(\text{kin})$. Then, obviously, the t-integral of $L = L(\text{kin}) + L(\text{extra})$ is invariant. We claim that a solution is

$$L(\text{extra}) = \frac{i}{2}(\partial_k g_{ij})(\psi^i \dot{\varphi}^j \psi^k). \qquad \text{(D.6)}$$

To verify this claim, note that we need not vary $\partial_k g_{ij}(\varphi)$, because it would yield $(\partial_l \partial_k g_{ij})\,(-i\epsilon\psi^l)$ and since $\psi^l\psi^k$ is antisymmetric in l, k, this variation would vanish. The variation of ψ^i cancels $\psi^k \dot{\varphi}^i \dot{\varphi}^j$ in (D.5). The variation of $\dot{\varphi}^j$ in $L(\text{extra})$ cancels the $\psi\psi\dot{\psi}$ term in $\delta L(\text{kin})$. Finally, the variations of ψ^k in $L(\text{extra})$ cancel the $\dot{\varphi}^k \psi^i \dot{\varphi}^j$ term in $\delta L(\text{kin})$.

Hence, $I \equiv \int L\, dt$ with

$$L = \frac{1}{2} g_{ij} \dot{\varphi}^i \dot{\varphi}^j + \frac{i}{2}\psi^i (g_{ij}\dot{\psi}^j + \partial_k g_{ij} \dot{\varphi}^j \psi^k) \qquad \text{(D.7)}$$

is invariant under the transformation laws in (D.3). We can rewrite the terms with fermions in a way which has a geometrical meaning,

$$\psi^i (g_{ij}\dot{\psi}^j + \partial_k g_{ij}\dot{\varphi}^j \psi^k) = \psi^i g_{ij}(\dot{\psi}^j + \dot{\varphi}^k \Gamma^j_{kl}\psi^l) \equiv \psi^i g_{ij}\frac{D}{dt}\psi^j \qquad \text{(D.8)}$$

where Γ^j_{kl} is the Christoffel symbol. The derivative $\frac{D}{dt}\psi^j$ transforms as a contravariant vector under general diffeomorphisms $x^i \to x^i + \xi^i(x)$, as we shall shortly discuss.

To obtain the Noether current for supersymmetry, we let ϵ become a local parameter, and repeat the evaluation of δL. The terms proportional to $\dot{\epsilon}$ then yield the Noether current [77]. From (D.4) we obtain one such term, namely when d/dt in the first term inside the square brackets acts on ϵ. The last term in (D.4) was partially integrated, so it does not yield a $\dot{\epsilon}$ term. Another term with $\dot{\epsilon}$ might seem to come from varying $\dot{\varphi}^j$ in (D.6), but this contribution vanishes as it is proportional to $\partial_k g_{ij}\psi^i\psi^j$. Hence, we find that the Noether current for supersymmetry is proportional to

$$j_N = g_{ij}\dot{\varphi}^i\psi^j. \tag{D.9}$$

It is now clear that the other way to proceed mentioned under (D.5), namely replacing $-i\epsilon\psi^k$ by $\delta\varphi^k$ in $\delta L(\text{kin})$, would not have worked. It would have led to an $L(\text{extra}) = \frac{1}{2}\partial_k g_{ij}\varphi^k\dot{\varphi}^i\dot{\varphi}^j$ but whereas variation of φ^k would have canceled the first term in $\delta L(\text{kin})$ (by construction), the other variations would not have been canceled. One might have expected this, since a coordinate φ^k is not a tensor in general relativity (in contrast to $\dot{\varphi}^k$ or $\delta\varphi^k$), so that the action should not contain undifferentiated φ (except in $g_{ij}(\varphi)$).

The difficult step was to find the correct $L(\text{extra})$. One might have guessed this result by noting that the terms in parentheses in (D.7) form a covariant derivative

$$\psi^i(g_{ij}\dot{\psi}^j + \partial_k g_{ij}\dot{\varphi}^j\psi^k) = \psi_j\frac{D}{dt}\psi^j, \qquad \psi_j = g_{ji}\psi^i$$
$$\frac{D}{dt}\psi^j = \dot{\psi}^j + \dot{\varphi}^k\Gamma^j_{kl}\psi^l, \quad \Gamma^j_{kl} = \frac{1}{2}g^{jm}\Big(\partial_k g_{lm} + \partial_l g_{km} - \partial_m g_{kl}\Big). \tag{D.10}$$

The covariant derivative $\frac{D}{Dt}\psi^j$ is indeed a contravariant vector under infinitesimal general coordinate transformation in spacetime (because Γ^j_{kl} is a connection: $\delta\Gamma^j_{kl} = \partial_k\partial_l\xi^j + \cdots$)

$$\varphi^i \;\to\; \varphi^i + \xi^i(\varphi), \quad \psi^i \;\to\; \psi^i + \frac{\partial\xi^i}{\partial\varphi^j}\psi^j$$
$$\frac{D}{dt}\psi^j \;\to\; \frac{D}{dt}\psi^j + \frac{\partial\xi^j}{\partial\varphi^k}\frac{D}{dt}\psi^k. \tag{D.11}$$

A less insightful, but still correct way to obtain the result in (D.7) would have been to write down all possible candidates for $L(\text{extra})$, with arbitrary coefficients, and then fixing these coefficients by requiring that, up to partial integrations, $\delta L(\text{extra}) = -\delta L(\text{kin})$. However, one would like to have a method which guarantees success even if one is not clever enough to use such tricks or patient enough to do much algebra, and such a method is the superspace method.

The superspace method

In $d = 1$ $N = 1$ superspace, one has one bosonic coordinate which we call t (because $d = 1$) and one fermionic coordinate θ (because $N = 1$). The

θ are Grassmann variables, $\{\theta, \theta\} = 2\theta\theta = 0$, and by definition θ is real, $(\theta)^\dagger = \theta$. In this superspace, we consider superfields, i.e. fields depending on t and θ.

We consider superfields with an index, $\phi^i(t, \theta)$. Expanding in powers of θ, there are only two terms since $\theta^2 = 0$, and we define

$$\phi^i(t, \theta) = \varphi^i(t) + i\beta\theta\psi^i(t) \tag{D.12}$$

where β is a real constant to be fixed later. The factor of i is again needed in order for $\phi^i(t, \theta)$ to be real, and we require that $\phi^i(t, \theta)$ be real because $\varphi^i(t)$ is real.

We shall now obtain the transformation law of $\phi^i(t, \theta)$ under supersymmetry and construct an invariant action involving superfields ϕ^i by using an approach called the "coset method" which is also used in more complicated cases (the $d = 4, N = 1$ or $N = 2$ cases in particular). This approach starts from a superalgebra which is at the basis of the whole approach. In our case we shall use the superalgebra of supersymmetric quantum mechanics, with Hamiltonian H and supersymmetry generator Q,

$$\{Q, Q\} = 2H, \qquad [Q, H] = 0. \tag{D.13}$$

In fact, $[Q, H] = 0$ follows from $\{Q, Q\} = 2Q^2 = 2H$. We shall assume that H and Q are hermitian, $(H)^\dagger = H$ and $(Q)^\dagger = Q$. Given any superalgebra, one first *deduces* how supercoordinates transform under supersymmetry. Then one finds how superfields transform under supersymmetry.

To deduce how the supercoordinates x and θ transform, one takes a group element of the form

$$g(t, \theta) = \mathrm{e}^{itH+\theta Q} \tag{D.14}$$

where $g^\dagger = g^{-1}$ (at least formally because $(\theta Q)^\dagger = Q^\dagger\theta^\dagger = Q\theta = -\theta Q$). We multiply from the left by a group element

$$h(\alpha, \epsilon) = \mathrm{e}^{i\alpha H+\epsilon Q} \tag{D.15}$$

and we work to linear order in α and ϵ ("h near the origin"). Since the product can again be written as an exponent, with multiple commutators of ϵH and ϵQ in the exponent, we have

$$h(\alpha, \epsilon)g(t, \theta) = g(t + \delta t, \theta + \delta\theta) \tag{D.16}$$

where δt and $\delta\theta$ are linear in α and ϵ. Using the Baker–Campbell–Hausdorff formula for the bosonic objects $i\alpha H, \epsilon Q, itH$ and θQ we find

$$\mathrm{e}^{i\alpha H+\epsilon Q}\ \mathrm{e}^{itH+\theta Q} = \mathrm{e}^{i(t+\alpha)H+(\theta+\epsilon)Q+\frac{1}{2}[\epsilon Q, \theta Q]}. \tag{D.17}$$

No further commutators are needed, since

$$[\epsilon Q, \theta Q] = -\epsilon\theta\{Q, Q\} = -2\epsilon\theta H \tag{D.18}$$

which commutes with all generators (H and Q). Hence,

$$\delta t = \alpha + i\epsilon\theta, \qquad \delta\theta = \epsilon. \tag{D.19}$$

To check that this indeed forms a representation of the superalgebra, we rewrite this as

$$\delta t = [t, i\alpha H + \epsilon Q], \qquad \delta\theta = [\theta, i\alpha H + \epsilon Q] \tag{D.20}$$

and by equating (D.19) and (D.20) we find the generators H and Q in the supercoordinate representation

$$H = i\frac{\partial}{\partial t}, \quad Q = -\frac{\partial}{\partial\theta} - i\theta\frac{\partial}{\partial t}. \tag{D.21}$$

As one may check, one indeed has a representation of the superalgebra in terms of differential operators. For example,

$$\{Q, Q\} = 2H. \tag{D.22}$$

We now **declare** $\phi^i(t, \theta)$ to be scalar superfields. By this we mean the same as, for example, in ordinary quantum mechanics: the transformation of the fields is induced by the transformation of the (super) coordinates

$$\phi'^i(t', \theta') = \phi^i(t, \theta). \tag{D.23}$$

Putting $\phi' = \phi + \delta\phi$ and $(t', \theta') = (t, \theta) + (\delta t, \delta\theta)$, we obtain

$$\begin{aligned}
\delta\phi &= \left(-\delta t\frac{\partial}{\partial t} - \delta\theta\frac{\partial}{\partial\theta}\right)\phi \\
&= \left[-\alpha\frac{\partial}{\partial t} - \epsilon\left(\frac{\partial}{\partial\theta} + i\theta\frac{\partial}{\partial t}\right)\right]\phi \\
&= [i\alpha H + \epsilon Q, \phi].
\end{aligned} \tag{D.24}$$

From (D.20) and (D.24) we see that coordinates and fields transform contragradiently, again a well-known result from ordinary quantum mechanics (and just a consequence of the definition $\phi'(t', \theta') = \phi(t, \theta)$).

We have now obtained the supersymmetry generator

$$Q = -\left(\frac{\partial}{\partial\theta} + i\theta\frac{\partial}{\partial t}\right). \tag{D.25}$$

Let us check whether the components $\varphi^i(t)$ and $\psi^i(t)$ transform as in (D.3),

$$\begin{aligned}\delta\phi^i = [\epsilon Q, \phi^i] &= -\epsilon\Big(\frac{\partial}{\partial\theta} + i\theta\frac{\partial}{\partial t}\Big)\Big[\varphi^i(t) + i\beta\theta\psi^i(t)\Big] \\ &= \delta\varphi^i(t) + i\beta\theta\delta\psi^i(t).\end{aligned} \tag{D.26}$$

Equating terms with and without θ yields

$$\begin{aligned}\delta\varphi^i(t) &= -\epsilon i\beta\psi^i(t) \\ i\beta\theta\delta\psi^i(t) &= -\epsilon i\theta\dot{\varphi}^i(t) = i\theta\epsilon\dot{\varphi}^i.\end{aligned} \tag{D.27}$$

It is clear that for $\beta = 1$ we retrieve the transformation rules of $\varphi^i(t)$ and $\psi^i(t)$ which we obtained in (D.3). Hence, so far we have obtained the following results:

$$\begin{aligned}\phi^i &= \varphi^i(t) + i\theta\psi^i(t) \\ \epsilon Q &= -\epsilon\Big(\frac{\partial}{\partial\theta} + i\theta\frac{\partial}{\partial t}\Big) \\ \delta\phi^i &= [\epsilon Q, \phi^i].\end{aligned} \tag{D.28}$$

For the construction of invariant actions it is useful to have an operator which anticommutes with Q. Such an operator is given by

$$D = \frac{\partial}{\partial\theta} - i\theta\frac{\partial}{\partial t}. \tag{D.29}$$

It is formally obtained by the same steps as Q, but using right multiplication for h and g. Since left and right multiplication commute, $(h_1 g)h_2 = h_1(gh_2)$, it follows that Q and D anticommute. Let us check explicitly that $\{D, Q\} = 0$. This follows by writing out all terms as

$$\begin{aligned}-\{D, Q\} &= \Big\{\frac{\partial}{\partial\theta} - i\theta\frac{\partial}{\partial t}\,,\, \frac{\partial}{\partial\theta} + i\theta\frac{\partial}{\partial t}\Big\} \\ &= \Big\{\frac{\partial}{\partial\theta}\,,\, i\theta\Big\}\frac{\partial}{\partial t} + \Big\{-i\theta\,,\, \frac{\partial}{\partial\theta}\Big\}\frac{\partial}{\partial t} = 0.\end{aligned} \tag{D.30}$$

We can now at once write down a set of invariant actions for ϕ^i. We claim that for any m and n the following action is invariant under (D.28):

$$I = \int dt\, d\theta \Big(D^m\phi^i\Big)\Big(D^n\phi^j\Big) g_{ij}(\phi). \tag{D.31}$$

To see why $\delta I = 0$, note that $\delta g_{ij}(\phi) = [\epsilon Q, g_{ij}(\phi)]$, because of (D.28), and

$$\delta(D^m\phi^i) \equiv D^m\delta\phi^i = D^m\Big(\epsilon Q\phi^i\Big) = \Big(\epsilon Q(D^m\phi^i)\Big) \tag{D.32}$$

where we used in the last step that $\epsilon QD = D\epsilon Q$. Hence,

$$\delta\Big(D^m\phi^i D^n\phi^j\ g_{ij}(\phi)\Big) = \Big(\epsilon Q[D^m\phi^i D^n\phi^j g_{ij}(\phi)]\Big). \tag{D.33}$$

(We write $\delta\phi^i = (\epsilon Q\phi^i)$ to indicate that the differential operator Q acts on ϕ^i but not beyond ϕ^i. We could equivalently write $\delta\phi^i = [\epsilon Q, \phi^i]$.) Now,

$$\int dt\, d\theta\ \epsilon QL = 0 \tag{D.34}$$

for any L, because the d/dt in Q yields zero (assuming, as we always do, that all functions vanish at $t = \pm\infty$), while the $\partial/\partial\theta$ in Q also gives zero due to the following property:

Theorem: $$\int d\theta\, \frac{\partial}{\partial\theta} f(\theta) = 0 \quad \text{for any } f(\theta). \tag{D.35}$$

The proof of this theorem is trivial: $f(\theta) = f_0 + \theta f_1$ and $\int d\theta\, \theta = 1$ but $\int d\theta = 0$, so $\int d\theta \frac{\partial}{\partial\theta}\, f(\theta) = \int d\theta\, f_0 = 0$.

We now use the superspace formalism to construct the $N=1$ and $N=2$ nonlinear sigma models which play a central role in the second part of this book.

The N = 1 model

This model is used for the calculation of chiral anomalies of spin-$\frac{1}{2}$ fields. We have seen that for any m and n the action $I = \int dt\, d\theta\, (D^m\phi^i)(D^n\phi^j)g_{ij}(\phi)$ is a supersymmetric action. Which m and n should we take? We want an action which contains, to begin with, the kinetic term $\frac{1}{2}g_{ij}(\varphi)\dot\varphi^i\dot\varphi^j$. Since $D = \partial/\partial\theta - i\theta\partial/\partial t$ satisfies

$$\begin{aligned} D^2 &= -i\frac{\partial}{\partial t}, \quad iD^2\phi^i = \dot\varphi^i + i\theta\dot\psi^i \\ D\phi^i &= \left(\frac{\partial}{\partial\theta} - i\theta\frac{\partial}{\partial t}\right)\left(\varphi^i + i\theta\psi^i\right) = i\psi^i - i\theta\dot\varphi^i \end{aligned} \tag{D.36}$$

we see that for $m = 1$, $n = 2$ (or $m = 2$, $n = 1$) we obtain this kinetic term. So we take as an action

$$\begin{aligned} I &= \alpha\int dt\, d\theta (D\phi^i)(D^2\phi^j)g_{ij}(\phi) \\ &= \alpha\int dt\, d\theta\left[i\psi^i - i\theta\dot\varphi^i\right]\left[-i\dot\varphi^j + \theta\dot\psi^j\right] \\ &\quad\times\left[g_{ij}(\varphi) + i\theta\psi^k\,\partial_k g_{ij}(\varphi)\right] \end{aligned} \tag{D.37}$$

where α is a constant we shall soon fix. To obtain a nonzero result for the θ integral, we need only the terms proportional to θ. There are only three such terms, and we find

$$L = -\alpha\dot{\varphi}^i\dot{\varphi}^j g_{ij} - i\alpha\psi^i\dot{\psi}^j g_{ij} - i\alpha\psi^i\dot{\varphi}^j\psi^k\partial_k g_{ij}(\varphi). \tag{D.38}$$

(In obtaining this result, we moved θ to the left of all ψ functions, which causes some minus signs.) For $\alpha = -\frac{1}{2}$, this is indeed the action of (D.7) obtained from the Noether method, but now the last term comes out automatically. In Euclidean space, we find, putting $t = -it_E$

$$L_E = \frac{1}{2}g_{ij}\dot{\varphi}^i\dot{\varphi}^j + \frac{1}{2}\psi^i g_{ij}\left(\dot{\psi}^j + \dot{\varphi}^l\Gamma^j_{lk}\psi^k\right). \tag{D.39}$$

Another way to see that one needs the combination $D^2\phi^i D\phi^j$ in the action is to use dimensional arguments. Since $[dt] = -1, [d\theta] = +\frac{1}{2}$ (note that $[\theta] = -\frac{1}{2}$ but $\int d\theta\,\theta = 1), [D] = [\partial/\partial\theta] = \frac{1}{2}$ and $[\phi^i] = [\varphi^i] = -\frac{1}{2}$ (because $\int dt\,\dot{\varphi}^i\dot{\varphi}^j$ should be dimensionless), we see that the action in (D.31) is dimensionless provided $-1 + \frac{1}{2} + (m+n)\frac{1}{2} + 2(-\frac{1}{2}) = 0$, hence $m + n = 3$. For $m = 0$ and $m = 3$ one obtains $(D^3\phi^i)\phi^j g_{ij}(\phi)$, which is not a tensor in target space, hence only $(D^2\phi^i)(D\phi^j)g_{ij}(\phi)$ is allowed.

One can write down a supersymmetric extension of the $A_i(x)\dot{x}^i$ coupling, namely $\int d\theta A_i(\phi)(D\phi^i)$. One can also try to add a potential term $V(\phi)$ to the action. Then,

$$\int dt\,d\theta\; V(\phi) = \int dt\,d\theta\,\left[V(\varphi) + i\theta\psi^k\partial_k V\right] = \int dt\; i\psi^k(t)\partial_k V(\varphi). \tag{D.40}$$

It follows that the resulting potential is fermionic, so not very interesting for most applications. However, the next model allows useful potentials.

The $N = 2$ model

This model is used for the calculation of chiral anomalies of selfdual antisymmetric tensor fields, and for the Euler index theorem. If one uses a $d = 1,\ N = 2$ superspace approach with coordinates $t, \theta, \bar{\theta} \equiv (\theta)^\dagger$, where of course $\{\theta, \bar{\theta}\} = 0$, then a suitable action is

$$\begin{aligned} I &= \int dt\,d\theta\,d\bar{\theta}(D\phi^i)(\bar{D}\phi^j)g_{ij}(\phi) \\ D &= \left(\frac{\partial}{\partial\bar{\theta}} - i\theta\frac{\partial}{\partial t}\right), \quad \bar{D} = \left(\frac{\partial}{\partial\theta} - i\bar{\theta}\frac{\partial}{\partial t}\right) \\ Q &= -\left(\frac{\partial}{\partial\bar{\theta}} + i\theta\frac{\partial}{\partial t}\right), \quad \bar{Q} = -\left(\frac{\partial}{\partial\theta} + i\bar{\theta}\frac{\partial}{\partial t}\right). \end{aligned} \tag{D.41}$$

The susy of this action follows immediately from the observation that D and $\bar{D}$ anticommute with Q and $\bar{Q}$. One could also choose a real basis, with $D_+ = \partial/\partial\theta^+ - i\theta^+\partial_t$ and $D_- = \partial/\partial\theta^- - i\theta^-\partial_t$, where $\theta^+ + i\theta^- = \sqrt{2}\theta$ and $\theta^+ - i\theta^- = \sqrt{2}\bar{\theta}$. Then $\sqrt{2}D = D_+ + iD_-$ and $\sqrt{2}\bar{D} = D_+ - iD_-$, and the action is written as

$$\int dt\, d\theta^+ d\theta^-\; g_{ij} D_+\phi^i D_-\phi^j . \tag{D.42}$$

The underlying superalgebra is now

$$\begin{aligned}
&\{Q, \bar{Q}\} = -\{D, \bar{D}\} = 2H \\
&[H, D] = [H, \bar{D}] = [H, Q] = [H, \bar{Q}] = 0 \\
&\{Q, Q\} = \{\bar{Q}, \bar{Q}\} = \{D, D\} = \{\bar{D}, \bar{D}\} = 0.
\end{aligned} \tag{D.43}$$

In the coordinate representation, $H = i\partial/\partial t$. Again $[H, D] = [H, \bar{D}] = [H, Q] = [H, \bar{Q}] = 0$ follows from the definition $2H = \{Q, \bar{Q}\}$. Most importantly, D and $\bar{D}$ anticommute with Q and $\bar{Q}$. The reader may apply the coset formalism discussed above to derive (D.41).

Dimensional arguments reveal that one needs one D and one $\bar{D}$ in the action (two D would yield zero as $D\phi^i$ and $D\phi^j$ anticommute). Putting

$$\phi^i = \varphi^i + i\theta\bar{\psi}^i + i\bar{\theta}\psi^i + \bar{\theta}\theta F^i \tag{D.44}$$

one finds after integration over θ and $\bar{\theta}$

$$\begin{aligned}
L = {} & g_{ij}(\dot{\varphi}^i\dot{\varphi}^j + F^iF^j) + ig_{ij}\left(\psi^i\frac{D}{dt}\bar{\psi}^j + \bar{\psi}^j\frac{D}{dt}\psi^i\right) \\
& - (\partial_k\partial_l g_{ij})(\psi^i\bar{\psi}^j\psi^k\bar{\psi}^l) - 2\bar{\psi}^i\Gamma^l_{ij}\psi^j F_l.
\end{aligned} \tag{D.45}$$

The transformation rules $\delta\phi = [\bar{\epsilon}Q, \phi] + [\epsilon\bar{Q}, \phi]$ preserve the reality of ϕ and yield

$$\begin{aligned}
\delta\varphi^i &= -i\bar{\epsilon}\psi^i - i\epsilon\bar{\psi}^i, \qquad & \delta F^i &= -\bar{\epsilon}\dot{\psi}^i + \epsilon\dot{\bar{\psi}}^i \\
\delta\psi^i &= \dot{\varphi}^i\epsilon + iF^i\epsilon, & \delta\bar{\psi}^i &= -\dot{\varphi}^i\bar{\epsilon} + iF^i\bar{\epsilon}.
\end{aligned} \tag{D.46}$$

Substituting the algebraic field equation $F^i = \bar{\psi}^j\Gamma^i_{jk}\psi^k$, the fields ψ^i and $\bar{\psi}^i$ transform as follows:

$$\begin{aligned}
&\delta(\epsilon)\psi^i + (\delta(\epsilon)\varphi^j)\Gamma^i_{jl}\psi^l = \dot{\varphi}^i\epsilon \\
&\delta(\bar{\epsilon})\bar{\psi}^i + (\delta(\bar{\epsilon})\varphi^j)\Gamma^i_{jl}\bar{\psi}^l = \dot{\varphi}^i\bar{\epsilon}.
\end{aligned} \tag{D.47}$$

The left-hand sides transform covariantly (as contravariant vectors) under general coordinate transformation $\varphi^i \to \varphi^i + \xi^i(\varphi)$. The left-hand side defines a covariant variation, similar to a covariant derivative, see (D.10).

The terms with $\partial_k g_{lm}$ covariantize the ψ derivatives. Eliminating F^i one obtains in the action $(\bar{\psi}\Gamma\psi)^2$ terms which covariantize $\partial_k\partial_l g_{ij}$ to a full Riemann tensor (see Appendix A). This yields, adding an overall factor of $\frac{1}{2}$,

$$L = \frac{1}{2}g_{ij}\dot{\varphi}^i\dot{\varphi}^j + \frac{i}{2}g_{ij}\left(\psi^i\frac{D}{dt}\bar{\psi}^j + \bar{\psi}^i\frac{D}{dt}\psi^j\right) - \frac{1}{8}R_{ijkl}(\Gamma)\psi^i\bar{\psi}^j\psi^k\bar{\psi}^l. \qquad \text{(D.48)}$$

Decomposing $\psi = (\psi_1 + i\psi_2)/\sqrt{2}$ and $\bar{\psi} = (\psi_1 - i\psi_2)/\sqrt{2}$ one obtains the same action in terms of Majorana fermions,

$$L = \frac{1}{2}g_{ij}\dot{\varphi}^i\dot{\varphi}^j + \frac{i}{2}g_{ij}\psi^i_\alpha\left(\dot{\psi}^j_\alpha + \dot{\varphi}^k\Gamma^j_{kl}\psi^l_\alpha\right) - \frac{1}{8}R_{ijkl}(\Gamma)\psi^i_\alpha\psi^j_\alpha\psi^k_\beta\psi^l_\beta \qquad \text{(D.49)}$$

with $R_{ijk}{}^l(\Gamma) = \partial_i\Gamma^l_{jk} + \Gamma^l_{im}\Gamma^m_{jk} - (i \leftrightarrow j)$. To cast the 4-fermion term in this form, one may use the cyclic identity for the Riemann tensor. Note that $R_{ijk}{}^a(\Gamma) = R_{ij}{}^a{}_k(\omega)$. In Euclidean space

$$L_E = \frac{1}{2}g_{ij}\dot{\varphi}^i\dot{\varphi}^j + \frac{1}{2}g_{ij}\psi^i_\alpha\frac{D}{dt}\psi^j_\alpha + \frac{1}{8}R_{abcd}(\Gamma)\psi^a_\alpha\psi^b_\alpha\psi^c_\beta\psi^d_\beta. \qquad \text{(D.50)}$$

The potential term which we write as $(-i)W(\phi)$ to make it real now yields

$$\int dt\, d\theta\, d\bar{\theta}\; W(\phi) = \int dt\left[F^k\partial_k W(\varphi) - \psi^i\bar{\psi}^j\partial_i\partial_j W(\varphi)\right]. \qquad \text{(D.51)}$$

Eliminating F^k by its field equation

$$F^k = g^{kl}\partial_l W \qquad \text{(D.52)}$$

yields in t-space a positive-definite bosonic potential

$$V = \frac{1}{2}g^{ij}\partial_i W(\varphi)\partial_j W(\varphi). \qquad \text{(D.53)}$$

Finally, we return to our promise to discuss the changes that occur if one uses ψ^a instead of ψ^i. We begin with the $N=1$ model. We recall that $\psi^a = e^a_i(\varphi)\psi^i$. The transformation rule of ψ^a becomes

$$\delta\psi^a = e^a_i\delta\psi^i + (\delta e^a_i)\psi^i = e^a_i\dot{\varphi}^i\epsilon + \delta\varphi^j\partial_j e^a_i\psi^i. \qquad \text{(D.54)}$$

Next, we use the vielbein postulate

$$\partial_j e^a_i = -\omega_j{}^a{}_b e^b_i + \Gamma^k_{ji}e^a_k. \qquad \text{(D.55)}$$

Because $\delta\varphi^j$ contains ψ^j, the Christoffel term cancels when inserted into (D.54), and we find

$$\delta\psi^a + \delta\varphi^j \omega_j{}^a{}_b \psi^b = e_i^a \dot{\varphi}^i \epsilon. \tag{D.56}$$

The left-hand side looks like a covariant derivative (with d/dt replaced by δ) and is indeed locally Lorentz covariant, just as for the right-hand side. For curved indices one would expect

$$\delta\psi^i + \delta\varphi^j \Gamma^i_{jk} \psi^k = \dot{\varphi}^i \epsilon \tag{D.57}$$

but the term with a Christoffel symbol cancels, see (D.3).

In the action for the $N=1$ model we have found the covariant derivative, see (D.8),

$$g_{ij}\psi^i \left(\frac{D}{dt}\psi^j \right) = g_{ij}\psi^i \left(\dot{\psi}^j + \dot{\varphi}^l \Gamma^j_{lk} \psi^k \right). \tag{D.58}$$

It is straightforward to check that if one replaces the curved ψ^i by flat ψ^a one finds the corresponding covariant derivatives

$$\psi^a \left(\frac{D}{dt}\psi^a \right) = \psi^a \left(\dot{\psi}^a + \dot{\varphi}^j \omega_j{}^a{}_b \psi^b \right). \tag{D.59}$$

For the $N = 2$ model one finds covariant derivatives of ψ and $\bar{\psi}$ (both for curved or flat indices), and in the transformation rules one now finds pullback terms both in the flat and the curved case. The reason one now also finds a nonvanishing covariantizing term in $\delta\psi^i$ for curved indices is that this term is now of the form $\bar{\psi}\Gamma\psi$ instead of $\psi\Gamma\psi$.

Appendix E

Nonlinear susy sigma models for internal symmetries

In the main text, we were led to conjecture the existence of an extension of the usual $N = 1$ nonlinear supersymmetric σ-model for quantum mechanics which contains a term proportional to

$$c^*_A \, (T_\alpha)^A{}_B \, c^B \, A_i{}^\alpha(\varphi). \tag{E.1}$$

Here the antihermitian matrices T_α generate a Lie algebra in the same representation as the spin-$\frac{1}{2}$ fields of the original spacetime quantum field theory, while $c^*_A(t)$ and $c^B(t)$ are now anticommuting functions, which we call an antighost and a ghost, respectively. In this section we shall construct this extension.

The usual $N=1$ susy nonlinear σ-model in $(0+1)$-dimensional Minkowski time is given by

$$L = \frac{1}{2} g_{ij}(\varphi)\dot{\varphi}^i\dot{\varphi}^j + \frac{i}{2}\psi^a(\dot{\psi}^a + \dot{\varphi}^k \omega_k{}^a{}_b \psi^b). \tag{E.2}$$

Hermiticity requires the factor of i. We introduced vielbein fields $e_i{}^a(\varphi)$ satisfying $e_i{}^a e_j{}^b \eta_{ab} = g_{ij}$ and defined $\psi^a = e_i{}^a \psi^i$. Then (E.2) follows from (D.6) and (D.7) by using the "vielbein postulate" as in (D.58) and (D.59)

$$\partial_i e_j{}^a - \Gamma^k_{ij} e_k{}^a + \omega_i{}^a{}_b e_j{}^b = 0. \tag{E.3}$$

If we had chosen ψ^i instead of ψ^a to work with, there would also have been a metric g_{ij} in front of the fermionic terms.

We now require that the quantum mechanical Hamiltonian H should be a representation of the regulator $\not{D}\not{D}$ with $\not{D} = \gamma^a e_a{}^i(\varphi) D_i$ and

$$D_i = \left(\partial_i + \frac{1}{4}\omega_{iab}\gamma^a\gamma^b + A_i{}^\alpha T_\alpha\right). \tag{E.4}$$

We have absorbed the Yang–Mills coupling constant in $A_i{}^\alpha$. In the main text we have seen that the covariant derivative is proportional to $g_{ij}\dot{x}^j$.

Here we denote $\dot{x}^i$ by $\dot{\varphi}^i$, hence we anticipate that the conjugate momentum for φ^i should be given by

$$p_i = g_{ij}\dot{\varphi}^j + \frac{i}{2}\omega_{iab}\psi^a\psi^b + iA_i{}^\alpha\Big(c^*T_\alpha c\Big) \tag{E.5}$$

where the ghost field c^* is the hermitian conjugate of c, so $c^* = (c)^*$. Inverting this relation we obtain

$$\begin{aligned}\dot{\varphi}^i &= g^{ij}\Big(p_j - \frac{i}{2}\omega_{jab}\psi^a\psi^b - iA_j{}^\alpha c^*T_\alpha c\Big)\\ &= -ig^{ij}\Big(\hbar\frac{\partial}{\partial\varphi^j} + \frac{1}{2}\omega_{jab}\psi^a\psi^b + A_j{}^\alpha c^*T_\alpha c\Big).\end{aligned} \tag{E.6}$$

To obtain the correct anticommutator for c^* and c after canonical quantization, $\{c^A, c^*_B\} = \hbar\delta^A_B$, we add the kinetic term $ic^*_A\dot{c}^A$. The normalization of this kinetic term is such that $p_A = \frac{\partial}{\partial\dot{c}^A}L$ satisfies the quantum anticommutator $\{p_A, c^B\} = -\hbar i\delta_A{}^B$. Furthermore, from (E.2) we find $\{\psi^a(t), \psi^b(t)\} = \hbar\delta^{ab}$. Hence, we must construct an action which leads to (E.5).

These considerations lead us to consider the following action:

$$\begin{aligned}L = &\frac{1}{2}\Big[g_{ij}\dot{\varphi}^i\dot{\varphi}^j + i\psi^a(\dot{\psi}^a + \dot{\varphi}^k\omega_k{}^a{}_b\psi^b)\Big]\\ &+ ic^*_A\dot{c}^A + i\dot{\varphi}^kA_k{}^\alpha(\varphi)(c^*T_\alpha c).\end{aligned} \tag{E.7}$$

Since T_α are antihermitian and $c^* = (c)^*$, the action is hermitian. As it turns out, this action is not yet supersymmetric. To find the terms which complete it, we shall analyze how it transforms. We already know that the c-independent part is invariant, so we study the c-dependent terms.

We begin by varying the φ fields in $L(c)$, using the known susy rule for $\delta\varphi$,

$$\begin{aligned}\delta\varphi^i &= -i\epsilon\psi^i, \qquad \delta\psi^i = \dot{\varphi}^i\epsilon\\ \psi^a &= \psi^i e_i{}^a(\varphi), \quad e_i{}^a e_j{}^b\delta_{ab} = g_{ij}.\end{aligned} \tag{E.8}$$

For this variation we find

$$\delta L(c) = \Big[i\frac{d}{dt}\Big(-i\epsilon\psi^k\Big)A_k{}^\alpha(\varphi) + i\dot{\varphi}^k\Big(\partial_l A_k{}^\alpha\Big)\Big(-i\epsilon\psi^l\Big)\Big]\Big(c^*T_\alpha c\Big). \tag{E.9}$$

Partially integrating the first term, we produce an ordinary curl of the vector field $A_k{}^\alpha$, plus a t-derivative of c^*Tc. The latter we can cancel by suitable susy laws for c^*_A and c^B, such that they also combine to give a t-derivative of c^*Tc. Clearly,

$$\begin{aligned}\delta c^A &= +i\epsilon\psi^k A_k{}^\alpha(T_\alpha)^A{}_B c^B\\ \delta c^*_A &= -i\epsilon\psi^k A_k{}^\alpha c^*_B(T_\alpha)^B{}_A\end{aligned} \tag{E.10}$$

does the job since, using a partial integration, we obtain

$$\begin{aligned}\delta\Big(ic^*\dot{c}\Big) &= ic^*\frac{d}{dt}\Big(i\epsilon\psi^k\, A_k{}^\alpha T_\alpha c\Big) + i\Big(-i\epsilon\psi^k A_k{}^\alpha c^* T_\alpha\Big)\dot{c} \\ &= \epsilon\psi^k A_k{}^\alpha \frac{d}{dt}\Big(c^* T_\alpha c\Big)\end{aligned} \tag{E.11}$$

which indeed cancels the terms in (E.9) proportional to $\frac{d}{dt}(c^* T_\alpha c)$ obtained by partially integrating the first term in (E.9).

The remainder reads

$$\delta L(c) = \dot{\varphi}^k \epsilon\psi^l \Big(\partial_l A_k{}^\alpha - \partial_k A_l{}^\alpha\Big)\Big(c^* T_\alpha c\Big). \tag{E.12}$$

Substituting the rules for $\delta c^*{}_A$ and δc^B into the last term of (E.7) yields a commutator of two T_α matrices and completes the ordinary curl in (E.12) to a nonabelian curl. Hence, at this point we have

$$\begin{aligned}\delta L(c) &= \dot{\varphi}^k(\epsilon\psi^l) F_{lk}{}^\alpha (c^* T_\alpha c) \\ F_{lk}{}^\alpha &= \partial_l A_k{}^\alpha - \partial_k A_l{}^\alpha + f^\alpha{}_{\beta\gamma} A_l{}^\beta A_k{}^\gamma.\end{aligned} \tag{E.13}$$

To cancel this $\delta L(c)$, we observe that the combination $\dot{\varphi}^k\epsilon$ is equal to $\delta\psi^k$. (As in Appendix D, we could also consider the combination $\epsilon\psi^l$ as coming from $\delta\varphi^l$, but this would lead to bare φ^l in the new term in the action, which as we have already seen does not work.) Hence, we add the following extra term to the action:

$$L(\text{extra}) = \frac{1}{2}\psi^k\psi^l F_{kl}{}^\alpha \Big(c^* T_\alpha c\Big). \tag{E.14}$$

(We need the factor of $\frac{1}{2}$ since the variation of ψ^k and ψ^l both give a $\epsilon\dot{\varphi}$ term.)

At this point we have canceled all variations of $L(c)$, but we still have to vary F and c^*Tc in L(extra). These variations cancel by themselves, due to the Bianchi identity for $F_{kl}{}^\alpha$ as we now demonstrate. The variations of the φ fields in $F_{kl}{}^\alpha$ produce $(\partial_i F_{kl}{}^\alpha)(-i\epsilon\psi^i)$. Variation of $c^* T_\alpha c$ yields $i\epsilon\psi^i A_i{}^\beta c^*[T_\alpha, T_\beta]c$. These two variations combine into

$$\begin{aligned}\delta L(\text{extra}) &= \frac{1}{2}\psi^k\psi^l\Big(-i\epsilon\psi^i\Big)[\partial_i F_{kl}{}^\alpha + f^\alpha{}_{\beta\gamma} A_i{}^\beta F_{kl}{}^\gamma]\Big(c^* T_\alpha c\Big) \\ &= -\frac{i\epsilon}{2}\Big(\psi^i\psi^k\psi^l\Big)\Big(D_i F_{kl}{}^\alpha\Big)\Big(c^* T_\alpha c\Big)\end{aligned} \tag{E.15}$$

where $D_i F_{jk} = \partial_i F_{jk} + [A_i, F_{jk}]$ for $F_{ij} = F_{ij}{}^\alpha T_\alpha$. This indeed vanishes since $\psi^i\psi^k\psi^l$ is totally antisymmetric, while $D_i F_{kl}$ + two cyclic terms = 0 due to the Bianchi identity.

We conclude that

$$L(\text{ghost}) = ic^*_A \dot{c}^A + \left(i\dot{\varphi}^i A_i{}^\alpha + \frac{1}{2}\psi^i\psi^j F_{ij}{}^\alpha\right)\left(c^*_A (T_\alpha)^A{}_B c^B\right) \tag{E.16}$$

is supersymmetric by itself under

$$\begin{aligned} \delta c^A &= i\epsilon\psi^i A_i{}^\alpha (T_\alpha)^A{}_B c^B, & \delta c^*_A &= -i\epsilon\psi^i A_i{}^\alpha c^*_B (T_\alpha)^B{}_A \\ \delta\varphi^i &= -i\epsilon\psi^i, & \delta\psi^i &= \dot{\varphi}^i\epsilon. \end{aligned} \tag{E.17}$$

The susy Noether charge Q for the model consisting of the sum of (E.2) and (E.16) is unchanged. To see this, we repeat the analysis of varying $L(\text{ghost})$ in (E.16), this time with local $\epsilon = \epsilon(t)$. Nowhere do we pick up an $\dot{\epsilon}$ term, and hence $L(\text{ghost})$ is even locally supersymmetric. Therefore $\delta L(\text{total}) = -i\dot{\epsilon} g_{ij}\dot{\varphi}^i\psi^j$ as before, and

$$Q = g_{ij}\dot{\varphi}^i\psi^j = \psi^a e_{ai}\dot{\varphi}^i. \tag{E.18}$$

We can write the ghost part of the action in a more covariant form as

$$\begin{aligned} L(\text{ghost}) &= \left[ic^*_A \frac{D}{Dt}c^A + \frac{1}{2}\psi^i\psi^j c^* F_{ij} c\right] \\ \frac{D}{Dt}c^A &= \frac{d}{dt}c^A + \dot{\varphi}^i (A_i c)^A, \quad A_i = A_i{}^\alpha\, T_\alpha. \end{aligned} \tag{E.19}$$

In this form, the local Yang–Mills (gauge) invariance of the action under

$$\delta A_i{}^\alpha = \partial_i\lambda^\alpha + f^\alpha{}_{\beta\gamma} A_i{}^\beta\lambda^\gamma \quad \text{with } \partial_i\lambda^\alpha \equiv \frac{\partial}{\partial\varphi^i}\lambda^\alpha(\varphi) \tag{E.20}$$

and $\delta c^A = -\lambda^\alpha (T_\alpha)^A{}_B c^B, \delta c^*_A = c^*_B (T_\alpha)^B{}_A \lambda^\alpha$ becomes manifest. For example, the $c^* F c$ term varies to give

$$\lambda^\gamma\left(-c^*[T_\beta, T_\gamma]cF^\beta + f^\alpha{}_{\beta\gamma}F^\beta c^* T_\alpha c\right) \tag{E.21}$$

which clearly vanishes. Furthermore, the covariant derivative $\frac{D}{Dt}c^A$ indeed transforms like c^A itself.

The Yang–Mills symmetry could have been used to anticipate the term with A_i in the ghost action, but the term with F_{ij} (a so-called Pauli term) is typical for supersymmetry and is not required by Yang–Mills symmetry. In a similar way, it follows that the action is invariant under general coordinate transformations and local Lorentz transformations in target space, with $\delta\varphi^i = \xi^i(\varphi)$, $\delta\dot{\varphi}^i = [\partial\xi^i(\varphi)/\partial\varphi^j]\dot{\varphi}^j, \delta\psi^a = \lambda^a{}_b(\varphi)\psi^b, \delta g_{ij} = \partial_i\xi^k g_{kj} + (i \leftrightarrow j)$.

Appendix F

Gauge anomalies for exceptional groups

In the main text we showed that gravitational and gauge anomalies cancel in 10 dimensions for $N = 1$ supergravity coupled to Yang–Mills theory if the gauge group G is either $SO(32)$ or $E_8 \times E_8$. None of the other "classical groups" ($SO(n)$, $SU(n)$ and $Sp(n)$) was allowed. We now complete this analysis by discussing the exceptional groups, namely G_2, F_4, E_6, E_7 and E_8. As we have shown in the main text, cancellation of gravitational anomalies allows only Lie groups with 496 generators. There are clearly many products of simple Lie algebras with this number of generators. In particular, there are semisimple Lie algebras with one or more exceptional groups as simple factors. However, we can at once rule out these exceptional groups if we study one-loop hexagon graphs with six gauge fields all belonging to the same exceptional Lie group, and if factorization of the kind discussed in the main text does not occur. In four dimensions gauge anomalies are proportional to the symmetrized trace of three generators, $d_{abc}(R) = \mathrm{tr}(T_a\{T_b, T_c\})$, where T_a are the generators of the gauge group in a representation R, and thus real or pseudoreal representations do not carry anomalies [187]. A representation R can only carry an anomaly if $d_{abc}(R)$ is nonvanishing, and this is only possible if there exists a cubic Casimir operator for the group. In 10 dimensions gauge anomalies are proportional to the symmetrized trace of six generators, and then real or pseudoreal representations can carry anomalies. Now anomalies can only be present if there exists a sixth-order Casimir operator for the group, and even if it exists, it may still happen that for a particular representation the value of the sixth-order Casimir operator is zero or factorizes into a product of lower-dimensional Casimir operators.

What follows is amusing group theory. Readers who are somewhat rusty in their G_2 or F_4 may brush up their knowledge of exceptional groups by working their way through the discussions below. There is of course an enormous literature on this subject and our aim is to give a complete account of those results which we need. For a clear introduction to Lie algebra theory we recommend [191], while Casimir invariants for arbitrary representations and groups are in considerable detail discussed in [192].

G_2

Do anomalies cancel for G_2? This group has 14 generators, and the fundamental representation (at the same time the defining representation) is the **7**.[1] We consider the maximal subgroup $SU(3)$. The **14** of G_2 decomposes under $SU(3)$ into $\mathbf{8} + \mathbf{3} + \mathbf{3}^*$. In fact, the **7** of G_2 is most easily defined by first decomposing it under $SU(3)$ as $\mathbf{7} \to \mathbf{3} + \mathbf{3}^* + \mathbf{1}$ and then defining the action of $SU(3)$ (with parameters $\lambda^\alpha{}_\beta$) and the extra generators (with parameters $\bar{\sigma}_\alpha$ and σ^α) on the $\mathbf{7} = \{x^\alpha, \bar{x}_\alpha, y\}$ in manifestly $SU(3)$ covariant form [188],

$$\begin{aligned}
\delta x^\alpha &= \lambda^\alpha{}_\beta x^\beta + \frac{1}{\sqrt{2}} \epsilon^{\alpha\beta\gamma} \bar{\sigma}_\beta \bar{x}_\gamma + \sigma^\alpha y \\
\delta \bar{x}_\alpha &= \bar{\lambda}_\alpha{}^\beta \bar{x}_\beta + \frac{1}{\sqrt{2}} \epsilon_{\alpha\beta\gamma} \sigma^\beta x^\gamma + \bar{\sigma}_\alpha y \\
\delta y &= -a(\bar{\sigma}_\alpha x^\alpha + \sigma^\alpha \bar{x}_\alpha) \quad (a = 1 \text{ see below}).
\end{aligned} \tag{F.1}$$

We defined $\bar{x}_\alpha = (x^\alpha)^*$, $\bar{\lambda}_\alpha{}^\beta = (\lambda^\alpha{}_\beta)^*$ and $\bar{\sigma}_\alpha = (\sigma^\alpha)^*$, and y is real. By rescaling of y and σ^α we can achieve that only a has to be fixed. The dimensions of representations and the number of parameters always refer to real quantities. For example, the **7** has seven real dimensions ($x + \bar{x}$, $-i(x - \bar{x})$ and the real y), and there are **14** real parameters ($\lambda + \bar{\lambda}$, $-i(\lambda - \bar{\lambda})$, $\sigma + \bar{\sigma}$ and $-i(\sigma - \bar{\sigma})$). The antihermiticity of the $SU(3)$ generators $\lambda^\alpha{}_\beta$ requires that λ be equal to $-\lambda^\dagger$, namely $\lambda^\beta{}_\alpha = -\bar{\lambda}_\alpha{}^\beta$, and, furthermore, they are traceless, $\lambda^\alpha{}_\alpha = 0$.

These transformations form a closed algebra. For the $SU(3)$ commutators one has

$$[\delta(\lambda_2), \delta(\lambda_1)] = \delta([\lambda_1, \lambda_2]) \tag{F.2}$$

while[2]

$$[\delta(\lambda), \delta(\sigma)] = \delta(\sigma' = -\lambda\sigma). \tag{F.3}$$

[1] For any group the defining representation is the representation one uses to define the group, whereas the fundamental representation has by definition the property that its tensor products yield all other representations. Thus for $SO(n)$ with odd n the vector representation is the defining representation, but the spinor representation is the fundamental representation.

[2] To prove (F.3) one needs the "Schouten identity" $\lambda^\alpha{}_\beta \epsilon^{\beta\gamma\delta} = -\lambda^\gamma{}_\beta \epsilon^{\alpha\beta\delta} - \lambda^\delta{}_\beta \epsilon^{\alpha\gamma\beta}$ and the antihermiticity relation $\lambda^\alpha{}_\beta = -\bar{\lambda}_\beta{}^\alpha$. The former follows by antisymmetrizing the four contravariant indices $\alpha, \beta, \gamma, \delta$.

Finally,

$$[\delta(\sigma_2), \delta(\sigma_1)] = \delta(\lambda'^{\alpha}{}_{\beta}) + \delta\Big(\sigma'^{\alpha} = \sqrt{2}\epsilon^{\alpha\beta\gamma}\bar{\sigma}_{1\beta}\bar{\sigma}_{2\gamma}\Big)$$

$$\lambda'^{\alpha}{}_{\beta} = \frac{3}{2}(\sigma_1^{\alpha}\bar{\sigma}_{2\beta} - \sigma_2^{\alpha}\bar{\sigma}_{1\beta}) - \frac{1}{2}\delta_{\beta}^{\alpha}(\bar{\sigma}_1 \cdot \sigma_2 - \bar{\sigma}_2 \cdot \sigma_1). \qquad \text{(F.4)}$$

Closure of the (σ, σ) commutator fixes $a = 1$, and this then defines the group G_2 as well as the defining representation **7**.

The decomposition of the generators of G_2 into the **8** of $SU(3)$ and the remaining **6** is not a symmetric decomposition (because of the term with $\delta(\sigma'^{\alpha})$ in (F.4)). A maximally noncompact version of an algebra always yields a symmetric decomposition, and has the property that the number of noncompact generators minus the number of compact generators is equal to its rank ($8-6=2$ for G_2). Such a decomposition is also called a Cartan decomposition [189]. The decomposition into the subgroup $SU(3)$ and six coset generators is not a Cartan decomposition because the subgroup $SU(3)$ does not have six generators. For G_2 a Cartan decomposition is, for example, into the generators of $SU(2) \times SU(2)$ and the $(\mathbf{2}, \mathbf{4})$ coset generators. The **7** decomposes under $SU(2) \times SU(2)$ into $(\mathbf{1}, \mathbf{3}) + (\mathbf{2}, \mathbf{2})$ [190]. One can find the maximal regular subalgebras of G_2 from its extended Dynkin diagram [191],

$G_2' =$ 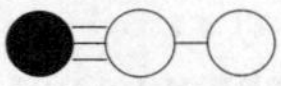 $\qquad G_2 =$ 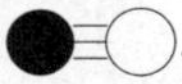 . (F.5)

Deleting one dot, one finds the maximal (rank-2) regular subalgebras $SU(3)$, $SU(2) \times SU(2)$ and of course G_2 itself.

Since G_2 has only two Casimir operators, C_2 and C_6, we know that

$$\mathrm{Tr}F^6 = \alpha\, \mathrm{tr}F^6 + \beta(\mathrm{tr}F^2)^3 \qquad \text{(F.6)}$$

with α and β to be computed. The trace Tr is over the adjoint representation **14**, while the trace tr is over the fundamental **7**. If it were to turn out that α vanishes, anomaly cancellation would be possible. To compute α and β we choose two particular generators of the subgroup $SU(3)$, namely λ_3 and λ_8, and evaluate the trace relation (F.6) on each of them. We specify $SU(3)$ by the following generators λ_k normalized to $\mathrm{tr}\lambda_k\lambda_l = -\frac{1}{2}\delta_{kl}$:

$$\lambda_1 = -\frac{i}{2}\begin{pmatrix} 0 & 1 & 0 \\ 1 & 0 & 0 \\ 0 & 0 & 0 \end{pmatrix} \qquad \lambda_2 = -\frac{i}{2}\begin{pmatrix} 0 & -i & 0 \\ i & 0 & 0 \\ 0 & 0 & 0 \end{pmatrix}$$

$$\lambda_3 = -\frac{i}{2}\begin{pmatrix} 1 & 0 & 0 \\ 0 & -1 & 0 \\ 0 & 0 & 0 \end{pmatrix} \qquad \lambda_4 = -\frac{i}{2}\begin{pmatrix} 0 & 0 & 1 \\ 0 & 0 & 0 \\ 1 & 0 & 0 \end{pmatrix}$$

$$\lambda_5 = -\frac{i}{2}\begin{pmatrix} 0 & 0 & -i \\ 0 & 0 & 0 \\ i & 0 & 0 \end{pmatrix} \quad \lambda_6 = -\frac{i}{2}\begin{pmatrix} 0 & 0 & 0 \\ 0 & 0 & 1 \\ 0 & 1 & 0 \end{pmatrix}$$

$$\lambda_7 = -\frac{i}{2}\begin{pmatrix} 0 & 0 & 0 \\ 0 & 0 & -i \\ 0 & i & 0 \end{pmatrix} \quad \lambda_8 = -\frac{i}{2\sqrt{3}}\begin{pmatrix} 1 & 0 & 0 \\ 0 & 1 & 0 \\ 0 & 0 & -2 \end{pmatrix}. \tag{F.7}$$

The first three generators λ_1, λ_2 and λ_3 define an $SU(2)$ subgroup of $SU(3)$, while (λ_4, λ_5) and (λ_7, λ_6) form doublets under this $SU(2)$:[3]

$$\begin{aligned}
&[\lambda_3, \lambda_1] = \lambda_2; \qquad [\lambda_3, \lambda_2] = -\lambda_1; \qquad [\lambda_3, \lambda_4] = \frac{1}{2}\lambda_5 \\
&[\lambda_3, \lambda_5] = -\frac{1}{2}\lambda_4; \quad [\lambda_3, \lambda_6] = -\frac{1}{2}\lambda_7; \quad [\lambda_3, \lambda_7] = \frac{1}{2}\lambda_6 \\
&[\lambda_8, \lambda_4] = \frac{\sqrt{3}}{2}\lambda_5; \quad [\lambda_8, \lambda_5] = -\frac{\sqrt{3}}{2}\lambda_4; \quad [\lambda_8, \lambda_6] = \frac{\sqrt{3}}{2}\lambda_7 \\
&[\lambda_8, \lambda_7] = -\frac{\sqrt{3}}{2}\lambda_6.
\end{aligned} \tag{F.8}$$

We now evaluate the trace Tr on **8**+**3**+**3*** and the trace tr on **3**+**3***+**1**. We find for the generator λ_3 that λ_3^2 is diagonal on all states. It is then easy to obtain

$$\mathrm{Tr}(\lambda_3)^6 = \left(-2 - \frac{1}{16}\right) + \left(\frac{-2-2}{64}\right) = -2 - \frac{1}{8} \tag{F.9}$$

$$\mathrm{tr}(\lambda_3)^6 = -\frac{4}{64} \qquad \mathrm{tr}(\lambda_3)^2 = -1 \tag{F.10}$$

where we have used to fact that $(\lambda_3)^2$ vanishes on y and equals $-\frac{1}{4}I$ on two of the three states in **3** and **3***. As a check one may rederive these relations using (6.250) for $SU(3)$. Hence from λ_3 we learn that $-\frac{17}{8} = \alpha(-\frac{1}{16}) + \beta(-1)$.

Similarly, we find for the generator λ_8

$$\mathrm{Tr}(\lambda_8)^6 = \left(-\frac{3}{4}\right)^3 4 - \left(\frac{1}{12}\right)^3 (66+66)$$

$$\mathrm{tr}(\lambda_8)^6 = \left(-\frac{1}{12}\right)^3 (66+66) \qquad \mathrm{tr}(\lambda_8)^2 = -\frac{1}{12}(6+6). \tag{F.11}$$

Hence λ_8 tells us that $-\frac{27}{16} - \frac{11}{144} = \alpha(-\frac{11}{144}) + \beta(-1)$. There is no solution of both equations with $\alpha = 0$, and since α is nonzero, $\mathrm{Tr}F^6$ does not factorize for G_2. Hence, this group produces gauge anomalies which cannot be canceled by a counterterm.

[3] An $SO(3)$ subgroup of $SU(3)$ is generated by λ_4, λ_5 and $\frac{1}{2}\lambda_3 + \frac{1}{2}\sqrt{3}\lambda_8$.

F_4

Next, we consider the group F_4. It has 52 generators, the defining representation is **26** and $SO(9)$ is a maximal regular subalgebra. (A regular subalgebra H of G has roots which are a subset of the roots of G, and Cartan generators which are a linear combination of the Cartan generators of G. If it is maximal, the rank of H is equal to the rank of G.) Spinors ψ^α of $SO(9)$ have 16 components and $SO(9)$ has 36 generators, so we expect that under $SO(9)$ the **26** and the **52** decompose as follows:

$$\begin{aligned}\mathbf{26} &\to \mathbf{9}+\mathbf{16}+\mathbf{1}\\ \mathbf{52} &\to \mathbf{36}+\mathbf{16}.\end{aligned} \tag{F.12}$$

So the **26** consists of a vector v^i, a spinor ψ^α and a scalar s, all of which are real. We define F_4 by its action on the **26** [188]. Manifest $SO(9)$ covariance only allows

$$\begin{aligned}\delta v^i &= \Lambda^i{}_j v^j + \lambda^\alpha \Gamma^i_{\alpha\beta}\psi^\beta\\ \delta\psi^\alpha &= \frac{1}{4}\Lambda^{ij}(\Gamma_{ij})^\alpha{}_\beta\psi^\beta + v^i(\Gamma_i)^\alpha{}_\beta\lambda^\beta + s\lambda^\alpha\\ \delta s &= a\lambda^\alpha\psi_\alpha \quad (a=3,\ \text{see below}).\end{aligned} \tag{F.13}$$

Orthogonal groups leave $y^i\delta_{ij}x^j$ invariant, hence $\delta_{ik}\Lambda^k{}_j+\Lambda^k{}_i\delta_{kj}=0$. Then $\Lambda_{ij}=-\Lambda_{ji}$ where $\Lambda_{ij}=\delta_{ik}\Lambda^k{}_j$ are the 36 real parameters of $SO(9)$, and λ^α are the 16 extra real parameters. Again only a has to be fixed. Since the Dirac matrices (with $\{\Gamma_i,\Gamma_j\}=2\delta_{ij}$) in nine Euclidean dimensions can be taken to be real and symmetric 16×16 matrices [77, 148], the **26** is real. In fact, the charge conjugation matrix is the unit matrix since $C\Gamma^i=\Gamma^{i,T}C$ for $C=1$. (In odd dimensions one has either $C\Gamma^i=-\Gamma^{i,T}C$ or $C\Gamma^i=+\Gamma^{i,T}C$, but not both possibilities. Here, clearly one has the + sign.) Because $C=1$, the matrices $\Gamma^i_{\alpha\beta}$ and $(\Gamma^i)^\alpha{}_\beta$ in (F.13) are the same.

To show that (F.13) defines a Lie algebra, we must show that the commutators close. Of course the subalgebra of $SO(9)$ closes, $[\delta(\Lambda_2),\delta(\Lambda_1)]=\delta([\Lambda_1,\Lambda_2])$. Also $[\delta(\Lambda),\delta(\lambda)]$ is, as expected, equal to $\delta(\lambda'=-\frac{1}{4}\Lambda^{jk}\Gamma_{jk}\lambda)$. The crucial question is whether $[\delta(\lambda_2),\delta(\lambda_1)]$ closes. On s and v^i one easily establishes that this commutator is equal to an $SO(9)$ rotation with composite parameter $\Lambda'_{ij}=2\lambda_1\Gamma_{ij}\lambda_2$. On ψ a Fierz rearrangement yields

$$[\delta(\lambda_2),\delta(\lambda_1)]\psi=\frac{1}{16}(\lambda_2\Gamma^iO^I\Gamma_i\lambda_1+a\lambda_2O^I\lambda_1)O_I\psi-(1\leftrightarrow 2). \tag{F.14}$$

Only $O^I\sim\Gamma^{jkl}$ and $O^I\sim\Gamma^{jk}$ contribute (the rest, I and Γ^i and Γ^{ijkl}, are symmetric matrices); the contribution of the former cancels if one chooses $a=3$, and then the contribution of the latter yields the correctly normalized $SO(9)$ rotation. Hence, (F.13) defines a Lie algebra, namely F_4.

The decomposition of the generators of F_4 into generators of $SO(9)$ and coset generators is not a Cartan decomposition because that would require 24 subgroup generators and 28 coset generators; however, it still is a symmetric decomposition (because two λ transformations only produce an $SO(9)$ rotation). Hence, not every symmetric decomposition is a Cartan decomposition. (The reverse is, however, true: every Cartan decomposition is a symmetric decomposition.) Of course, 24 is the number of generators of $SU(5)$, but 28 is not the sum of the dimensions of a set of irreducible representations of $SU(5)$ (except if one has many $SU(5)$ singlets). In fact, we may use a so-called extended Dynkin diagram [191] to find all maximal regular subalgebras of F_4,

$F_4' =$ 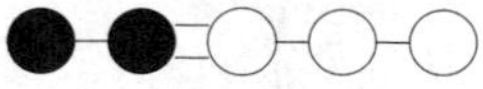 $\qquad F_4 =$ 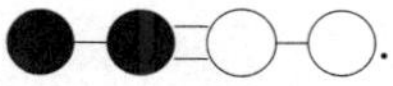. (F.15)

Deleting any point gives the following set of maximal regular subalgebras[4] of F_4

$$SO(9); \quad SU(2) \times SU(4); \quad SU(3) \times SU(3); \quad Sp(6) \times SU(2). \tag{F.16}$$

The last one yields the Cartan decomposition $F_4/Sp(6) \times SU(2)$, and the **28**-dimensional coset is indeed a representation of $Sp(6) \times SU(2)$, namely **28** = (**14**, **2**) (where **14** is the antisymmetric symplectic-traceless representation $t^{ij} = -t^{ji}$ of $Sp(6)$).

The group F_4 has four Casimir operators (because it has rank 4), namely C_2, C_6, C_8, C_{12}. So, as in the case of G_2, we know that

$$\mathrm{Tr} F^6 = \alpha \, \mathrm{tr} F^6 + \beta \, (\mathrm{tr} F^2)^3 \tag{F.17}$$

and the issue is whether $\alpha = 0$.

We need two particular generators of $SO(9)$ to be able to fix α and β. As such, we take a rotation in the 1–2 plane of R^9, and a simultaneous rotation in the 1–2 plane and the 1–3 plane. These generators have the following form in the defining representation of $SO(9)$:

$$T_I = \begin{pmatrix} 0 & 1 & 0 & . \\ -1 & 0 & 0 & . \\ 0 & 0 & 0 & . \\ . & . & . & . \end{pmatrix}_{9\times 9} \qquad T_{II} = \begin{pmatrix} 0 & 1 & 1 & . \\ -1 & 0 & 0 & . \\ -1 & 0 & 0 & . \\ . & . & . & . \end{pmatrix}_{9\times 9} . \tag{F.18}$$

In the spinor representation of $SO(9)$ they are given by $\frac{1}{2}\gamma^1\gamma^2$ and $\frac{1}{2}\gamma^1\gamma^2 + \frac{1}{2}\gamma^1\gamma^3$, respectively.

[4]Actually, it is known [190] that for the exceptional groups there are five regular subalgebras obtained in this way which are not maximal, and one of these exceptions is for F_4 where $SU(2) \times SU(4)$ is contained in $SO(9)$.

We evaluate the trace Tr on the **36** and the **16** of $SO(9)$, and the trace tr on the **9**, **16** and **1** of $SO(9)$. One obtains, using (6.245) for the **36** of $SO(9)$

$$\mathrm{Tr}(T_I)^6 = -14 - \frac{16}{64}$$
$$\mathrm{tr}(T_I)^6 = -2 - \frac{16}{64} = -\frac{9}{4}, \quad \mathrm{tr}(T_I)^2 = -2 - \frac{16}{4} = -6. \qquad \text{(F.19)}$$

Hence the first equation for α and β is $-\frac{57}{4} = \alpha(-\frac{9}{4}) + \beta(-216)$.

The computation of the traces with T_{II} is simplified by noting that for the vector representation

$$T_{II}^2 = \begin{pmatrix} -2 & 0 & 0 & . \\ 0 & -1 & 0 & . \\ 0 & 0 & -1 & . \\ . & . & . & 0 \end{pmatrix}_{9\times 9} \qquad \text{(F.20)}$$

while for the spinor representation $T_{II}^2 = -\frac{1}{2} I_{16\times 16}$. Now one finds, again using (6.245) to compute $\mathrm{Tr}\, T_{II}^6$ on the adjoint representation **36**,

$$\mathrm{Tr}(T_{II})^6 = -23(-10) + 15(-4)6 - \frac{16}{8} = -132$$
$$\mathrm{tr}(T_{II})^6 = -10 - \frac{16}{8} = -12, \quad \mathrm{tr}(T_{II})^2 = -4 - \frac{16}{2}. \qquad \text{(F.21)}$$

Hence the second equation for α and β is $-132 = \alpha(-12) + \beta(-12)$. Again, there is no solution to (F.17) with $\alpha = 0$, hence also for F_4 gauge anomalies do not cancel.

E_6

The group E_6 has 78 generators, and the fundamental representation is the **27**. Suitable subgroups and the corresponding decompositions are [190]

$$\begin{aligned} SO(10)\times U(1){:}\ & \mathbf{78} \to \mathbf{45}(0) + \mathbf{16}(-3) + \mathbf{16}^*(3) + \mathbf{1}(0) \\ & \mathbf{27} \to \mathbf{16}(1) + \mathbf{10}(-2) + \mathbf{1}(4) \\ Usp(8){:}\ & \mathbf{78} \to \mathbf{36} + \mathbf{42} \\ & \mathbf{27} \to \mathbf{27} \\ F_4{:}\ & \mathbf{78} \to \mathbf{52} + \mathbf{26} \\ & \mathbf{27} \to \mathbf{26} + \mathbf{1}. \end{aligned} \qquad \text{(F.22)}$$

None of these is a maximal regular subalgebra, as one may deduce from the extended Dynkin diagram for E_6

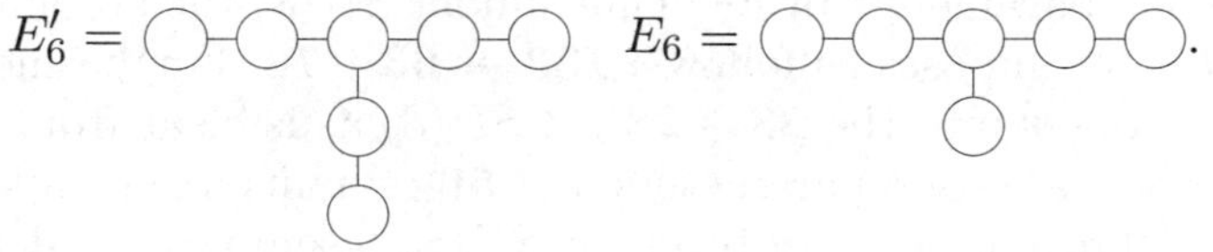

(F.23)

In the literature the coset $E_6/Usp(8)$ has been studied in detail [193, 194], so let us choose $Usp(8)$ as the subgroup of E_6. The corresponding decomposition of E_6 is a Cartan decomposition. The group E_6 acts on the **27** as

$$\delta z^{\alpha\beta} = \Lambda^{\alpha}{}_{\gamma} z^{\gamma\beta} + \Lambda^{\beta}{}_{\gamma} z^{\alpha\gamma} + \Sigma^{\alpha\beta}{}_{\gamma\delta} z^{\gamma\delta} \tag{F.24}$$

where $z^{\alpha\beta} = -z^{\beta\alpha}$ with $\alpha, \beta = 1, 8$ and $z^{\alpha\beta}\Omega_{\alpha\beta} = 0$ with $\Omega_{\alpha\beta}$ being the metric of $Sp(8)$. The generators of $Usp(n)$ in the defining representation are the antihermitian matrices $\Lambda^{\alpha}{}_{\beta}$ preserving $z^{\alpha}\Omega_{\alpha\beta}y^{\beta}$. Hence, $\Lambda^{\alpha}{}_{\beta} = -(\Lambda^{\beta}{}_{\alpha})^*$ and $\Lambda^{\gamma}{}_{\alpha}\Omega_{\gamma\beta} \equiv \Lambda_{\beta\alpha} = -\Omega_{\alpha\gamma}\Lambda^{\gamma}{}_{\beta} = \Lambda_{\alpha\beta}$. This shows that there are 36 generators. For $n = 2$ one can easily check that $Usp(2) = SU(2)$, but for higher n the dimension of $SU(n)$ is larger then that of $Usp(n)$. Furthermore, $\Sigma_{\alpha\beta\gamma\delta} \equiv \Sigma^{\alpha'\beta'}{}_{\gamma\delta}\Omega_{\alpha'\alpha}\Omega_{\beta'\beta}$ is totally antisymmetric, traceless with respect to $\Omega^{\alpha\beta}$ and satisfies the reality condition $(\Sigma^{\alpha\beta\gamma\delta})^* = \Sigma_{\alpha\beta\gamma\delta}$ where $\Sigma^{\alpha\beta\gamma\delta}\Omega_{\alpha\alpha'}\Omega_{\delta\delta'} \equiv \Sigma_{\alpha'\beta'\gamma'\delta'}$ with $\Omega^{\alpha\beta}\Omega_{\gamma\beta} = \delta^{\alpha}_{\gamma}$. $\Sigma^{\alpha\beta}{}_{\gamma\delta}$ are the 42 generators of the coset $E_6/Usp(8)$.

There are three Casimir operators with rank ≤ 6, namely C_2, C_5 and C_6, but only C_2 and C_6 play a role in the decomposition of $\mathrm{Tr}F^6$

$$\mathrm{Tr}F^6 = a\,\mathrm{tr}F^6 + b\,(\mathrm{tr}F^2)^3. \tag{F.25}$$

We leave the proof that a is nonvanishing as an exercise. (Hint: two suitable 8×8 matrices in the defining representation of $Usp(8)$ are the matrices with $i\sigma_3$ or $i\sigma_1$ in the first 2×2 block and for $\mathrm{Tr}T^6$ over the **36** one may use (6.245) with + signs instead of − signs.)

E_7

The group E_7 has 133 generators and $SU(8)$ is a maximal subgroup. The fundamental representation is the **56**, spanned by antisymmetric x^{ij} and $x_{ij} = (x^{ij})^*$ with $i, j = 1, 8$. E_7 is defined by its action on the **56** as follows [195]:

$$\begin{aligned}\delta x^{ij} &= \Lambda^{i}{}_{k} x^{kj} + \Lambda^{j}{}_{k} x^{ik} + \frac{1}{4!}\epsilon^{ijklmnop}\Sigma_{mnop}\bar{x}_{kl} \\ \delta \bar{x}_{ij} &= \bar{\Lambda}_{i}{}^{k}\bar{x}_{kj} + \bar{\Lambda}_{j}{}^{k}\bar{x}_{ik} + \Sigma_{ijkl}x^{kl}.\end{aligned} \tag{F.26}$$

As before $\bar{x}_{ij} = (x^{ij})^*$ and $\bar{\Lambda}_{i}{}^{k} = (\Lambda^{i}{}_{k})^*$. The $\Lambda^{i}{}_{k}$ yield the 63 real parameters of $SU(8)$, hence they are antihermitian and traceless, $\bar{\Lambda}_{i}{}^{k} = -\Lambda^{k}{}_{i}$, and $\Lambda^i{}_i = 0$, while Σ_{ijkl} are totally antisymmetric and satisfy the self-duality relation $({}^*\Sigma)^{ijkl} \equiv (\Sigma_{ijkl})^* = \frac{1}{4!}\epsilon^{ijklmnop}\Sigma_{mnop}$, yielding the remaining 70 real parameters of E_7. Thus, under $SU(8)$ the adjoint representation of E_7 decomposes as follows: $\mathbf{133} \to \mathbf{63} + \mathbf{70}$. Furthermore, the $\mathbf{56}$ of E_7 decomposes into the $\mathbf{28} + \mathbf{28}^*$ of $SU(8)$ as is clear from (F.26). (If we only allow real group parameters, the $\mathbf{56}$ remains irreducible under $SU(8)$.) This definition of E_7 in terms of $SU(8)$ resembles the definition of G_2 in terms of $SU(3)$, but $E_7/SU(8)$ yields a Cartan decomposition.

The relevant Casimir operators are in this case C_2 and C_6, so

$$\mathrm{Tr}F^6 = a\,\mathrm{tr}F^6 + b\,(\mathrm{tr}F^2)^3 \tag{F.27}$$

where the trace "Tr" is over the adjoint representation of E_7 and the trace "tr" is over the fundamental representation of E_7.

Decomposing the adjoint and fundamental representation of E_7 with respect to $SU(8)$, and denoting the resulting $SU(8)$ representations by their dimensionality we find from (F.27)

$$\mathrm{Tr}_{\mathbf{63}}F^6 + \mathrm{tr}_{\mathbf{70}}F^6 = 2a\mathrm{tr}_{\mathbf{28}}F^6 + 8b(\mathrm{tr}_{\mathbf{28}}F^2)^3. \tag{F.28}$$

Again we choose suitable generators for F, namely any linear combination A of the seven independent generators which are diagonal (and imaginary) in the fundamental representation $\mathbf{8}$ of $SU(8)$. We denote these entries by a_i with $i = 1, 8$, and then have the constraint $\sum a_i = 0$.

The trace of A^6 in the adjoint representation of $SU(8)$ follows from (6.250),

$$\mathrm{Tr}_{\mathbf{63}}A^6 = 16\sum a_i^6 + 30\left(\sum a_i^2\right)\left(\sum a_j^4\right) - 20\left(\sum a_i^3\right)^2. \tag{F.29}$$

All sums run over $1 \le i \le 8$ and $1 \le j \le 8$.

For the trace of A^6 in the $\mathbf{70}$ of $SU(8)$ we use the fact that states are labeled[5] by $i < j < k < l$ and A is represented on these states by

[5] As the 70 independent components of Σ_{ijkl} we can take the real parts R_{ijkl} and the imaginary parts I_{ijkl} with i, j, k, l running from 1 to 7. The self-duality relation expresses R_{ijk8} and I_{ijk8} in terms of these 70 components. The matrix A maps R_{ijkl} into $-(\alpha_i + \alpha_j + \alpha_k + \alpha_l)I_{ijkl}$ and I_{ijkl} into $(\alpha_i + \alpha_j + \alpha_k + \alpha_l)R_{ijkl}$, where the diagonal entries $a_1, \ldots, a_8$ of A, satisfying $a_1 + \cdots + a_8 = 0$, have been written as $a_j = i\alpha_j$ with real α_j. Then A^2 is separately diagonal on the 35 states R_{ijkl} and the 35 states I_{ijkl}. The trace of A^{2p} becomes

$$\begin{aligned}\mathrm{tr}_{\mathbf{70}}A^{2p} &= (-2)\sum_{1\le i<j<k<l\le 7}(\alpha_i + \alpha_j + \alpha_k + \alpha_l)^{2p} \\ &= \sum_{1\le i<j<k<l\le 8}(a_i + a_j + a_k + a_l)^{2p}\end{aligned}$$

because the terms with an index α_8 give the same contribution as the terms without an index α_8.

$(a_i + a_j + a_k + a_l)$. We rewrite the restricted sum as a combination of unrestricted sums

$$\sum_{i<j<k<l} (a_i + a_j + a_k + a_l)^6 = \frac{1}{24}\Bigg[\sum_{i,j,k,l} (a_i + a_j + a_k + a_l)^6 - 6 \sum_{i=j,k,l} (2a_i + a_k + a_l)^6 + 3 \sum_{i=j,k=l} (2a_i + 2a_k)^6 + 8 \sum_{i=j=k,l} (3a_i + a_l)^6 - 6 \sum_{i=j=k=l} (4a_i)^6\Bigg]. \quad \text{(F.30)}$$

As a check one may verify that the number of terms on the left-hand side is equal to that on the right-hand side. This expression can be expanded into terms with $\sum a_i^6$, $(\sum a_i^4)(\sum a_j^2)$ and $(\sum a_i^2)^3$. Since A depends on seven arbitrary parameters, there is enough resolving power to treat these invariants as independent.

For the traces of A^6 and A^2 over **28** of $SU(8)$ one easily finds

$$\mathrm{tr}_{\mathbf{28}} A^6 = \frac{1}{2}\Big[\sum_{i,j} (a_i + a_j)^6 - \sum_{i=j} (2a_i)^6\Big]$$
$$\mathrm{tr}_{\mathbf{28}} A^2 = \frac{1}{2}\Big[\sum_{i,j} (a_i + a_j)^2 - \sum_{i=j} (2a_i)^2\Big]. \quad \text{(F.31)}$$

Finally, we collect all terms with $\sum a_i^6$. There are no terms of this kind proportional to b, and our aim is to show that a is nonvanishing. Thus, we assume that a vanishes, and show that this leads to a contradiction, namely we show that the sum of all terms in $\mathrm{Tr}_{133} A^6$ which are proportional to $\sum a_i^6$ does not vanish. From the **63** we find a coefficient 16, while the **70** yields the following coefficient:

$$\frac{1}{24}\Big[4 - 6(2^6 + 2) + 3(2^6 + 2^6) + 8(3^6 + 1) - 6(4^6)\Big] = \frac{1}{24}\Big[-3 \cdot 2^7 + 3 \cdot 2^7 + 2^3 \cdot 3^6 - 3 \cdot 2^{13}\Big] = 3^5 - 2^{10}. \quad \text{(F.32)}$$

Clearly a is nonvanishing, and hence the group E_7 also leads to gauge anomalies in ten dimensions.

E_8

The group E_8 has 248 generators, and $SO(16)$ is a maximal subgroup. This **248** adjoint representation of E_8 splits into the adjoint representation **120** of $SO(16)$ and one of the two spinor representations $\mathbf{128}_s$ and $\mathbf{128}_c$ of $SO(16)$. It does not matter which spinor representation one chooses, hence

$$\mathbf{248} \underset{SO(16)}{\longrightarrow} \mathbf{120} + \mathbf{128} \quad \text{(F.33)}$$

We take as carrier space the adjoint representation since this is for E_8 also the defining representation, and the representation with the lowest dimension. So as carrier space we take an antisymmetric tensor $\phi^{ij} = -\phi^{ji}(i,j = 1,16)$ and a chiral spinor $\psi^\alpha(\alpha = 1,128)$. The **248** parameters of E_8 are also decomposed into the **120** parameters $\Lambda^i{}_j$ of $SO(16)$ and **128** further parameter λ^α. It is then fairly obvious what the transformation rules are [188]

$$\begin{aligned}\delta\phi^{ij} &= \Lambda^i{}_k\phi^{kj} + \Lambda^j{}_k\phi^{ik} + \lambda^\alpha\Gamma^{ij}_{\alpha\beta}\psi^\beta \\ \delta\psi^\alpha &= \frac{1}{4}\Lambda^{ij}(\Gamma_{ij})^\alpha{}_\beta\psi^\beta + \phi^{ij}(\Gamma_{ij})^\alpha_\beta\lambda^\beta.\end{aligned} \quad \text{(F.34)}$$

By rescaling one can eliminate other constants in these terms. Clearly both ψ^α and λ^α have the same chirality. The Dirac matrices in 8 dimensions, and therefore also in 16 dimensions, are real, symmetric and off-diagonal. It follows that the charge conjugation matrix $(C_+)_{\alpha\beta}$, defined by $C_+\Gamma_i C_+^{-1} = +\Gamma_i^T$, is the unit matrix, $C_+ = I$. This explains why there is no "bar" on λ in the term $\lambda\Gamma^{ij}\psi$. So all indices i, j, α, β are raised and lowered with Kronecker deltas. Also $\phi^{ij}, \Lambda^i{}_j, \lambda^\alpha$ and ψ^α are all real. Note that λ^α are commuting objects (they are just parameters) and also ψ^α will be taken to be commuting.

To prove that the transformation rules define a Lie algebra, we must show that

(i) $[\delta(\Lambda_1), \delta(\Lambda_2)] = \delta[\Lambda_2, \Lambda_1]$ which is just the statement that $SO(16)$ is a Lie algebra.

(ii) $[\delta(\lambda), \delta(\Lambda)] = \delta[\lambda']$ with $\lambda' = \frac{1}{4}\Lambda_{ij}\Gamma^{ij}\lambda$. This states that the coset generators corresponding to the parameters λ form a representation of the subgroup. This is clearly true in our case.

(iii) $[\delta(\lambda_1), \delta(\lambda_2)] = \delta[\Lambda']$ with $\Lambda'_{ij} = 4\lambda_2\Gamma_{ij}\lambda_1$.

Because $SO(16)$ is a maximal subgroup of E_8, the structure of (i), (ii) and (iii) should hold (it is a Cartan decomposition).

Consider first the last commutator on ϕ^{ij}

$$[\delta(\lambda_2), \delta(\lambda_1)]\phi^{ij} = \lambda_1^\alpha\Gamma^{ij}_{\alpha\beta}(\phi^{kl}\Gamma_{kl})^\beta{}_\gamma\lambda_2^\gamma - 1 \leftrightarrow 2. \quad \text{(F.35)}$$

One can write the sum of both terms as a commutator of $[\Gamma^{ij}, \Gamma_{kl}]$, and this proves (iii) when acting on ϕ^{ij}. All that is left is to prove that (iii) also holds when acting on ψ^α. One finds

$$[\delta(\lambda_2), \delta(\lambda_1)]\psi^\alpha = (\lambda_2\Gamma^{ij}\psi)(\Gamma_{ij}\lambda_1) - 1 \leftrightarrow 2. \quad \text{(F.36)}$$

We next make a Fierz rearrangement which couples to λ's together

$$= +\sum_I \left(\frac{1}{256}\right)(\lambda_2 O^I\lambda_1)(\Gamma_{ij}O^I\Gamma^{ij}\psi) - 1 \leftrightarrow 2 \quad \text{(F.37)}$$

(with a plus sign since the λ's are commuting spinors). Here O^I are all 2^{16} products of Dirac matrices normalized to $TrO^IO^J = 256\ \delta^{IJ}$.

Because the λ's are chiral, only products of an even number of Γ_i matrices contributes. Moreover, the matrices $\Gamma_{17}, \Gamma^i\Gamma_{17}, \Gamma^{ij}\Gamma_{17}\ldots$ give the same result as $I, \Gamma^i, \Gamma^{ij}\ldots$ so we only need to consider products with none, one, ..., eight Dirac matrices. But $\lambda_2\lambda_1, \lambda_2\Gamma_{klmn}\lambda_1$ and $\lambda_2\Gamma_{i_1\ldots i_8}\lambda_1$ are symmetric in λ_2 and λ_1, so only $O^I \sim \Gamma^{ij}$ and $O^I \sim \Gamma^{i_1\ldots i_6}$ can contribute. But now there is an amusing identity

$$\Gamma^{ij}\Gamma^{i_1\ldots i_6}\Gamma_{ij} = 0. \tag{F.38}$$

Indeed, there are $\binom{6}{2}$ choices of ij from the set $i_1\ldots i_6$, $\binom{10}{2}$ choices of ij lies outside this same set, and 6×10 ways of choosing i from $i_{1\ldots i6}$ and j from the complement. The first two cases yield $+\Gamma^{i_1\ldots i_6}$, while the last case yields $-\Gamma^{i_1\ldots i_6}$. This proves the identity.

We have arrived at

$$\begin{aligned}\sum_{ij}\sum_{I}\frac{1}{256}(\lambda_2O^I\lambda_1)(\Gamma_{ij}O^I\Gamma^{ij}\psi) - 1\leftrightarrow 2\\ = 4\left(-\frac{1}{2}\right)\frac{1}{256}\sum_{ij}\sum_{kl}(\lambda_2\Gamma^{kl}\lambda_1)(\Gamma_{ij}\Gamma^{kl}\Gamma^{ij}\psi).\end{aligned} \tag{F.39}$$

The factor 4 is needed to include $O^I = \Gamma^{ij}\Gamma_{17}$ and the terms due to $1\leftrightarrow 2$, and the factor $-1/2$ is needed since $\Gamma^{12}\Gamma^{12} = -1$ and we sum over $k > l$ and $k < l$. The same arguments as used before now show that $\sum_{ij}\Gamma_{ij}\Gamma^{kl}\Gamma^{ij} = -128\Gamma^{kl}$ (since $2\binom{2}{2} + 2\binom{14}{2} - 2\times 2\times 14 = 128$). Thus we obtain $(\lambda_2\Gamma^{kl}\lambda_1)\Gamma^{kl}\psi$, which confirms $[\delta(\lambda_2), \delta(\lambda_1)]$ also for ψ^α. This proves the existence of E_8, and provides a simple explicit representation.

$E_8 \times E_8$

We mentioned in the main text that also for the gauge group $E_8 \times E_8$ all anomalies in 10-dimensional simple supergravity coupled to Yang–Mills theory can be canceled by a counterterm. We give the details here.

From (6.288) we see that we must factorize $\mathrm{Tr}\tilde{F}^6$ in order to have a chance of canceling the anomalies. As we have already discussed in the main text, for E_8 $\mathrm{Tr}\tilde{F}^6$ factorizes. In fact, $\mathrm{Tr}\tilde{F}^4$ also factorizes. Consequently, also for $E_8 \times E_8$ these traces factorize. This guarantees that the anomaly factorizes. We first prove this factorization and then explicitly construct the counterterm. Hence, $E_8 \times E_8$ is as good a candidate as $SO(32)$.

For E_8 we have the following relations:

$$\begin{aligned}\mathrm{Tr}\,F^6 &= \frac{1}{7200}(\mathrm{Tr}\,F^2)^3\\ \mathrm{Tr}\,F^4 &= \frac{1}{100}(\mathrm{Tr}\,F^2)^2.\end{aligned} \tag{F.40}$$

To derive these relations, we consider a particular generator A of E_8 and compute $\mathrm{Tr}A^6$, $\mathrm{Tr}A^4$ and $\mathrm{Tr}A^2$ separately. To find a suitable generator of E_8, we note that $SO(16)$ is a maximal regular subalgebra of E_8, and the adjoint representation **248** of E_8 decomposes under this $SO(16)$ into the adjoint **120** of $SO(16)$ and the spinor representation **128** of $SO(16)$. We then act with A separately onto the **120** part and the **128** part of **248**. Let the $SO(16)$ generator be a rotation in the x–y plane of R^{16}; then the spinor representation[6] is given by the 128×128 matrix $\frac{1}{2}\gamma^1\gamma^2$. So the generator A lies in the $SO(16)$ subgroup of E_8, and its representation in the vector representation and in the spinor representation is given by

$$A_{vector} = \begin{pmatrix} 0 & 1 & 0 & \cdot \\ -1 & 0 & 0 & \cdot \\ 0 & 0 & 0 & \cdot \\ \cdot & \cdot & \cdot & \cdot \end{pmatrix}_{16\times 16} \qquad A_{spinor} = \left(\frac{1}{2}\gamma^1\gamma^2\right)_{128\times 128.} \tag{F.41}$$

From applying (6.245) to $SO(16)$ we find $\mathrm{Tr}A^2 = (n-2)\mathrm{tr}A^2_{vector} = 14(-2)$ on the **120** and from $(\frac{1}{2}\gamma^1\gamma^2)^2 = -\frac{1}{4}I$ one obtains $\mathrm{Tr}A^2_{spinor} = -\frac{1}{4}128$ on the **128**. Together $\mathrm{Tr}A^2 = -60$ for E_8. For $\mathrm{Tr}A^4$, in the same way, one finds $\mathrm{Tr}A^4 = 14(+2) + \frac{1}{16}128 = 36$. Finally, $\mathrm{Tr}A^6 = 14(-2) - \frac{1}{64}128 = -30$. With these results one obtains (F.40).

For $E_8 \times E_8$ one has $\mathrm{Tr}\, F^n = \dim E_8\, (\mathrm{Tr_I}\, F^n + \mathrm{Tr_{II}}\, F^n)$, where $\mathrm{Tr_I} F^n$ refers to the trace in the first E_8 and $\mathrm{Tr_{II}}\, F^n$ refers to the second E_8. Furthermore, $\dim E_8$ equal 248. Then,

$$\mathrm{Tr}\, F^6 = \frac{\dim E_8}{7200}\Big[(\mathrm{Tr_I}\, F^2)^3 + (\mathrm{Tr_{II}}\, F^2)^3\Big]. \tag{F.42}$$

This can be factorized using $x^3 + y^3 = (x+y)(x^2 - xy + y^2)$

$$\begin{aligned}\mathrm{Tr}\, F^6 = {} & \frac{\dim E_8}{7200}\Big[\mathrm{Tr_I}\, F^2 + \mathrm{Tr_{II}}\, F^2\Big] \\ & \times\Big[(\mathrm{Tr_I}\, F^2)^2 - (\mathrm{Tr_I}\, F^2)(\mathrm{Tr_{II}}\, F^2) + (\mathrm{Tr_{II}}\, F^2)^2\Big].\end{aligned} \tag{F.43}$$

For $\mathrm{Tr}F^4$ one finds

$$\begin{aligned}\mathrm{Tr}\, F^4 &= \dim E_8\Big(\mathrm{Tr_I}\, F^4 + \mathrm{Tr_{II}}\, F^4\Big) \\ &= \frac{\dim E_8}{100}\Big[(\mathrm{Tr_I}\, F^2)^2 + (\mathrm{Tr_{II}}\, F^2)^2\Big].\end{aligned} \tag{F.44}$$

[6] The Dirac matrices in 16 dimensions are 256×256 matrices, which can be chosen to be block off-diagonal. Then $\frac{1}{2}\gamma^1\gamma^2$ is block diagonal with one 128×128 block for chiral spinors and the other 128×128 block for antichiral spinors. These two spinor representations are inequivalent, just as in the case of $SO(8)$. In the text, by $\frac{1}{2}\gamma^1\gamma^2$ we mean one of these 128×128 blocks.

Finally, one has of course

$$\mathrm{Tr}\, F^2 = \dim E_8 \Big(\mathrm{Tr_I}\, F^2 + \mathrm{Tr_{II}}\, F^2 \Big). \tag{F.45}$$

The first question to be settled is whether the factorization in (6.293) holds for $E_8 \times E_8$. Substitution of (F.42), (F.44) and (F.45) into (6.293) shows that it does hold. The counterterm is now constructed in the same way as for $SO(32)$, see Section 6.9. The 12-form from which the descent equation starts is given by (6.294). Note that the traces $\mathrm{Tr}F^n$ in this formula refer to traces over $E_8 \times E_8$. The counterterm is then given by

$$\Delta\mathcal{L} = BX_8 + \Big(\mathrm{tr}R^2 - \frac{1}{30}\mathrm{Tr}F^2\Big) X_7 \tag{F.46}$$

where $dX_7 = X_8$ and

$$X_8 = \frac{1}{8}\mathrm{tr}R^4 + \frac{1}{32}(\mathrm{tr}R^2)^2 - \frac{1}{240}\mathrm{tr}R^2\mathrm{Tr}F^2 + \frac{1}{24}\mathrm{Tr}F^4 - \frac{1}{7200}(\mathrm{Tr}F^2)^2. \tag{F.47}$$

References

[1] L. Alvarez-Gaumé and E. Witten, *Nucl. Phys.* B **234** (1983) 269.

[2] W. A. Bardeen and B. Zumino, *Nucl. Phys.* B **244** (1984) 421.

[3] P. van Nieuwenhuizen, *Anomalies in Quantum Field Theory, Leuven Notes in Mathematical and Theoretical Physics*, series B, vol. 3 (Leuven: Leuven University Press, 1988).

[4] M. B. Green and J. H. Schwarz, *Phys. Lett.* B **149** (1984) 117.

[5] J. L. Gervais and A. Jevicki, *Nucl. Phys.* **110** (1976) 93. The appendix of this paper contains a very clear and short derivation of the extra terms of order $\hbar^2$ in the action for the path integral. Extra order $\hbar^2$ terms were earlier found by S. F. Edwards and Y. V. Gulyaev, *Proc. Roy. Soc. London* A **279** (1964) 269, who studied the change from Cartesian to polar coordinates in path integrals, and D. W. Laughlin and L. S. Schulman, *J. Math. Phys.* **12** (1971) 2520, who found extra terms in the potential for path integrals in curved space. See also refs. [6, 7, 8, 9].

[6] J. Schwinger, *Phys. Rev.* **127** (1962) 324; *Phys. Rev.* **130** (1963) 406. In these articles the commutation relations of the stress tensor and Lorentz currents of Yang–Mills theory are studied in a canonical formalism in the Coulomb gauge, and extra terms of order $\hbar^2$ are found to be needed to close the algebra.

[7] R. P. Treat, *Phys. Rev.* D **10** (1975) 3145. Quantum Yang–Mills theory is formulated in terms of background field theory, and the order $\hbar^2$ terms of [6] (produced by double commutators) are given in gauge-covariant form.

[8] N. H. Christ and T. D. Lee, *Phys. Rev.* D **22** (1980) 939. In this article Yang–Mills theory is first quantized in the $A_0^a = 0$ gauge. There are then no extra terms in the Hamiltonian in this gauge. Next, a transformation is made to the radiation (Coulomb) gauge and extra $\hbar^2$ terms are found to be present. These terms are nonlocal (in space, not in time), and some of them are smooth while other contains Dirac delta functions with space-dependent arguments.

[9] T. D. Lee, *Particle Physics and Introduction to Field Theory, Contemporary Concepts in Physics*, vol. 1 (New York: Harwood Academic Publishers, 1981). This textbook contains a clear and explicit derivation of the $\hbar^2$ terms found in [8].

[10] G. 't Hooft and M. J. G. Veltman, *Nucl. Phys.* B **44** (1972) 189; C. G. Bollini and J. J. Giambiagi, *Phys. Lett.* B (1972) **40** 566; see also: J. F. Ashmore, *Lett. Nuovo Cimento* **4** (1972) 289. G. M. Cicuta and E. Montaldi, *Lett. Nuovo Cim.* **4** (1972) 329.

[11] H. Kleinert and A. Chervyakov, *Phys. Lett.* B **464** (1999) 257 (hep-th/9906156).

[12] K. Fujikawa, *Phys. Rev. Lett.* **42** (1979) 1195; *Phys. Rev.* D **21** (1980) 2848, *Phys. Rev.* D **22** (1980) 1499 (erratum) and *Phys. Rev.* D **29** (1984) 285 (chiral anomalies); *Phys. Rev. Lett.* **44** (1980) 1733, *Phys. Rev.* D **23** (1981) 2262 (conformal anomalies). See also [99]. The twiddled fields were first mentioned by S. Hawking, *Commun. Math. Phys.* **55** (1977) 133, where he noticed that the normalization constant in front of the path integral remains unchanged under a scale transformation of the metric if one "assumes that the measure in the path integral over all configuration of the field ϕ is defined not on a scalar field but on a scalar density of weight $\frac{1}{2}$". Fujikawa showed that the anomalies arise from the Jacobian of the path integral and noted that using $\tilde{\phi} = g^{1/4}\phi$ as a path integral variable one obtains the correct Weyl anomaly for a scalar field. He also noted that with these variables the Jacobian for Einstein transformations automatically vanishes. In [13] it was shown that if one uses $g^\alpha\phi$ with $\alpha \neq \frac{1}{4}$ as a path integral variable one finds a particular linear combination of Weyl and Einstein anomalies.

[13] A. Diaz, W. Troost, P. van Nieuwenhuizen and A. Van Proeyen, *Int. J. Mod. Phys.* A **4** (1989) 3959; M. Hatsuda, W. Troost, A. Van Proeyen and P. van Nieuwenhuizen, *Nucl. Phys.* B **335** (1990) 166; F. Bastianelli, A. Van Proeyen and P. van Nieuwenhuizen, *Phys. Lett.* B **253** (1991) 67; E. Laenen and P. van Nieuwenhuizen, *Ann. Phys.* **207** (1991) 77.

[14] C. Becchi, A. Rouet and R. Stora, *Phys. Lett.* B **52** (1974) 344; *Commun. Math. Phys.* **42** (1975) 127; *Ann. Phys.* **98** (1976) 287; I. V. Tyutin, Lebedev Institute, internal report FIAN 39/1975 (unpublished).

[15] B. S. DeWitt, *Rev. Mod. Phys.* **29** (1957) 377. In this article DeWitt pioneered path integrals in curved space. He generalized to curved space a proposal by Pauli [16] to write the transition element for small times as a product of the exponent of the classical action and the square root of the Van Vleck–Morette determinant [17]. Pauli had shown that this approximation of the transition element satisfies a Schrödinger equation with extra $\hbar^2$ terms, and DeWitt found that for the case of curved space these extra terms are proportional to the scalar curvature. There is much overlap with our Chapter 2, and his work has led to an enormous literature on this subject.

[16] W. Pauli, *Feldquantisierung*, *Lecture Notes*, Zürich (1950–1951), appendix; also in *Selected Topics in Field Quantization*, chapter 6 (Cambridge, MA: MIT Press, 1973). Pauli was the first to approximate Feynman's path

integrals for short times by the product of the classical action and the Van Vleck–Morette one-loop determinant [17]. P. Jordan, *Z. f. Physik* **38** (1926) 513, studied generating functions $S(\alpha, q)$ for canonical transformations from α, β to p, q which yielded hermitian momenta, but the application of Hamilton–Jacobi theory to determine the short time transition element is due to Pauli.

[17] J. H. Van Vleck, *Proc. Nat. Acad. Sci.* **14** (1928) 178; C. Morette, *Phys. Rev.* **81** (1951) 848.

[18] B. S. DeWitt, *Relativity, Groups and Topology*, ed. B. S. DeWitt and C. DeWitt (New York: Gordon and Breach, 1964); and *Relativity, Groups and Topology II*, ed. B. S. DeWitt and R. Stora (Amsterdam: North-Holland, 1984).

[19] B. S. DeWitt, *Supermanifolds*, 2nd edition (Cambridge: Cambridge University Press, 1992).

[20] M. S. Marinov, *Phys. Rep.* **60** (1980) 1.

[21] L. S. Schulman, *Techniques and Applications of Path Integration* (New York: J. Wiley and Sons, 1981).

[22] F. Langouche, D. Roekaerts and E. Tirapegui, *Functional Integration and Semiclassical Expansions, Mathematics and Its Applications*, vol. 10 (Dordrecht: Reidel, 1982).

[23] B. Sakita, *Quantum Theory of Many-variable Systems and Fields* (Singapore: World Scientific, 1985).

[24] H. Kleinert, *Path Integrals in Quantum Mechanics* (Singapore: World Scientific, 1995); *Path Integrals in Quantum Mechanics, Statistics, Polymer Physics, and Financial Markets* (Singapore: World Scientific, 2004).

[25] M. Chaichian and A. Demichev, *Path Integrals in Physics*, vol. I (Bristol: Institute of Physics Publishing, 2001).

[26] B. DeWitt, *The Global Approach to Quantum Field Theory* (Oxford: Oxford University Press, 2003).

[27] M. G. Schmidt and C. Schubert, *Phys. Lett.* B **318** (1993) 438 (hep-th/9309055); C. Schubert, *Phys. Rep.* **355** (2001) 73 (hep-th/0101036).

[28] F. Bastianelli and A. Zirotti, *Nucl. Phys.* B **642** (2002) 372 (hep-th/0205182).

[29] F. Bastianelli, O. Corradini and A. Zirotti, *Phys. Rev.* D **67** (2003) 104 009 (hep-th/0211134).

[30] F. Bastianelli, P. Benincasa and S. Giombi, *JHEP* **0504** (2005) 010 (hep-th/0503155).

[31] K. S. Cheng, *J. Math. Phys.* **13** (1972) 1723.

[32] L. Parker, *Phys. Rev.* D **19** (1979) 438.

[33] C. DeWitt-Morette, A. Maheshwari and B. Nelson, *Phys. Rep.* C **50** (1979) 256.

[34] K. Kuchař, *J. Math. Phys.* **24** (1983) 2122.

[35] L. Alvarez-Gaumé, *Commun. Math. Phys.* **90** (1983) 161; *J. Phys.* A **16** (1983) 4177.

[36] L. Alvarez-Gaumé, Supersymmetry and index theory, in *Supersymmetry, NATO Advanced Study Institute*, ed. K. Dietz *et al.* (New York: Plenum, 1985).

[37] S. Cecotti and L. Girardello, *Phys. Lett.* B **110** (1982) 39.

[38] P. Windey, *Acta Phys. Pol.* B **15** (1984) 435; D. Friedan and P. Windey, *Nucl. Phys.* B **234** [FS11] (1984) 395.

[39] F. Bastianelli, *Nucl. Phys.* B **376** (1992) 113 (hep-th/9112035).

[40] F. Bastianelli and P. van Nieuwenhuizen, *Nucl. Phys.* B **389** (1993) 53 (hep-th/9208059).

[41] T. D. Lee and C. N. Yang, *Phys. Rev.* **128** (1962) 885; see also the discussion in pages 62–63 in E. S. Abers and B. W. Lee, *Phys. Rep.* **9** (1973) 1.

[42] J. F. Colombeau, *Bull. AMS* **23** (1990) 251; J. F. Colombeau, *Multiplication of Distributions*, *Lecture Notes in Mathematics*, vol. 1532 (Berlin: Springer-Verlag).

[43] A. Litvintsev and P. van Nieuwenhuizen, hep-th/0010051.

[44] P. M. Dirac, *Phys. Z. Sov.* **3** (1933) 64.

[45] R. P. Feynman, *Rev. Mod. Phys.* **20** (1948) 367.

[46] R. P. Feynman and A. R. Hibbs, *Quantum Mechanics and Path Integrals* (New York: McGraw-Hill, 1965).

[47] J. Govaerts, *Hamiltonian Quantization and Constrained Dynamics*, *Leuven Notes in Mathematical and Theoretical Physics* (Leuven: Leuven University Press, Leuven, 1991).

[48] L. Faddeev and A. Slavnov, *Gauge Fields: an Introduction to Quantum Theory*, 2nd edition (Redwood City, MA: Addison-Wesley, 1991).

[49] C. Itzykson and J. B. Zuber, *Quantum Field Theory* (New York: McGraw-Hill, 1980).

[50] J. Zinn-Justin, *Quantum Field Theory and Critical Phenomena* (Oxford: Oxford University Press, 1996).

[51] M. Henneaux and C. Teitelboim, *Quantization of Gauge Systems* (Princeton, NJ: Princeton University Press, 1992).

[52] G. 't Hooft, On peculiarities and pit falls in path integrals, in *Lectures Given at 7th International Conference on Path Integrals*, Antwerpen, Belgium, 2002, hep-th/0208054.

[53] See, for example, H. Goldstein, *Classical Mechanics* (Reading, MA: Addison-Wesley, 1950).

[54] J. Mehra, *The Beat of a Different Drum – The Life and Science of Richard Feynman* (Oxford: Clarendon Press, 1994).

[55] E. Schrödinger, *Ann. Phys. (Leipzig)* **79** (1926) 734.

[56] F. A. Berezin, *Theor. Math. Phys.* **6** (1971) 194 and *The Method of Second Quantization* (New York: Academic Press, 1966).

[57] J. de Boer, B. Peeters, K. Skenderis and P. van Nieuwenhuizen, *Nucl. Phys.* B **446** (1995) 211 (hep-th/9504097).

[58] J. de Boer, B. Peeters, K. Skenderis and P. van Nieuwenhuizen, *Nucl. Phys.* B **459** (1996) 631 (hep-th/9509158).

[59] B. Peeters and P. van Nieuwenhuizen, hep-th/9312147.

[60] B. Peeters and P. van Nieuwenhuizen, *Proceedings of HEP Conference*, Marseille, France, 22–28 July, 1993.

[61] K. Skenderis and P. van Nieuwenhuizen, hep-th/9401024.

[62] J. de Boer, B. Peeters, K. Skenderis and P. van Nieuwenhuizen, *Proceedings of Workshop on Gauge Theories, Applied Supersymmetry and Quantum Gravity*, Leuven, July 10–14, 1995; and *Strings'95*, USC, March 13–18, 1995; hep-th/9511141.

[63] J. de Boer, B. Peeters, K. Skenderis and P. van Nieuwenhuizen, *Proceedings for International Europhysics Conference on High-energy Physics (HEP 95)*, Brussels, Belgium, 27 July–2 August, 1995.

[64] A. Hatzinikitas, K. Schalm and P. van Nieuwenhuizen, *Nucl. Phys.* B **518** (1998) 424 (hep-th/9711088).

[65] K. Schalm and P. van Nieuwenhuizen, *Phys. Lett.* B **446** (1999) 247 (hep-th/9810115).

[66] H. Weyl, *Z. f. Physik* **46** (1927) 1, and *The Theory of Groups and Quantum Mechanics*, page 275 (New York: Dover, 1950). Weyl used his "Weyl correspondence" to convert any classical function $f(p, q)$ into an operator with definite ordering of p and q, but as such it was only one of many ad hoc rules. The Weyl correspondence is discussed clearly in S. R. de Groot and L. G. Suttorp, *Foundations of Electrodynamics* (Amsterdam: North-Holland, 1972).

[67] F. A. Berezin, *Theor. Math. Phys.* **6** (1971) 141. In this paper the special role of Weyl ordering for the transition from the operatorial formalism to path integrals is discussed. This connection is established by Fourier transformation. The results are entirely equivalent to our Appendices B and C.

[68] M. Mizrahi, *J. Math. Phys.* **16** (1975) 2201. This paper also discusses the Weyl transform from an operator $A(\hat{p}, \hat{q})$ to a function and back, and applies this to path integrals. The correct order $\hbar^2$ terms in the phase-space action are obtained.

[69] M. Sato, *Prog. Theor. Phys.* **58**, 1262 (1977).

[70] P. Salomonson, *Nucl. Phys.* B **121** (1977) 433.

[71] P. T. Matthews, *Phys. Rev.* **76** (1949) 686.

[72] Y. Nambu, *Progr. Theor. Phys.* **7**, 131 (1952).

[73] F. Bastianelli, K. Schalm and P. van Nieuwenhuizen, *Phys. Rev.* D **58** (1998) 044002 (hep-th/9801105).

[74] C. Bernard and A. Duncan, *Phys. Rev.* D **11**, (1975) 848.

[75] P. van Nieuwenhuizen and A. Waldron, *Phys. Lett.* B **389** (1996) 29 (hep-th/9608174).

[76] P. Solomonson and J. van Holten, *Nucl. Phys.* B **196** (1982) 509.

[77] P. van Nieuwenhuizen, *Phys. Rep.* **68** (1981) 189.

[78] A. Das and M. Kaku, *Phys. Rev.* D **18** (1978) 4540.

[79] L. van Hove, *Nucl. Phys.* B **207** (1982) 15.

[80] A. Das, *Physica* A **158** (1989) 1.

[81] F. Bastianelli and O. Corradini, *Phys. Rev.* D **60** (1999) 044014 (hep-th/9810119).

[82] F. Bastianelli, *Int. J. Mod. Phys.* D **3** (1994) 145 (hep-th/9308041).

[83] F. Bastianelli, *Proceedings for Path Integrals from peV to PeV*, Firenze, Italy, hep-th/9810143.

[84] H. Decker, *Physica* A **103** (1980) 586.

[85] F. Bastianelli, O. Corradini and P. van Nieuwenhuizen, *Phys. Lett.* B **494** (2000) 161 (hep-th/0008045).

[86] F. Bastianelli, O. Corradini and P. van Nieuwenhuizen, *Phys. Lett.* B **490** (2000) 154 (hep-th/0007105).

[87] F. Bastianelli and O. Corradini, *Phys. Rev.* D **63** (2001) 065005 (hep-th/0010118).

[88] F. Bastianelli and N.D. Hari Dass, *Phys. Rev.* D **64** (2001) 047701 (hep-th/0104234).

[89] F. Bastianelli, O. Corradini and A. Zirotti, *JHEP* **0401** (2004) 023 (hep-th/0312064).

[90] H. Kleinert and A. Chervyakov, *Phys. Lett.* B **464**, 257 (1999) (hep-th/9906156); *Phys. Lett.* B **477**, 373 (2000) (quant-ph/9912056); *Phys. Lett.* A **273**, 1 (2000) (quant-ph/0003095); *Int. J. Mod. Phys.* A **17**, 2019 (2002) (quant-ph/0208067).

[91] H. Kleinert and A. Chervyakov, *Int. J. Mod. Phys.* A **18** (2003) 5521 (quant-ph/0301081).

[92] R.P. Feynman, *Phys. Rev.* **80** (1950) 440.

[93] J. Schwinger, *Phys. Rev.* **82** (1951) 664.

[93a] M. B. Halpern, A. Jevicki and P. Senjanovic, *Phys. Rev.* **D 16** (1977) 2476; M. B. Halpern and W. Siegel, *Phys. Rev.* **D 16** (1977) 2486.

[94] G. 't Hooft, *Nucl. Phys.* B **75** (1974) 461.

[95] A.M. Polyakov and P.B. Wiegmann, *Phys. Lett.* B **131** (1983) 121; *Phys. Lett.* B **141** (1984) 223.

[96] J. Wess and B. Zumino, *Phys. Lett.* B **37** (1971) 95; S.P. Novikov, *Usp. Mat. Nauk.* **37N5** (1982) 3; E. Witten, *Commun. Math. Phys.* **92** (1984) 455.

[97] H. Ooguri, K. Schoutens, A. Sevrin and P. van Nieuwenhuizen, *Commun. Math. Phys.* **145** (1992) 515; K. Schoutens, A. Sevrin and P. van Nieuwenhuizen, *Proceedings of Strings and Symmetries 1991*, Stony Brook, NY, 1991.

[98] A.M. Polyakov, in *Gauge Fields and Strings*, chapter 9 (New York: Harwood, 1987).

[99] K. Fujikawa and H. Suzuki, *Path Integrals and Quantum Anomalies* (Oxford: Oxford University Press, 2004).

[100] P. H. Frampton, D. R. T. Jones, P. van Nieuwenhuizen and S. C. Zhang, *Quantum Field Theory and Quantum Statistics, Festschrift for E. S. Fradkin*, ed. I. A. Batalin, C. J. Isham and G. A. Volkovisky (Bristol: Hilger), page 379.

[101] H. Römer and P. van Nieuwenhuizen, *Phys. Lett.* B **162** (1985) 290.

[102] R. Floreanini and R. Jackiw, *Phys. Rev. Lett.* **59** (1987) 1873.

[103] M. Henneaux and C. Teitelboim, *Phys. Lett.* B **206** (1988) 650.

[104] F. Bastianelli and P. van Nieuwenhuizen, *Phys. Rev. Lett.* **63** (1989) 728.

[105] P. Pasti, D. P. Sorokin and M. Tonin, *Phys. Rev.* D **52** (1995) 4277 (hep-th/9506109).

[106] K. Lechner, *Nucl. Phys.* B **537** (1999) 361 (hep-th/9808025).

[107] A. Ceresole, P. Pizzochero and P. van Nieuwenhuizen, *Phys. Rev.* D **39** (1989) 1567; J. Kowalski-Glikman and P. van Nieuwenhuizen, *Int. J. Mod. Phys.* A **6** (1991) 4077.

[108] Z. Koba and G. Takeda, *Progr. Theor. Phys.* **3** (1948) 98; T. Tati and S. Tomonaga, *Progr. Theor. Phys.* **4** (1948) 391.

[109] S. Tomonaga, *Phys. Rev.* **74** (1948) 224.

[110] H. Fukuda and Y. Miyamoto, *Progr. Theor. Phys.* **4** (1949) 235 (L) and 347.

[111] J. Steinberger, *Phys. Rev.* **76** (1949) 1180.

[112] W. Pauli and F. Villars, *Rev. Mod. Phys.* **21** (1949) 434. See also, S. N. Gupta, *Proc. Phys. Soc.* A **66** (1953) 129.

[113] H. Fukuda, Y. Miyamoto, T. Miyazima and S. Tomonaga, *Progr. Theor. Phys.* **4** (1949) 385.

[114] S. L. Adler, Anomalies to all orders, hep-th/0405040; in *Fifty Years of Yang–Mills Theory*, G. 't Hooft editor (World Sc.)

[115] M. Gell-Mann and M. Levy, *Nuovo Cimento* **16** (1960) 705.

[116] B. W. Lee, *Chiral Dynamics* (New York: Gordon and Breach, 1972).

[117] J. Bernstein, M. Gell-Mann and L. Michel, *Nuovo Cimento* **16** (1960) 560.

[118] J. Bell and R. Jackiw, *Nuovo Cimento* A **60** (1969) 47.

[119] S. L. Adler, *Phys. Rev.* **177** (1969) 2426.

[120] D. G. Sutherland, *Nucl. Phys.* B **2** (1967) 433; M. Veltman, *Proc. Roy. Phys. Soc.* A **301** (1967) 107.

[121] W. Bardeen, unpublished.

[122] L. Rosenberg, *Phys. Rev.* **129** (1963) 2786.

[123] C. Bouchiat, J. Iliopoulos and Ph. Meyer, *Phys. Lett.* B **38** (1972) 519; D. J. Gross and R. Jackiw, *Phys. Rev.* D **6** (1972) 477.

[124] T. Kimura, *Prog. Theor. Physics* **42** (1969) 1191; R. Delburgo and A. Salam, *Phys Lett.* B **40** (1972) 381; T. Eguchi and P. G. O. Freund, *Phys. Rev. Lett.* **37** (1976) 1251.

[125] S. M. Christensen and M. J. Duff, *Phys. Lett.* B **76**, 571 (1978); *Nucl. Phys.* B **154** (1979) 301; N. K. Nielsen, M. T. Grisaru, H. Römer and P. van Nieuwenhuizen, *Nucl. Phys.* B **140** (1978) 477.

[126] R. Endo and T. Kimura, *Prog. Theor. Phys.* **63** (1980) 683; R. Endo and M. Takao, *Phys. Lett.* B **161** (1985), 155.

[127] S. R. Coleman and R. Jackiw, *Ann. Phys.* **67** (1971) 552.

[128] D. M. Capper and M. J. Duff, *Nuovo Cimento* A **23** (1974) 173.

[129] S. Deser, M. J. Duff and C. J. Isham, *Nucl. Phys.* B **111** (1976) 45.

[130] M. J. Duff, *Nucl. Phys.* B **125** (1977) 334; L. Bonora, P. Cotta-Ramusino and C. Reina, *Phys. Lett.* B **126** (1983) 305; L. Bonora, P. Pasti and M. Bregola, *Class. Quant. Grav.* **3** (1986) 635.

[131] G. 't Hooft and M. J. G. Veltman, *Ann. H. Poincaré Phys. Theor.* A **20** (1974) 69 (pure gravity and coupling to scalars); S. Deser and P. van Nieuwenhuizen, *Phys. Rev. Lett.* **32** (1974) 245 and *Phys. Rev.* D **10** (1974) 401 (coupling to Maxwell fields); S. Deser and P. van Nieuwenhuizen, *Phys. Rev.* D **10** (1974) 411 and *Lett. Nuovo Cimento* **11** (1974) 218 (coupling to fermions); S. Deser, H.-S. Tsao and P. van Nieuwenhuizen, *Phys. Rev.* D **10** (1974) 3337 (coupling to Yang–Mills fields); S. Deser, M. T. Grisaru, P. van Nieuwenhuizen and C. C. Wu, *Phys. Lett.* B **58** (1975) 355 (scale dependence in quantum gravity).

[132] M. J. Duff and P. van Nieuwenhuizen, *Phys. Lett.* B **94** (1980) 179.

[133] W. Siegel, *Phys. Lett.* B **103** (1981) 107; M. T. Grisaru, N. K. Nielsen, W. Siegel and D. Zanon, *Nucl. Phys.* B **247** (1984) 157.

[134] S. M. Christensen and M. J. Duff, *Phys. Lett.* B **76** (1978) 571; *Nucl. Phys.* B **154** (1979) 301.

[135] S. L. Adler, *Lectures on Elementary Particles and Quantum Field Theory*, ed. S. Deser *et al.* (Reading, MA: MIT Press, 1970).

[136] S. Treiman, R. Jackiw, B. Zumino and E. Witten, *Current Algebra and Anomalies* (Singapore: World Scientific, 1985).

[137] R. A. Bertelmann, *Anomalies in Quantum Field Theory* (Oxford: Oxford University Press, 2000).

[138] L. Alvarez-Gaumé and M. A. Vazquez-Mozo, *Gravitation and Quantizations*, ed. B. Julia and J. Zinn-Justin (Amsterdam: North-Holland, 1995) (hep-th/9212006).

[139] M. J. Duff, *Class. Quant. Grav.* **11** (1994) 1387 (hep-th/9308075).

[140] C. A. Scrucca and M. Serone, *Int. J. Mod. Phys.* A **19** (2004) 2579 (hep-th/0403163).

[141] K. Fujikawa, S. Ojima and S. Yajima, *Phys. Rev.* D **34** (1986) 3223.

[142] L. Alvarez-Gaumé and P. Ginsparg, *Nucl. Phys.* B **243** (1984) 449.

[143] L. Alvarez-Gaumé and P. Ginsparg, *Ann. Phys.* **161** (1985) 423 (erratum *Ann. Phys.* **171** (1986) 233).

[144] L. N. Chang and H. T. Nieh, *Phys. Rev. Lett.* **53** (1984) 21. In this article the existence of local Lorentz anomalies in $4k + 2$ dimensions is shown by evaluating the Fujikawa Jacobian $\mathrm{tr}\sigma_{ab}a_{2k+1}$, where a_{2k+1} are the Schwinger–Seeley–DeWitt coefficients [145] obtained from the heat kernel method and zeta function regularization.

[145] J. Schwinger, *Phys. Rev.* **82** (1951) 664; B. S. DeWitt, *Relativity, Groups and Topology*, ed. B. S. DeWitt and C. DeWitt (New York: Gordon and Breach, 1964) and *Relativity, Groups and Topology II*, ed. B. S. DeWitt and R. Stora (Amsterdam: North-Holland 1984); J. S. Dowker and R. Critchley, *Phys. Rev.* D **13** (1976) 3224; S. W. Hawking, *Commun. Math. Phys.* **55** (1977) 133. For a review, see B. Schroer, *Acta Phys. Austriaca Suppl.* **19** (1978) 155.

[146] S. Yajima and T. Kimura, *Phys. Lett.* B **173**, 154 (1986).

[147] J. von Neumann, *Math Ann.* **104** (1931) 104.

[148] P. van Nieuwenhuizen, *Relativity, Groups and Topology II, Lectures at 1983 Les Houches School*, ed. B. S. DeWitt and R. Stora (Amsterdam: North-Holland, 1984).

[149] F. Gliozzi, J. Scherk and D. I. Olive, *Nucl. Phys.* B **122** (1977) 253.

[150] H. T. Nieh, *Phys. Rev. Lett.* **53** (1984) 2219. In this article it is noted that in flat space chiral spinors in four dimensions coupled to $U(1)$ gauge fields have Lorentz anomalies. They are obtained by evaluating $\mathrm{tr}\sigma_{ab}\gamma^5 a_2(x)$ as in [144]. In the standard model they cancel separately for quarks and leptons.

[151] R. Endo, *Prog. Theor. Phys.* **78** (1987) 440; *Prog. Theor. Phys.* **80** (1988) 311; *Phys. Lett.* B **202** (1988) 105.

[152] K. Fujikawa, *Nucl. Phys.* B **226** (1983) 437; K. Fujikawa and O. Yasuda, *Nucl. Phys.* B **245** (1984) 436; K. Fujikawa, M. Tomiya and O. Yasuda, *Z. Phys.* C **28** (1985) 289; K. Fujikawa, *Quantum Gravity and Cosmology*, ed. H. Sato and T. Inami (Singapore: World Scientific, 1986).

[153] B. Zumino and R. Stora, unpublished; B. Zumino, *Relativity, Groups and Topology II, Lectures at 1983 Les Houches School*, ed. B. S. DeWitt and R. Stora (Amsterdam: North-Holland, 1984); B. Zumino, Y.-S. Wu and A. Zee, *Nucl. Phys.* B **239** (1984) 477.

[154] F. Bastianelli and P. van Nieuwenhuizen, *Phys. Lett.* B **217** (1989) 98.

[155] F. Bastianelli, *Phys. Lett.* B **254** (1991) 427.

[156] E. Kähler, *Rend. Math.* **21** (3.4) (1962) 425.

[157] T. Banks, Y. Dothan and D. Horn, *Phys. Lett.* B **117** (1982) 413.

[158] C. Lanczos, *Z. Phys.* **57** (1929) 447, 474, 484.

[159] M. B. Green, J. H. Schwarz and E. Witten, *Superstring Theory*, vol. I (Cambridge: Cambridge University Press, 1987).

[160] R. Delbourgo and T. Matsuki, *J. Math. Phys.* **26** (1985) 1334.

[161] M. B. Green, J. H. Schwarz and E. Witten, *Superstring Theory*, vol. II (Cambridge: Cambridge University Press, 1987).

[162] E. Sezgin and A. Salam, *Phys. Scripta* **32** (1985) 32; S. J. Gates and H. Nishino, *Phys. Lett.* B **157** (1985) 157; E. Sezgin, *Czech. J. Phys.* B **37** (1987) 465.

[163] S. Randjebar Daemi, A. Salam, E. Sezgin and A. Strathdee, *Phys. Lett.* B **151** (1985) 351 (for a model with group $E_7 \times E_6 \times U(1)$); E. Bergshoeff,

A. Salam and E. Sezgin, *Phys. Lett.* B **173** (1986) 73; E. Bergshoeff and M. Rakowski, *Phys. Lett.* B **191** (1987) 399.

[164] E. Bergshoeff, A. Van Proeyen and E. Sezgin, *Nucl. Phys.* B **264** (1986) 653.

[165] A. Sen, *Phys. Rev.* D **53** (1996) 6725.

[166] E. Bergshoeff, M. de Roo, B. de Wit and P. van Nieuwenhuizen, *Nucl. Phys.* B **195** (1982) 97; G. F. Chapline and N. Manton, *Phys. Lett.* B **120** (1983) 105.

[167] M. F. Atiyah and I. M. Singer, *Ann. Math.* **87** (1968) 485, 546; *Ann. Math.* **93** (1971) 1, 139; M. F. Atiyah and G. B. Segal, *Ann. Math.* **87** (1968) 531.

[168] E. Witten, *Nucl. Phys.* B **202** (1982) 253.

[169] A. K. Waldron, *Phys. Rev.* D **53** (1996) 5692 (hep-th/9511148).

[170] N. K. Nielsen and P. van Nieuwenhuizen, *Phys. Rev.* D **38** (1988) 3183.

[171] L. Alvarez-Gaumé, D. Z. Freedman and S. Mukhi, *Ann. Phys.* **134** (1981) 85; P. S. Howe, G. Papadopoulos and K. S. Stelle, *Nucl. Phys.* B **269** (1988) 26.

[172] F. Bastianelli, S. Frolov and A. A. Tseytlin, *JHEP* **0002** (2000) 013 (hep-th/0001041).

[173] U. Theis and P. van Nieuwenhuizen, *Class. Quant. Grav.* **18** (2001) 5469 (hep-th/0108204).

[174] N. K. Nielsen and P. van Nieuwenhuizen, *Phys. Rev.* D **38** (1988) 3183.

[175] M. J. Strassler, *Nucl. Phys.* B **385** (1992) 145 (hep-ph/9205205).

[176] D. H. Friedan, *Ann. Phys.* **163** (1985) 318.

[177] P. S. Howe, G. Papadopoulos and K. S. Stelle, *Nucl. Phys.* B **296** (1988) 26.

[178] A. A. Tseytlin, *Phys. Lett.* B **223** (1989) 165.

[179] J. de Boer, K. Schalm and J. Wijnhout, hep-th/0310150.

[180] H. Kluberg-Stern and J. B. Zuber, *Phys. Rev.* D **12** (1975) 482.

[181] D. Fliegner, P. Haberl, M. G. Schmidt and C. Schubert, *Ann. Phys.* **264** (1998) 51 (hep-th/9707189).

[182] K. Schalm and P. van Nieuwenhuizen, *Phys. Lett.* B **446** (1999) 247 (hep-th/9810115).

[183] E. Witten, *Commun. Math. Phys.* **100** (1985) 197 (see, in particular, page 218).

[184] S. Weinberg, *Gravitation and Cosmology*, section 12.5 (New York: Wiley, 1972).

[185] G. M. Gavazzi, *J. Math. Phys.* **30** (1989) 2904.

[186] P. van Nieuwenhuizen, Supersymmetry, supergravity, superspace and BRST symmetry in a simple model (hep-th/0408179).

[187] H. Georgi and S. L. Glashow, *Phys. Rev.* D **6** (1972) 429.

[188] B. S. DeWitt and P. van Nieuwenhuizen, *J. Math. Phys.* **23** (1982) 1953.

[189] A. O. Barut and A. Raczka, *Theory of Group Representations and Applications* (Singapore: World Scientific, 1986), pages 39–43.

[190] R. Slansky, *Phys. Rep.* **79** (1981) 1.

[191] H. Georgi, *Lie Algebras in Particle Physics, Frontiers in Physics* (New York: Perseus, 1999).

[192] T. van Ritbergen, A. N. Schellekens and J. A. M. Vermaseren, *Int. J. Mod. Phys.* A **14** (1999) 41 (hep-ph/98023760).

[193] E. Cremmer, J. Scherk and J. H. Schwarz, *Phys. Lett.* B **84** (1979) 83.

[194] E. Sezgin and P. van Nieuwenhuizen, *Nucl. Phys.* B **195** (1982) 325.

[195] E. Cremmer and B. Julia, *Nucl. Phys.* B **159** (1979) 141.

Index